The Economics of Technical Change

The International Library of Critical Writings in Economics

Series Editor: Mark Blaug

Professor Emeritus, University of London
Professor Emeritus, University of Buckingham
Visiting Professor, University of Exeter

This series is an essential reference source for students, researchers and lecturers in economics. It presents by theme an authoritative selection of the most important articles across the entire spectrum of economics. Each volume has been prepared by a leading specialist who has written an authoritative introduction to the literature included.

A full list of published and future titles in this series is printed at the end of this volume.

The Economics of Technical Change

Edited by

Edwin Mansfield

Director of the Center for Economics and Technology
University of Pennsylvania, US

and

Elizabeth Mansfield

Teaching Fellow in Counseling Psychology
Northwestern University, US

An Elgar Reference Collection

[redacted]

Published by
Edward Elgar Publishing Limited
Gower House
Croft Road
Aldershot
Hants GU11 3HR
England

Edward Elgar Publishing Company
Old Post Road
Brookfield
Vermont 05036
USA

British Library Cataloguing in Publication Data
Economics of Technical Change. –
(International Library of Critical
Writings in Economics;No.31)
I. Mansfield, Edwin II. Mansfield,
Elizabeth III. Series
338.9

Library of Congress Cataloguing in Publication Data
The Economics of technical change / edited by Edwin Mansfield and
Elizabeth Mansfield.
p. cm. — (The International library of critical writings in
economics; 31) (An Elgar reference collection)
Includes bibliographical references and index.
1. Technological innovations—Economic aspects. I. Mansfield,
Edwin. II. Mansfield, Elizabeth. III. Series. IV. Series: An
Elgar reference collection.
HC79.T4E268 1993
338′.064—dc20

93–22664
CIP

ISBN 1 85278 783 X

Printed in Great Britain by Galliard (Printers) Ltd, Great Yarmouth

Contents

Acknowledgements

The editors and publishers wish to thank the following who have kindly given permission for the use of copyright material.

American Economic Association for articles: Paul A. David (1985), 'Clio and the Economics of QWERTY', *American Economic Review*, **75** (2), May, 332–7; Zvi Griliches (1986), 'Productivity, *R&D*, and Basic Research at the Firm Level in the 1970s', *American Economic Review*, **76** (1), March, 141–54; Jeffrey I. Bernstein and M. Ishaq Nadiri (1988), 'Interindustry *R&D* Spillovers, Rates of Return, and Production in High-Tech Industries', *American Economic Review*, **78** (2), May, 429–34; Gene M. Grossman and Elhanan Helpman (1990), 'Comparative Advantage and Long-Run Growth', *American Economic Review*, **80** (4), September, 796–815; Suzanne Scotchmer (1991), 'Standing on the Shoulders of Giants: Cumulative Research and the Patent Law', *Journal of Economic Perspectives*, **5** (1), Winter, 29–41; Janusz A. Ordover (1991), 'A Patent System for Both Diffusion and Exclusion', *Journal of Economic Perspectives*, **5** (1), Winter, 43–60; F.M. Scherer (1992), 'Schumpeter and Plausible Capitalism', *Journal of Economic Literature*, **XXX** (3), September, 1416–33.

Basil Blackwell Ltd. for articles: D.J. Teece (1977), 'Technology Transfer by Multinational Firms: The Resource Cost of Transferring Technological Know-How', *Economic Journal*, **87**, June, 242–61; Partha Dasgupta and Joseph Stiglitz (1980), 'Industrial Structure and the Nature of Innovative Activity', *Economic Journal*, **90**, June, 266–93; Edwin Mansfield, Mark Schwartz and Samuel Wagner (1981), 'Imitation Costs and Patents: An Empirical Study', *Economic Journal*, **91**, December, 907–18; Wesley M. Cohen and Daniel A. Levinthal (1989), 'Innovation and Learning: The Two Faces of R&D', *Economic Journal*, **99**, September, 569–96.

Brookings Institution for article: Richard C. Levin, Alvin K. Klevorick, Richard R. Nelson and Sidney G. Winter (1987), 'Appropriating the Returns from Industrial Research and Development', *Brookings Papers on Economic Activity*, **3** (Special Issue), 783–831.

Econometric Society for article: Edwin Mansfield (1961), 'Technical Change and the Rate of Imitation', *Econometrica*, **29** (4), October, 741–66.

Elsevier Science Publishers B.V. for articles: Robert M. Solow (1957), 'Technical Change and the Aggregate Production Function', *Review of Economics and Statistics*, **XXXIX**, August, 312–20; Richard R. Nelson (1961), 'Uncertainty, Learning, and the Economics of Parallel Research and Development Efforts', *Review of Economics and Statistics*, **XLIII**, November, 351–64.

MIT Press Journals for articles: Raymond Vernon (1966), 'International Investment and International Trade in the Product Cycle', *Quarterly Journal of Economics*, **LXXX**, May, 190–207; Edwin Mansfield, John Rapoport, Anthony Romeo, Samuel Wagner and George Beardsley (1977), 'Social and Private Rates of Return from Industrial Innovations', *Quarterly Journal of Economics*, **XCI**, May, 221–40.

Oxford University Press for article: P.A. Geroski (1990), 'Innovation, Technological Opportunity, and Market Structure', *Oxford Economic Papers*, New Series, **42**, July, 586–602.

Review of Economic Studies Ltd. for articles: Kenneth J. Arrow (1962), 'The Economic Implications of Learning by Doing', *Review of Economic Studies*, **XXIX**, June, 155–73; Jennifer F. Reinganum (1981), 'On the Diffusion of New Technology: A Game-Theoretic Approach', *Review of Economic Studies*, **XLVIII**, July, 395–405.

University of Chicago Press for articles: Zvi Griliches (1958), 'Research Costs and Social Returns: Hybrid Corn and Related Innovations', *Journal of Political Economy*, **LXVI**, October, 419–31; Richard R. Nelson (1959), 'The Simple Economics of Basic Scientific Research', *Journal of Political Economy*, **XLVII**, June, 297–306; Nathan Rosenberg (1970), 'Economic Development and the Transfer of Technology: Some Historical Perspectives', *Technology and Culture*, **11** (4), October, 550–75; Michael L. Katz and Carl Shapiro (1986), 'Technology Adoption in the Presence of Network Externalities', *Journal of Political Economy*, **94** (4), August, 822–41; Paul M. Romer (1990), 'Endogenous Technological Change', *Journal of Political Economy*, **98** (5), October, S71–S102.

Every effort has been made to trace all the copyright holders but if any have been inadvertently overlooked the publishers will be pleased to make the necessary arrangement at the first opportunity.

In addition the publishers wish to thank the library of the London School of Economics and Political Science, the Marshall Library, Cambridge University and the Photographic Unit of the University of London Library for their assistance in obtaining these articles.

Introduction*

The purpose of this book is to provide a selection of well-known articles covering major aspects of the economics of technological change. These articles are concerned with the following seven broad topics: (1) technical change and economic growth; (2) social and private returns from research and development; (3) market structure and technical change; (4) intellectual property rights; (5) the diffusion of innovations; (6) international technology transfer; and (7) the management of technology.

Technical Change and Economic Growth

Pioneering studies of the relationship between technological change and economic growth – by Robert Solow, Moses Abramowitz, Solomon Fabricant, and others – occurred in the mid-1950s. The opening paper in this book – Solow's classic paper, part of the work for which he received the Nobel Prize – was extremely influential. Assuming that there were constant returns to scale, that capital and labour were paid their marginal products, and that technological change was neutral, Solow attempted to estimate the rate of technological change for the American nonfarm economy during the period 1909–1949. His findings suggested that, for the period as a whole, the average rate of technological change – the increase in the amount of output that could be derived from a fixed amount of inputs – was about 1.5 per cent per year. On the basis of these findings, Solow concluded that about 90 per cent of the increase in output per capita during this period was attributable to technological change.

Essentially, the basic methodology used in the early studies was the following: economists, who view the total output of the economy as being due to various inputs of productive services into the productive process, began by specifying these inputs as labour and capital, and by attempting to estimate their contribution to the measured growth of output. Whatever portion of the measured growth of output could not be explained by these inputs was attributed to technological change. The crudeness of this procedure is obvious. The resulting measure of the effect of technological change does not isolate the effects of technological change alone. In addition, it contains the effects of whatever inputs are excluded, which, depending on the study, may be economies of scale, improved allocation of resources, changes in product mix, increases in education, or improved health and nutrition of workers.

Many studies have built on this early work of Solow and others. Among the most interesting papers in the 1990s have been those by Paul Romer and by Gene Grossman and Elhanen Helpman that are reprinted here. In Romer's model, growth occurs because of technological change due to investment decisions made by profit-maximizing agents. Technology as an input is regarded as a non-rival, partially excludable good, and the equilibrium is one with

*This introduction draws on the senior editor's previous writings in this area, some of which are cited in the papers reprinted in this volume.

monopolistic competition. Romer concludes that the stock of human capital determines the growth rate, that too little human capital will be devoted to research in equilibrium, and that integration into world markets will raise growth rates.

Grossman and Helpman analyse a dynamic, two-country model of trade and growth. The determinants of R&D are studied, where research results in designs for new intermediate products and in making further research less costly. Cross-country differences in efficiency at R&D and manufacturing are assumed to exist. Among their findings are that growth in world resources or improvements in R&D efficiency need not increase the rate of steady-state growth if these developments take place predominantly in the nation with a comparative disadvantage in research and development.

Social and Private Returns from Research and Development

The social rate of return from an investment in new technology is the interest rate realized by society as a whole from that investment – the payoff to society. A high social rate of return indicates that society's resources are being used effectively, and that more should be devoted to such investments if the rate of return remains high.

In 1958, Zvi Griliches estimated the social rate of return from the investment in R&D in the case of hybrid corn, an important agricultural innovation. The internal rate of return was very high, about 40 per cent, which was expected since hybrid corn was known in advance to be a very successful innovation. Griliches's study, reprinted in this volume, was followed by a variety of other studies of agricultural R&D by Robert Evenson, Vernon Ruttan, and others. In general, the results seemed to indicate that social rates of return from agricultural R&D have been very substantial.

The first attempt to measure the social and private returns from investments in industrial innovations was a study conducted by myself and others, the results of which were published in 1977. This paper, reproduced in this book, indicated that the median social rate of return from the investment in our sample of innovations was 56 per cent, a very high figure. This high rate of return was borne out in two subsequent studies commissioned by the National Science Foundation to replicate our study. Based on separate samples of 20 innovations each, the subsequent studies found the median social rate of return to be 70 per cent and 99 per cent, respectively. It is important to note that all three of these samples included many very minor innovations and failures as well as successes. An attempt was made to obtain reasonably representative samples.

While the social returns from innovative activities are very high, this does not mean that the private returns – the returns to the innovating firm – are also high. Although the median social rate of return in our study was 56 per cent, the median private rate of return was only 25 per cent. One of the main reasons why the private rate of return from innovative activity is so much lower than the social rate of return is that the innovator frequently finds it difficult to appropriate the returns from the innovation. Many of these benefits accrue to imitators, who often obtain information quickly concerning the detailed nature and operation of new products and processes.

There are many channels through which information on innovation spreads – the movement of personnel from one firm to another, informal communications networks among engineers and scientists working at various firms, and professional meetings at which information is exchanged. In some industries, input suppliers and customers are important channels (since they pass on a great deal of relevant information), patent applications are scrutinized very carefully, and reverse engineering is carried out. In other industries, the diffusion process is accelerated by the fact that firms do not go to great lengths to keep such information secret, partly because they believe that it would be futile.

While information on innovative products and processes leaks out relatively quickly, this does not mean that imitation will occur equally as quickly. It often takes considerable time to invent around patents (if they exist), to develop prototypes, to alter or build plant and equipment, and to engage in the manufacturing and marketing startup activities required to introduce an imitative product or process. Nonetheless, the basic information concerning the nature and operation of the innovation is of great importance to the innovator's rivals. And because information of this sort is likely to find its way into the hands of rivals within a year or two, this has obvious and important implications concerning the incentives for innovation.

In 1988, Jeremy Bernstein and M. Ishaq Nadiri, in the next paper reprinted here, studied the spillovers from R&D. They conclude that:

> the social rate of return on R&D capital in the non-electric machinery industry was 2 to 3 times the private rate of return... In [electrical products and transportation equipment], the social rate was greater than the private rate, but only by about 10 to 20 per cent... The R&D capital from scientific instruments generated substantial spillover effects... In fact, the social rate of return was around 10 times the private rate.

In 1959, Richard Nelson, in a paper reprinted here, analysed the economic aspects of basic research. He concludes that:

> The problem of getting enough resources to flow into basic research is basically the classical external-economy problem. External economies result from two facts: first, that research results often are of little value to the firm that sponsors the research, though of great value to another firm, and, second, that research results often cannot be quickly patented.

Indeed, as later papers indicate, they may not be patentable at all.

In the following paper, Zvi Griliches studies the relationship between R&D expenditure and productivity growth at the firm level during the 1970s. He concludes that:

> evidence has been presented for the proposition that R&D contributes significantly to productivity growth, that the basic research component of it does so even more strongly, and that privately financed R&D expenditures have a significantly larger effect on private productivity and profitability than federally financed R&D.

With regard to the last finding, it is important, of course, to note that much federally financed R&D is not aimed at increasing private productivity.

Market Structure and Technical Change

Joseph Schumpeter and John Kenneth Galbraith have argued that a perfectly competitive economy is likely to be inferior in a dynamic sense to an economy including many imperfectly competitive industries (that is, monopolies, oligopolies, etc.). In their view, although a perfectly competitive economy will satisfy the conditions for static economic efficiency, it will not result in as high a rate of technological change and productivity increase as an imperfectly competitive economy. To illustrate their arguments, suppose that we compare the performance of two economies, one perfectly competitive, the other imperfectly competitive. Assume that they start off with comparable technology in 1994, but that the rate of technological change is higher in the imperfectly competitive economy than in the perfectly competitive economy, so that the annual rate of productivity change is 3 per cent in the imperfectly competitive economy and 2 per cent in the perfectly competitive economy. If the quantity of inputs is the same in each economy and constant over time, it follows that:

$$Q(t) = Q_0(1.02)^t$$
$$Q'(t) = Q'_0(1.03)^t$$

where $Q(t)$ is the output in year t in the perfectly competitive economy, $Q'(t)$ is the output in year t in the imperfectly competitive economy, and t is measured in years from 1994.

Assume too that, because of its static inefficiencies, the imperfectly competitive economy produces 98 per cent as much as the perfectly competitive economy when their technology levels are the same. Thus $Q'_0 = 0.98Q_0$, and:

$$Q(t) = Q_0(1.02)^t$$
$$Q'(t) = 0.98Q_0(1.03)^t.$$

Hence it follows that:

$$Q'(t) \div Q(t) = 0.98(1.03 \div 1.02)^t,$$

which implies that $Q'(t) \div Q(t)$ will be greater than one when t is greater than about 2. Thus, despite its being relatively inefficient in a static sense, the imperfectly competitive economy out-produces the perfectly competitive economy from about 1996 on, because it has the higher rate of technological change and productivity increase.

In 1980, Partha Dasgupta and Joseph Stiglitz published a paper, reprinted here, that analysed the relationship between industrial structure and innovative activity. Among their conclusions were the following:

> (1) Even when one regards industrial structure to be endogenous, theoretical considerations are consistent with the empirical finding that when the degree of concentration in industries is small, industry-wide R&D is positively correlated with concentration... (2) High degrees of concentration are by themselves not an evidence of lack of effective competition... (3) When the degree of concentration in industries with free-entry is small, R&D effort per firm (and therefore cost reduction) is often positively correlated with concentration. Moreover, the degree of concentration is positively correlated with the degree of monopoly.

Paul Geroski, in the next paper reprinted here, carries out an empirical study of the Schumpeterean hypothesis based on data for the United Kingdom. He concludes:

> that monopoly appears to inhibit the response to a given level of post-innovation returns, and that the indirect effects on innovativeness are relatively small. There is, in short, almost no support in the data for popular Schumpeterean assertions about the role of actual monopoly in stimulating progressiveness.

In his 1992 paper, F.M. Scherer surveys Schumpeter's views and summarizes much of the relevant theoretical and empirical work that was carried out to test and extend Schumpeter's hypotheses. Among his conclusions are that

> [the] only simple conclusion stemming from this and much other theoretical research stimulated by Schumpeter's original conjectures is that the links between market structure, innovation, and economic welfare are extremely complex... Most... research supports a conclusion that Schumpeter overstated the advantages of large, monopolistic corporations as engines of technological change. But doubts have intruded, and important policy relevant questions remain unanswered.

Intellectual Property Rights

Intellectual property consists chiefly of patents, plant breeders' rights, copyrights, trademarks, and trade secrets. Economists have focused more attention on patents than other forms of intellectual property. The following arguments have been used to justify the existence of the patent system:

1. Patents are viewed as an important incentive to get the inventor to put in the work required to produce an invention. Particularly for the individual inventor, patent protection is claimed to be a strong incentive.
2. Patents are viewed as a major incentive to get firms to carry out the further work and make the necessary investment in pilot plants and other items that are needed to bring the invention to commercial use. If an invention became public property when made, a firm might be unwilling to incur the costs and risks involved in experimenting with a new process or product, since another firm could watch, take no risks, and duplicate the process or product if it were successful.
3. It is said that the patent laws result in inventions being disclosed earlier than otherwise, the result being that other inventions are facilitated by the earlier dissemination of the information.

Nonetheless, not all economists believe that the patent system is beneficial. Some stress that patents have been used to establish monopoly positions in many industries, and say that patents are not really important as incentives for innovation because long lead times ensure that most of the innovative profits from many types of innovations can be obtained before imitators have a chance to enter the market. Others argue that new knowledge is not used as widely under the patent system as it should be, from the viewpoint of static efficiency, because the price of the information should, according to static welfare economics, be set equal to its marginal cost, which is often practically zero. But the fly in the ointment is that this, of course, would provide no incentive for invention. In essence, a nation's patent laws must reflect a balancing of incentives for inventors and rapid diffusion of new technology.

Industrialized countries and the developing countries have often differed in their attitudes toward intellectual property rights. To the developing countries, such rights give inventors and innovators an undesirable monopoly on advanced technology that can be employed to raise prices and to impose unwarranted restrictions on the use of the technology. To them, the strong enforcement of intellectual property rights would do little to aid their own development; instead, it would tend to hinder their attempts to raise per capita income. A view commonly expressed in developing countries is that knowledge should be made available at minimal cost to everyone, since it is a common property of all, and that because the development of the relatively impoverished countries of the world is a goal that benefits everyone, the technology needed by these countries should be given to them at a low cost.

On the other hand, the industrialized countries argue that intellectual property rights must be respected to provide a fair return to the private investors who take the considerable risks involved in developing and commercializing a new technology. Unless such returns are available, the incentives for inventive and innovative activity will be impaired, to the detriment of all nations, rich or poor. Also, the industrialized countries sometimes assert that the establishment of stronger intellectual property rights would help to promote indigenous technological and innovative activities in the developing countries, although it is generally conceded that this is only one of many relevant factors influencing the indigenous rate of innovation.

Suzanne Scotchmer, in her essay reprinted here, focuses attention on the use of patent protection and cooperative agreements among firms to protect incentives for cumulative research. She concludes that:

> patent policy is a very blunt instrument trying to solve a very delicate problem. Its bluntness derives largely from the narrowness of what patent breadth can depend on, namely the realized values of the technologies. As a consequence, the prospects for fine-tuning the patent system seem limited, which may be an argument for more public sponsorship of basic research.

Janusz Ordover, in the following paper, argues that:

> weak patent protection need not be inimical to economic growth and, conversely, that strong patent protection need not be an enemy of diffusion. Appropriately structured patent law and antitrust rules can together ensure incentives for R&D and also induce cooperation among firms in diffusing R&D results through licensing and other means. At the same time, cooperation among firms at the R&D stage can counterbalance weak patent protection by internalizing spillovers from ongoing R&D programs, and such cooperation may also produce additional spillovers from the existing knowledge.

Turning to empirical studies, Mark Schwartz, Samuel Wagner, and I published a paper in 1981 looking at the effects of patents on imitation costs. For a sample of 48 product innovations, the average ratio of imitation cost to innovation cost was about 0.65, and the average ratio of imitation time to innovation time was about 0.70. Within four years of their introduction, about 60 per cent of the patented successful innovations in our sample were imitated. Nonetheless, patent protection generally increased imitation costs, particularly in the pharmaceutical industry.

Richard Levin, Alvin Klevorick, Richard Nelson, and Sidney Winter carried out a subsequent study in which they surveyed over 600 individuals representing 130 lines of business. Their paper, reprinted here, concludes that:

> The patent system and related institutions to protect intellectual property... are not the only nor necessarily the primary barriers that prevent general access to what would otherwise be pure public goods. Lead time accrues naturally to the innovator, even in the absence of any deliberate effort to enhance its protective effect. Secrecy, learning advantages, and sales and service efforts can provide additional protection, though they require the innovator's deliberate effort. The survey confirmed that these other means of appropriation are typically more important than the patent system. Hence in examining a proposed adjustment of the patent system or related institutions, it is important to recognize that the incremental effect of the policy change depends on the protection other mechanisms provide.

The Diffusion of Innovations

When a new process or product is first introduced, the diffusion process (the process by which the use of an innovation spreads) begins. The diffusion of a new technique is often a slow process, as indicated by my 1961 paper on this topic reprinted here. Measuring from the date of first commercial application, it generally took more than ten years for all of the major American firms in the bituminous coal, steel, railroad, and brewing industries to begin using a sample of important new techniques. The diffusion process often conforms to an S-shaped curve. My paper presents a model that helps to explain this fact, as well as the determinants of the rate of diffusion.

Based on studies of a number of industries, it is clear that firms where the expected returns from the innovation are highest tend to be quickest to introduce an innovation. Also, holding other factors constant, firms with younger and better-educated managers tend to be quicker to introduce new techniques – or at least, this seems to be the case in industries where firms are small. Firms also differ greatly with regard to the intrafirm rate of diffusion – the rate at which, once it has begun to use the new technique, a firm substitutes it for older methods. A considerable amount of this variation can be explained by differences among firms in the profitability of the innovation, the size of the firm, and its liquidity. Also, there is a tendency for late starters to 'catch up', i.e. firms that are slow to begin using an innovation tend to substitute it for older techniques more rapidly than those that are quick to begin using it. It is also relevant to note that the same sort of process occurs on the international scene: countries that are slow to begin using an innovation tend to substitute it for older techniques more rapidly than countries that are quick to begin using it. The reasons for this tendency, both at the firm and national levels, seem clear enough.

In 1985, Paul David published an essay, reprinted here, on the factors leading to the adoption of the dominant keyboard arrangement on typewriters. He observes that:

> There may be many eligible candidates for supremacy, and from an *ex ante* vantage point we cannot say with corresponding certainty which among the contending colors – or rival keyboard arrangements – will be the one to gain eventual dominance. That part of the story is likely to be governed by 'historical accidents'...

Michael Katz and Carl Shapiro, in the next paper reprinted here, studied the adoption of technology in industries where network externalities are important. They emphasize the role of a sponsor, defined as an entity that has property rights to the technology and thus is willing to make investments to promote it. They conclude that:

> (1) compatibility tends to be undersupplied by the market, but excessive standardization can occur; (2) in the absence of sponsors, the technology superior *today* has a strategic advantage and is likely to dominate the market; (3) when one of two rival technologies is sponsored, that technology has a strategic advantage and may be adopted even if it is inferior; (4) when two competing technologies both are sponsored, the technology that will be superior *tomorrow* has a strategic advantage.

In the following paper, Jennifer Reinganum represents the choice of the date of adoption of a new technology as a two-person nonzero-sum game:

> Under Nash behavior we find that there exists a pair of equilibria in pure strategies (for identical firms). Each Nash equilibrium involves one firm adopting relatively early with the other adopting at a relatively late date, despite the facts that information is perfect and players are identical. For non-identical firms, a Nash equilibrium always exists and is asymmetric so long as the net value of being first is positive.

International Technology Transfer

International technology transfer plays a major role in the process by which nations and firms grow and achieve economic success. A considerable amount of technology is transferred without any appreciable payment for it. For example, scientists and engineers exchange information at international meetings, and one country's scientists and engineers read the publications of other countries' scientists and engineers. Another important channel of international technology transfer is the export of goods. The mere existence or availability of a good in a foreign country may result in the transfer of technology, since the good may provide information to the importers of the good. Thus, the export of advanced computers to a particular country may result in technology transfer. Also, the country may gain technology because the exporters of the good will help the country to use the good efficiently. For example, they may help train workers and others – and this training is a form of technology transfer.

Still another way of transferring technology is through the establishment and utilization of overseas subsidiaries. In a great many cases, firms have become multinational because they have wanted to exploit a technological lead. In some cases, the only way that a firm could introduce its innovation into a foreign market was through the establishment of overseas production facilities. In establishing and utilizing foreign subsidiaries, multinational firms transfer technology in a variety of ways. They train operatives and managers, communicate information and capabilities to engineers and technicians, help the users of their products to use them more effectively, and help suppliers to upgrade their technology.

Firms with significant new products or processes may also engage in licensing agreements with foreigners covering patents, trademarks, franchise, technical assistance, and so on. Licensing agreements often call for the licensee to pay a certain percentage of its sales to the licensor, plus, in some cases, a flat fee for technical help. Further, still another way in which technology can be transferred is through the formation of a joint venture, an operation owned jointly by the firm with the technology and a firm or agency of the host country.

In a paper reprinted here, Raymond Vernon analyses international investment and international trade. His emphasis is on the timing of innovation, the effects of scale economies, and the roles of uncertainty in affecting trade patterns. He emphasizes the advantages of the US market and of US firms as sources of innovation. As the demand for a new product expands, a certain amount of standardization takes place. Eventually, US producers establish production units in other countries. The 1990s are not the 1960s, but Vernon's analysis was very influential. His paper was an early attempt to integrate technological change with the analysis of international investment and international trade.

In the next paper, Nathan Rosenberg provides an historian's view of the transfer of technology, both among and within countries. He points out that:

> One of the advantages of not taking the lead in economic development is that, once an objective has been reached and clearly demarcated, other and easier routes to attain that objective may become obvious. Or, to put the point a little differently, there is no reason that the optimal path in the development of a new technology is the same as the optimal path for transferring and adapting that technology, once it has been developed.

David Teece, in the following paper, is concerned with the size and determinants of the costs of transferring technology. He stresses that technology is by no means a set of blueprints that is usable at nominal cost to all. In his sample of international transfers of technology, transfer costs averaged about 20 per cent of total project costs. The number of previous manufacturing startups, the age of the technology, and the number of years of manufacturing experience of the transferee have a statistically significant effect on transfer costs.

The Management of Technology

The management of technology entails a great deal more than the establishment of an R&D laboratory that produces a lot of good technical output. A central problem facing a firm that attempts to be innovative is to effect a proper coupling between R&D, on the one hand, and marketing and production on the other. Many R&D projects are designed without sufficient understanding of market and production realities. Many marketing and production people are unnecessarily impervious to the good ideas produced by R&D people. To try to reduce these problems, some firms promote frequent contacts between R&D people and people in other parts of the firm, there being considerable evidence that person-to-person contacts are the most effective way of transferring ideas and technology.

Firms also try to break down resistance – in the R&D department and elsewhere – to ideas stemming from outside the firm. Even in R&D-intensive industries like chemicals and pharmaceuticals, a large proportion of the innovations are based on inventions made outside the innovating firm. Companies where problems and tasks are broken down into specialties, and where there is a strict vertical chain of command, seem to be less likely to be innovative than companies where there is no strictly defined hierarchy, where communication resembles consultation rather than command, and where individuals have to perform tasks in the light of their knowledge of the tasks of the whole organization.

One of the classic analytical tools in the management of technology is the learning curve. In 1962, Kenneth Arrow published a classic paper, reprinted here, on this topic. He suggests:

> an endogenous theory of the changes in knowledge which underlie intertemporal and international shifts in production functions. The acquisition of knowledge is what is usually termed 'learning', and we might perhaps pick up some clues from the many psychologists who have studied this phenomenon . . . I do not think that the picture of technical change as a vast and prolonged process of learning about the environment in which we operate is in any way a far-fetched analogy . . . .

Research and development is characterized by considerable uncertainty, thus prompting decision makers to consider whether they should run parallel R&D efforts. In the next paper, Richard Nelson shows:

> that if there is considerable uncertainty regarding the best way to achieve a particular objective, and if it is possible to reduce uncertainty substantially in the early stages of a development effort, the parallel running of several alternatives may be a very good strategy.

In practice, parallel R&D efforts often are used with success. For example, the IBM Corporation has reported that it has often benefited substantially from the use of parallel R&D efforts.

In the final paper, Wesley Cohen and Daniel Leventhal point out that firms rely on R&D to provide the capability to exploit externally available information, as well as to do the work leading directly to product and process innovations:

> Recognition of the dual role of R&D suggests that factors that affect the character and ease of learning will affect firms' incentives to conduct R&D. We hypothesized that such factors include the degree to which knowledge is targeted to a firm's needs, and more generally, the character of knowledge within each of the scientific and technological fields upon which innovation depends.

Without question, R&D is to a large extent an activity directed at gaining a window on the external world, and should be modelled as such by economists. Also, this fact should be taken into account in the management of technology.

Part I
Technical Change and Economic Growth

[1]

TECHNICAL CHANGE AND THE AGGREGATE PRODUCTION FUNCTION *

Robert M. Solow

IN this day of rationally designed econometric studies and super-input-output tables, it takes something more than the usual "willing suspension of disbelief" to talk seriously of the aggregate production function. But the aggregate production function is only a little less legitimate a concept than, say, the aggregate consumption function, and for some kinds of long-run macro-models it is almost as indispensable as the latter is for the short-run. As long as we insist on practicing macro-economics we shall need aggregate relationships.

Even so, there would hardly be any justification for returning to this old-fashioned topic if I had no novelty to suggest. The new wrinkle I want to describe is an elementary way of segregating variations in output per head due to technical change from those due to changes in the availability of capital per head. Naturally, every additional bit of information has its price. In this case the price consists of one new required time series, the share of labor or property in total income, and one new assumption, that factors are paid their marginal products. Since the former is probably more respectable than the other data I shall use, and since the latter is an assumption often made, the price may not be unreasonably high.

Before going on, let me be explicit that I would not try to justify what follows by calling on fancy theorems on aggregation and index numbers.[1] Either this kind of aggregate economics appeals or it doesn't. Personally I belong to both schools. If it does, I think one can draw some crude but useful conclusions from the results.

Theoretical Basis

I will first explain what I have in mind mathematically and then give a diagrammatic exposition. In this case the mathematics seems simpler. If Q represents output and K and L represent capital and labor inputs in "physical" units, then the aggregate production function can be written as:

$$Q = F(K,L;t). \qquad (1)$$

The variable t for time appears in F to allow for technical change. It will be seen that I am using the phrase "technical change" as a shorthand expression for *any kind of shift* in the production function. Thus slowdowns, speedups, improvements in the education of the labor force, and all sorts of things will appear as "technical change."

It is convenient to begin with the special case of *neutral* technical change. Shifts in the production function are defined as neutral if they leave marginal rates of substitution untouched but simply increase or decrease the output attainable from given inputs. In that case the production function takes the special form

$$Q = A(t)f(K,L) \qquad (1a)$$

and the multiplicative factor $A(t)$ measures the cumulated effect of shifts over time. Differentiate (1a) totally with respect to time and divide by Q and one obtains

$$\frac{\dot{Q}}{Q} = \frac{\dot{A}}{A} + A\frac{\partial f}{\partial K}\frac{\dot{K}}{Q} + A\frac{\partial f}{\partial L}\frac{\dot{L}}{Q}$$

where dots indicate time derivatives. Now define $w_k = \frac{\partial Q}{\partial K}\frac{K}{Q}$ and $w_L = \frac{\partial Q}{\partial L}\frac{L}{Q}$ the relative shares of capital and labor, and substitute in the above equation (note that $\partial Q/\partial K = A\,\partial f/\partial K$, etc.) and there results:

$$\frac{\dot{Q}}{Q} = \frac{\dot{A}}{A} + w_K\frac{\dot{K}}{K} + w_L\frac{\dot{L}}{L}. \qquad (2)$$

* I owe a debt of gratitude to Dr. Louis Lefeber for statistical and other assistance, and to Professors Fellner, Leontief, and Schultz for stimulating suggestions.

[1] Mrs. Robinson in particular has explored many of the profound difficulties that stand in the way of giving any precise meaning to the quantity of capital ("The Production Function and the Theory of Capital," *Review of Economic Studies*, Vol. 21, No. 2), and I have thrown up still further obstacles (ibid., Vol. 23, No. 2). Were the data available, it would be better to apply the analysis to some precisely defined production function with many precisely defined inputs. One can at least hope that the aggregate analysis gives some notion of the way a detailed analysis would lead.

From time series of $\dot{Q}/Q$, w_k, $\dot{K}/K$, w_L and $\dot{L}/L$ or their discrete year-to-year analogues, we could estimate $\dot{A}/A$ and thence $A(t)$ itself. Actually an amusing thing happens here. Nothing has been said so far about returns to scale. But if all factor inputs are classified either as K or L, then the available figures always show w_K and w_L adding up to one. Since we have assumed that factors are paid their marginal products, this amounts to assuming the hypotheses of Euler's theorem. The calculus being what it is, we might just as well assume the conclusion, namely that F is homogeneous of degree one. This has the advantage of making everything come out neatly in terms of intensive magnitudes. Let $Q/L = q$, $K/L = k$, $w_L = 1 - w_K$; note that $\dot{q}/q = \dot{Q}/Q - \dot{L}/L$ etc., and (2) becomes

$$\frac{\dot{q}}{q} = \frac{\dot{A}}{A} + w_K \frac{\dot{k}}{k}. \qquad (2a)$$

Now all we need to disentangle the technical change index $A(t)$ are series for output per man hour, capital per man hour, and the share of capital.

So far I have been assuming that technical change is neutral. But if we go back to (1) and carry out the same reasoning we arrive at something very like (2a), namely

$$\frac{\dot{q}}{q} = \frac{1}{F}\frac{\partial F}{\partial t} + w_k \frac{\dot{k}}{k} \qquad (2b)$$

It can be shown, by integrating a partial differential equation, that if $\dot{F}/F$ is independent of K and L (actually under constant returns to scale only K/L matters) then (1) has the special form (1a) and shifts in the production function are neutral. If in addition $\dot{F}/F$ is constant in time, say equal to a, then $A(t) = e^{at}$ or in discrete approximation $A(t) = (1 + a)^t$.

The case of neutral shifts and constant returns to scale is now easily handled graphically. The production function is completely represented by a graph of q against k (analogously to the fact that if we know the unit-output isoquant, we know the whole map). The trouble is that this function is shifting in time, so that if we observe points in the (q,k) plane, their movements are compounded out of movements along the curve and shifts of the curve. In Chart 1, for instance, every ordinate on the curve for $t = 1$ has been multiplied by the same factor to give a neutral upward shift of the production function for period 2. The problem is to estimate this shift from knowledge of points P_1 and P_2. Obviously it would be quite misleading to fit a curve through raw observed points like P_1, P_2 and others. But if the shift factor for each point of time can be estimated, the observed points can be corrected for technical change, and a production function can then be found.[2]

CHART 1

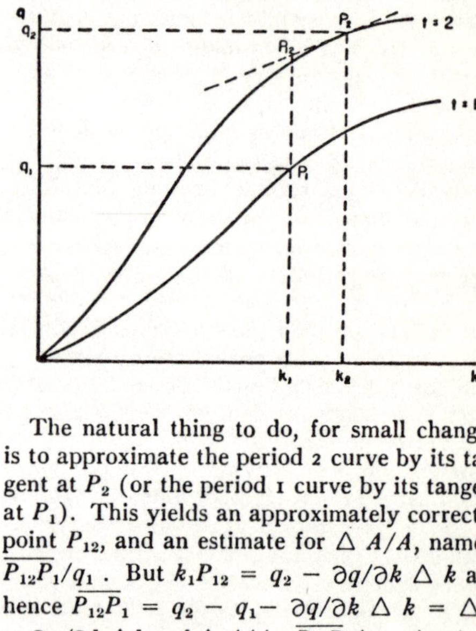

The natural thing to do, for small changes, is to approximate the period 2 curve by its tangent at P_2 (or the period 1 curve by its tangent at P_1). This yields an approximately corrected point P_{12}, and an estimate for $\Delta A/A$, namely $\overline{P_{12}P_1}/q_1$. But $k_1P_{12} = q_2 - \partial q/\partial k \, \Delta k$ and hence $\overline{P_{12}P_1} = q_2 - q_1 - \partial q/\partial k \, \Delta k = \Delta q - \partial q/\partial k \, \Delta k$ and $\Delta A/A = \overline{P_{12}P_1}/q_1 = \Delta q/q - \partial q/\partial k \, (k/q) \, \Delta k/k = \Delta q/q - w_k \, \Delta k/k$ which is exactly the content of (2a). The not-necessarily-neutral case is a bit more complicated, but basically similar.

[2] Professors Wassily Leontief and William Fellner independently pointed out to me that this "first-order" approximation could in principle be improved. After estimating a production function corrected for technical change (see below), one could go back and use it to provide a second approximation to the shift series, and on into further iterations.

An Application to the U.S.: 1909–1949

In order to isolate shifts of the aggregate production function from movements along it, by use of (2a) or (2b), three time series are needed: output per unit of labor, capital per unit of labor, and the share of capital. Some rough and ready figures, together with the obvious computations, are given in Table 1.

The conceptually cleanest measure of aggregate output would be real net national product. But long NNP series are hard to come by, so I have used GNP instead. The only difference this makes is that the share of capital has to include depreciation. It proved possible to restrict the experiment to private non-farm economic activity. This is an advantage (a) because it skirts the problem of measuring government output and (b) because eliminating agriculture is at least a step in the direction of homogeneity. Thus my q is a time series of real private non-farm GNP per man hour, Kendrick's valuable work.

The capital time series is the one that will really drive a purist mad. For present purposes, "capital" includes land, mineral deposits, etc. Naturally I have used Goldsmith's estimates (with government, agricultural, and consumer durables eliminated). Ideally what one would like to measure is the annual flow of capital services. Instead one must be content with a less utopian estimate of the stock of capital goods in existence. All sorts of conceptual problems arise on this account. As a single example, if the capital stock consisted of a million identical machines and if each one as it wore out was replaced by a more durable machine of the same annual capacity, the stock of capital as measured would surely increase. But the maximal flow of capital services would be constant. There is nothing to be done about this, but something must be done about the fact of idle capacity. What belongs in a production function is capital in use, not capital in place. Lacking any reliable year-by-year measure of the utilization of capital I have simply reduced the Goldsmith figures by the fraction of the labor force unemployed in each year, thus assuming that labor and capital always suffer unemployment to the same percentage. This is undoubtedly wrong, but probably gets closer to the truth than making no correction at all.[3]

CHART 2

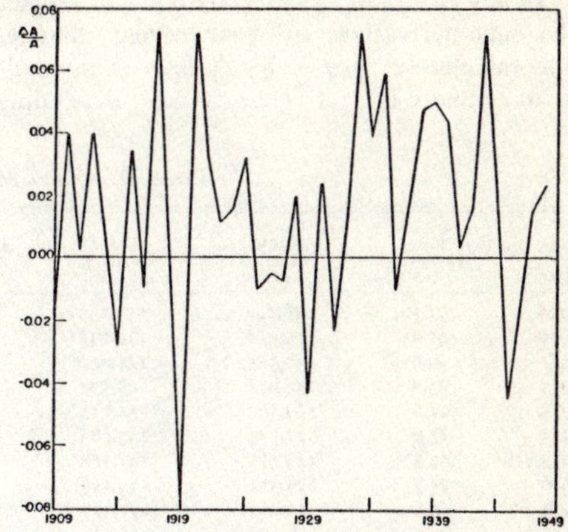

The share-of-capital series is another hodgepodge, pieced together from various sources and ad hoc assumptions (such as Gale Johnson's guess that about 35 per cent of non-farm entrepreneurial income is a return to property). Only after these computations were complete did I learn that Edward Budd of Yale University has completed a careful long-term study of factor shares which will soon be published. It seems unlikely that minor changes in this ingredient would grossly alter the final results,

CHART 3

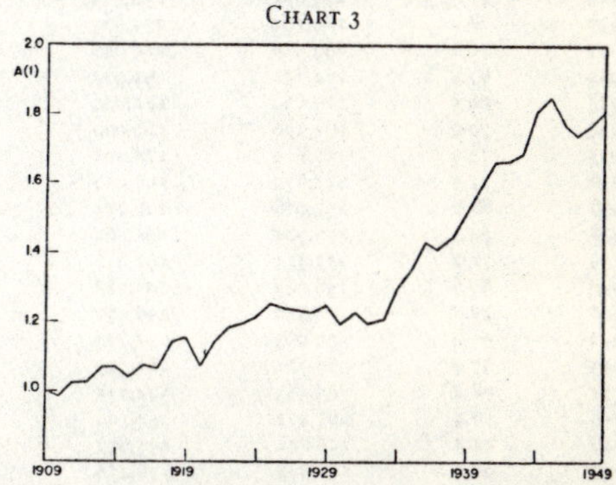

[3] Another factor for which I have not corrected is the changing length of the work-week. As the work-week shortens, the intensity of use of existing capital decreases, and the stock figures overestimate the input of capital services.

but I have no doubt that refinement of this and the capital time-series would produce neater results.

In any case, in (2a) or (2b) one can replace the time-derivatives by year-to-year changes and calculate $\Delta q/q - w_k \Delta k/k$. The result is an estimate of $\Delta F/F$ or $\Delta A/A$, depending on whether these relative shifts appear to be neutral or not. Such a calculation is made in Table 1 and shown in Chart 2. Thence, by arbitrarily setting $A(1909) = 1$ and using the fact that $A(t+1) = A(t)\ (1 + \Delta A(t)/A(t))$ one can successively reconstruct the $A(t)$ time series, which is shown in Chart 3.

TABLE 1. — DATA FOR CALCULATION OF $A(t)$

Year	% labor force employed (1)	Capital stock ($ mill.) (2)	Col. 1 x Col. 2 (3)	Share of property in income (4)	Priv. nonfarm GNP per manhour (5)	Employed capital per manhour (6)	$\Delta A/A$ (7)	$A(t)$ (8)
1909	91.1	146,142	133,135	.335	$.623	$2.06	−.017	1.000
1910	92.8	150,038	139,235	.330	.616	2.10	.039	.983
1911	90.6	156,335	141,640	.335	.647	2.17	.002	1.021
1912	93.0	159,971	148,773	.330	.652	2.21	.040	1.023
1913	91.8	164,504	151,015	.334	.680	2.23	.007	1.064
1914	83.6	171,513	143,385	.325	.682	2.20	−.028	1.071
1915	84.5	175,371	148,188	.344	.669	2.26	.034	1.041
1916	93.7	178,351	167,115	.358	.700	2.34	−.010	1.076
1917	94.0	182,263	171,327	.370	.679	2.21	.072	1.065
1918	94.5	186,679	176,412	.342	.729	2.22	.013	1.142
1919	93.1	189,977	176,869	.354	.767	2.47	−.076	1.157
1920	92.8	194,802	180,776	.319	.721	2.58	.072	1.069
1921	76.9	201,491	154,947	.369	.770	2.55	.032	1.146
1922	81.7	204,324	166,933	.339	.788	2.49	.011	1.183
1923	92.1	209,964	193,377	.337	.809	2.61	.016	1.196
1924	88.0	222,113	195,460	.330	.836	2.74	.032	1.215
1925	91.1	231,772	211,198	.336	.872	2.81	−.010	1.254
1926	92.5	244,611	226,266	.327	.869	2.87	−.005	1.241
1927	90.0	259,142	233,228	.323	.871	2.93	−.007	1.235
1928	90.0	271,089	243,980	.338	.874	3.02	.020	1.226
1929	92.5	279,691	258,714	.332	.895	3.06	−.043	1.251
1930	88.1	289,291	254,865	.347	.880	3.30	.024	1.197
1931	78.2	289,056	226,042	.325	.904	3.33	.023	1.226
1932	67.9	282,731	191,974	.397	.879	3.28	.011	1.198
1933	66.5	270,676	180,000	.362	.869	3.10	.072	1.211
1934	70.9	262,370	186,020	.355	.921	3.00	.039	1.298
1935	73.0	257,810	188,201	.351	.943	2.87	.059	1.349
1936	77.3	254,875	197,018	.357	.982	2.72	−.010	1.429
1937	81.0	257,076	208,232	.340	.971	2.71	.021	1.415
1938	74.7	259,789	194,062	.331	1.000	2.78	.048	1.445
1939	77.2	257,314	198,646	.347	1.034	2.66	.050	1.514
1940	80.6	258,048	207,987	.357	1.082	2.63	.044	1.590
1941	86.8	262,940	228,232	.377	1.122	2.58	.003	1.660
1942	93.6	270,063	252,779	.356	1.136	2.64	.016	1.665
1943	97.4	269,761	262,747	.342	1.180	2.62	.071	1.692
1944	98.4	265,483	261,235	.332	1.265	2.63	.021	1.812
1945	96.5	261,472	252,320	.314	1.296	2.66	−.044	1.850
1946	94.8	258,051	244,632	.312	1.215	2.50	−.017	1.769
1947	95.4	268,845	256,478	.327	1.194	2.50	.016	1.739
1948	95.7	276,476	264,588	.332	1.221	2.55	.024	1.767
1949	93.0	289,360	269,105	.326	1.275	2.70	...	1.809

NOTES AND SOURCES:

Column (1): Percentage of labor force employed. 1909–26, from Douglas, *Real Wages in the United States* (Boston and New York, 1930), 460. 1929–49, calculated from *The Economic Almanac*, 1953–54 (New York, 1953), 426–28.

Column (2): Capital Stock. From Goldsmith, *A Study of Saving in the United States*, Vol. 3 (Princeton, 1956), 20–21, sum of columns 5, 6, 7, 9, 12, 17, 22, 23, 24.

Column (3): (1) x (2).

Column (4): Share of property in income. Compiled from *The Economic Almanac*, 504–505; and Jesse Burkhead, "Changes in the Functional Distribution of Income," *Journal of the American Satistical Association*, Vol. 48 (June 1953), 192–219. Depreciation estimates from Goldsmith, 427.

Column (5): Private nonfarm GNP per man hour, 1939 dollars. Kendrick's data, reproduced in *The Economic Almanac*, 490.

Column (6): Employed capital per man hour. Column (3) divided by Kendrick's man hour series, *ibid.*

Column (7): $\Delta A/A = \Delta (5)/(5) - (4) \times \Delta (6)/(6)$.

Column (8): From (7).

I was tempted to end this section with the remark that the $A(t)$ series, which is meant to be a rough profile of technical change, at least looks reasonable. But on second thought I decided that I had very little prior notion of what would be "reasonable" in this context. One notes with satisfaction that the trend is strongly upward; had it turned out otherwise I would not now be writing this paper. There are sharp dips after each of the World Wars; these, like the sharp rises that preceded them, can easily be rationalized. It is more suggestive that the curve shows a distinct levelling-off in the last half of the 1920's. A sustained rise begins again in 1930. There is an unpleasant sawtooth character to the first few years of the $\Delta A/A$ curve, which I imagine to be a statistical artifact.

The Outlines of Technical Change

The reader will note that I have already drifted into the habit of calling the curve of Chart 2 $\Delta A/A$ instead of the more general $\Delta F/F$. In fact, a scatter of $\Delta F/F$ against K/L (not shown) indicates no trace of a relationship. So I may state it as a formal conclusion that over the period 1909–49, shifts in the aggregate production function netted out to be approximately neutral. Perhaps I should recall that I have defined neutrality to mean that the shifts were pure scale changes, leaving marginal rates of substitution unchanged at given capital/labor ratios.

Not only is $\Delta A/A$ uncorrelated with K/L, but one might almost conclude from the graph that $\Delta A/A$ is essentially constant in time, exhibiting more or less random fluctuations about a fixed mean. Almost, but not quite, for there does seem to be a break at about 1930. There is some evidence that the average rate of progress in the years 1909–29 was smaller than that from 1930–49. The first 21 relative shifts average about 9/10 of one per cent per year, while the last 19 average 2¼ per cent per year. Even if the year 1929, which showed a strong downward shift, is moved from the first group to the second, there is still a contrast between an average rate of 1.2 per cent in the first half and 1.9 per cent in the second. Such *post hoc* splitting-up of a period is always dangerous. Perhaps I should leave it that there is some evidence that technical change (broadly interpreted) may have accelerated after 1929.

The over-all result for the whole 40 years is an average upward shift of about 1.5 per cent per year. This may be compared with a figure of about .75 per cent per year obtained by Stefan Valavanis-Vail by a different and rather less general method, for the period 1869–1948.[4] Another possible comparison is with the output-per-unit-of-input computations of Jacob Schmookler,[5] which show an increase of some 36 per cent in output per unit of input between the decades 1904–13 and 1929–38. Our $A(t)$ rises 36.5 per cent between 1909 and 1934. But these are not really comparable estimates, since Schmookler's figures include agriculture.

As a last general conclusion, after which I will leave the interested reader to his own impressions, over the 40 year period output per man hour approximately doubled. At the same time, according to Chart 2, the cumulative upward shift in the production function was about 80 per cent. It is possible to argue that about one-eighth of the total increase is traceable to increased capital per man hour, and the remaining seven-eighths to technical change. The reasoning is this: real GNP per man hour increased from \$.623 to \$1.275. Divide the latter figure by 1.809, which is the 1949 value for $A(t)$, and therefore the full shift factor for the 40 years. The result is a "corrected" GNP per man hour, net of technical change, of \$.705. Thus about 8 cents of the 65 cent increase can be imputed to increased capital intensity, and the remainder to increased productivity.[6]

Of course this is not meant to suggest that the observed rate of technical progress would have persisted even if the rate of investment had been much smaller or had fallen to zero. Obviously much, perhaps nearly all, innovation must be embodied in new plant and equipment to be realized at all. One could imagine this process taking place without net capital for-

[4] S. Valavanis-Vail, "An Econometric Model of Growth, U.S.A. 1869–1953," *American Economic Review, Papers and Proceedings*, XLV (May 1955), 217.

[5] J. Schmookler, "The Changing Efficiency of the American Economy, 1869–1938," this REVIEW (August 1952), 226.

[6] For the first half of the period, 1909–29, a similar computation attributes about one-third of the observed increase in GNP per man-hour to increased capital intensity.

mation as old-fashioned capital goods are replaced by the latest models, so that the capital-labor ratio need not change systematically. But this raises problems of definition and measurement even more formidable than the ones already blithely ignored. This whole area of interest has been stressed by Fellner.

For comparison, Solomon Fabricant[7] has estimated that over the period 1871–1951 about 90 per cent of the increase in output per capita is attributable to technical progress. Presumably this figure is based on the standard sort of output-per-unit-of-input calculation.

It might seem at first glance that calculations of output per unit of resource input provide a relatively assumption-free way of measuring productivity changes. Actually I think the implicit load of assumptions is quite heavy, and if anything the method proposed above is considerably more general.

Not only does the usual choice of weights for computing an aggregate resource-input involve something analogous to my assumption of competitive factor markets, but in addition the criterion output ÷ a weighted sum of inputs would seem tacitly to *assume* (a) that technical change is neutral and (b) that the aggregate production function is *strictly* linear. This explains why numerical results are so closely parallel for the two methods. We have already verified the neutrality, and as will be seen subsequently, a strictly linear production function gives an excellent fit, though clearly inferior to some alternatives.[8]

[7] S. Fabricant, "Economic Progress and Economic Change," *34th Annual Report of the National Bureau of Economic Research* (New York, 1954).

[8] For an excellent discussion of some of the problems, see M. Abramovitz "Resources and Output Trends in the U.S. since 1870," *American Economic Review, Papers and Proceedings*, XLVI (May 1956), 5–23. Some of the questions there raised could in principle be answered by the method used here. For example, the contribution of improved quality of the labor force could be handled by introducing various levels of skilled labor as separate inputs. I owe to Prof. T. W. Schultz a heightened awareness that a lot of what appears as shifts in the production function must represent improvement in the quality of the labor input, and therefore a result of real capital formation of an important kind. Nor ought it be forgotten that even straight technical progress has a cost side.

The Aggregate Production Function

Returning now to the aggregate production function, we have earned the right to write it in the form (1a). By use of the (practically unavoidable) assumption of constant returns to scale, this can be further simplified to the form

$$q = A(t)f(k,1), \qquad (3)$$

which formed the basis of Chart 1. It was there noted that a simple plot of q against k would give a distorted picture because of the shift factor $A(t)$. Each point would lie on a different member of the family of production curves. But we have now provided ourselves with an estimate of the successive values of the shift factor. (Note that this estimate is quite *independent* of any hypothesis about the exact shape of the production function.) It follows from (3) that by plotting $q(t)/A(t)$ against $k(t)$ we reduce all the observed points to a *single* member of the family of curves in Chart 1, and we can then proceed to discuss the shape of $f(k,1)$ and reconstruct the aggregate production function. A scatter of q/A against k is shown in Chart 4.

CHART 4

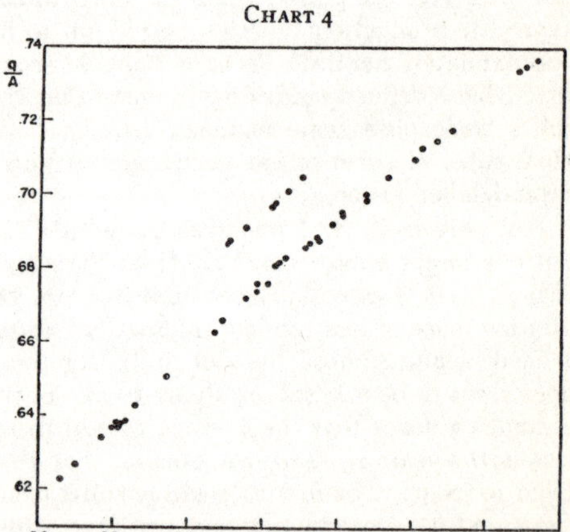

Considering the amount of *a priori* doctoring which the raw figures have undergone, the fit is remarkably tight. Except, that is, for the layer of points which are obviously too high. These maverick observations relate to the seven last years of the period, 1943–49. From the way they lie almost exactly parallel to the main scatter, one is tempted to conclude that in 1943 the aggregate production function simply shifted. But the whole earlier procedure was designed to purify those points from shifts in the function, so that way out would seem to be closed. I suspect the explanation may lie in some systematic incomparability of the capital-in-use series. In particular during the war there was almost certainly a more intensive use of capital services through two- and three-shift operation than the stock figures would show, even with the crude correction that has been applied. It is easily seen that such an underestimate of capital inputs leads to an overestimate of productivity increase. Thus in effect each of the affected points should really lie higher and toward the right. But further analysis shows that, for the orders of magnitude involved, the net result would be to pull the observations closer to the rest of the scatter.

At best this might account for 1943–1945. There remains the postwar period. Although it is possible that multi-shift operation remained fairly widespread even after the war, it is unlikely that this could be nearly enough to explain the whole discrepancy.[9] One might guess that accelerated amortization could have resulted in an underestimate of the capital stock after 1945. Certainly other research workers, notably Kuznets and Terborgh, have produced capital stock estimates which rather exceed Goldsmith's at the end of the period. But for the present, I leave this a mystery.

In a first version of this paper, I resolutely let the recalcitrant observations stand as they were in a regression analysis of Chart 4, mainly because such casual amputation is a practice I deplore in others. But after some experimentation it seemed that to leave them in only led to noticeable distortion of the results. So, with some misgivings, in the regressions that follow I have omitted the observations for 1943–1949. It would be better if they could be otherwise explained away.

Chart 4 gives an inescapable impression of curvature, of persistent but not violent diminishing returns. As for the possibility of approaching capital-saturation, there is no trace on this gross product level, but even setting aside all other difficulties, such a scatter confers no particular license to guess about what happens at higher K/L ratios than those observed.

As for fitting a curve to the scatter, a Cobb-Douglas function comes immediately to mind, but then so do several other parametric forms, with little to choose among them.[10] I can't help feeling that little or nothing hangs on the choice of functional form, but I have experimented with several. In general I limited myself to two-parameter families of curves, linear in the parameters (for computational convenience), and at least capable of exhibiting diminishing returns (except for the straight line, which on this account proved inferior to all others).

The particular possibilities tried were the following:

$$q = \alpha + \beta k \qquad (4a)$$
$$q = \alpha + \beta \log k \qquad (4b)$$
$$q = \alpha - \beta/k \qquad (4c)$$
$$\log q = \alpha + \beta \log k \qquad (4d)$$
$$\log q = \alpha - \beta/k. \qquad (4e)$$

Of these, (4d) is the Cobb-Douglas case; (4c and e) have upper asymptotes; the semi-logarithmic (4b) and the hyperbolic (4c) must cross the horizontal axis at a positive value of k and continue ever more steeply but irrelevantly downward (which means only that some positive k must be achieved before any output is forthcoming, but this is far outside the range of observation); (4e) begins at the origin with a phase of increasing returns and ends with a phase of diminishing returns — the point of inflection occurs at $k = \beta/2$ and needless to say all our observed points come well to the right of this.

The results of fitting these five curves to the scatter of Chart 4 are shown in Table 2.

The correlation coefficients are uniformly so high that one hesitates to say any more than

[9] It is cheering to note that Professor Fellner's new book voices a suspicion that the postwar has seen a substantial increase over prewar in the prevalence of multi-shift operation. See *Trends and Cycles in Economic Activity* (New York, 1956), 92.

[10] A discussion of the same problem in a different context is to be found in Prais and Houthakker, *The Analysis of Family Budgets* (Cambridge, England, 1955), 82–88. See also S. J. Prais, "Non-Linear Estimates of the Engel Curves," *Review of Economic Studies*, No. 52 (1952–53), 87–104.

TABLE 2

Curve	a	β	r
4a	.438	.091	.9982
b	.448	.239	.9996
c	.917	.618	.9964
d	−.729	.353	.9996
e	−.038	.913	.9980

that all five functions, even the linear one, are about equally good at representing the general shape of the observed points. From the correlations alone, for what they are worth, it appears that the Cobb-Douglas function (4d) and the semilogarithmic (4b) are a bit better than the others.[11]

Since all of the fitted curves are of the form $g(y) = a + \beta\, h(x)$, one can view them all as linear regressions and an interesting test of goodness of fit proposed by Prais and Houthakker (*ibid.*, page 51) is available. If the residuals from each regression are arranged in order of increasing values of the independent variable, then one would like this sequence to be disposed "randomly" about the regression line. A strong "serial" correlation in the residuals, or a few long runs of positive residuals alternating with long runs of negative residuals, would be evidence of just that kind of smooth departure from linearity that one would like to catch. A test can be constructed using published tables of critical values for runs of two kinds of elements.

[11] It would be foolhardy for an outsider (or maybe even an insider) to hazard a guess about the statistical properties of the basic time series. A few general statements can be made, however. (a) The natural way to introduce an error term into the aggregate production function is multiplicatively: $Q = (1 + u)F(K,L;t)$. In the neutral case it is apparent that the error factor will be absorbed into the estimated $A(t)$. Then approximately the error in $\Delta A/A$ will be $\Delta u/1 + u$. If u has zero mean, the variance of the estimated $\Delta A/A$ will be approximately $2(1 - \rho)$ var u, where ρ is the first autocorrelation of the u series. (b) Suppose that marginal productivity distribution doesn't hold exactly, so that $K/Q \partial Q/\partial K = w_k + v$, where now v is a random deviation and w_k is the share of property income. Then the error in the estimated $\Delta A/A$ will be $v\, \Delta k/k$, with variance $(\Delta k/k)^2$ var v. Since K/L changes slowly, the multiplying factor will be very small. The effect is to bias the estimate of $\Delta A/A$ in such a way as to lead to an overestimate when property receives less than its marginal product (and k is increasing). (c) Errors in estimating $A(t)$ enter in a relatively harmless way so far as the regression analysis is concerned. Errors of observation in k will be more serious and are likely to be large. The effect will of course be to bias the estimates of β downward.

This has been done for the linear, semilogarithmic, and Cobb-Douglas functions. The results strongly confirm the visual impression of diminishing returns in Chart 4, by showing the linear function to be a systematically poor fit. As between (4b) and (4d) there is little to choose.[12]

A Note on Saturation

It has already been mentioned that the aggregate production function shows no signs of levelling off into a stage of capital-saturation. The two curves in Table 2 which have upper asymptotes (c and e) happen to locate that asymptote at about the same place. The limiting values of q are, respectively, .92 and .97. Of course these are both true asymptotes, approached but not reached for any finite value of k. It could not be otherwise: no analytic function can suddenly level off and become constant unless it has always been constant. But on the other hand, there is no reason to expect nature to be infinitely differentiable. Thus any conclusions extending beyond the range actually observed in Chart 4 are necessarily treacherous. But, tongue in cheek, if we take .95 as a guess at the saturation level of q, and use the *linear* function (4a) (which will get there first) as a lower-limit guess at the saturation level for k, it turns out to be about 5.7, more than twice its present value.

But all this is in terms of *gross output*, whereas for analytic purposes we are interested in the *net* productivity of capital. The difference between the two is depreciation, a subject about which I do not feel able to make guesses. If there were more certainty about the meaning of existing estimates of depreciation, especially over long periods of time, it would have been better to conduct the whole analysis in terms of net product.

However, one can say this. Zero net marginal productivity of capital sets in when gross marginal product falls to the "marginal rate of depreciation," i.e. when adding some capital adds only enough product to make good the depreciation on the increment of capital itself. Now in recent years NNP has run a bit over

[12] The test statistic is R, the total number of runs, with small values significant. For (4a), $R = 4$; for (4b), $R = 13$. The 1% critical value in both cases is about 9.

90 per cent of GNP, so capital consumption is a bit under 10 per cent of gross output. From Table 1 it can be read that capital per unit of output is, say, between 2 and 3. Thus annual depreciation is between 3 and 5 per cent of the capital stock. Capital-saturation would occur whenever the gross marginal product of capital falls to .03–.05. Using (4b), this would happen at K/L ratios of around 5 or higher, still well above anything ever observed.[13]

Summary

This paper has suggested a simple way of segregating shifts of the aggregate production function from movements along it. The method rests on the assumption that factors are paid their marginal products, but it could easily be extended to monopolistic factor markets. Among the conclusions which emerge from a crude application to American data, 1909–49, are the following:

1. Technical change during that period was neutral on average.

2. The upward shift in the production function was, apart from fluctuations, at a rate of about one per cent per year for the first half of the period and 2 per cent per year for the last half.

3. Gross output per man hour doubled over the interval, with 87½ per cent of the increase attributable to technical change and the remaining 12½ per cent to increased use of capital.

4. The aggregate production function, corrected for technical change, gives a distinct impression of diminishing returns, but the curvature is not violent.

[13] And this is under relatively pessimistic assumptions as to how technical change itself affects the rate of capital consumption. A warning is in order here: I have left Kendrick's GNP data in 1939 prices and Goldsmith's capital stock figures in 1929 prices. Before anyone uses the β's of Table 2 to reckon a yield on capital or any similar number, it is necessary to convert Q and K to a comparable price basis, by an easy calculation.

[2]

Endogenous Technological Change

Paul M. Romer

University of Chicago

Growth in this model is driven by technological change that arises from intentional investment decisions made by profit-maximizing agents. The distinguishing feature of the technology as an input is that it is neither a conventional good nor a public good; it is a nonrival, partially excludable good. Because of the nonconvexity introduced by a nonrival good, price-taking competition cannot be supported. Instead, the equilibrium is one with monopolistic competition. The main conclusions are that the stock of human capital determines the rate of growth, that too little human capital is devoted to research in equilibrium, that integration into world markets will increase growth rates, and that having a large population is not sufficient to generate growth.

I. Introduction

Output per hour worked in the United States today is 10 times as valuable as output per hour worked 100 years ago (Maddison 1982). In the 1950s, economists attributed almost all the change in output per hour worked to technological change (Abramovitz 1956; Kendrick 1956; Solow 1957). Subsequent analysis raised our estimates of

Prepared for the conference "The Problem of Economic Development: Exploring Economic Development through Free Enterprise," held at the State University of New York at Buffalo, May 1988. I have benefited from the comments of many seminar and conference participants and two discussants (Rob Vishny, Buffalo, May 1988, and Dale Jorgenson, National Bureau of Economic Research Economic Fluctuations meeting, July 1988). Discussions with Gary Becker, Karl Shell, Robert Lucas, Gene Grossman, and Elhanan Helpman were especially helpful. Research assistance was provided by Danyang Xie. The original work was supported by National Science Foundation grant SES-8618325. It was revised while I was a visitor at the Center for Advanced Study in the Behavioral Sciences and supported by NSF grant BNS87-00864.

[*Journal of Political Economy*, 1990, vol. 98, no. 5, pt. 2]

the importance of increases in the effective labor force and the effective stock of capital in generating growth in output per worker (Jorgenson, Gollop, and Fraumeni 1987), but technological change has surely been important as well. The raw materials that we use have not changed, but as a result of trial and error, experimentation, refinement, and scientific investigation, the instructions that we follow for combining raw materials have become vastly more sophisticated. One hundred years ago, all we could do to get visual stimulation from iron oxide was to use it as a pigment. Now we put it on plastic tape and use it to make videocassette recordings.

The argument presented in this paper is based on three premises. The first is that technological change—improvement in the instructions for mixing together raw materials—lies at the heart of economic growth. As a result, the model presented here resembles the Solow (1956) model with technological change. Technological change provides the incentive for continued capital accumulation, and together, capital accumulation and technological change account for much of the increase in output per hour worked.

The second premise is that technological change arises in large part because of intentional actions taken by people who respond to market incentives. Thus the model is one of endogenous rather than exogenous technological change. This does not mean that everyone who contributes to technological change is motivated by market incentives. An academic scientist who is supported by government grants may be totally insulated from them. The premise here is that market incentives nonetheless play an essential role in the process whereby new knowledge is translated into goods with practical value. Our initial understanding of electromagnetism arose from research conducted in academic institutions, but magnetic tape and home videocassette recorders resulted from attempts by private firms to earn a profit.

The third and most fundamental premise is that instructions for working with raw materials are inherently different from other economic goods. Once the cost of creating a new set of instructions has been incurred, the instructions can be used over and over again at no additional cost. Developing new and better instructions is equivalent to incurring a fixed cost. This property is taken to be the defining characteristic of technology.

Most models of aggregate growth, even those with spillovers or external effects, rely on price-taking behavior. But once these three premises are granted, it follows directly that an equilibrium with price taking cannot be supported. Section II of the paper starts by showing why this is so. It also indicates which of the premises is dropped in growth models that do depend on price-taking behavior. The argument in this section is fundamental to the motivation for the particular model of monopolistic competition that follows, but it is more general than the model itself.

In the specific model outlined in Section III, a firm incurs fixed design or research and development costs when it creates a new good. It recovers those costs by selling the new good for a price that is higher than its constant cost of production. Since there is free entry into this activity, firms earn zero profit in a present value sense.

The conclusions of the model follow directly from this specification. On the basis of results from the static theory of trade with differentiated goods (see, e.g., Helpman and Krugman 1985), one should expect that fixed costs lead to gains from increases in the size of the market and therefore to gains from trade between different countries. Perhaps the most interesting feature of the equilibrium calculated for the model constructed here is that increases in the size of the market have effects not only on the level of income and welfare but also on the rate of growth. Larger markets induce more research and faster growth.

The analysis also suggests why population is not the right measure of market size and why the presence of a large domestic market in countries such as China or India is not a substitute for trade with the rest of the world. The growth rate is increasing in the stock of human capital, but it does not depend on the total size of the labor force or the population. In a limiting case that may be relevant for historical analysis and for the poorest countries today, if the stock of human capital is too low, growth may not take place at all.

These implications of the model are taken up briefly in the final sections of the paper. Section III describes the functional forms that are used to describe the preferences and the technology for the model. It defines an equilibrium that allows for both monopolistic competition and external effects arising from knowledge spillovers. Section IV offers a brief intuitive description of a balanced growth equilibrium for the model. Section V formally characterizes the equilibrium. Section VI describes the welfare properties of the equilibrium. Section VII discusses the connection implied by the model between trade, research, and growth. Algebraic details of the derivations are placed in the Appendix.

II. Rivalry, Excludability, and Nonconvexities

Economists studying public finance have identified two fundamental attributes of any economic good: the degree to which it is rivalrous and the degree to which it is excludable (Cornes and Sandler 1986). Rivalry is a purely technological attribute. A purely rival good has the property that its use by one firm or person precludes its use by another; a purely nonrival good has the property that its use by one firm or person in no way limits its use by another. Excludability is a function of both the technology and the legal system. A good is excludable if the owner can prevent others from using it. A good such as the code for a computer program can be made excludable by means of a legal system that prohibits copying or by means of encryption and copy protection schemes.

Conventional economic goods are both rivalrous and excludable. They are privately provided and can be traded in competitive markets. By definition, public goods are both nonrival and nonexcludable. Because they are nonexcludable, they cannot be privately provided or traded in markets. Public goods can be introduced into a model of price-taking behavior by assuming the existence of a government that can levy taxes. Basic scientific research is an example of a public good that could be provided in this way and that is relevant for modeling growth.

Rivalry and excludability are closely linked because most rival goods are excludable. (A parking space in a shopping center parking lot is an example of a good that is effectively nonexcludable because the cost of enforcing excludability is too high relative to the value of the good.) The interesting case for growth theory is the set of goods that are nonrival yet excludable. The third premise cited in the Introduction implies that technology is a nonrival input. The second premise implies that technological change takes place because of the actions of self-interested individuals, so improvements in the technology must confer benefits that are at least partially excludable. The first premise therefore implies that growth is driven fundamentally by the accumulation of a partially excludable, nonrival input.

To evaluate these claims, it helps to have a specific case in mind. The example of a nonrival input used in what follows is a design for a new good. The vast majority of designs result from the research and development activities of private, profit-maximizing firms. A design is, nonetheless, nonrival. Once the design is created, it can be used as often as desired, in as many productive activities as desired.

In this sense, a design differs in a crucial way from a piece of human capital such as the ability to add. The design is nonrival but the ability to add is not. The difference arises because the ability to add is inherently tied to a physical object (a human body) whereas the design is not.[1] The ability to add is rivalrous because the person who

[1] The original version of this paper used the terms "embodied" and "disembodied" to refer to the difference between an intangible such as the ability to add, which is tied to a specific person, and an intangible such as a design, which is not. This choice of terminology is not used in this revision because embodiment has another meaning in growth theory and because the notion of rivalry already exists in the public finance literature.

possesses this ability cannot be in more than one place at the same time; nor can this person solve many problems at once. As noted above, rivalry leads to a presumption that human capital is also excludable. Thus human capital can be privately provided and traded in competitive markets. In contrast, the design is nonrival because it is independent of any physical object. It can be copied and used in as many different activities as desired.

Like any scientific concept, nonrivalry is an idealization. It is sometimes observed that a design cannot be a nonrival good because it is itself tied to the physical piece of paper or the physical computer disk on which it is stored. What is unambiguously true about a design is that the cost of replicating it with a drafter, a photocopier, or a disk drive is trivial compared to the cost of creating the design in the first place. This is not true of the ability to add. Training the second person to add is as costly as training the first. For simplicity, the arguments here will treat designs as idealized goods that are not tied to any physical good and can be costlessly replicated, but nothing hinges on whether this is literally true or merely close to being true.

Nonrivalry has two important implications for the theory of growth. First, nonrival goods can be accumulated without bound on a per capita basis, whereas a piece of human capital such as the ability to add cannot. Each person has only a finite number of years that can be spent acquiring skills. When this person dies, the skills are lost, but any nonrival good that this person produces—a scientific law; a principle of mechanical, electrical, or chemical engineering; a mathematical result; software; a patent; a mechanical drawing; or a blueprint—lives on after the person is gone. Second, treating knowledge as a nonrival good makes it possible to talk sensibly about knowledge spillovers, that is, incomplete excludability. These two features of knowledge—unbounded growth and incomplete appropriability—are features that are generally recognized as being relevant for the theory of growth. What thinking about nonrivalry shows is that these features are inextricably linked to nonconvexities.

If a nonrival input has productive value, then output cannot be a constant-returns-to-scale function of all its inputs taken together. The standard replication argument used to justify homogeneity of degree one does not apply because it is not necessary to replicate nonrival inputs. Suppose that a firm can invest 10,000 hours of engineering time to produce a design for a 20-megabyte hard disk drive for computers. Suppose that it can produce a total of 2 trillion megabytes of storage per year (i.e., 100,000 units of the drive) if it builds a $10 million factory and hires 100 workers. If it merely replicates the rival inputs—the factory and the workers—it can double its output to 4 trillion megabytes of storage per year.

Suppose that the firm could have invested 20,000 hours of engineering time in the design work instead of 10,000 hours and, by doing so, could have designed a 30-megabyte hard disk drive that could be manufactured with the same factory and workers. When the firm doubles all its inputs, it uses a 20,000-hour design, two factories, and 200 workers and produces 6 trillion megabytes of storage per year, three times the original output.

More formally, if $F(A, X)$ represents a production process that depends on rival inputs X and nonrival inputs A, then by a replication argument, it follows that $F(A, \lambda X) = \lambda F(A, X)$. This replication argument assumes that X is an exhaustive list of the rival inputs. Because the focus here is on national economies, the argument neglects integer problems that may be relevant for a small market that gets stuck between n and $n + 1$ plants. The fact that it may not be possible to actually replicate all the inputs in the list X has no bearing on this argument about the properties of $F(\cdot)$.

If A is productive as well, it follows that F cannot be a concave production function because $F(\lambda A, \lambda X) > \lambda F(A, X)$. Because of the properties of homogeneous functions, it also follows that a firm with these kinds of production possibilities could not survive as a price taker. If disk drives sold for marginal cost, annual revenue for the firm would just equal interest payments on the capital and wage payments to workers. More generally, since $F(A, X) = X \cdot (\partial F/\partial X)(A, X)$, it follows that

$$F(A, X) < A \cdot \frac{\partial F}{\partial A}(A, X) + X \cdot \frac{\partial F}{\partial X}(A, X).$$

If all inputs were paid their value marginal product, the firm would suffer losses.

This point has been made many times before (Schumpeter 1942; Arrow 1962*b*; Shell 1966, 1967, 1973; Nordhaus 1969; Wilson 1975). Previous growth models have avoided this difficulty in various ways. Solow (1956) treats A as an exogenously provided public input (i.e., an input that is both nonexcludable and nonrival). Shell (1966, 1967) treats it as a public input that is provided by the government. In each case, the factor A receives no compensation, and every individual firm is assumed to be free to exploit the entire stock of A. These models are consistent with the first premise, that technological change drives growth, and the third, that the technology is a nonrival good, but they are inconsistent with the second premise. They both deny the role that private, maximizing behavior plays in generating technological change.

In an attempt to make the evolution of A responsive to market incentives, Arrow (1962*a*) assumed that an increase in K necessarily leads to an equiproportionate increase in knowledge through "learning by doing," but he still treats knowledge as a public good. Lucas (1988) assumed in effect that it is production of human capital rather than physical capital that generates this nonrival, nonexcludable good. Both of these papers make the production of a nonrival, nonexcludable good an unintentional side effect of the production of a conventional good.

The learning-by-doing formulation has the advantage that it makes the rate of accumulation of nonrival knowledge endogenous, but it is unsatisfactory because it takes the strict proportionality between knowledge and physical capital or knowledge and education as an unexplained and exogenously given feature of the technology. It preserves the public-good character of knowledge assumed by Solow and Shell but makes it a public good that is privately provided as a side effect. Like the other public-good formulations, it rules out the possibility that firms make intentional investments in research and development.

This formulation has the additional difficulty that it is not robust. The nonrival input produced through learning by doing must be completely nonexcludable. If it were even partially excludable, Dasgupta and Stiglitz (1988) show that decentralized equilibrium with many firms would not be sustainable.

In a partial equilibrium model of an industry in which firms face upward-sloping cost curves, Shell (1973) proposed a model with price taking in which expenditure on research was compensated out of quasi rents. Griliches (1979), again in an industry setting, made this formulation more explicit. He assumed that the production function takes the form $F(A_N, A_E, X)$, where A_E represents an excludable part of the benefits of research and development and A_N represents the nonexcludable part. Since A_E is excludable, it is accumulated intentionally. The nonexcludable part A_N is created as a side effect of producing A_E. He also assumed that the function $F(\cdot)$ is homogeneous of degree one in X and A_E taken together.

In an aggregate model of growth, I made the same kind of assumption (Romer 1986). To make the dynamic analysis in this paper simple, I reduced the dynamic model to one with a single-state-variable model by assuming that the excludable good A_E that the firm produces intentionally is used in fixed proportions with physical capital. As a result, the model ends up having dynamics similar to those of Arrow's learning-by-doing model, and the mathematical equations can be interpreted equally well in terms of learning by doing that is incidental to capital production.

The advantage of the interpretation that knowledge is compensated out of quasi rents is that it allows for intentional private investments in research and development. The difficulty is that it violates the logic of the replication argument. If the input A_E is truly the result of research and development, it is a nonrival good. In this case, the function $F(\cdot)$ must be homogeneous of degree one in X alone and the Dasgupta and Stiglitz (1988) argument applies. If a nonrival input is even partially excludable, nonconvexities are present and a decentralized equilibrium cannot be sustained. Stated in terms of cost curves, the replication argument implies that long-run cost curves are horizontal when all rival inputs are treated as variable. What appear to be quasi rents are merely competitive returns to rival factors that are in fixed supply. These quasi rents cannot be used to compensate both the innovation activity and the rival fixed factors. Even if these factors are in fixed supply at the aggregate level, they presumably have alternative uses and will not be supplied to an activity if they are not paid their marginal product.

An alternative approach to growth theory with price-taking competition is to dispute the first and third premises, that technological change drives growth and that knowledge about the technology is a nonrival input. Human capital models such as those presented by King and Rebelo (1987), Jones and Manuelli (1988), Rebelo (1988), and Becker, Murphy, and Tamura (this issue) treat all forms of intangible knowledge as being analogous to human capital skills that are rivalrous and excludable. There is no nonrival input like the technology and, hence, no nonconvexities or spillovers.

The only way to accept all three premises described in the Introduction is to return to the suggestion of Schumpeter (1942) and explicitly introduce market power. Shell (1973) described a model with a single monopolist who invests in technological change, but as he recognizes, it is difficult to give an aggregate interpretation to a model dominated by a single firm. In a recent paper (Romer 1987), I presented a model with market power but also with free entry and many firms. It builds on the model of monopolistic competition in consumption goods formulated by Dixit and Stiglitz (1977), applied in a dynamic setting by Judd (1985), used in a dynamic model of trade by Grossman and Helpman (1989*c*), and extended to differentiated inputs in production by Ethier (1982). The specification used here differs from that in my earlier paper primarily because it emphasizes the importance of human capital in the research process. The earlier model showed that scale is an important determinant of the rate of growth. The analysis here shows that the correct measure of scale is not population but human capital.

III. Description of the Model

The four basic inputs in this model are capital, labor, human capital, and an index of the level of the technology. Capital is measured in units of consumption goods. Labor services L are skills such as eye-hand coordination that are available from a healthy physical body. They are measured by counts of people. As used here, human capital H is a distinct measure of the cumulative effect of activities such as formal education and on-the-job training.

The concept of human capital as years of education or training that are person specific is close to the one used in labor market contexts (e.g., Heckman 1976; Rosen 1976). It corresponds to the practice in growth accounting applications that take account of changes in the quality of the labor force due to changes in observables such as the level of education and experience (see, e.g., Gollop and Jorgenson 1980). This concept of human capital is more limited than the notion used in theoretical models of growth based on unlimited human capital accumulation such as those presented by King and Rebelo (1987), Lucas (1988), and Becker et al. (this issue). These models implicitly combine a notion of knowledge that can outlive any individual with a labor market notion of human capital that does not.

The model used here separates the rival component of knowledge, H, from the nonrival, technological component, A. Because it has an existence that is separate from that of any individual, A can grow without bound. In the specific formulation used below, each new unit of knowledge corresponds to a design for a new good, so there is no conceptual problem measuring A. It is a count of the number of designs.

The formal model of the economy has three sectors. The research sector uses human capital and the existing stock of knowledge to produce new knowledge. Specifically, it produces designs for new producer durables. An intermediate-goods sector uses the designs from the research sector together with forgone output to produce the large number of producer durables that are available for use in final-goods production at any time. In practice, one might expect research on a new design and the production of the new good to take place within the same firm, and nothing in the analysis here rules this out; design work can take place either internally or in a separate firm that sells its patent to the firm that will produce the actual good. A final-goods sector uses labor, human capital, and the set of producer durables that are available to produce final output. Output can be either consumed or saved as new capital.

To keep the dynamic analysis simple and highlight the effects of interest, several simplifying assumptions are used. The first is that the population and the supply of labor are both constant. This rules out an analysis of fertility, labor force participation, or variation in hours worked per worker. The second is that the total stock of human capital in the population is fixed and that the fraction supplied to the market is also fixed. Thus the supply of the aggregate factors L and H

is fixed. The assumption on H is made largely for technical reasons. The dynamic analysis is greatly simplified by restricting attention to equilibria with constant growth rates. In a stationary population in which people have finite lives, the only feasible constant growth rate for total years of education or experience is zero. A more complicated dynamic analysis could consider the effects of the kinds of increases in H and L that have been observed historically, but even in this kind of analysis, H must ultimately approach an upper bound.

The other simplifying assumptions are extreme assumptions on factor intensities. One has already been made implicitly. Assuming that capital can be accumulated as forgone output is equivalent to assuming that capital goods are produced in a separate sector that has the same technology as the final-output sector. Forgoing consumption is then equivalent to shifting resources from the consumption sector into the capital sector. Also, the plausible assertion that research is relatively human capital– and knowledge-intensive is translated into an extreme specification in which only knowledge and human capital are used to produce new designs or knowledge. Labor and capital do not enter at all. These kinds of restrictions will reduce the analysis of the dynamics of this system to a system of equations that can be explicitly solved by doing algebra. Presumably, a relaxation of these assumptions that preserves the factor intensity orderings used here would not change the basic dynamics of the model.

Final output Y in this model is expressed as a function of physical labor L, human capital devoted to final output H_Y, and physical capital. The unusual feature of the production technology assumed here is that it disaggregates capital into an infinite number of distinct types of producer durables. For now, let these durables be indexed by an integer i. (Soon, the index i will be assumed to be a continuous variable instead of a discrete one to avoid integer constraints.) Only a finite number of these potential inputs, the ones that have already been invented and designed, are available for use at any time. Thus if $x = \{x_i\}_{i=1}^{\infty}$ is the list of inputs used by a firm that produces final output, there is some value A such that $x_i = 0$ for all $i \geq A$. Because A changes as new producer durables are invented, it is important to be able to describe final output as a stationary function of all conceivable input lists.

In this kind of environment, a simple functional form for output is the following extension of the Cobb-Douglas production function:

$$Y(H_Y, L, x) = H_Y^{\alpha} L^{\beta} \sum_{i=1}^{\infty} x_i^{1-\alpha-\beta}. \qquad (1)$$

This production function differs from the usual production function only in its assumption about the degree to which different types of capital goods are substitutes for each other. In the conventional specification, total capital K is implicitly defined as being proportional to the sum of all the different types of capital. This definition implies that all capital goods are perfect substitutes. One additional dollar of capital in the form of a truck has the same effect on the marginal productivity of mainframe computers as an additional dollar's worth of computers. Equation (1) expresses output as an additively separable function of all the different types of capital goods so that one additional dollar of trucks has no effect on the marginal productivity of computers.

Among the various types of capital goods, one can imagine pairs of inputs that are close substitutes (trucks and trains), pairs that are complements (computers and communications networks), and many pairs that fall somewhere in between. The conventional formulation explores the case in which all durables are perfect substitutes. The model here considers the case in which all durables have additively separable effects on output. An investigation of complementarity as well as of mixtures of types of substitutability is left for future work.

Because the production function in equation (1) is homogeneous of degree one, output in the final-goods sector can be described in terms of the actions of a single, aggregate, price-taking firm. The sector that produces producer durables, however, cannot be described by a representative firm. There is a distinct firm i for each durable good i. A firm must purchase or produce a design for good i before commencing production. Once it owns the design, the firm can convert η units of final output into one durable unit of good i. As in the standard one-sector model, the formal specification here describes the sector that produces capital goods as a black box that takes final output in on one side and gives capital goods out of the other side. The correct interpretation of this formal description is that the forgone consumption is never manufactured. The resources that would have been used to produce the forgone output are used instead to manufacture capital goods. It is possible to exchange a constant number of consumption goods for each unit of capital goods if the production function used to manufacture capital goods has exactly the same functional form as the production function used to manufacture consumption goods.

Once a firm has produced a design for durable i, it can obtain an infinitely lived patent on that design. If the firm manufactures $x(i)$ units of the durable, it rents those durables to final-output firms for a rental rate $p(i)$. Since firm i will be the only seller of capital good i, it will face a downward-sloping demand curve for its good. Since the durables are assumed not to depreciate, the value of one unit of durable i is the present discounted value of the infinite stream of rental income that it generates.

There are many equivalent institutional arrangements that can support any given equilibrium. In what follows, the firm that owns the patent on the design for durable i is assumed to be the only one that manufactures it, but this is only a convenience. Whether the owner of the patent manufactures the good itself or licenses others to do so, it can extract the same monopoly profit. Design of new durables and manufacturing could take place within the same firm, but it is easier to describe the equilibrium if the research and development department is treated as a separate firm and designs are transferred for an explicit price. It is also easier to assume that the firm that buys a design and manufactures a differentiated producer durable rents its durables instead of selling them outright. In particular, this shows that there are market mechanisms that avoid the usual durable-goods-monopoly problem. Provided that the manufacturer of the durable could commit to levels of output, nothing would change if the durables were sold instead of rented. The analysis is further simplified by assuming that the durables do not depreciate. Adding depreciation would merely add a familiar term to the user cost of capital.

In parallel with the usual one-sector model and in conformity with national income accounting conventions, it is useful to define an accounting measure of total capital K as cumulative forgone output. Thus $K(t)$ evolves according to the rule

$$\dot{K}(t) = Y(t) - C(t), \tag{2}$$

where $C(t)$ denotes aggregate consumption at time t. Because it takes η units of forgone consumption to create one unit of any type of durable, this accounting measure K is related to the durable goods that are actually used in production by the rule $K = \eta \sum_{i=1}^{\infty} x_i = \eta \sum_{i=1}^{A} x_i$.

Thus H and L are fixed, and K grows by the amount of forgone consumption. It remains to specify the process for the accumulation of new designs, that is, for the growth of $A(t)$. As noted above, research output depends on the amount of human capital devoted to research. It also depends on the stock of knowledge available to a person doing research. If designs were treated as discrete indivisible objects that are not produced by a deterministic production process, the production technology for designs would have to take explicit account of both integer constraints and uncertainty. There is no doubt that both indivisibility and uncertainty are important at the micro level and over short periods of time. The simplifying assumption made here is that neither is crucial to a first-pass analysis of technological change at the aggregate level. Henceforth, the index i

for the different types of goods is treated as a continuous variable, and the sum in equation (1) is replaced by an integral:

$$Y(H_Y, L, x) = H_Y^{\alpha} L^{\beta} \int_0^{\infty} x(i)^{1-\alpha-\beta}\, di. \qquad (1')$$

(Here x must be interpreted as a function from an appropriately defined function space.) With this formal structure, the output of new designs produced by researcher j can be written as a continuous, deterministic function of the inputs applied. If the researcher possesses an amount of human capital H^j and has access to a portion A^j of the total stock of knowledge implicit in previous designs, the rate of production of new designs by researcher j will be $\delta H^j A^j$, where δ is a productivity parameter.

This formulation with a continuum of goods is close to that used by Judd (1985) in his discussion of patents. He also studies the process whereby goods are introduced as new patents are produced. The model here corresponds to Judd's discussion of the case in which patents are infinitely lived. The main differences are that Judd treats the differentiated goods as consumption goods rather than producer durables and uses a form of exogenous technological change to generate growth of productivity in the research sector. Here, growth in A by itself increases the productivity of human capital in the research sector.

Although other assumptions about secrecy and property rights could be considered, the equilibrium here is based on the assumption that anyone engaged in research has free access to the entire stock of knowledge. This is feasible because knowledge is a nonrival input. All researchers can take advantage of A at the same time. The output of researcher j is therefore $\delta H^j A$. If we sum across all people engaged in research, the aggregate stock of designs evolves according to

$$\dot{A} = \delta H_A A, \qquad (3)$$

where H_A has the obvious interpretation of total human capital employed in research.

Equation (3) contains two substantive assumptions and two functional form assumptions. The first substantive assumption is that devoting more human capital to research leads to a higher rate of production of new designs. The second is that the larger the total stock of designs and knowledge is, the higher the productivity of an engineer working in the research sector will be. According to this specification, a college-educated engineer working today and one working 100 years ago have the same human capital, which is measured in terms of years of forgone participation in the labor market. The engineer working today is more productive because he or she can take advantage of all the additional knowledge accumulated as design problems were solved during the last 100 years.

The two functional form assumptions are that the output of designs is linear in each of H_A and A when the other is held constant. These assumptions, like the exclusion of the inputs L and $x(\cdot)$, are made largely for analytical convenience. Linearity in H_A is not important for the dynamic properties of the model, but weakening this assumption would require a more detailed specification of how income in the research sector is allocated to the participants.

Linearity in A is what makes unbounded growth possible, and in this sense, unbounded growth is more like an assumption than a result of the model. In what follows, it will become clear that the marginal product of human capital H_Y employed in the manufacturing sector grows in proportion to A. If A were replaced in equation (3) by some concave function of A—that is, if the marginal productivity of human capital in the research sector does not continue to grow in proportion to A—human capital employed in research would shift out of research and into manufacturing as A becomes larger. This will cause the rate of growth to slow down. Whether opportunities in research are actually petering out, or will eventually do so, is an empirical question that this kind of theory cannot resolve. The specification here, in which unbounded growth at a constant rate is feasible, was chosen because there is no evidence from recent history to support the belief that opportunities for research are diminishing. Moreover, linearity in A is convenient analytically, and assumptions about what will happen in the far future for values of A that are very large relative to the current level have very little effect on the question of interest: How do other variables in the model affect the rate of growth of A?

The crucial feature of the specification used here is that knowledge enters into production in two distinct ways. A new design enables the production of a new good that can be used to produce output. A new design also increases the total stock of knowledge and thereby increases the productivity of human capital in the research sector. The owner of a design has property rights over its use in the production of a new producer durable but not over its use in research. If an inventor has a patented design for widgets, no one can make or sell widgets without the agreement of the inventor. On the other hand, other inventors are free to spend time studying the patent application for the widget and learn knowledge that helps in the design of a wodget. The inventor of the widget has no ability to stop the inventor of a wodget from learning from the design of a widget. This means that the benefits from the first productive role for a design are completely excludable, whereas the benefits from the second are completely nonexcludable. In an overall sense, this means that the nonrival design inputs are partially excludable.

In theory and in practice, there is always some ambiguity about what constitutes a design for a new and different good and what constitutes a copy of an existing design. In the model, this ambiguity is artificially resolved by the form of the production function Y. This functional form implies that new goods are never close substitutes for existing goods since all the producer durables enter into production in an additively separable fashion. (See Pascoa [1986] for a discussion of this general property of "no neighboring goods.") Although it greatly simplifies the analysis, this is not a realistic feature of the model. In particular, it rules out the possibility of obsolescence. Nonetheless, the general results here should be robust to more careful modeling of the nature of the interaction between different specialized producer durables. What matters for the results is that the knowledge is a nonrival good that is partially excludable and privately provided.

At the aggregate level, H_A and H_Y are related by the constraint $H_Y + H_A = H$. According to these equations, any person can devote human capital to either the final-output sector or the research sector. Implicitly this formulation neglects the fact that L and H are supplied jointly. To take the equations used here literally, one must imagine that there are some skilled persons who specialize in human capital accumulation and supply no labor.

To fix notation for prices, let spot prices at any point in time be measured in units of current output and let r denote the interest rate on loans denominated in goods. Let P_A denote the price of new designs, and let w_H denote the rental rate per unit of human capital. Because goods can be converted into capital one for one, the spot price for capital is one and its rate of return is r. Because of the assumption that anyone engaged in research can freely take advantage of the entire existing stock of designs in doing research to produce new designs, it follows from equation (3) that P_A and w_H are related by $w_H = P_A \delta A$.

Once a design has been produced, a large number of potential suppliers of the new good bid for the right to do so. Each of these firms takes the price P_A for designs, the price of one for capital goods, and the interest rate as given, but if it begins production it sets prices to maximize profits. Formally, it helps to let the rental price $p(i)$ for the ith durable lie in the range $\mathbb{R}_+ \cup \{\infty\}$ for its durable good. If no firm produces good i, its price can be understood to be $p(i) = \infty$.

Faced with a price list $\{p(i): i \in \mathbb{R}_+\}$ for all the producer durables, including infinite prices for the durables that have not been invented yet, the representative final-output firm chooses a profit-maximizing quantity $x(i)$ for each durable. Because it is a constant-returns-to-scale firm, its input demands are defined only after the scale of operation is pinned down. Let L and H_Y be the total amounts of labor and human capital that are used in the production of final-output goods. (The split of total H between H_Y and H_A remains to be determined.) Given values for H_Y and L, it is possible to derive the aggregate demand for the durables from a maximization problem that is conditional on them:

$$\max_x \int_0^\infty [H_Y^\alpha L^\beta x(i)^{1-\alpha-\beta} - p(i)x(i)]di.$$

Differentiating under the integral sign leads to an inverse demand function

$$p(i) = (1 - \alpha - \beta)H_Y^\alpha L^\beta x(i)^{-\alpha-\beta}. \tag{4}$$

(There is an important technical issue about what it means to the final-goods producer if prices change on a set of values of i that has measure zero. Because of the symmetry in this model, eq. [4] can readily be derived by a limiting argument. For a general discussion of this issue, see Pascoa [1986, 1990].)

The demand curve in equation (4) is what the producer of each specialized durable takes as given in choosing the profit-maximizing price to set. Faced with given values of H_Y, L, and r, a firm that has already incurred the fixed-cost investment in a design will choose a level of output x to maximize its revenue minus variable cost at every date:

$$\begin{aligned}\pi &= \max_x p(x)x - r\eta x \\ &= \max_x (1 - \alpha - \beta)H_Y^\alpha L^\beta x^{1-\alpha-\beta} - r\eta x.\end{aligned} \tag{5}$$

The flow of rental income is $p(x)$ times x. The cost is the interest cost on the ηx units of output needed to produce x durables. To keep the analysis simple, it is convenient to assume that the capital is putty-putty, so that the firm can solve this problem at every point in time, converting units of durables back into general capital, and avoid the interest cost if it decides to supply fewer units. At any date, the only sunk cost is the initial expenditure on the design. This assumption is harmless because the demand for durables is stationary in equilibrium, so no disinvestment ever takes place.

The monopoly pricing problem specified in equation (5) is that of a firm with constant marginal cost that faces a constant elasticity demand curve. The resulting monopoly price is a simple markup over marginal cost, where the markup is determined by the elasticity of demand, $\bar{p} = r\eta/(1 - \alpha - \beta)$. The flow of monopoly profit is $\pi = (\alpha + \beta)\bar{p}\bar{x}$, where $\bar{x}$ is the quantity on the demand curve (4) implied by the price $\bar{p}$.

Each producer of specialized durables must rent its output to a large number of final-goods producers that can operate at any scale. By assumption, it is not possible for the producer to monitor the use of its durables. As a result, price discrimination is not feasible. The best the firm can do is charge the simple monopoly price.

The decision to produce a new specialized input depends on a comparison of the discounted stream of net revenue and the cost P_A of the initial investment in a design. Because the market for designs is competitive, the price for designs will be bid up until it is equal to the present value of the net revenue that a monopolist can extract. At every date t, it must therefore be true that

$$\int_t^{\infty} e^{-\int_t^{\tau} r(s)ds}\pi(\tau)d\tau = P_A(t). \tag{6}$$

If P_A is constant (as it will be in the equilibrium described below), this condition can be put in a more intuitive form. Differentiating with respect to time t yields

$$\pi(t) - r(t)\int_t^{\infty} e^{-\int_t^{\tau} r(s)ds}\pi(\tau)d\tau = 0.$$

Substituting in the expression for P_A from equation (6) yields

$$\pi(t) = r(t)P_A. \tag{6'}$$

This equation says that at every point in time, the instantaneous excess of revenue over marginal cost must be just sufficient to cover the interest cost on the initial investment in a design. (This formulation of the intertemporal zero profit constraint is taken from Grossman and Helpman [1989*c*].)

The solution of the model for a balanced growth equilibrium given in Section V shows that the technology described above implies a negatively sloped linear relation between the rate of growth of output and the rate of return on investment. To close the model, it remains to specify preferences that imply a parallel relation between the rate of growth of consumption and the marginal rate of intertemporal substitution. This relation is easily derived for Ramsey consumers with discounted, constant elasticity preferences:

$$\int_0^\infty U(C)e^{-\rho t}dt, \quad \text{with } U(C) = \frac{C^{1-\sigma} - 1}{1 - \sigma} \quad \text{for } \sigma \in [0, \infty).$$

The implied intertemporal optimization condition for a consumer faced with a fixed interest rate r is that $\dot{C}/C = (r - \rho)/\sigma$. Preferences enter the solution of the model only through this relation between the consumption growth rate and the interest rate.

The consumers are endowed with fixed quantities of labor L and human capital H that are supplied inelastically. At time 0, consumers own the existing durable-goods-producing firms, and the net revenues of these firms are paid to consumers as dividends. Final-goods firms earn zero profits and own no assets, so they can be ignored in the specification of endowments.

An equilibrium for this model will be paths for prices and quantities such that (i) consumers make savings and consumption decisions taking interest rates as given; (ii) holders of human capital decide whether to work in the research sector or the manufacturing sector taking as given the stock of total knowledge A, the price of designs P_A, and the wage rate in the manufacturing sector w_A; (iii) final-goods producers choose labor, human capital, and a list of differentiated durables taking prices as given; (iv) each firm that owns a design and manufactures a producer durable maximizes profit taking as given the interest rate and the downward-sloping demand curve it faces, and setting prices to maximize profits; (v) firms contemplating entry into the business of producing a durable take prices for designs as given; and (vi) the supply of each good is equal to the demand.

IV. Discussion of the Model

A reasonable intuition for the behavior of this model can be inferred by considering the Solow (1956) model, in which the evolution of A is given exogenously, and the Uzawa (1965) model, in which the evolution of A is determined by the allocation of resources between a research sector and a final-goods sector. For a fixed amount of A, and therefore a fixed set of producer durables, the model is almost identical to the Solow model. Because of the symmetry in the model, all the durable goods that are available are supplied at the same level, henceforth denoted as $\bar{x}$. If they were not, it would be possible to increase profits in the producer durable sector by reducing the output of high-output firms and diverting the capital released in this way to low-output goods. Since A determines the range of durables that can be produced and since η units of capital are required per unit of durable goods, it is possible to solve for $\bar{x}$ from the equation $K = \eta A\bar{x}$. Then output Y can be written as

$$
\begin{aligned}
Y(H_A, L, x) &= H_Y^{\alpha}L^{\beta}\int_0^{\infty} x(i)^{1-\alpha-\beta}\,di \\
&= H_Y^{\alpha}L^{\beta}A\bar{x}^{1-\alpha-\beta} \\
&= H_Y^{\alpha}L^{\beta}A\left(\frac{K}{\eta A}\right)^{1-\alpha-\beta} \qquad (7)\\
&= (H_YA)^{\alpha}(LA)^{\beta}(K)^{1-\alpha-\beta}\eta^{\alpha+\beta-1}.
\end{aligned}
$$

The last line of this equation shows that the model behaves just like the neoclassical model with labor and human capital augmenting technological change. In particular, it exhibits the usual diminishing returns to capital accumulation. Given the assumed form of preferences, a fixed level of A will lead to an equilibrium with a steady state in which the level of K is determined by the requirement that the marginal product of capital is equal to the discount rate. If A grew at an exogenously specified exponential rate, the economy would converge to a path on which K grows at the same exponential rate as A, just as it does in the Solow model. Along the transition path, the ratio of K to A would change, which implies that r and $\bar{x}$ would change as well. Along the balanced growth path, r, $\bar{x}$, and the ratio of K to A are all constant.

The nonconvexity evident in the expression for final output as a function of the primary inputs of the model (H, L, K, and A) is supported in a decentralized equilibrium that relies on monopolistic competition. In contrast, the nonconvexity present in equation (3) describing output of designs is supported through competition with external effects that arise from knowledge spillovers. In each case, the nonconvexity arises because the nonrival good A is an input in production. In the final-output sector, A matters indirectly because of its effects on the availability of the new $x(\cdot)$ goods. In the research sector, A enters directly.

Both spillovers and price setting seem essential to capturing the features of knowledge in a model of growth. There is little doubt that much of the value to society of any given innovation or discovery is not captured by the inventor, and any model that missed these spillovers would miss important elements of the growth process. Yet it is still the case that private, profit-maximizing agents make investments in the creation of new knowledge and that they earn a return on these investments by charging a price for the resulting goods that is greater than the marginal cost of producing the goods.

V. Solution of the Model for a Balanced Growth Equilibrium

The strategy for characterizing the model that is followed here is to solve for an equilibrium in which the variables A, K, and Y grow at constant exponential rates. This is generally referred to as a balanced growth equilibrium. The intuition from the Solow model suggests that such an equilibrium will exist if A grows at a constant exponential rate. The intuition from the Uzawa model suggests that it is possible for A to grow at an exponential rate because equation (3) for $\dot{A}$ is linear in A. It will grow at a constant rate if the amount of human capital H_A that is devoted to research stays constant. Verifying that a balanced growth equilibrium exists therefore reduces to the problem of showing that prices and wages are such that H_Y and H_A remain constant as Y, K, C, and A grow.

By focusing only on balanced growth paths, the analysis neglects the transient dynamics that arise when the economy starts from a ratio of K to A that differs from the ratio that is maintained along the balanced growth path. One should be able to study convergence to the balanced growth ratio of K to A using the tools used for studying the Solow and Uzawa models, but this analysis is not attempted here.

The first step in the solution of the model is to derive the relation between the growth rate of output and the rate of return on investment. Coupled with the relation between interest rates and growth rates implied by the preference side of the model, this will determine the rate of growth and interest rate. Given the intuition from the Solow model and the results derived in my earlier model of differentiated inputs (Romer 1987), it follows that along the balanced growth path, the ratio of K to A should be constant, which implies that $\bar{x}$ is constant as well. Because of the accumulation of both K and A, the wage paid for human capital in the final-output sector will grow in proportion to A, but by equation (3), the productivity of human capital in research also grows in proportion to A. Since the productivity of human capital grows at the same rate in both sectors, H_Y and H_A will remain constant if the price P_A for new designs is constant.

Figure 1 illustrates the behavior of the inputs in the model. All the producer durables that have been designed up to time t are used at the level $\bar{x}$ that indicates the height of the rectangle. The width is the measure of the number of designs or durables in use, $A(t)$. The area $A(t)\bar{x}$ is equal to total capital divided by η. Over time, $\bar{x}$ remains constant and A grows at a constant exponential rate.

It remains to check that this description of a balanced growth path is consistent with all the equilibrium conditions. As noted in the dis-

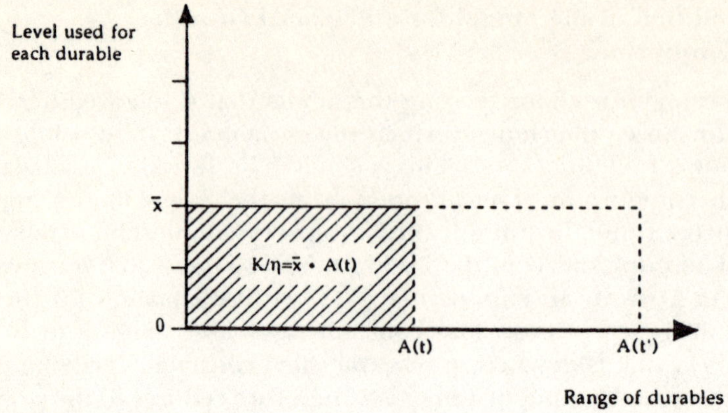

FIG. 1.—Producer durables used in production at time t and $t' > t$

cussion following the monopoly profit problem (5), the flow of profit that can be extracted by the seller of any particular durable input is $\pi = (\alpha + \beta)\bar{p}\bar{x}$. Since the present discounted value of this stream of profit must equal the price P_A of the design, it follows that

$$P_A = \frac{1}{r}\pi = \frac{\alpha + \beta}{r}\bar{p}\bar{x} = \frac{\alpha + \beta}{r}(1 - \alpha - \beta)H_Y^{\alpha}L^{\beta}\bar{x}^{1-\alpha-\beta}. \quad (8)$$

The last equality in this expression follows by using equation (4) to evaluate $\bar{p}$ in terms of $\bar{x}$.

The condition determining the allocation of human capital between the final-output and research sectors says that the wages paid to human capital in each sector must be the same. In the final-output sector, the wage for human capital is its marginal product. Since human capital receives all the income from the research sector, the wage there is $P_A\delta A$. To equalize returns to human capital in both sectors, $H_Y = H - H_A$ must be chosen so that

$$\begin{aligned} w_H = P_A\delta A &= \alpha H_Y^{\alpha-1}L^{\beta}\int_0^{\infty}\bar{x}^{1-\alpha-\beta}\,di \\ &= \alpha H_Y^{\alpha-1}L^{\beta}A\bar{x}^{1-\alpha-\beta}. \end{aligned} \quad (9)$$

Substituting P_A from equation (8) into equation (9) and simplifying yield

$$H_Y = \frac{1}{\delta}\frac{\alpha}{(1 - \alpha - \beta)(\alpha + \beta)}r. \quad (10)$$

For a fixed value of $H_A = H - H_Y$, the implied exponential growth rate for A is δH_A. From the monopoly pricing problem, we know that

$\bar{x}$ is constant if r is. The intuition from the Solow model suggests, and examination of the expression for final output

$$Y = H_Y^{\alpha} L^{\beta} \int_0^{\infty} \bar{x}^{1-\alpha-\beta} di = H_Y^{\alpha} L^{\beta} A \bar{x}^{1-\alpha-\beta}$$

from equation (7) shows, that output grows at the same rate as A if L, H_Y, and $\bar{x}$ are fixed. If $\bar{x}$ is fixed, then K must grow at the same rate as A, because total usage of capital is $A\bar{x}\eta$. Let g denote the growth rate of A, Y, and K. Since K/Y is a constant, the ratio

$$\frac{C}{Y} = 1 - \frac{\dot{K}}{Y} = 1 - \frac{\dot{K}}{K}\frac{K}{Y}$$

must also be constant. The common growth rate g for all these variables is therefore

$$g = \frac{\dot{C}}{C} = \frac{\dot{Y}}{Y} = \frac{\dot{K}}{K} = \frac{\dot{A}}{A} = \delta H_A.$$

Together with equation (10), the constraint $H_Y = H - H_A$ implies a relation between the growth rate g and the interest rate r:

$$g = \delta H_A = \delta H - \frac{\alpha}{(1-\alpha-\beta)(\alpha+\beta)} r, \tag{11}$$

which can be simplified as

$$g = \delta H_A = \delta H - \Lambda r, \tag{11'}$$

where Λ is a constant that depends on the technology parameters α and β,

$$\Lambda = \frac{\alpha}{(1-\alpha-\beta)(\alpha+\beta)}. \tag{12}$$

Implicitly, the allocation of H between the two sectors is constrained by the requirement that H_A must be nonnegative and that H_Y can be no larger than H. This implies that g is nonnegative. If this constraint is binding, equation (10) will hold as an inequality.

To close the model, it remains to impose the relation between the growth rate g and the interest rate r implied by the preference side of the model, $g = \dot{C}/C = (r - \rho)/\sigma$. Combined with equation (11), this gives an expression for g in terms of the fundamentals of the model,

$$g = \frac{\delta H - \Lambda\rho}{\sigma\Lambda + 1}, \tag{13}$$

where Λ is as defined in equation (12).

The expression for the growth rate suggests a minor technical restriction. For the integral in the consumer's preferences to be finite, the rate of growth of current utility $(1 - \sigma)g$ must be less than the discount rate ρ. Thus, for $\sigma \in [0, 1)$, $(1 - \sigma)\delta H/(\Lambda + 1)$ must be less than the discount rate ρ. If this does not hold, the integral can be infinite and some kind of overtaking criterion must be used to describe the behavior of the consumer.

VI. Interpretation and Welfare Properties of the Solution

Almost all the content of the model is contained in equation (11), which summarizes the effects of the technological side of the model, including the effects of imperfect competition in the market for producer durables. Part of the intuition behind equation (11) is easy to grasp. The opportunity cost of human capital is the wage income that can be earned instantaneously in the manufacturing sector. The return to investing human capital in research is a stream of net revenue that a design generates in the future. If the interest rate is larger, the present discounted value of the stream of net revenue will be lower. Less human capital will be allocated to research, and the rate of growth will be lower.

Other aspects of the intuition behind equation (11) are more subtle. The surprising feature of equation (11) is that neither L nor the parameter η is present. Equation (4) shows that an increase in L increases the demand faced by each monopolistic firm selling a durable good. From the monopoly problem (5), it is clear that a reduction in η reduces the costs of the monopolist and increases output $\bar{x}$. In either case, the stream of net revenue generated by a new design increases. Nonetheless, the amount of human capital devoted to research increases neither when L increases nor when η falls.

In a partial equilibrium analysis, one would always expect any change that increases the return to an activity to increase the allocation of resources to that activity. What a general equilibrium analysis emphasizes is that any intervention that increases the return to one activity can very well increase the return to some other activity that competes with the first activity for resources. An increase in L or a reduction in η (which increases $\bar{x}$) raises the return to human capital employed in manufacturing at the same time that it raises the return to human capital in research. For the functional forms used here, these two effects exactly cancel. The allocation of human capital between research and manufacturing does not change when L and η change.

This kind of exact cancellation is not a robust feature of the model.

If slightly different functional forms were used, the net effect of an increase in L or a decrease in η could be either to increase or to reduce the amount of human capital used in the research sector. For example, in an extension of this model (Romer 1990), I show how an increase in L could reduce research effort and the rate of growth. The correct inference from this model is that the effect of an increase in L or a decrease in η on the rate of growth is ambiguous, something that a priori theorizing cannot resolve.

This ambiguity contrasts with the strong and robust implication that reductions in the interest rate will speed up growth. Equivalently, from equation (12), it follows that any change in the preference parameters that acts to reduce the interest rate (an increase in patience captured by a decrease in the discount rate ρ, or an increase in the intertemporal rate of substitution, as captured by a decrease in σ) will increase research and growth. This implication follows directly from the assumption that the benefits of research come largely in the future and that the costs are incurred immediately.

From a policy point of view, the difference between a reduction in η and a reduction in the equilibrium interest rate is very important. A direct subsidy to investment in physical capital financed by a lump-sum tax is mathematically equivalent to a reduction in η. (An illustration of this easy result is given in Romer [1989].) On the basis of an intuition developed for the one-sector model, economists often identify the marginal product of physical capital with the market interest rate. If they were the same, anything that increases the capital stock and reduces the marginal product of physical capital would have the same growth-enhancing effect as a reduction in the interest rate. The intuition that subsidies to physical capital accumulation will also speed up growth in A was reinforced by the first generation of endogenous growth models. In Arrow's (1962*a*) model of learning by doing and in my first model (Romer 1986), the rate of growth of A was forced by assumption to be the same as the rate of growth of K. As a result, an intervention such as an investment tax credit that increased the accumulation of K necessarily increased the accumulation of A as well.

The model presented here shows that when the decision to invest in physical capital is uncoupled from the decision to invest in research, the effects of a subsidy to physical capital are quite different from the effects of a reduction in the market interest rate. If the fundamental policy problem is that we have too many lawyers and MBAs and not enough engineers, a subsidy to physical capital accumulation is a weak, and possibly counterproductive, policy response.

In the previous paper with differentiated producer durables (Romer 1987), I found that an increase in scale measured by L would increase the rate of growth. Here, an increase in L has no effect.

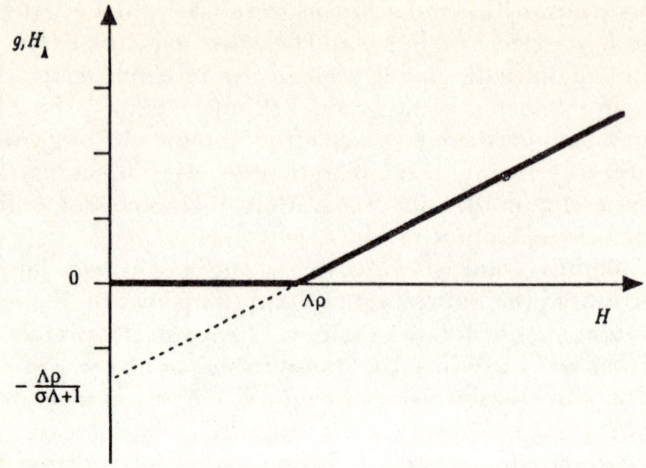

FIG. 2.—Growth rate and amount of human capital in research as a function of total human capital (for $\delta = 1$).

Instead, an increase in scale as measured by total human capital H has the effect of speeding up the rate of growth. This effect is illustrated in figure 2, which plots the rate of growth and the amount of human capital used in research as a function of total human capital. Both of these models have an underlying form of increasing returns in research. As a result, an increase in a scale variable induces an increase in the rate of growth. Human capital is the relevant scale variable in this model because it is the input that is used most intensively in research.

That the research sector in this model exhibits increasing returns is clear from equation (4). If the research technology exhibited constant returns to scale, a doubling of both the human capital and the stock of knowledge would leave the marginal product of human capital in research unchanged. Under the specification used here, a doubling of both leads to an increase in the marginal product of human capital in research. As a result, a permanent increase in the total stock of human capital in the population leads to an increase in the ratio of A to K and a more than proportional increase in the amount of human capital that is devoted to the research sector, as illustrated in figure 2. This implication is of interest from both a historical and a cross-sectional point of view. It is surely the case that the total level of human capital and the fraction of human capital devoted to research are higher now than they were at any time in the past. Moreover, the fraction of human capital devoted to research is apparently highest in the most developed countries of the world.

An extreme possibility illustrated in figure 2 is that if the total level of human capital is too small, stagnation may arise. If H is too low, the nonnegativity constraint on H_A is binding and growth does not take place. In this case, all the feasible growth rates for A are too small relative to the discount rate to justify the sacrifice in current output necessary for growth to take place. This result offers one possible way to explain the wide variation in growth rates observed among countries and the fact that in some countries growth in income per capita has been close to zero. This explanation is reminiscent of the explanation for the absence of growth in prehistoric time that is offered by some historians and anthropologists: civilization, and hence growth, could not begin until human capital could be spared from the production of goods for immediate consumption. This model cannot offer a complete explanation for these observations because it treats the stock of H (and of L) as given, but it does suggest directions for further work.

The effects of a subsidy to capital can be contrasted with a policy designed to encourage research. A subsidy to employment in the research sector that is financed through lump-sum taxes has the same effects on growth as an increase in the productivity parameter δ in equation (3). In the long run, the subsidy will cause an increase in the growth rate, a fall in P_A, and a reduction in $\bar{x}$ and in the ratio of K to A.

There are two reasons to expect that too little human capital is devoted to research. The most obvious reason is that research has positive external effects. An additional design raises the productivity of all future individuals who do research, but because this benefit is nonexcludable, it is not reflected at all in the market price for designs. The second and an equally important reason why too little human capital is devoted to research is that research produces an input that is purchased by a sector that engages in monopoly pricing. The markup of price over marginal cost forces a wedge between the marginal social product of an input used in this sector and its market compensation. Equation (7) shows that a new design (i.e., an additional unit of A) increases output at every date by an amount $H_Y^{\alpha}L^{\beta}\bar{x}^{1-\alpha-\beta}$; from the calculation in (5), the producer of a design captures only a fraction $1 - \alpha - \beta$ of this net benefit to society.

Both of these effects cause human capital to be undercompensated. The marginal product of capital in the manufacturing sector is equal to the wage w_H, but the marginal product in the research sector is higher than the wage because the price of the patent captures only part of the social value of the patent. From equation (9), (10), or (11), it follows that an increase in total human capital would lead to an increase in the amount of human capital employed in the research sector. As a result, in equilibrium, the marginal value of an additional unit of human capital is higher than the market wage. For simplicity, the model here has treated the stock of total human capital as being exogenously fixed, so undercompensating human capital has no welfare effects. In a more realistic model in which human capital is accumulated endogenously, the supply will be too low. As a result, a second-best policy for a government that cannot affect the allocation of human capital between different sectors would be to subsidize the production of human capital.

Within the confines of the model, the social optimum can be achieved by subsidizing the accumulation of A. Demonstrating this result rigorously starting from arbitrary initial conditions forces the analysis to depart from consideration of balanced growth paths. Any intervention designed to move an economy from one balanced growth path to another must consider the transition dynamics along the way, and an explicit analysis of these dynamics is beyond the scope of this paper. It is easy, however, to compare the growth rate along the balanced growth path that emerges from the equilibrium with the one that would emerge from the solution to a social planning problem. Because of the symmetry between the different producer durables in this economy, the optimal level of $x(i)$ is the same for all i between zero and A. This level is related to K and A by the constraint that $K = \eta x A$. Using this to express x in terms of K and A, we can therefore write the social planning problem for this economy as

$$\max \int_0^\infty \frac{C^{1-\sigma} - 1}{1 - \sigma} e^{-\rho t} dt,$$

subject to

$$\dot{K} = \eta^{\alpha+\beta-1} A^{\alpha+\beta} H_Y^{\alpha} L^{\beta} K^{1-\alpha-\beta} - C,$$

$$\dot{A} = \delta H_A A,$$

$$H_Y + H_A \leq H.$$

As is shown in the Appendix, the balanced growth solution to the first-order necessary conditions for this problem has a growth rate g^* that is given by

$$g^* = \frac{\delta H - \Theta\rho}{\Theta\sigma + (1 - \Theta)}, \tag{14}$$

where $\Theta = \alpha/(\alpha + \beta)$. The coefficient Λ from equation (12) is equal to this coefficient Θ times the markup from the monopoly sector, $1/(1 - \alpha - \beta)$. This accounts for part of the difference between the equilibrium and the socially optimal rate of growth. However, there is an additional effect because the term in the denominator in equation (14) also replaces the constant 1 from equation (13) with the expression $1 - \Theta$. This change reflects the effect of correcting for the external effects associated with the production of new ideas. Both of these changes in the expression for the rate of growth—the substitution of Θ for Λ and the substitution of $1 - \Theta$ for 1—cause the socially optimal allocation of human capital to research to be higher, and this causes the socially optimal rate of growth to be higher.

VII. Growth, Trade, and Research

The final observations about this model pertain to its implications for growth, trade, and research. These can be seen most simply by comparing the balanced growth equilibrium for two identical closed economies that operate in isolation with the balanced growth equilibrium that would obtain if the economies had always been fully integrated. In isolation, the common growth rate is given by g from equation (13), with H set equal to the amount of human capital in each country. In the second case, the growth rate is found by replacing H by total worldwide human capital, $2H$. Both the fraction H_A/H of total worldwide human capital devoted to research and the rate of growth increase.

This thought experiment suggests why a decision to engage in trade may be important even for a country that has a large population, such as China or India. If access to a large number of workers or consumers were all that mattered, having a large population would be a good substitute for trade with other nations. The model here suggests that what is important for growth is integration not into an economy with a large number of people but rather into one with a large amount of human capital. Many of the details of trade between different economies of this kind remain to be worked out,[2] but since growth seems to be correlated with the degree of integration into worldwide markets but not closely related to population size or density, the results from this model seem promising.

The most direct test of this implication of the model would come from a controlled experiment in which the level of research activity was monitored both before and after a country was opened to trade with the rest of the world. Sokoloff (1988) reports historical data on a natural experiment that comes close to this test. He finds cross-sectional variation that supports the model: counties in the United States in the early nineteenth century that had access to navigable waterways had higher rates of patenting than counties that did not.

[2] Since the first draft of this paper was written, Grossman and Helpman (1989*a*, 1989*b*) have explored detailed models of trade and growth with this kind of underlying technology.

More convincingly, he shows that over time, the introduction of water transportation (because of either the construction of a new canal or the dredging of a river) was followed by a sharp increase in the rate of patenting in counties adjacent to the waterway.

In subsequent work, Sokoloff and Khan (1989) examine the time-series variability in patenting for specific individuals. They find that there was a substantial group of people with broad general knowledge who moved in and out of research in response to aggregate disturbances (primarily business cycles and a trade embargo by the British) just as one would expect if human capital that could be used to do research had alternative uses in manufacturing and commercial trade. They conclude that at least in the nineteenth century, there was a fairly elastic short-run supply of human capital for use in research. Combined with the prior evidence that patenting activity does respond to changes in the size of the market, this offers some assurance that the basic mechanisms described in the model are relevant for historical experience.

VIII. Conclusions

The model presented here is essentially the one-sector neoclassical model with technological change, augmented to give an endogenous explanation of the source of the technological change. The most robust welfare conclusion from the model is that because research projects exchange current costs for a stream of benefits in the future, the rate of technological change is sensitive to the rate of interest. Although all the research is embodied in capital goods, a subsidy to physical capital accumulation may be a very poor substitute for direct subsidies that increase the incentive to undertake research. In the absence of feasible policies that can remove the divergence between the social and private returns to research, a second-best policy would be to subsidize the accumulation of total human capital.

The most interesting positive implication of the model is that an economy with a larger total stock of human capital will experience faster growth. This finding suggests that free international trade can act to speed up growth. It also suggests a way to understand what it is about developed economies in the twentieth century that permitted rates of growth of income per capita that are unprecedented in human history. The model also suggests that low levels of human capital may help explain why growth is not observed in underdeveloped economies that are closed and why a less developed economy with a very large population can still benefit from economic integration with the rest of the world.

Appendix

Calculation of the Balanced Growth Social Optimum

To derive the necessary conditions for the social optimization problem given in the text, construct the current-valued Hamiltonian:

$$\mathcal{H} = \frac{C^{1-\sigma}}{1-\sigma} + \lambda[\eta^{\alpha+\beta-1}A^{\alpha+\beta}(H - H_A)^{\alpha}L^{\beta}K^{1-\alpha-\beta} - C] + \mu\delta H_A A.$$

The necessary conditions follow by maximizing $\mathcal{H}$ with respect to the control variables C and H_A, and from the equations for the evolution of the multipliers λ and μ:

$$\dot{\lambda} = \rho\lambda - \frac{\partial \mathcal{H}}{\partial K}, \quad \dot{\mu} = \rho\mu - \frac{\partial \mathcal{H}}{\partial A}.$$

The first-order condition for maximizing $\mathcal{H}$ with respect to C gives the usual expression relating marginal utility and the multiplier λ:

$$C^{-\sigma} = \lambda. \tag{A1}$$

If the symbol Δ is used to represent the term $\eta^{\alpha+\beta-1}A^{\alpha+\beta}(H - H_A)^{\alpha}L^{\beta}K^{1-\alpha-\beta}$ from the Hamiltonian, the first-order condition for maximizing $\mathcal{H}$ with respect to H_A can be written as

$$\Delta = (H - H_A)\frac{\delta\mu}{\alpha\lambda}A. \tag{A2}$$

Then with equation (A2), the evolution equation for μ can be simplified to yield

$$\frac{\dot{\mu}}{\mu} = \rho - \delta\left(\frac{\alpha+\beta}{\alpha}H - \frac{\beta}{\alpha}H_A\right). \tag{A3}$$

For a balanced growth equilibrium, it must be the case that $\dot{\mu}/\mu = \dot{\lambda}/\lambda$ and that $\dot{C}/C = \dot{A}/A$. With equation (A1), these can be combined to yield $-\sigma(\dot{A}/A) = \dot{\mu}/\mu$. Then combining the evolution equation for A, $\dot{A}/A = \delta H_A$, with equation (A3) gives an equation in H, H_A, and the basic parameters of the model:

$$-\sigma\delta H_A = \rho - \delta\left(\frac{\alpha+\beta}{\alpha}H - \frac{\beta}{\alpha}H_A\right). \tag{A4}$$

Equation (14) in the text can be derived from this equation by solving for H_A and using the fact that the growth rate g is given by $g = \delta H_A$.

References

Abramovitz, Moses. "Resource and Output Trends in the United States since 1870." *A.E.R. Papers and Proc.* 46 (May 1956): 5–23.

Arrow, Kenneth J. "The Economic Implications of Learning by Doing." *Rev. Econ. Studies* 29 (June 1962): 155–73. (*a*)

———. "Economic Welfare and the Allocation of Resources for Invention." In *The Rate and Direction of Inventive Activity: Economic and Social Factors,* edited by Richard Nelson. Princeton, N.J.: Princeton Univ. Press (for NBER), 1962. (*b*)

Becker, Gary S.; Murphy, Kevin M.; and Tamura, Robert. "Human Capital, Fertility, and Economic Growth." *J.P.E.*, this issue.

Cornes, Richard, and Sandler, Todd. *The Theory of Externalities, Public Goods, and Club Goods.* Cambridge: Cambridge Univ. Press, 1986.

Dasgupta, Partha, and Stiglitz, Joseph E. "Learning-by-Doing, Market Structure and Industrial and Trade Policies." *Oxford Econ. Papers* 40 (June 1988): 246–68.

Dixit, Avinash K., and Stiglitz, Joseph E. "Monopolistic Competition and Optimum Product Diversity." *A.E.R.* 67 (June 1977): 297–308.

Ethier, Wilfred J. "National and International Returns to Scale in the Modern Theory of International Trade." *A.E.R.* 72 (June 1982): 389–405.

Gollop, Frank M., and Jorgenson, Dale W. "U.S. Productivity Growth by Industry." In *New Developments in Productivity Measurement and Analysis,* edited by John W. Kendrick and Beatrice N. Vaccara. Chicago: Univ. Chicago Press (for NBER), 1980.

Griliches, Zvi. "Issues in Assessing the Contribution of Research and Development to Productivity Growth." *Bell J. Econ.* 10 (Spring 1979): 92–116.

Grossman, Gene M., and Helpman, Elhanan. "Comparative Advantage and Long Run Growth." Working Paper no. 2809. Cambridge, Mass.: NBER, January 1989. (*a*).

———. "Growth and Welfare in a Small Open Economy." Working Paper no. 2970. Cambridge, Mass.: NBER, May 1989. (*b*)

———. "Product Development and International Trade." *J.P.E.* 97 (December 1989): 1261–83. (*c*)

Heckman, James J. "A Life-Cycle Model of Earnings, Learning, and Consumption." *J.P.E.* 84, no. 4, pt. 2 (August 1976): S11–S44.

Helpman, Elhanan, and Krugman, Paul R. *Market Structure and Foreign Trade: Increasing Returns, Imperfect Competition, and the International Economy.* Cambridge, Mass.: MIT Press, 1985.

Jones, Larry, and Manuelli, Rodolfo. "A Model of Optimal Economic Growth." Working paper. Stanford, Calif.: Stanford Univ., 1988.

Jorgenson, Dale W.; Gollop, Frank M.; and Fraumeni, Barbara M. *Productivity and U.S. Economic Growth.* Cambridge, Mass.: Harvard Univ. Press, 1987.

Judd, Kenneth L. "On the Performance of Patents." *Econometrica* 53 (May 1985): 567–85.

Kendrick, John W. "Productivity Trends: Capital and Labor." *Rev. Econ. and Statis.* 38 (August 1956): 248–57.

King, Robert G., and Rebelo, Sergio. "Business Cycles with Endogenous Growth." Working paper. Rochester, N.Y.: Univ. Rochester, 1987.

Lucas, Robert E., Jr. "On the Mechanics of Economic Development." *J. Monetary Econ.* 22 (July 1988): 3–42.

Maddison, Angus. *Phases of Capitalist Development.* Oxford: Oxford Univ. Press, 1982.

Nordhaus, William D. *Invention, Growth, and Welfare: A Theoretical Treatment of Technological Change.* Cambridge, Mass.: MIT Press, 1969.

Pascoa, Mario. "Monopolistic Competition and No-Neighboring-Goods." Working Paper no. 86-14. Philadelphia: Univ. Pennsylvania, Center Analytic Res. Econ. and Soc. Sci., 1986.

———. "Noncooperative Equilibrium and Chamberlinian Monopolistic Competition." *J. Econ. Theory* (1990), in press.

Rebelo, Sergio. "Long Run Policy Analysis and Long Run Growth." Working paper. Rochester, N.Y.: Univ. Rochester, 1988.

Romer, Paul M. "Increasing Returns and Long-Run Growth." *J.P.E.* 94 (October 1986): 1002–37.

———. "Growth Based on Increasing Returns Due to Specialization." *A.E.R. Papers and Proc.* 77 (May 1987): 56–62.

———. "Capital Accumulation in the Theory of Long Run Growth." In *Modern Business Cycle Theory*, edited by Robert J. Barro. Cambridge, Mass.: Harvard Univ. Press, 1989.

———. "Capital, Labor, and Productivity." *Brookings Papers Econ. Activity* (1990), in press.

Rosen, Sherwin. "A Theory of Life Earnings." *J.P.E.* 84, no. 4, pt. 2 (August 1976): S45–S67.

Schumpeter, Joseph A. *Capitalism, Socialism, and Democracy.* New York: Harper, 1942.

Shell, Karl. "Toward a Theory of Inventive Activity and Capital Accumulation." *A.E.R. Papers and Proc.* 56 (May 1966): 62–68.

———. "A Model of Inventive Activity and Capital Accumulation." In *Essays on the Theory of Optimal Economic Growth*, edited by Karl Shell. Cambridge, Mass.: MIT Press, 1967.

———. "Inventive Activity, Industrial Organisation and Economic Growth." In *Models of Economic Growth*, edited by James A. Mirrlees and Nicholas H. Stern. New York: Wiley, 1973.

Sokoloff, Kenneth L. "Inventive Activity in Early Industrial America: Evidence from Patent Records, 1790–1846." *J. Econ. Hist.* 48 (December 1988): 813–50.

Sokoloff, Kenneth L., and Khan, Zorina. "The Democratization of Invention during Early Industrialization: Evidence from the United States, 1790–1846." Working paper. Los Angeles: Univ. California, 1989.

Solow, Robert M. "A Contribution to the Theory of Economic Growth." *Q.J.E.* 70 (February 1956): 65–94.

———. "Technical Change and the Aggregate Production Function." *Rev. Econ. and Statis.* 39 (August 1957): 312–20.

Uzawa, Hirofumi. "Optimum Technical Change in an Aggregative Model of Economic Growth." *Internat. Econ. Rev.* 6 (January 1965): 18–31.

Wilson, Robert. "Informational Economies of Scale." *Bell J. Econ.* 6 (Spring 1975): 184–95.

[3]

Comparative Advantage and Long-Run Growth

By GENE M. GROSSMAN AND ELHANAN HELPMAN*

We construct a dynamic, two-country model of trade and growth in which endogenous technological progress results from the profit-maximizing behavior of entrepreneurs. We study the role that the external trading environment and that trade and industrial policies play in the determination of long-run growth rates. Cross-country differences in efficiency at R&D versus manufacturing (i.e., comparative advantage) bear importantly on the growth effects of economic structure and commerical policies. (JEL *411, 111, 621)*

What role do the external trading environment and commercial policy play in the determination of *long-run* economic performance? This central question of international economics has received surprisingly little attention in the theoretical literature over the years.

Previous research on trade and growth has adopted the neoclassical framework to focus on factor accumulation in the open economy. (See the surveys by Ronald Findlay, 1984, and Alasdair Smith, 1984.) This research largely neglects the effects of trade structure on rates of growth, however, addressing instead the reverse causation from growth and accumulation to the pattern of trade.[1] The direction that the research followed almost surely can be ascribed to the well-known property of the standard neoclassical growth model with diminishing returns to capital that (endogenous) growth in per capita income dissipates in the long run. For this reason, the familiar models that incorporate investment only in capital equipment seem ill-suited for analysis of long-run growth.

The available evidence collected since the seminal work of Robert Solow (1957) also leads one to look beyond capital accumulation for an explanation of growth. Exercises in growth accounting for a variety of countries generally find that increases in the capital to labor ratio account for considerably less than half of the last century's growth in per capita incomes.[2] Although econometric efforts to explain the residual have been somewhat disappointing (see, for example, Zvi Griliches, 1979), professional opinion and common sense continue to impute much of this residual to improvements in technology.[3] We share the view, expressed by Paul Romer (1986, 1990), that a full understanding of growth in the long run requires appreciation of the economic determinants of the accumulation of knowledge.

In this paper we draw on the pioneering work by Romer to construct a model that highlights the roles of scale economies and technological progress in the growth process. As in Romer's work, our model im-

*Grossman: Woodrow Wilson School, Princeton University, Princeton, NJ 08544. Helpman: Department of Economics, Tel Aviv University, Ramat Aviv, Israel. We are grateful to Avinash Dixit, Carl Shapiro, and Lars Svensson for their comments on an earlier draft, and to the National Science Foundation and the Alfred P. Sloan Foundation for financial support. Grossman also thanks the World Bank and Helpman thanks the International Monetary Fund for providing support and a stimulating environment during part of our work on this project. These organizations are not responsible for the views expressed herein.

[1]An exception is Max Corden (1971), who studies how the opening up of trade affects the speed of transition to the steady state in a two-factor neoclassical growth model with fixed savings propensities.

[2]See Angus Maddison (1987) for a careful, recent exercise in growth accounting.

[3]The benefits of education and experience undoubtedly contribute part of the explanation for the growth residual. See, for example, Robert Lucas (1988) and Gary Becker and Kevin Murphy (1990) for growth models that highlight the role of human capital accumulation as a source of growth.

plies an endogenous rate of long-run growth in per capita income, and we study its economic determinants. Our primary contribution lies in casting the growth process in a two-country setting. We provide, for the first time, a rigorous analysis linking long-run growth rates to trade policies and other international economic conditions. Moreover, we show that recognition of cross-country differences in economic structure impinge upon conclusions about the long-run effects of domestic shocks and policies.

Our model incorporates the essential insights from Romer (1990), although we introduce some differences in detail. The building blocks are an R&D sector that produces designs or blueprints for new products using primary resources and previously accumulated knowledge, an intermediate-goods sector consisting of oligopolistic producers of differentiated products, and a consumer-goods sector in each country that produces a country-specific final output using labor and intermediate inputs. As in Wilfred Ethier (1982), total factor productivity in final production increases when the number of available varieties of differentiated inputs grows. Thus, resources devoted to R&D contribute over time to productivity in the production of final goods, as well as to the stock of scientific and engineering knowledge.

The new elements in our analysis stem from the assumed presence of cross-country differences in the effectiveness with which primary resources can perform different activities, that is, *comparative advantage*. For simplicity we specify a one-primary-factor model, and allow the productivity of this factor in the three activities to vary internationally. Similar results could be derived from a multifactor model with interindustry differences in factor intensities (see Grossman and Helpman, 1989d, and Grossman, 1989). In any event, we find that many comparative dynamic results hinge on a comparison across countries of parameters reflecting efficiency in R&D relative to efficiency in manufacturing the goods that make use of the knowledge generated by R&D. The effects of policy in a single country, of accumulation of primary resources in a single country, and of a shift in world tastes toward the final output of one of the countries all depend upon the identity of the country in which the change originates in relation to the international pattern of comparative advantage.

We describe the economic setting in Section I below. Then, in Section II, we derive the dynamic equilibrium of the world economy and calculate two reduced-form equations that determine the steady-state growth rate. In Section III, we investigate the structural determinants of long-run growth. There, the implications for growth of variations in consumer preferences, primary-input coefficients in one or both countries, and stocks of available primary resources are considered. Section IV contains policy analysis. We study barriers and inducements to trade in consumer goods and subsidies to research and development. The analysis is extended in Section V to incorporate lags in the dissemination of knowledge and asymmetries in the speed of diffusion within and between countries. We use the extended model to reconsider the effects of trade policies on the steady-state rate of growth. Finally, Section VI provides a brief summary of our findings.

I. The Model

We study a world economy comprising two countries. Each country engages in three productive activities: the production of a final good, the production of a continuum of varieties of differentiated middle products (i.e., intermediate inputs), and research and development (R&D). A single primary factor is used in production, and is taken to be in fixed and constant supply in each country. Although we refer to this factor as "labor," we have in mind an aggregate of irreproducible resources that for any given level of technical know-how limits aggregate output.

At a point in time, output of final goods in country i is given by

$$Y_i = BA_i L_{Yi}^{1-\beta}\left[\int_0^n x_i(\omega)^\alpha d\omega\right]^{\beta/\alpha},$$

$$0 < \alpha, \beta < 1,$$

where L_{Yi} represents employment in the final-goods sector, $x_i(\omega)$ denotes the input of middle product ω, A_i is a country-specific productivity parameter, and n is (the measure of) the number of varieties of middle products available at that time.[4] This production function exhibits constant returns to scale for given n, but an increase in the measure of varieties of middle products raises total factor productivity. This specification, which we borrow from Ethier (1982), captures the notion that an increasing degree of *specialization* generates technical efficiency gains. The economy's potential for augmenting the degree of specialization by developing new middle products implies the existence of dynamic scale economies at the industry level that are external to the individual final-good-producing firms.

Competition in the final-goods sectors ensures marginal-cost pricing. Hence, by appropriate choice of the constant B, producer prices satisfy

$$(1)\quad p_{Yi}=\left(\frac{w_i}{A_i}\right)^{1-\beta}\left[\int_0^n p_X(\omega)^{1-\varepsilon}d\omega\right]^{\beta/(1-\varepsilon)},$$

$$\varepsilon=\frac{1}{1-\alpha}>1,$$

where w_i is the wage in country i and $p_X(\omega)$ is the price of variety ω. Final-good producers worldwide pay the same prices for (freely traded) middle products.

At every moment in time, the existing producers of middle products engage in oligopolistic price competition. The producer of a variety ω in country i chooses $p_X(\omega)$ to maximize profits,

$$\pi_i(\omega)=[p_X(\omega)-w_i a_{LXi}]\times\frac{p_X(\omega)^{-\varepsilon}}{\int_0^n p_X(\omega)^{1-\varepsilon}d\omega}\beta\Sigma_i p_{Yi}Y_i,$$

where a_{LXi} is the unit labor requirement for production of intermediates in country i. This expression for profits comprises the product of profits per unit (in square brackets) and derived demand for variety ω, where the latter incorporates the assumption that neither the prices of competing products nor the value of final production varies with $p_X(\omega)$. The first-order condition for a profit maximum implies the usual fixed-markup pricing rule,

$$(2)\qquad \alpha p_X(\omega)=w_i a_{LXi}.$$

It is clear from (2) that varieties originating from the same country bear the same price. Letting p_{Xi} represent the price of a variety produced in country i and n_i be the number of intermediate inputs produced there, equations (1) and (2) imply

$$(3)\quad p_{Yi}=\left(\frac{w_i}{A_i}\right)^{1-\beta}\left(\sum_j n_j p_{Xj}^{1-\varepsilon}\right)^{\beta/(1-\varepsilon)},$$

$$(4)\qquad \alpha p_{Xi}=w_i a_{LXi}.$$

With these prices, profits per firm can be expressed as

$$(5)\qquad \pi_i=(1-\alpha)p_{Xi}X_i/n_i,$$

where X_i is aggregate output of intermediates in country i (n_i times per-firm output) and is given by

$$(6)\qquad X_i=\frac{n_i p_{Xi}^{-\varepsilon}}{\sum_j n_j p_{Xj}^{1-\varepsilon}}\beta\left(\sum_j p_{Yj}Y_j\right).$$

As in our 1989c paper (see also Kenneth Judd, 1985), resources devoted to research generate "blueprints" that expand the measure of differentiated products. Research outlays are made by private, profit-maximizing entrepreneurs, who appropriate some of the benefits from a new technological innovation in the form of a stream of oligopoly profits. We assume that innovators receive indefinite patent protection, but that blueprints are not tradable, so that all profits must be derived from production in

[4] Here, and henceforth, we omit time arguments when no confusion is caused by doing so.

the country in which a middle product has been developed. Then, with free entry by entrepreneurs, if resources are devoted to R&D in country i at time t, the present value of future operating profits from producing there—discounted to time t—must equal the current cost of R&D. We denote R&D costs by $c_{ni}(t)$ and write the zero-profit condition as

$$\int_t^{\infty} e^{-[R(\tau)-R(t)]}\pi_i(\tau)\,d\tau = c_{ni}(t),$$

where $R(t)$ is the cumulative interest factor from time 0 to time t ($R(0)=1$). Differentiating this condition with respect to t, we find

$$\frac{\pi_i + \dot{c}_{ni}}{c_{ni}} = \dot{R}. \tag{7}$$

Equation (7) expresses a standard no-arbitrage condition. Recognizing that $c_{ni}(t)$ represents the value of an input-producing firm in country i at time t, (7) equates the instantaneous rate of return on shares in such a firm (the sum of dividends and capital gains) to the rate of interest.

We follow Romer (1990) in assuming that R&D generates a second output, which takes the form of a contribution to the stock of disembodied knowledge. Knowledge here includes all general scientific information, as well as some forms of engineering data with more widespread applicability, generated in the course of developing marketable products. Knowledge contributes to the productivity of further research efforts by reducing the amount of labor needed for an inventor to develop a new product. Due to the more general and non-patentable nature of this product of the R&D effort, appropriation of the resulting returns by the creator seems problematic. We assume to begin with that general knowledge disseminates immediately and costlessly throughout the world. This approximates a situation in which information spreads through technical journals, professional organizations, and interpersonal commercial contacts, and where literature, scientists, and business-people move freely across international borders (see Luigi Pasinetti, 1981, ch. 11). We relax this assumption by introducing lags in the dissemination of knowledge in Section V.

With these knowledge spillovers in mind, we specify our R&D technology as follows. If L_{ni} units of labor engage in research in country i, they generate a flow of new products $\dot{n}_i$ given by

$$\dot{n}_i = L_{ni}K/a_{Lni}, \tag{8}$$

where K is the current stock of knowledge and a_{Lni} is a country-specific productivity parameter. We take the stock of knowledge to be proportional to cumulative experience in R&D; that is, there are no diminishing returns to research in adding to scientific understanding. By choosing units for K so that the factor of proportionality is unity, we have $K=n$ and

$$\dot{K} = \sum_i L_{ni}K/a_{Lni}. \tag{9}$$

Since knowledge has been assumed to be a free input to each individual entrepreneur, the cost of product development in country i can be written as

$$c_{ni} = w_i a_{Lni}/n. \tag{10}$$

We turn now to the demand side of the model. Consumers worldwide share identical, homothetic preferences. They view the final goods produced in the two countries as imperfect substitutes. We represent preferences by a time-separable intertemporal utility function

$$U_t = \int_t^{\infty} e^{-\rho(\tau-t)} \times \log u[y_1(\tau), y_2(\tau)]\,d\tau, \tag{11}$$

where ρ is the subjective discount rate and $y_i(\tau)$ is consumption of final goods from country i in period τ. The instantaneous sub-utility function $u(\cdot)$ is nondecreasing, strictly quasi-concave, and positively linearly homogeneous.

A typical consumer maximizes (11) subject to an intertemporal budget constraint. With $u(\cdot)$ linearly homogeneous, this optimization problem can be solved in two stages. First, the consumer maximizes static utility for a given level of expenditure at time τ, $E(\tau)$. The solution to this subproblem generates an indirect utility function, $v[p_{Y1}(\tau), p_{Y2}(\tau)]E(\tau)$, where p_{Yi} is the price of Y_i. In the absence of barriers to trade in final goods, these prices are common to consumers in the two countries. The second-stage problem involves choosing the time pattern of expenditures to maximize

$$(12) \quad V_t = \int_t^{\infty} e^{-\rho(\tau-t)}\{\log v[p_{Y1}(\tau), p_{Y2}(\tau)] + \log E(\tau)\}\, d\tau$$

subject to

$$(13) \quad \int_t^{\infty} e^{-[R(\tau)-R(t)]}E(\tau) \leq \int_t^{\infty} e^{-[R(\tau)-R(t)]}w(\tau)\, L\, d\tau + Z(t),$$

where $w(t)$ is the consumer's wage rate at time t, L is his labor supply, and $Z(t)$ is the value of his time t asset holdings. The interest factor in (13) is common to all individuals as a result of trade on the integrated world capital market, but the wage rate varies by country.

From the first-order conditions to this problem, we find that the optimal path for expenditure obeys

$$(14) \quad \frac{\dot{E}}{E} = \dot{R} - \rho.$$

Savings are used to accumulate either ownership claims in input-producing firms or riskless bonds issued by these same firms. Arbitrage ensures that the rates of return on these two assets are equal, and in equilibrium consumers are indifferent as to the composition of their portfolios.

II. Equilibrium Dynamics

During the course of the development of our model in the previous section, we provided some of the equilibrium conditions. For example, we derived pricing equations for goods and a no-arbitrage condition relating equilibrium asset returns. In this section we complete the list of equilibrium requirements by adding conditions that stipulate market clearing in factor and final-goods markets. We then derive and discuss a reduced-form system that describes equilibrium dynamics.

Static equilibrium in the markets for the two final goods implies

$$(15) \quad p_{Yi}Y_i = s_i E,$$

where $s_i(p_{Y1}, p_{Y2})$ is the share of world spending allocated to Y_i and E is world spending on consumer goods. The share function is, of course, homogeneous of degree zero. We establish below that relative commodity prices are constant in the vicinity of a steady state with active R&D sectors in both countries. For this reason, we take s_i to be constant in our subsequent analysis, and omit its functional dependence on relative prices.

The labor-market clearing conditions equate labor supply and labor demand in each country. Using (3) and Shephard's lemma, we see that final-goods producers demand $(1-\beta)p_{Yi}Y_i/w_i$ workers. The demand for labor by middle-products manufacturers is $a_{LXi}X_i$, while (8) and the fact that $K = n$ imply demand for labor by product developers of $(a_{Lni}/n)\dot{n}_i$. Hence,

$$(16) \quad (a_{Lni}/n)\dot{n}_i + a_{LXi}X_i + (1-\beta)p_{Yi}Y_i/w_i = L_i,$$

where L_i is the labor force available in country i.

Since we neglect here the monetary determinants of the price level, we may choose freely a time pattern for one nominal variable. It proves convenient to specify the numeraire as follows:

$$(17a)\qquad p_{X1} = n(a_{LX1}/a_{Ln1})^{1/\varepsilon}.$$

We show in Appendix A of our 1989a working paper that, with this normalization, a necessary condition for convergence to a steady state with positive R&D in both countries (i.e., nonspecialization) is

$$(17b)\qquad p_{X2} = n(a_{LX2}/a_{Ln2})^{1/\varepsilon}.$$

Together, (17a) and (17b) imply that relative prices of middle products are constant along the convergent path, which further implies with (4) the constancy of relative wages, and with (3) the constancy of relative prices of final goods. This last fact justifies our treatment of expenditure shares as fixed.

Let $g(t)$ denote the rate of growth of the number of products and the stock of knowledge; that is, $g \equiv \dot{n}/n = \dot{K}/K$. Then from (17) and (4) we see that prices of intermediates and wages grow at rate g, while from (10), product development costs are constant. Equations (5)–(7), (10), (15), and (17) imply

$$(18)\qquad X_i = \frac{n_i b_i^{1/\alpha}}{\Sigma_j n_j b_j}\frac{\beta E}{n}$$

and

$$(19)\qquad \dot{R} = \frac{1}{\varepsilon - 1}\frac{\beta E}{\Sigma_j n_j b_j},$$

where $b_i \equiv (a_{Lni}/a_{LXi})^{\alpha}$. The coefficients b_i will serve as our measures of *comparative advantage*. Country 1 enjoys comparative advantage in conducting R&D if and only if $b_1 < b_2$.

Since wages grow at the same rate as n, it proves convenient to define $e \equiv E/n$. Letting $\sigma_i \equiv n_i/n$ be the share of products manufactured in country i and noting that $g = \Sigma_i \dot{n}_i/n$, (16), (15), (17), (4), and (18) imply

$$(20)\qquad g = H - \frac{\beta e}{\sigma} - \frac{1-\beta}{\alpha}se,$$

where we have defined $H \equiv \Sigma_i L_i/a_{Lni}$, the total *effective* labor force, $\sigma \equiv \Sigma_i \sigma_i b_i$, a weighted average of the comparative advantage parameters with product shares as weights, and $s \equiv \Sigma_i s_i/b_i$. Observe that the parameter σ, which provides a useful summary of the static intersectoral resource allocation, grows (shrinks) over time if and only if the growth rate of the number of differentiated middle products in the country with comparative *disadvantage* in R&D exceeds (falls short of) that of the other country.

We are now prepared to derive two equations that describe the dynamic evolution of the world economy. From the definition of e, we have $\dot{e}/e = \dot{E}/E - g$, or, substituting (14), (19), and (20),

$$(21)\qquad \frac{\dot{e}}{e} = \frac{\beta e}{\alpha\sigma} + \frac{1-\beta}{\alpha}se - H - \rho.$$

Hence, the rate of increase of spending per middle product is larger the greater is spending per product and the smaller is the share of the country with comparative disadvantage in R&D in the total number of varieties.

Now, from the definition of the product shares σ_i, their rates of change are given by $\dot{\sigma}_i/\sigma_i = \dot{n}_i/n_i - \dot{n}/n$. Using (16) together with (17), (4), (18), and (20), we obtain

$$(22)\qquad \dot{\sigma}_i = h_i - \frac{s_i}{b_i}\frac{1-\beta}{\alpha}e - \sigma_i\left(H - \frac{1-\beta}{\alpha}se\right),$$

where $h_i \equiv L_i/a_{Lni}$ is effective labor in country i, and $\Sigma_i h_i = H$. Since the evolution of the two product shares are related by $\Sigma_i \dot{\sigma}_i = 0$, we can replace (22) by a single differential equation in σ. Making use of the fact that $\dot{\sigma} = \Sigma_i \dot{\sigma}_i b_i$, we find

$$(23)\qquad \dot{\sigma} = h - \frac{1-\beta}{\alpha}e - \sigma\left(H - \frac{1-\beta}{\alpha}se\right),$$

where $h \equiv \Sigma_i h_i b_i$.

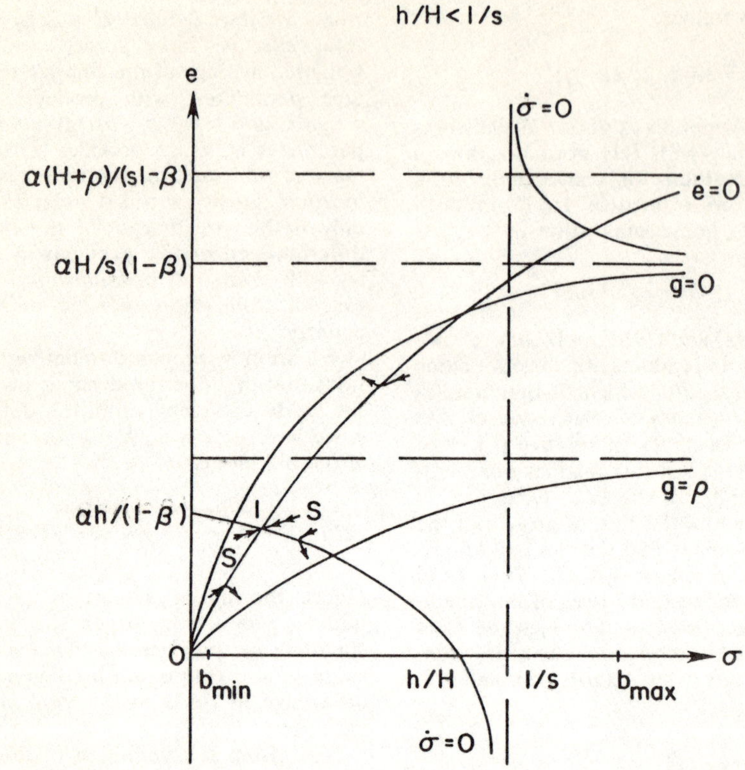

FIGURE 1

Equations (21) and (23) constitute an autonomous system of differential equations in e and σ. The solution to this system, together with (13), (20), and the definition of σ, provide a complete description of the evolution of spending and the number of products in each country. From these, the paths for outputs, employments and final-goods prices are easily derived. Thus, we shall use this two-equation system to analyze equilibrium dynamics.[5]

In Figures 1 and 2 we depict the stationary points for e. We draw the $\dot{e}=0$ locus as increasing and concave (see (21)). To understand the positive slope of this curve, observe from (19) that the interest rate can be expressed as

$$(24) \qquad \dot{R} = \beta e/\sigma(\varepsilon - 1).$$

Thus, an increase in spending per product increases the interest rate and (from (20)) reduces the rate of growth of n (the former because the profitability of R&D rises with

[5]A restriction that we have not explicitly taken into account to this point is that product development must be nonnegative in each country. This condition certainly is satisfied in the neighborhood of a steady-state equilibrium with positive growth, but it need not hold all along the convergent path to such an equilibrium. Thus, strictly speaking, the equilibrium dynamics that we describe below apply for sure only in the vicinity of a steady state.

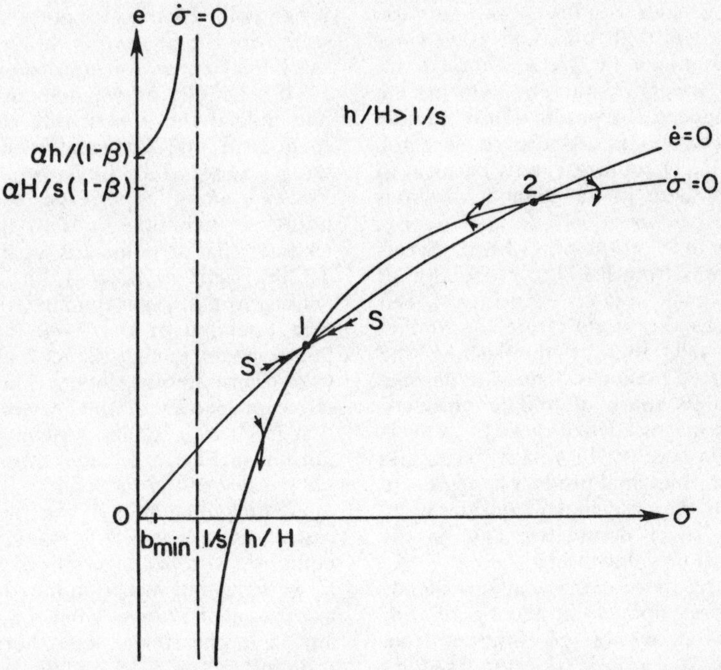

FIGURE 2

derived demand, the latter because more spending means less savings and hence less investment). Since an increase in the interest rate raises the rate of growth of nominal spending, and the rate of growth of e is just the difference between the rates of growth of E and n, it follows that an increase in e raises the growth rate of e. To compensate for this acceleration in spending per product, if e is to be stationary, σ must rise. An increase in σ lowers the interest rate and raises the rate of growth of n, thereby reducing the rate of growth of e.

Next, we distinguish two subcases depending on the relative sizes of h/H and $1/s$.[6] It can be shown that $h/H > 1/s$ if and only if $(b_2 - b_1)(h_2 b_2/s_2 - h_1 b_1/s_1) > 0$. Therefore, the case $h/H > 1/s$, depicted in Figure 2, applies, for example, if the shares of the two countries' final outputs are in proportion to their relative *effective* labor forces; that is, $h_1/s_1 = h_2/s_2$. But a bias in size relative to budget share of final output can reverse the inequality and hence the relationship between h/H and $1/s$. This gives us the case $h/H < 1/s$, shown in Figure 1, with which we begin the discussion.

In general, both $1/s$ and h/H must lie between b_{min} and b_{max}. The $\dot{\sigma} = 0$ curve in Figure 1 is everywhere downward sloping, crosses the horizontal axis at h/H, and is discontinuous at $\sigma = 1/s$. The slope of the curve is understood as follows. For $\dot{\sigma} = 0$, we must have $\dot{\sigma}_1 = 0$, which requires that the resources available for R&D in each country be just sufficient to preserve the country's *share* in the world's number of

[6]A borderline case arises when $b_1 = b_2$; that is, when comparative advantage is absent. Then $h/H = 1/s$, and $\sigma = b$ always. Convergence to the steady-state level of e is immediate in this case; see Grossman and Helpman (1989a).

varieties. Consider country 1 and suppose for concreteness that this country has comparative advantage in R&D. Then an increase in σ lowers σ_1, thereby reducing the resources needed for production of middle products. The fall in σ_1 also reduces the amount of R&D country 1 must perform to preserve its share in the number of products. *Ceteris paribus*, σ_1 would tend to rise. An increase in e, on the other hand, diverts resources away from R&D to production of middle and final products in country 1. But it also causes the world's rate of product growth to fall, thereby diminishing the amount of R&D country 1 must undertake to maintain its share of middle products. The relative magnitudes of these two effects depend upon country 1's relative size, and on the share of its final product in aggregate spending. In the case under consideration, the second effect dominates, and so the $\dot{\sigma}=0$ curve slopes downward.

In this case, there exists a unique steady state shown as point 1 in the figure. For initial values of σ not too different from that at point 1, a unique trajectory (saddle-path) converges to the steady state. This trajectory, labeled SS, fulfills all equilibrium requirements and satisfies the intertemporal budget constraint with equality. Along this trajectory (in the vicinity of the steady state), the interest rate and profit rate are declining (see (24)) and nominal expenditure E is rising. If the country with comparative advantage initially has a share of products that is smaller (larger) than its steady-state share, expenditure rises more slowly (rapidly) than the number of products.

The case depicted in Figure 2 arises when $h/H > 1/s$. Then the $\dot{\sigma}=0$ schedule slopes upward. If the curve intersects the $\dot{e}=0$ locus in the positive orthant at all, it must intersect it twice, as at points 1 and 2.[7] The lower point (point 1) represents the steady state with the higher rate of growth (growth rates increase as we move down along the $\dot{e}=0$ schedule, as we demonstrate below) and indeed the growth rate corresponding to point 2 may be negative. More importantly, as we show in Appendix B of our 1989a working paper, the equilibrium at point 1 exhibits saddle-path stability, whereas that at point 2 is locally unstable.[8] To the right of point 1, the saddle-path leading to that point remains trapped in the area bounded by the $\dot{e}=0$ locus and the line segment joining points 1 and 2, and is everywhere upward sloping. Thus, the qualitative properties of the dynamic trajectory that leads to a stable, positive-growth equilibrium in Figure 2 mimic those of the stable saddle-path in Figure 1.

For the remainder of this paper we shall restrict our discussion to stable steady-state equilibria with positive rates of growth. That is, we focus our attention on equilibria such as those at the points labeled 1 in Figures 1 and 2. In the steady state there occurs intraindustry trade in middle products and interindustry trade in consumer goods, with the long-run pattern of trade determined by comparative advantage, productivities in the two final-goods industries, and consumer preferences.

III. Determinants of Long-Run Growth

Our model generates an endogenous rate of long-run growth. We now are prepared to explore how economic structure and economic policy affect this growth rate. In this section, we derive the implications of sec-

[7] The geometry supports this claim, once we recognize that the $\dot{\sigma}=0$ curve asymptotes to the horizontal line at $\alpha H/s(1-\beta)$, whereas the $\dot{e}=0$ curve asymptotes to the horizontal line at $\alpha(H+\rho)/s(1-\beta)$. The algebra provides confirmation, as simple manipulation reveals that the steady-state growth rate solves a quadratic equation.

[8] We strongly suspect, however, that whenever there exist two positive-growth, steady-state equilibria in the admissible range, there also exists a third (saddle-path stable) steady-state equilibrium with zero growth. We have established the existence of such an equilibrium for some parameter values, but so far have been unable to construct a general existence proof. Since the equilibrium at point 1 in Figure 2 can only be reached if the initial value of σ is less than that at point 2, we suspect that initial values of σ in excess of that at point 2 (and perhaps only these) imply convergence to a steady state with zero growth.

toral productivity levels, country sizes, and demand composition for the steady-state growth rate. The influence of trade policies and of subsidies to R&D are treated in the next section.

We derive the long-run values of e and σ by setting $\dot{e}$ and $\dot{\sigma}$ to zero in (21) and (23). The steady-state magnitudes, $\bar{e}$ and $\bar{\sigma}$, solve

$$\frac{\beta\bar{e}}{\alpha\bar{\sigma}}+\frac{1-\beta}{\alpha}s\bar{e}=H+\rho; \tag{25}$$

$$\frac{1-\beta}{\alpha}\bar{e}(1-\bar{\sigma}s)+\bar{\sigma}H=h. \tag{26}$$

Whenever $1/s>h/H$, these equations provide at most one solution for $(\bar{e},\bar{\sigma})$ consistent with $\bar{g}>0$. When $1/s<h/H$, there may be two such solutions, in which case we select the stable equilibrium, that is, the one with the smaller values for $\bar{e}$ and $\bar{\sigma}$. Stability implies, in this latter case, that the $\dot{\sigma}=0$ curve intersects the $\dot{e}=0$ curve from below (see Appendix B of our 1989a working paper). We make use of this condition, namely,

$$\frac{\beta\bar{e}}{(H+\rho)\bar{\sigma}^2}>\frac{\alpha H-(1-\beta)s\bar{e}}{\bar{\sigma}H-h} \quad \text{for} \quad 1/s<h/H, \tag{27}$$

in signing the comparative-dynamics derivatives that follow.

The growth rate of the number of varieties in the steady-state equilibrium can be derived from the solution to (25) and (26), together with (20). From this, we can easily calculate the growth rate of output. In the steady state, nominal expenditure grows at rate $\bar{g}$, while (3) implies that p_{Yi} grows at rate $[1-\beta/(1-\varepsilon)]\bar{g}$. From these facts and (15), we deduce that final output grows at rate $\beta\bar{g}/(1-\varepsilon)$.

It is worth noting at this point that the steady-state equations (25) and (26), as well as the equation for $\bar{g}$, do not rely on our assumption of perfect capital mobility. In the absence of capital mobility, the steady state would be the same as long as consumers worldwide share identical preferences (and therefore common subjective discount rates).[9]

It is instructive to begin the discussion with the case in which neither country exhibits comparative advantage in conducting R&D; that is, $b_1=b_2=b$. This case has $h=bH$ and $s=1/b$. Then (25) and (26) provide a unique solution for $\bar{e}$ and $\bar{\sigma}$, which upon substitution into (20) yields the long-run growth rate

$$\bar{g}=\frac{\beta(H+\rho)}{\varepsilon}-\rho. \tag{28}$$

This equilibrium growth rate shares much in common with that derived by Romer (1990) for a closed economy. In particular, the growth rate rises with effective labor H and declines with the subjective discount rate. Our measure of effective labor adjusts raw labor for productivity in R&D (recall that $H=\Sigma_i L_i/a_{Lni}$), so greater effectiveness in research in either country, as well as a larger world labor force, necessarily means faster growth. Long-run growth does not, however, depend upon coefficients that determine absolute productivity in the intermediate or final goods sectors (such as A_i or a_{LXi}). Nor do properties of the instantaneous utility function $u(\cdot)$, including the product composition of final demand, play any role in the determination of $\bar{g}$. As we shall see presently, all these features (except for the absence of an effect of A_i on $\bar{g}$) are special to a world without any comparative advantage.

Consider next the case with $1/s>h/H$. The curves $\dot{e}=0$ and $\dot{\sigma}=0$ in Figure 3 describe the initial situation, with a unique initial steady state at point 1. Now suppose

[9] The cases of perfect and imperfect capital mobility do differ in their implications for the steady-state share of each country in aggregate spending E. However, as should be clear from (25) and (26), the cross-country composition of E does not matter for the issues taken up in the present section.

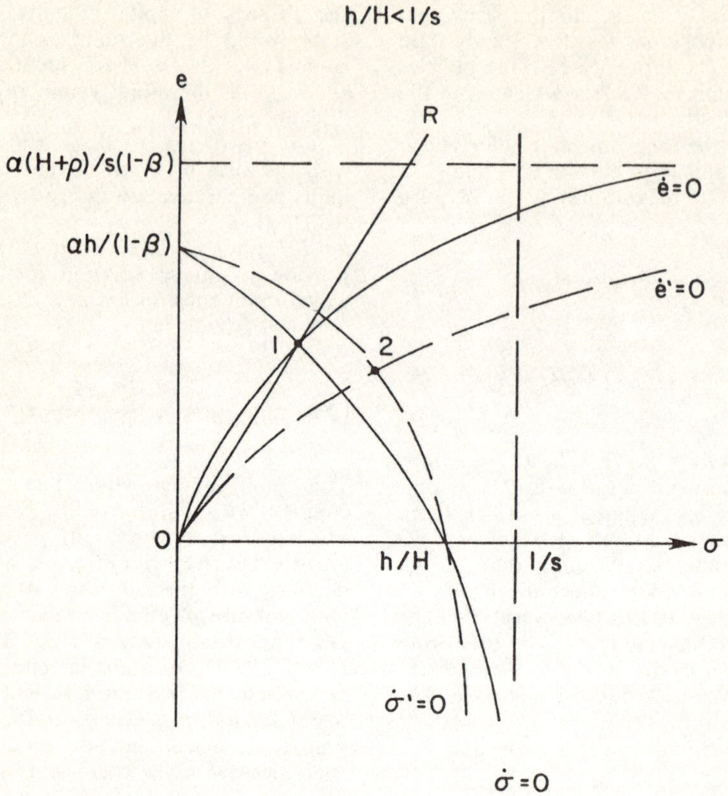

FIGURE 3

that preferences change so that s increases. This corresponds to a shift in tastes in favor of the final good produced by the country with comparative advantage in performing R&D. From (25), we see that the $\dot{e}=0$ curve shifts down, say to $\dot{e}'=0$ in the figure. Equation (26) implies that the $\dot{\sigma}=0$ schedule shifts out (in the positive orthant) to $\dot{\sigma}'=0$. The new steady state occurs at a point such as 2. But observe that all points on $\dot{e}'=0$ to the right of its intersection with ray OR are characterized by slower steady-state growth than at point 1. This claim follows from (20) and (25), whence

$$\bar{g} = \frac{\beta\bar{e}}{(\varepsilon-1)\bar{\sigma}} - \rho. \tag{29}$$

Since the intersection of $\dot{\sigma}'=0$ and $\dot{e}'=0$ necessarily lies to the right of the intersection of the latter curve with OR, we have established that an increase in s reduces steady-state growth.

When tastes shift unexpectedly toward the final good of the country with comparative advantage in R&D, resources there must be reallocated to satisfy the relatively higher consumer demand. A process begins whereby labor there shifts out of R&D and the manufacture of middle products. Products accumulate more slowly in this country than in the other, and over time its share of middle products falls (i.e., σ rises). Output per middle product changes by the same proportion in both countries (see (18)). So, in the new steady state, the country with comparative disadvantage in R&D is responsible for a relatively larger share of the world's innovation, with adverse consequences for the common steady-state growth rate. Of course, the opposite conclusion applies when s falls. Moreover, the same results obtain at stable equilibrium points when $1/s < h/H$ (see our 1989a working paper). We have thus proven the following:

PROPOSITION 1: *Stronger relative demand for the final good of the country with comparative advantage in R&D lowers the long-run share of that country in the number of middle products and slows long-run growth of the world economy. In the absence of comparative advantage in R&D, the long-run growth rate is independent of the relative demand for final goods.*

Next we consider the dependence of growth on the sizes of the effective labor forces. Effective labor may grow without affecting cross-country comparative advantage either because the stock of irreproducible resources expands, or because the productivity of labor in all uses (or in R&D and intermediate-good production) rises equiproportionately. In the first experiment, suppose that both countries experience equiproportionate, once-and-for-all increases in the sizes of their effective labor forces. We have already seen that this change would augment world growth in the absence of comparative advantage. In our 1989a working paper we prove that the same result carries over to a world with intercountry differences in relative productivities at R&D and manufacturing. We establish the following there:

PROPOSITION 2: *An equiproportionate, once-and-for-all increase in the effective labor forces of both countries accelerates long-run growth.*

Greater resources generate faster growth in our model, as in Romer (1990), essentially because dynamic scale economies characterize long-run production.

We investigate next the effects of an increase in the effective labor force of a single country. Conceptually, it proves convenient to decompose this change into two elements. First, we increase h and H by the same percentage amount. This percentage is chosen to equal the product of the share of the expanding country in the world's effective stock of labor times the percentage increase in effective labor force that the expanding country actually experiences. This accounts for the total percentage change in H when H_i changes. Then we must adjust h with H fixed to arrive at the appropriate change in h.

As an intermediate step, let us consider the effects of an increase in h alone. This corresponds to an increase in the effective labor of the country with comparative disadvantage in R&D, and a decrease in the effective labor of the other, so that the sum remains constant. This imaginary reshuffling of the world's resources shifts the $\dot{\sigma} = 0$ schedule upward when $1/s > h/H$, and downward otherwise. In either case, the $\dot{e} = 0$ curve is unaffected and the new steady-state point lies on this curve to the right of the original point. Noting (29), this proves the following:

LEMMA 1: *A reallocation of resources between countries that maintains a constant world stock of effective labor raises the long-run growth rate and increases the long-run product share of the relatively R&D efficient country if and only if the share of this country in effective labor increases.*

When the effective labor force of only country 1 (say) increases, h rises by proportionately more or less than H, according to whether country 1 has comparative disadvantage or advantage in R&D. If country 1 has comparative advantage in R&D, then both the uniform increase in H and h and the adjustment (lowering) of h, that together comprise the effect of an increase in H_1, serve to accelerate world growth. But if country 1 has comparative disadvantage in R&D, the two effects work in opposite directions. The increase in resources, by Proposition 2, speeds growth; but the reallocation of given resources, by Lemma 1, slows growth. The net effect is ambiguous, as the numerical examples that we present in Appendix D of our 1989a working paper serve to demonstrate. We have established the following:

PROPOSITION 3: *The long-run growth rate is higher the larger is the effective labor force of the country with comparative advantage in R&D. A larger effective labor force in the country with comparative disadvantage in R&D may be associated with faster or slower growth, depending upon the extent of productivity differences. In the absence of comparative advantage, long-run growth is faster the larger is the effective labor force of either country.*

These results emphasize the novel features of growth in a world with distinct countries and intercountry differences in relative productivities. They also suggest that findings reported by Paul Krugman (1990) may be somewhat special. A country need not enjoy faster growth by joining the integrated world economy, if the country enjoys substantial comparative advantage in R&D. Moreover, growth in resources or improvements in the productivity of existing resources do not guarantee faster long-run growth in a world equilibrium with free trade. If resources expand or become more efficient in the country with comparative disadvantage in R&D, then the resulting intersectoral reallocation of resources worldwide might slow innovation and growth everywhere.

IV. Economic Policy

In this section we discuss the effects of tariffs, export subsidies, and R&D subsidies on long-run growth. In order to do so, it is necessary for us to introduce the relevant policy parameters into the equations that describe instantaneous and steady-state equilibrium. To avoid repetition of the detailed arguments presented in Section I, we present here only the necessary modifications of the model and then explain their implications for the steady-state conditions. We restrict attention to small taxes and subsidies; this restriction facilitates exposition, as the channels through which economic policies affect long-run growth can be seen more clearly. We confine our analysis of trade policies to those that impede or encourage trade in final goods.

The introduction of taxes and subsidies to the model necessitates consideration of the government's budget. As usual, we assume that the government collects and redistributes net revenue by lump-sum taxes and subsidies. In a static framework, this specification suffices to determine completely the government's budgetary policy. But in a dynamic framework the budget need not balance period by period, so budgetary policy in general must specify the intertemporal pattern of lump-sum collections and transfers. However, with perfectly foresighted and infinitely lived agents, our model exhibits the Barro-Ricardo neutrality property. Hence, we need not concern ourselves with the intertemporal structure of budget deficits so long as the present value of the government's net cash flow equals zero.

The presence of the aforementioned policies modifies the decision problem for consumers in country 1 in two ways. First, we replace the price of good i in (12) by $T_i p_{Yi}$, where $T_1 = 1$. With this formulation, p_{Yi} remains the producer price of final good i, $T_2 > 1$ represents a tariff in country 1 on imports of consumer goods, and $T_2 < 1$ represents a subsidy by country 2 on exports of final output.[10] Second, we add the present value of net taxes to the right-hand side of (13) as a lump-sum addition to consumer wealth. The amount of this collection or redistribution will differ across countries according to their policies.

These modifications do not affect (14), which continues to describe the optimal intertemporal pattern of expenditures for consumers worldwide as a function of the pattern of equilibrium interest rates. In a steady state with $\dot{e} = 0$, (14) reduces to

$$\dot{R} = \bar{g} + \rho. \tag{30}$$

[10]The effects of a country 2 import tariff and a country 1 export subsidy can be derived symmetrically, so we neglect these policies here and leave the maximand for consumers in country 2 as before.

Notice that (30) implies that in any steady state in which countries grow at the same rate, long-run equalization of interest rates obtains. This property of our model holds irrespective of the presence or absence of international capital mobility and the presence or absence of tariffs or export subsidies on final goods and subsidies to research and development.

Turning to the production side, our policies do not alter equations (3)–(8) describing pricing and output relationships in the intermediate and final-goods sectors and the technology for knowledge creation. However, R&D subsidies do change the private cost of R&D. We replace (10) by

$$(10') \qquad c_{ni} = w_i a_{Lni} / nS_i,$$

where $S_i > 1$ represents subsidization of research costs in country i. It proves convenient to redefine our numeraire to normalize for the effect of the R&D subsidy on the price of intermediate inputs in country 1. Our new normalization dictates a modified equation for the price of intermediates produced in country 2 as well. Together, these relationships, which replace (17a) and (17b), can be written as

$$(17') \qquad p_{Xi} = n(S_i a_{LXi} / a_{Lni})^{1/\varepsilon}.$$

As for the market-clearing conditions, the factor-markets equation (16) is not affected, but we must replace (15) by

$$(15') \qquad p_{Yi} Y_i = \frac{s_{i1} E_1}{T_i} + s_{i2} E_2,$$

where E_i denotes aggregate spending by consumers in country i, and the shares of spending devoted to good i by residents of country 1 and country 2 are $s_{i1} = s_i(p_{Y1}, p_{Y2}T_2)$ and $s_{i2} = s_i(p_{Y1}, p_{Y2})$, respectively. Although import tariffs and export subsidies on final goods do not affect steady-state producer prices of final output in our model,[11] the direct response of spending shares in country 1 to changes in trade policy must now be taken into account. Moreover, R&D subsidies, if introduced at different rates in the two countries, will affect the steady-state value of p_{Y1}/p_{Y2}, and may therefore influence the long-run spending shares in both countries.

This completes the necessary modifications of the equilibrium relationships. We can now use the extended model to derive the equations describing steady-state equilibrium in the presence of policy intervention. In a steady state, employment in the R&D sector is given by $a_{Lni}\dot{n}_i/n = a_{Lni}\bar{g}\bar{\sigma}_i$. Making use of (4), (5), (7), (30), (10′), and (17′) (which together imply $\dot{c}_{ni} = 0$ in a steady state), we find employment in the manufacture of middle products equal to $a_{Lni}\bar{\sigma}_i(\bar{g} + \rho)(\varepsilon - 1)/S_i$. Substitution of these terms into (16) yields the steady-state labor market-clearing condition,

$$(31) \qquad \bar{g}\bar{\sigma}_i + \frac{(\varepsilon-1)(\bar{g}+\rho)}{S_i}\bar{\sigma}_i + \frac{1-\beta}{\alpha b_i S_i^{1/\varepsilon}}\bar{q}_i = h_i,$$

where $q_i \equiv p_{Yi}Y_i/n$. Next, from (4)–(7), (30), and (17′), we obtain

$$(32) \qquad (\varepsilon-1)(\bar{g}+\rho)\left(\sum_i \frac{\bar{\sigma}_i b_i}{S_i^{\alpha}}\right) - \beta\sum_i \bar{q}_i = 0.$$

Naturally, we also require

$$(33) \qquad \sum_i \bar{\sigma}_i = 1.$$

Finally, (15′) implies

$$(34) \qquad \bar{q}_i = \frac{\bar{s}_{i1}\bar{e}_1}{T_i} + \bar{s}_{i2}\bar{e}_2.$$

It is straightforward, now, to verify that (31)–(34) imply (25) and (26) when $T_i = S_i = 1$ for $i = 1,2$ (with $\bar{e} = \Sigma_i \bar{e}_i$). This provides a

[11] This statement can be verified using equations (3), (4), and (17′).

consistency check on the extended model with policy instruments.

We consider trade policies first. From (34), the ratio $\bar{q}_1/\bar{q}_2$ satisfies

$$(35) \qquad \frac{\bar{q}_1}{\bar{q}_2} = \frac{\bar{s}_{11}\bar{e}_1 + \bar{s}_{12}\bar{e}_2}{\bar{s}_{21}\bar{e}_1/T_2 + \bar{s}_{22}\bar{e}_2}.$$

Now, for given expenditure levels $\bar{e}_i$, equations (31)–(33) and (35)—which constitute a system of five equations—provide a solution for $(\bar{g}, \bar{\sigma}_1, \bar{\sigma}_2, \bar{q}_1, \bar{q}_2)$. In this system, the trade policy parameters appear only in (35). Therefore, the long-run effects of trade policy depend only on their effects on $\bar{q}_1/\bar{q}_2$, taking into account the induced adjustment in the spending levels $\bar{e}_1$ and $\bar{e}_2$. Moreover, for small trade policies (i.e, with an initial value of $T_2 = 1$), the spending shares are equal across countries ($\bar{s}_{i1} = \bar{s}_{i2}$), so the effect on $\bar{q}_1/\bar{q}_2$ of changes in the cross-country composition of aggregate spending "washes out."

Further inspection of (35) reveals that an increase in T_2 starting from free trade with $T_2 = 1$ (i.e., a small import tariff in country 1) unambiguously raises $\bar{q}_1/\bar{q}_2$.[12] A tariff shifts demand by residents of country 1 toward home consumer products, and since relative producer prices do not change in the long run, steady-state relative quantities must adjust. The effect of this change on the steady state is qualitatively the same as for an exogenous increase in world preference for final good 1, such as we studied in the previous section when we varied s_1. Similarly, a small export subsidy in country 2 (a reduction in T_2 to a value slightly below one) biases country 1 demand in favor of foreign final output. So we may apply directly our results from Proposition 1 to state the following:

PROPOSITION 4: *A small import tariff or export subsidy on final goods reduces a country's steady-state share in middle products and R&D. It increases the rate of long-run growth in the world economy if and only if the policy-active country has comparative disadvantage in R&D.*

Commercial policies *do* affect long-run growth rates. They do so by shifting resources in the policy-active country out of the growth-generating activity (R&D) and into production in the favored sector. At the same time, a resource shift of the opposite kind takes place abroad in the dynamic general equilibrium. The net effect on world growth hinges on the identity of the country that favors its consumer-good industry. If import protection or export promotion is undertaken by the country that is relatively less efficient in conducting R&D, then growth accelerates; otherwise, growth decelerates.

Next, we investigate the effects of small subsidies to R&D, introduced from an initial position of *laissez faire*. For these policy experiments, $T_2 = 1$ before and after the policy change, so the expenditure levels $\bar{e}_i$ cancel from (35). Suppose, first, that both countries apply subsidies at equal *ad valorem* rates; that is, $S_1 = S_2 = S$. In this case, relative prices of final output do not change across steady states. Therefore, the spending shares $\bar{s}_{ij}$ do not change. In Appendix C of our 1989a working paper, we totally differentiate (31)–(33) and (35) with respect to S to prove the following:

PROPOSITION 5: *A small R&D subsidy by both countries at a common rate increases the rate of long-run growth in the world economy.*

This proposition is not surprising, and corresponds to a similar result for the closed

[12]The easiest way to see this is to write the right-hand-side of (35) as

$$(\bar{p}_{Y1}, \bar{p}_{Y2})[\phi_1(\bar{p}_{Y1}, T_2\bar{p}_{Y2})\bar{e}_1 + \phi_1(\bar{p}_{Y1}, \bar{p}_{Y2})\bar{e}_2]/$$
$$[\phi_2(\bar{p}_{Y1}, T_2\bar{p}_{Y2})\bar{e}_1 + \phi_2(\bar{p}_{Y1}, \bar{p}_{Y2})\bar{e}_2],$$

where $\phi_i(\cdot)$ is minus the partial derivative of $r(\cdot)$ from (1) with respect to its ith argument divided by $r(\cdot)$. Then an increase in T_2 with $\bar{p}_{Y1}/\bar{p}_{Y2}$ constant clearly raises demand for final good 1 in country 1 (the first component of the bracketed term in the numerator increases) and lowers the demand there for final good 2 (the first component of the bracketed term in the denominator falls).

economy derived by Romer (1990). Since R&D represents the only source of gains in per capita income in our model, stimulation of this activity promotes growth.

What is more interesting, perhaps, is the effect of a small R&D subsidy in a single country. As for bilateral subsidies, a unilateral subsidy promotes growth by bringing more resources into product development in the policy-active country. But now, relative final-good prices change, so the spending shares in (35) must be allowed to vary unless the utility function has a Cobb-Douglas form. Depending on whether the elasticity of substitution between final products exceeds or falls short of one, this induced change in the pattern of spending can be conducive to or detrimental to growth. Moreover, an R&D subsidy in a single country will alter the relative shares of the two countries in product development. If the subsidy is introduced by the country that is relatively less efficient at performing R&D, this effect too can impede growth. In Appendix D of our 1989a working paper we show, by means of a numerical example using a Cobb-Douglas utility function, that an R&D subsidy introduced by the country with comparative disadvantage in R&D might (but need not) reduce the world's growth rate. We also prove in Appendix C that, for the case of constant spending shares, an R&D subsidy must encourage growth if it is undertaken by the country with comparative advantage in R&D. Thus we have the following:

PROPOSITION 6: *The provision of a subsidy to R&D in one country increases long-run growth if spending shares on the two final goods are constant and the policy is undertaken by the country with comparative advantage in R&D. Otherwise, the long-run growth rate may rise or fall.*

Our results here on the long-run growth effects of government policy have no immediate normative implications, both because we perform only steady-state comparisons and because the initial *level* of utility may differ along alternative growth paths. But in our 1989b paper we conduct a complete welfare analysis for a small country with endogenous growth generated by technological progress as here. We find that the market determined rate of growth is suboptimally low due to the presence of the nonappropriable spillovers in the process of knowledge generation. An R&D subsidy that speeds growth improves welfare for the small country until some optimal growth rate is achieved. Further increases in the subsidy rate reduce welfare even as they accelerate growth. We show also that trade policy need not raise welfare, even if it successfully speeds growth. The explanation for this lies in the presence of a second distortion in our economy, one that stems from the pricing of middle products above marginal cost. The mark-up pricing practiced by intermediate producers gives rise to suboptimal use of middle products in the production of final goods. Trade policy that accelerates growth but reduces the output of middle products in the general equilibrium can be detrimental to welfare. Conversely, commercial policy that augments the output of middle products can improve welfare even if the growth implications of such policy are adverse.

These results do not carry over immediately to the current environment, though the considerations we found to be relevant in our 1989b paper are certainly applicable here as well. Normative analysis in the present two-country setting is complicated by the terms-of-trade effects that arise when policy alters the relative price of the final goods, changes the price of intermediates (if sectoral trade in middle products is not balanced), or varies the interest rate. Because of these various terms-of-trade effects we have been unable thus far to find simple conditions under which growth-enhancing policy by one government raises that country's welfare.

V. Lags in the Diffusion of Knowledge

We have assumed all along that research and development creates as a by-product an addition to the stock of knowledge that facilitates subsequent R&D. Moreover, we supposed that the knowledge so created becomes available immediately to scientists and engineers worldwide. We now relax the latter assumption, in recognition of the fact that privately created knowledge, even if nonappropriable, may enter the public domain via an uneven and time-consuming process. Also, since legal and cultural barriers may inhibit the free movement of people and ideas across national borders, we shall allow here for the possibility that information generated in one country disseminates more rapidly to researchers in the same country than it does to researchers in the trade partner country. After extending the model we will reconsider the effects of trade policies on the steady-state rate of growth.

In place of our earlier assumption that world knowledge accumulates exactly at the rate of product innovation (equation (9)), we suppose now that R&D expenditures contribute to country-specific stocks of knowledge according to

$$(9^{\dagger}) \quad K_i(t) = \lambda_h \int_{-\infty}^{t} e^{\lambda_h(\tau - t)} n_i(\tau) + \lambda_f \int_{-\infty}^{t} e^{\lambda_f(\tau - t)} n_j(\tau)\, d\tau,$$

where $K_i(t)$ is the stock of knowledge capital at time t in country i. With this specification, the contribution of a particular R&D project to general knowledge is spread over time. At the moment after completion of the project, none of its findings have percolated through the scientific and professional community. After an infinite amount of time has passed, the R&D project makes, as before, a unit contribution to knowledge. After finite time, the contribution lies between these extremes of zero and one, as given by the exponential lag structure in ($9^{\dagger}$). The parameters λ_h and λ_f (with $\lambda_h \geq \lambda_f$) distinguish within-country and cross-country rates of diffusion.

The introduction of lags in the diffusion of knowledge alters two of the fundamental equations of the model. First, (8) becomes

$$(8^{\dagger}) \qquad \dot{n}_i = L_{ni} K_i / a_{Lni}.$$

Second, we have in place of (10),

$$(10^{\dagger}) \qquad c_{ni} = w_i a_{Lni} / K_i.$$

In a steady state with $\dot{n}_1 = \dot{n}_2 = g$, we have $n_i(\tau) = n_i(t) e^{g(\tau - t)}$, so that

$$(36) \quad K_i(t) = \frac{\lambda_h}{\lambda_h + \bar{g}} n_i(t) + \frac{\lambda_f}{\lambda_f + \bar{g}} n_j(t)$$
$$= \left[\frac{\lambda_h}{\lambda_h + \bar{g}} \bar{\sigma}_i + \frac{\lambda_f}{\lambda_f + \bar{g}} \bar{\sigma}_j \right] n(t)$$
$$\equiv \mu_i(\bar{\sigma}_1, \bar{\sigma}_2, \bar{g}) n(t).$$

So in the steady state, knowledge in each country is proportional once again to the total number of middle products. But the factor of proportionality has become country-specific and endogenous. This means that the steady-state labor-input coefficient for R&D in country i, a_{Lni}/μ_i, also is endogenous; that is, relative productivity in R&D depends now not only on relative natural abilities in performing this activity, but also on relative cumulative experience in research, as summarized by the σ_i's. This consideration leads us to draw a distinction henceforth between *natural* and *acquired* comparative advantage in R&D.

From (36) we see that when $\lambda_h = \lambda_f \to \infty$ (i.e., when diffusion lags are very short), $\mu_1 = \mu_2 \to 1$, and the extended model reverts to the earlier formulation. For $\lambda_h = \lambda_f$ finite, $\mu_1 = \mu_2$, so that the ratio of the natural-plus-acquired productivity parameters for each country is the same as for the natural productivity parameters alone. In this case, the pattern of comparative advantage cannot be reversed by endogenous learning, and all results from before continue to apply. We concentrate here on cases in which the rates of diffusion are *unequal* but the difference between them is small.[13]

[13]A large difference between the within-country and across-country rates of diffusion may imply that, in the steady-state equilibrium, all R&D is carried out by one country. Such specialization, which is common in mod-

We derive the long-run effects of trade policy in the extended model using equations (31)–(33) and (35), but with $S_1 = S_2 = 1$ (no R&D subsidies), with b_i replaced by b_i/μ_i^a (natural plus acquired comparative advantage in place of just natural comparative advantage), and with h_i replaced by $h_i\mu_i$ (natural plus acquired effective labor in place of natural effective labor). For clarity of exposition, we shall also assume for the remainder of this section that the spending shares s_i are constant. This assumption is valid when $u(\cdot)$ takes a Cobb-Douglas form.

The new elements that diffusion lags introduce to the analysis of policy stem from the effects of *relative size* and *demand-side bias*. Before considering these new aspects, let us suppose that labor forces are equal and demand for the two final goods is symmetric. By totally differentiating the system of steady-state equations (see Appendix E of our 1989a working paper), we establish the following:

PROPOSITION 7: *Suppose* $L_1 = L_2$, $s_1 = s_2$, $a_{LX1} = a_{LX2}$, *and* $\lambda_h - \lambda_f > 0$, *but small. Then a tariff on imports of final goods in country i raises the long-run growth rate if and only if* $a_{Lni} > a_{Lnj}$.

In this case, the effects of acquired comparative advantage necessarily *reinforce* those of natural comparative advantage. The country that is relatively more productive in creating new blueprints will attain, in the steady-state equilibrium prior to the introduction of policy, a majority share of the world's middle products. By its greater concentration in R&D, it will gain more experience in research and attain a higher steady-state stock of knowledge. Thus, the effects of learning will augment its initial comparative advantage in R&D. When policy is introduced in one country or the other, the implications of the dynamic resource reallocation for the global efficiency of R&D will be all the more significant.

Now suppose that the two countries differ initially only in (effective) size, as measured by h_i. Recall that with equal rates of diffusion, a small tariff in either country does not affect the long-run rate of growth. Now, however, we find the following:

PROPOSITION 8: *Suppose* $b_1 = b_2$, $s_1 = s_2$ *and* $\lambda_h - \lambda_f > 0$, *but small. Then a small tariff on imports of final goods raises the long-run growth rate if and only if the policy is introduced by the country with the relatively smaller effective labor force.*

Here, the larger country will come to acquire comparative advantage in R&D, though it starts with none. The reason is as follows. With differential rates of diffusion, knowledge takes on the characteristics of a *local public good*. The larger country will have more (effective) scientists to benefit from this nonexcludable good as its share in world R&D exceeds one half. So it acquires over time a relatively larger knowledge base and hence a relatively more productive corps of researchers. Trade policy that serves to divert resources away from the R&D sector in the larger country once comparative advantage has been established must be detrimental to growth.

The effects of demand-size bias are similar. The country whose good is in relatively greater demand must devote relatively more of its resources to final-goods production. Thus, its R&D sector initially will be smaller. This country develops over time a comparative disadvantage in R&D, as its learning lags that in its trade partner country. Protection in this country will improve world efficiency of R&D and thereby speed growth.

Once we allow for lags in the diffusion of scientific knowledge and differential speeds of diffusion within versus between countries, we find a richer set of possibilities for the long-run effects of trade policy. Comparative advantage continues to play a critical role in determining whether policy in one country will speed or decelerate growth.

els with a national component to increasing returns to scale, necessarily occurs here if static preferences are Cobb-Douglas and $\lambda_f = 0$ (i.e., all spillovers are internal). Then the equations that we have developed to describe the steady-state equilibrium (which presume non-specialization in each country) would not be valid.

But comparative advantage now must be interpreted with care, because it reflects not only natural ability but also the (endogenous) benefits from cumulative experience.[14] Since steady-state productivity in R&D varies positively with the size of the R&D sector, all determinants of the equilibrium allocation of resources to this sector come to be important in the analysis of policy.

VI. Conclusions

In this paper, we have analyzed a dynamic, two-country model of trade and growth in which long-run productivity gains stem from the profit-maximizing behavior of entrepreneurs. We have studied the determinants of R&D, where research bears fruit in the form of designs for new intermediate products and in making further research less costly. New intermediate products permit greater specialization in the process of manufacturing consumer goods, thereby enhancing productivity in final production. In order to highlight the role of endogenous technological improvements as a source of growth we have abstracted entirely from factor accumulation. But Romer (1990) has shown that capital accumulation can be introduced into a model such as the one we have studied without affecting the analysis in any significant way.

The interesting features of our analysis arise because of the assumed presence of *cross-country differences* in efficiency at R&D and manufacturing. Considerations of comparative advantage in research versus manufacturing of intermediate goods bear importantly on the implications of economic structure and economic policy for long-run patterns of specialization and long-run rates of growth. We find, for example, that growth in world resources or improvements in R&D efficiency need not speed the rate of steady-state growth, if those changes occur predominantly in the country with comparative disadvantage in R&D.

Concerning policy, we find for the first time a link between trade intervention and long-run growth. Any (small) trade policy that switches spending toward the consumer good produced by the country with comparative advantage in R&D will cause long-run growth rates to decline. Subsidies to R&D will accelerate growth when applied at equal rates in both countries, but need not do so if introduced only in the country with comparative disadvantage in R&D. When knowledge spillovers occur with a time lag and diffusion is faster within the country of origin than across national borders, comparative advantage becomes endogenous. Once we recognize that comparative advantage can be *acquired* as well as natural, we find a role for country size and demand-size bias in determining the long-run effects of policy.

Our emphasis on comparative advantage in research and development highlights only one channel through which trade structure and commercial policy might affect long-run growth. In other contexts, the trade environment might influence the rate of accumulation of human capital or the rate at which a technologically lagging (less developed) country adopts for local use the existing off-the-shelf techniques of production. Investigation of the links between trade regime and these other sources of growth seems to us a worthy topic for future research.

[14] Endogenous comparative advantage also plays a central role in Krugman's (1987) analysis of commodity-specific learning-by-doing. There, as here, productivity increases with cumulative experience. But each good is produced in only one country in Krugman's model, so long-run comparative advantage is fully determined by the initial pattern of specialization.

REFERENCES

Becker, Gary S. and Murphy, Kevin M., "Economic Growth, Human Capital and Population Growth," *Journal of Political Economy*, forthcoming, 1990.

Corden, W. Max, "The Effects of Trade on the Rate of Growth," in J. Bhagwati et al., eds., *Trade, Balance of Payments, and Growth: Papers in Honour of Charles P. Kindleberger*, Amsterdam: North-Holland, 1971.

Ethier, Wilfred J., "National and International Returns to Scale in the Modern Theory of International Trade," *American Economic Review*, June 1982, *72*, 389–405.

Findlay, Ronald, "Growth and Development in Trade Models," in R. Jones and P. Kenen, eds., *Handbook of International Economics*, Amsterdam: North-Holland, 1984.

Griliches, Zvi, "Issues in Assessing the Contribution of Research and Development in Productivity Growth," *Bell Journal of Economics*, Summer 1979, *10*, 92–116.

Grossman, Gene M., "Explaining Japan's Innovation and Trade: A Model of Quality Competition and Dynamic Comparative Advantage," National Bureau of Economic Research Working Paper No. 3194, December 1989.

______ **and Helpman, Elhanan,** (1989a) "Comparative Advantage and Long-Run Growth," National Bureau of Economic Research Working Paper No. 2809, January 1989.

______ **and** ______, (1989b) "Growth and Welfare in a Small, Open Economy," Woodrow Wilson School Discussion Paper in Economics No. 145, June 1989.

______ **and** ______, (1989c) "Product Development and International Trade," *Journal of Political Economy*, December 1989, *97*, 1261–83.

______ **and** ______, (1989d) "Quality Ladders in the Theory of Growth," National Bureau of Economic Research Working Paper No. 3099, September 1989. (*Review of Economic Studies*, forthcoming.)

______, ______ **and Razin, A.,** eds., *International Trade and Trade Policy*, Cambridge, MA: MIT Press, forthcoming.

Judd, Kenneth, "On the Performance of Patents," *Econometrica*, May 1985, *53*, 567–85.

Krugman, Paul R., "The Narrow Moving Band, the Dutch Disease, and the Competitive Consequences of Mrs. Thatcher: Notes on Trade in the Presence of Dynamic Scale Economies," *Journal of Development Economics*, October 1987, *27*, 41–55.

______, "Endogenous Innovation, International Trade and Growth," *Journal of Political Economy*, forthcoming, 1990.

Lucas, Robert E. Jr., "On the Mechanics of Economic Development," *Journal of Monetary Economics*, July 1988, *22*, 3–42.

Maddison, Angus, "Growth and Slowdown in Advanced Capitalist Economies: Techniques of Quantitative Assessment," *Journal of Economic Literature*, June 1987, *25*, 649–98.

Pasinetti, Luigi, *Structural Change and Economic Growth*, Cambridge: Cambridge University Press, 1981.

Romer, Paul M., "Increasing Returns and Long-Run Growth," *Journal of Political Economy*, October 1986, *94*, 1002–37.

______, "Endogenous Technological Change," *Journal of Political Economy*, forthcoming, 1990.

Solow, Robert M., "Technical Change and the Aggregate Production Function," *Review of Economics and Statistics*, August 1957, *39*, 312–20.

Part II
Social and Private Returns from Research and Development

[4]

RESEARCH COSTS AND SOCIAL RETURNS: HYBRID CORN AND RELATED INNOVATIONS[1]

ZVI GRILICHES
University of Chicago

INTRODUCTION AND SUMMARY

Both private and public expenditures on "research and development" have grown very rapidly in the last decade. Quantitatively, however, we know very little about the results of these investments. We have some idea of how much we have spent but very little of what we got in return. We know almost nothing about the realized rate of return on these investments, though we feel intuitively that it must have been quite high. This article presents a first step toward answering some of these questions. However, all that is attempted here is to estimate the realized social rate of return, as of 1955, on public and private funds invested in hybrid-corn research, one of the outstanding technological successes of the century. The calculated rate of return is an *estimate*, subject to a wide margin of error, but it should provide us with an order of magnitude for the "true" social rate of return on expenditures on hybrid-corn research. Actually, I believe that my estimate is biased downward, for, whenever I had to choose among alternative assumptions, I chose the assumption that led to the lowest estimate. This estimate will not tell us the global rate of return on research expenditures, but even a modest step in that direction may be of some use.

The following procedure is used to arrive at the estimate: First, private and public research expenditures on hybrid corn, 1910–55, are estimated on the basis of a mail survey and other data. Then the annual gross social returns are estimated on the assumption that they are approximately equal to the value of the resulting increase in corn production plus a price-change adjustment. The additional cost of producing hybrid seed is subtracted from these gross returns to arrive at an annual flow of net social returns. Using first a 5 and then a 10 per cent rate of interest, I bring all costs and returns forward to 1955, when the books are closed on this development and a rate of return is computed. Research costs are expressed as a capital sum, and returns are converted into a perpetual flow. The estimated perpetual flow of returns is divided by the cumulated research expenditures to arrive at a rate of return that will equalize the present value of the flow of returns with the cumulated value of research expenditures. This procedure leads to the estimate that *at least* 700 per cent per year was being earned, as of 1955, on the average dollar invested in hybrid-corn research.

Since this is not the only way in which a rate of return could be computed from these data, some alternative ways of

[1] This article is an outgrowth of a larger study of the economics of hybrid corn. See my article, "Hybrid Corn: An Exploration in the Economics of Technological Change," *Econometrica*, October, 1957. I am indebted to A. C. Harberger, Martin J. Bailey, Lester G. Telser, and T. W. Schultz for valuable comments and to the National Science Foundation and the Social Science Research Council for financial support.

defining and estimating the social rate of return are explored briefly. Comparisons are also made with estimates of returns in some other areas of technological change. Finally, I discuss the limitations of the procedure used and the implications of the results. In particular, I shall emphasize that almost no normative conclusions can be drawn from these few estimates.

RESEARCH EXPENDITURES

Inbred lines and hybrids have been developed by state agricultural experiment stations, the United States Department of Agriculture (USDA), and private seed companies. The distinction between the first two developing agencies is mainly in the source of funds. Except for funds spent on research and coordinating activities at Beltsville, Maryland, most of the USDA funds were spent on co-operative corn-breeding research at various experiment stations.

A mail inquiry to ascertain expenditures on hybrid-corn research was sent to all the agricultural experiment stations, and usable data were obtained from twenty of them. The twenty states represented by these replies include most of the important corn states in the country. Expenditures of non-responding stations were estimated by setting the expenditures of each of them equal to the expenditures of a "similar" station.[2]

Data on USDA expenditures on "corn-production research: agronomic phases" beginning with 1931 were obtained from the Agricultural Research Service and extrapolated back to 1910. They overestimate the USDA contribution substantially, because they include various other aspects of corn research besides hybrid corn. Moreover, some of the USDA funds have already been counted in the expenditures of agricultural experiment stations.

The research expenditures of one of the major private seed companies for the years 0925–55 were extrapolated back to 1911 and divided by that firm's estimated share of the total market for hybrid seed corn to arrive at an estimate of the research expenditures of the "private" segment of the industry.[3]

The figures for 1955 may be used as an example of the numbers involved. I estimate that in 1955 the USDA spent about $300,000 on hybrid-corn research, the experiment stations about $650,000, and the private companies about $1,900,000.[4]

The historical research expenditure data, deflated by the Consumers Price Index (1955 = 100), are reproduced in column 1 of Table 1. In view of all the assumptions made, these figures should be taken with several grains of salt, the dosage increasing as one goes back into the past. In particular, for the years 1910–25, the figures are little more than guesses.

For the purpose of estimating the rate

[2] The pairing was made on the basis of geographic proximity and general information about the industry. For example, Indiana expenditures were assumed to equal the reported Illinois expenditures; Oklahoma's to equal Kentucky's, and so forth. These pairings probably overestimate the total expenditures on hybrid-corn research.

[3] This will again overestimate expenditures, because "public" hybrids make up 25–30 per cent of the total market, and research expenditures on these have already been counted once.

[4] In 1951 M.T. Jenkins, of the USDA, estimated the total annual expenditures on hybrid-corn research as follows: USDA, $220,000; states, $600,000; and private industry, $1,100,000. My own independent estimate for 1951 is: USDA, $190,000; states, $550,000; private industry, $1,300,000. The two totals are $1,920,000 and $2,040,000, respectively. The agreement is very close, considering how arbitrary some of my assumptions are (see M. T. Jenkins, "Corn Breeding Research—Whither Bound," *Proceedings of the Sixth Annual Hybrid Corn Industry–Research Conference* [Chicago: American Seed Trade Association, November, 1951], pp. 42–45).

of return on these expenditures, I assume that the public sector will continue to invest in hybrid-corn research at an annual rate of $1 million, and the private sector at an annual rate of $2 million. No incremental returns, however, will be ascribed to these expenditures. I assume them to be "maintenance" expenditures in face of a malevolent nature.

COST OF ADDITIONAL RESOURCES DEVOTED TO PRODUCTION OF HYBRID SEED

I assume that the price of hybrid seed, approximately $11 per bushel in 1955, measures adequately the value of resources devoted to its production. If there were no hybrid corn, farmers would use mainly home-produced open-pollinated seed, which I value at $1.50 per bushel.[5] The quantity of hybrid seed used annually was estimated by multiplying the reported corn acreage planted with hybrid seed by the average seeding rate of corn. Multiplying the result by $9.50, the difference between the price of hybrid and non-hybrid seed, and subtracting $2 million research expenditures, I get $90 million as my estimate of the additional resources currently devoted each year to hybrid-seed production.[6] Using the average 1939–48 corn acreage (90 million), the 1951 seeding rate (7.5 pounds per acre), and the percentage planted with hybrid seed, I computed the additional cost of hybrid seed for the years 1933–55 and subtracted this from the subsequent estimate of gross returns to arrive at a net return figure.[7]

THE VALUE OF HYBRID CORN TO SOCIETY

As everyone knows, hybrid corn increased corn yields. The figure most often quoted for this increase is 20 per cent. For my purpose, I assume that the superiority of hybrid over open-pollinated varieties is 15 per cent, the lower figure in most estimated ranges.[8] The value of this increase to "society" will be measured by the loss in total corn production that would have resulted if there were no hybrid corn. This hypothetical loss will be valued at the estimated equilibrium price of corn plus a price-change adjustment, a procedure equivalent to computing the loss in "consumer surplus" that would occur if hybrid corn were to "disappear."

The amount of this loss will depend on our assumptions about the relevant demand-and-supply elasticities. As will be seen from the formulas presented below, these elasticities have only a second-order effect, and hence different reasonable assumptions about them will affect the results very little. I assumed that the price elasticity of the demand for corn is approximately -0.5.[9] Since we

[5] This is somewhat higher than the market price of corn because of the better quality of seed corn and the labor that would go into its selection. Since open-pollinated seed is now quoted at about $3.00 to $4.00 a bushel, this assumption also contributes to an over-all overestimate of cost.

[6] This result is reached as follows: 80 million acres × 90 per cent in hybrids × (8.6/64) average seeding rate × $9.50 = $92 million. Subtracting $2 million research expenditures, we get $90 million (source: *Agricultural Statistics, 1955*).

[7] Throughout the period these computations use the average corn acreage planted in 1939–48; they disregard annual fluctuations in total corn acreage and seeding rates. For any year before 1956, the extra cost of seed equals the percentage planted with hybrid seed times $98 million (90 million × [7.5/64] × $9.50 − $2 million [research expenditures] = $98 million).

[8] For example: "Plant breeders conservatively estimate increase in yields of 15 to 20 per cent from using hybrid seed under field conditions. They expect about the same relative increases in both low—and high—yielding areas" (USDA, *Technology on the Farm* [Washington, 1940], p. 7).

[9] This figure is based on a USDA demand analysis (see R. J. Foote, J. W. Klein, and M. Clough, *The Demand and Price Structure for Corn and Total Feed Concentrates* [Technical Bull. 1061 (Washington: USDA, October, 1952)]).

know much less about the supply elasticity of corn, I shall first explore the consequences of two different extreme assumptions about it.

Let us assume, first, that in the long run the supply of corn is infinitely elastic; that is, we face long-run constant costs. The "disappearance" of hybrid corn would shift the supply curve upward by the percentage reduction in the yield

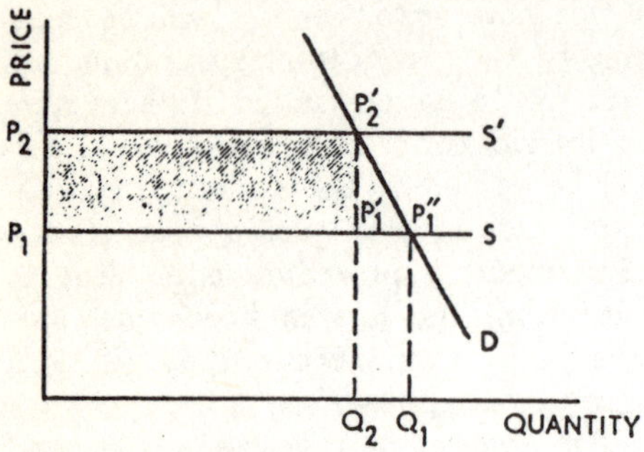

FIG. 1

of corn. The "loss" to society, in this case, is the total area under the demand curve between the new and the old supply curves. In Figure 1 this area is $P_1P_2P_2'P_1''$. This area can be interpreted as the increase in the total cost of producing the quantity Q_2 in the new situation, the rectangle $P_1P_2P_2'P_1'$, plus the loss in consumer surplus caused by the rise in price, the triangle $P_1'P_2'P_1''$. A linear approximation of this area is given by the formula

$$\text{Loss } 1 = kP_1Q_1(1 - \tfrac{1}{2}kn),$$

where k is the percentage change in yield (marginal cost and average cost), P_1 and Q_1 are, respectively, the previous equilibrium price of corn and quantity of corn produced, and n is the absolute value of the price elasticity of the demand for corn.

Alternatively, it could be assumed that the elasticity of the supply curve is zero. In this case, the loss is measured by the area $Q_2P_2'P_1''Q_1$ in Figure 2. Instead of assuming that the supply curve shifts upward, we now assume that it shifts k per cent to the left. The rectangle $Q_2P_1P_1''Q_1$ measures the loss in corn production at the old price P_1. The triangle $P_1'P_2'P_1''$ can be viewed as the additional loss in consumer surplus or as an adjustment for the increase in price from P_1 to P_2. The total loss is now given by the formula

$$\text{Loss } 2 = kP_1Q_1(1 + \tfrac{1}{2}kn).$$

It is easily seen that the second assumption leads to a higher estimate of the loss. It can be also shown that the two estimates bracket estimates implied by assuming other intermediate supply elasticities. The ratio of the loss under assumption 2 to the loss under assumption 1 is $(1 + \frac{1}{2}kn)/(1 - \frac{1}{2}kn)$. In our case, this ratio is approximately 1.07.[10] The difference between these two extreme assumptions implies only a 7 per

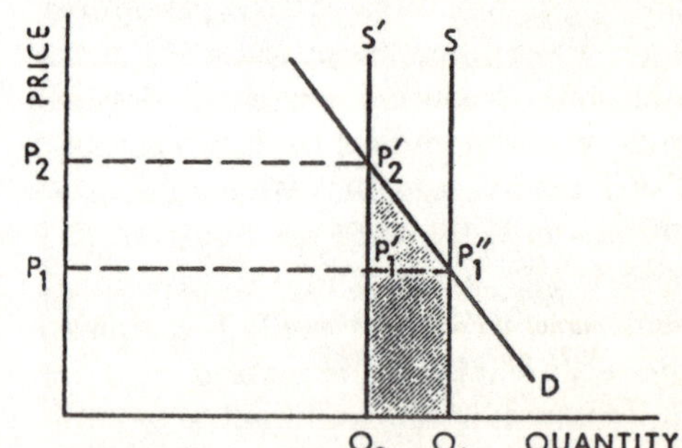

FIG. 2

cent difference in the final estimate of the total loss. Because this difference is so small and because I am striving for a lower-limit estimate, I have chosen

[10] Assuming k = 13, i.e., 15/115, and $n \doteq 0.5$, the ratio is $(1 + 0.5 \times 0.13 \times 0.5)/(1 - 0.5 \times 0.13 \times 0.5) = 1.07$.

the first assumption—that of an infinitely elastic long-run supply of corn.

To calculate the loss, we must assume an equilibrium price of corn. I shall use \$1.00 per bushel in 1955 dollars as a minimal estimate of the value of corn to society. The current price of corn about \$1.25, is affected by the esixtence of price-support programs and probably overestimates the social value of corn.[11]

Because not all corn acreage was or is planted with hybrids, I multiply the percentage shift k by h, the percentage of all corn acres planted with hybrids (loss $= hkPQ[1 - \frac{1}{2}hkn]$). This procedure disregards the fact that the acres first planted to hybrids were higher-yielding acres than those planted later, and hence the procedure underestimates total returns.

In estimating past returns from hybrid corn, I ignore annual fluctuations in prices and production, basing my computations on the average 1937–48 level of production of 2,900 million bushels[12] and a real price of corn of \$1.00 per bushel in 1955 dollars. On the returns side, only the percentage planted with hybrid seed varies over time. To calculate the annual flow of future returns, I assume that the average 1943–52 level of production—approximately 3,000 million bushels annually—will continue and that the percentage planted with hybrid seed will stabilize at 90. Both these assumptions are conservative and will result in an underestimate of returns.

Assuming that k, the relative shift in the supply curve, is 0.13 (15/115); that PQ is \$3,000 million; that n, the demand elasticity, is 0.5; and that h, the current and future fraction of all corn acres planted with hybrid seed, is 0.9, we can calculate the current and expected annual flow of gross social returns from hybrid corn as follows: $0.9 \times 0.13 \times \$3{,}000$ million $(1 - \frac{1}{2} \times 0.9 \times 0.13 \times 0.5) = 0.117 \times \$3{,}000$ million $(1 - 0.029) = 0.117 \times 0.971 \times \$3{,}000$ million $= \$341$ million. Subtracting the projected annual cost of hybrid-seed production and research—\$93 million—we get \$248 million as the current and expected annual flow of net social returns.

Similarly, returns for the past years, beginning with 1933, are calculated by multiplying 2,900 million 1955 dollars, the average total value of corn production, by the percentage of total corn acreage planted with hybrid seed in each year. Subtracting the estimated past costs of hybrid-seed production, we get the net social returns for the years 1933–55 shown in column 2 of Table 1.[13]

[11] An approximate formula for determining the price of corn in the absence of price supports is given by Marc Nerlove in "Estimates of the Elasticities of Supply of Selected Agricultural Commodities," *Journal of Farm Economics*, XXXVIII (May, 1956), 497: $dp/p_0 = (dq/q)/(n + e)$, where p_0 is the equilibrium price, n and e are the demand-and-supply elasticities, and dq is the quantity placed under loan. In recent years about 7 per cent of the annual corn crop, on the average, has been placed under loan with the Commodity Credit Corporation. The assumptions $n = 0.5$, $e = 0.2$, imply that the current price is about 10 per cent above the equilibrium price. The current price is about \$1.25 per bushel, which implies an "equilibrium" price of corn of about \$1.13. But this estimate does not take into account the impact on corn prices of the elimination of price supports on all other agricultural commodities. Taking this into account, the equilibrium price would be closer to \$1.00 per bushel. In any case, it is unlikely to be lower than \$1.00.

[12] This assumption was made to simplify the calculations. In the first part of the period, production was below this figure, and, since I use a relatively high rate of interest, this will result in an overestimate of returns. But the percentage planted with hybrids was also low then, while it was much higher during the period when production exceeded its average, and this will result in an underestimate of returns. On balance, the second effect should outweigh the first by a fair margin.

[13] These are calculated from the following formula: $h \times [0.13 \times \$2{,}900$ million $(1 - 1/2 \times 0.9 \times 0.13 \times 0.5) - 90$ million $\times (7.5/61) \times \$9.50 - \$2$ million$] = h \times \$268$ million. This procedure is an approximation, since h should also have entered

TABLE 1

HYBRID CORN: ESTIMATED RESEARCH EXPENDITURES AND NET SOCIAL RETURNS, 1910–55

(Millions of 1955 Dollars)

Year	Total Research Expenditures (Private and Public)	Net Social Returns*
1910	0.008	
1911	0.011	
1912	0.010	
1913	0.016	
1914	0.022	
1915	0.032	
1916	0.039	
1917	0.039	
1918	0.039	
1919	0.044	
1920	0.052	
1921	0.068	
1922	0.092	
1923	0.105	
1924	0.124	
1925	0.139	
1926	0.149	
1927	0.185	
1928	0.210	
1929	0.285	
1930	0.325	
1931	0.395	
1932	0.495	
1933	0.584	0.3
1934	0.564	1.1
1935	0.593	2.9
1936	0.661	8.3
1937	0.664	21.2
1938	0.721	39.9
1939	0.846	60.3
1940	1.090	81.7
1941	1.100	105.3
1942	1.070	124.3
1943	1.390	140.4
1944	1.590	158.7
1945	1.600	172.6
1946	1.820	184.7
1947	1.660	194.3
1948	1.660	203.7
1949	1.840	209.8
1950	2.060	209.0
1951	2.110	218.7
1952	2.180	226.7
1953	2.030	232.1
1954	2.270	234.2
1955	2.790	239.1
Annually after 1955	3.000	248.0

* Net of seed production cost but not net of research expenditures. Net social returns are zero before 1933.

into the second part of the first term of the formula, the "triangle." I neglect this. Because h is less than 1 and because that part is always subtracted, this procedure again underestimates total returns.

CALCULATION OF A RATE OF RETURN

Table 1 presents estimates of costs and returns. There are several ways in which these figures could be summarized and a rate of return calculated. My procedure is as follows: Consider the development closed as of 1955. Future expenditures will not increase returns, nor will there be an expansion of hybrid-corn acreage. Standing in 1955, I cumulate and bring forward to 1955 all past costs and returns at a reasonable *external* rate of interest. To explore the impact of two quite different rates of interest, I perform the calculations twice, using first a 5 per cent and then a 10 per cent rate of interest. Past research costs are cumulated and expressed as a capital sum. Past returns are cumulated to 1955, and a 5 or 10 per cent rate of return on these cumulated returns is projected into the future. The estimated flow of future net returns is added to the flow from past returns to arrive at a perpetual flow of net social returns from hybrid corn. This flow, divided by the cumulated research expenditures, gives us our estimate of the realized perpetual rate of return.

Table 2 presents the calculations that lead to my estimate of approximately $7.00 as the annual return in perpetuity, as of 1955, for every dollar that has been invested in hybrid-corn research. Actually, even if we ignore all past returns completely, the figure is still very high—approximately $4.00 annually (using the 5 per cent interest rate) for every research dollar.

This way of calculating a "rate of return" is not really different from a benefit-cost ratio calculation. It may be useful to bring out explicitly the relationship between these two concepts. The preceding rate of return is defined as follows: $r = 100\ (PR \times k + AFR)/RC$, where PR = cumulated past returns, k = the external rate of interest used to cumulate or discount returns, AFR = annual future returns, and RC = cumulated research costs.

A benefit-cost ratio from these same data would be $B/C = (PR + AFR/k)/RC$. Hence $r = 100\ k\ (B/C)$, and we can translate our calculation into a benefit-cost ratio, and vice versa. Using 5 and 10 per cent as the external rates of interest, the benefit-cost ratios for hybrid-corn research expenditures are 150 and 70, respectively.[14] When we recall that most Bureau of Reclamation watershed projects have ex post benefit-cost ratios of 1 or less, this does imply a certain misallocation of public resources.[15]

These calculations use an external rate of interest to bring all sums forward to 1955. It is reasonable to assume that the marginal productivity of capital in alternative investments is between 5 and 10 per cent and to use these rates as conversion factors for funds expended or earned at different dates. Alternatively, however, one could calculate an *internal* rate of return—that rate of interest which will equate the flow of costs and the flow of returns over time.[16] Such a rate has to be calculated using an iterative procedure, changing the rate used until the cumulated costs are equal to the discounted returns. The two procedures will give different answers when the time shape of costs differs markedly from the time shape of returns, as in our case. The internal rate of return on hybrid-corn research expenditures is between 35 and 40 per cent.[17] My objection to this particular procedure is that it values a dollar spent in 1910 at $2,300 in 1933. This does not seem very sensible to me. I prefer to value the 1910 dollar at a reasonable rate of return on some alternative social investment. Also, this procedure gives tremendous weight to the early expenditures, which are subject to the largest error of measurement. Actually, however, the two estimates are not very far apart. The estimate using an *external* rate of interest says that a dollar invested in hybrid-corn research earned 10 cents annually until 1955 and $7.00 annually thereafter. The *internal* rate estimate says that the dollar earned 40 cents annually throughout the whole period. If the average delay between investment and fruition is about ten years, then

TABLE 2

RATE OF RETURN ON HYBRID-CORN RESEARCH EXPENDITURES AS OF 1955

(Millions of Dollars)

	$r = 0.05$	$r = 0.10$
(1) Net cumulated past returns	4,405	6,542
(2) Past returns expressed as an annual flow	220	654
(3) Annual future gross returns	341	341
(4) Annual additional cost of production and research	93	93
(5) Total net annual returns, (2)+(3)−(4)	468	902
(6) Cumulated past research expenditures	63	131
(7) Rate of return 100 × (5)/(6)	3	689

[14] If before cumulating we were to subtract the research costs from net returns annually (that is, have in our denominator only research expenditures before 1934, the year when net returns began to exceed research costs), the benefit-cost ratios would be substantially higher (about 700 and 200, respectively), and so would also the rate of return as defined in the text.

[15] For an evaluation of public investments in watershed projects see E. F. Renshaw, *Toward Responsible Government* (Chicago: Idyia Press, 1957).

[16] I am indebted to Martin J. Bailey for suggesting this alternative way of calculating the rate of return.

[17] That is, 40 per cent is too high and 35 is too low. The iterative procedure was not carried further.

the two figures represent different ways of saying the same thing.[18]

LIMITATIONS

The estimate of 700 per cent is probably too low. At almost every point at which there was a choice of assumptions to be made, I have purposely chosen those that would result in a lower estimate. This is an attempt to arrive at an estimated lower limit of the social rate of return from hybrid corn.

Both the public and the private research expenditures are probably overstated substantially. In fact, the expenditure estimates supplied by some experiment stations are obviously too high. Of the USDA expenditures, perhaps less than half were devoted to hybrids. I did leave out all expenses incurred before 1910, but these in total could not have been more than a few thousand dollars. This should remind us, however, that the estimated rate of return is mainly a rate of return on applied rather than basic research. The basic idea of hybrid corn was developed between 1905 and 1920, with the help of very little money. The rate of return on this basic research, if it could be calculated, would be much higher. However, the idea had to be translated to commercial reality, and separate adaptable hybrids had to be developed for different areas. These are the activities reflected in my estimate of research cost.

The returns, on the other hand, are understated. The assumed price of corn of $1.00 per bushel in 1955 dollars and the assumption of a 15 per cent superiority of hybrids are both conservative and probably result in an underestimate of the real returns to society. I have also assumed that all past research has already borne all its fruit and that all future research on hybrids will result in no benefit whatsoever. Nor has credit been given for the impact of hybrid corn on other fields: the research on hybrid poultry and hybrid sorghum which it stimulated or the reduction of farmer resistance to new technology as a result of the spectacular success of hybrid corn.

Hence, as far as costs and returns from hybrid corn per se are concerned, the estimate is too low. One troublesome problem, however, remains to haunt us. Does it really make sense to calculate the rate of return on a successful "oil well"? What is the point of calculating the rate of return on one of the outstanding technological successes of the century? Obviously, it will be high. What we would like to have is an estimate that would also include the cost of all the "dry holes" that were drilled before hybrid corn was struck.

The estimate does include the cost of all the "dry holes" in hybrid-corn research itself. Hybrid corn was not a unique invention—it was an invention of a method of inventing. Many different combinations were tried before the right ones were found. One major seed company annually tests approximately fifteen hundred different combinations of inbred lines. Of these, at most three or four prove to be successful. The cost of the unsuccessful experiments is included in my estimate. What is excluded is investment in various other areas of agricultural research which has not borne fruit.

The problem of dry holes, however, can be reduced *ad absurdum*. What is the relevant segment for which an aggre-

[18] An annual flow of 40 cents discounted at 40 per cent has a present value of $1.00. An annual flow of $7.00, discounted at 40 per cent, will also have a present value of $1.00 if there is a lag of approximately ten years between the date of investment and the date at which the perpetual flow of returns begins.

gate rate of return is to be computed? Is it really reasonable to ascribe the cost of unsuccessful gold exploration to the oil industry? If one takes this kind of reasoning too seriously, there is only one rate that has any meaning—the rate of return on research for the economy as a whole.

Nevertheless, the rate of return on a successful innovation may be of some interest. In particular, it may be useful, ex ante, to break down the probable rate of return into two components: the rate of return if the development turns out to be a success and the probability that it will be a success. The approach outlined here is a way of estimating the first component. An estimate of the probability of success, however, must be made on the basis of data other than those presented in this article.

RETURNS IN SOME OTHER AREAS

T. W. Schultz has provided us with estimates of costs and returns of research for United States agriculture as a whole.[19] His data can be used to estimate the rate of return on agricultural research as a whole. Schultz gives an upper- and a lower-limit estimate of how much more input it would have taken to produce the 1950 output with 1940 techniques and inputs. His upper estimate is that it would have taken 18.5 per cent more input; his lower estimate is 3.7 per cent. Let us use Schultz's figures to estimate the perpetual gross annual returns, beginning with 1951, from the agricultural research that bore fruit between 1940 and 1950. Using the lower estimate, there would be a loss of 3.7 per cent in output if the new technology were to disappear. Taking the total annual value of farm output as $30 billion ($32 billion cash receipts from farm marketings in 1951, minus approximately 10 per cent to allow for the impact of the support programs), and assuming a price elasticity of demand for agricultural products of −.25 and an infinite supply elasticity, we get

$$k(1-\tfrac{1}{2}kn)\times\$30\text{ billion}=0.037\times(1-0.5\times0.037\times0.25)\times\$30\text{ billion}=0.037\times0.995\times\$30\text{ billion}=\$1{,}110\text{ million}.$$

Using the upper limit estimate of 18.5 per cent saving in inputs, we get

$$0.185\,(1-0.5\times0.185\times0.25)\times\$30\text{ billion}=\$5{,}430\text{ million}$$

as an upper-limit estimate of the gross annual social return from the technical change that occurred between 1940 and 1950.

Schultz also provides an estimate of total public expenditures on agricultural research for the years 1937–51. I assume that all these expenditures were used to produce the increase in output in 1951. This leaves out the contributions developed from funds spent before 1937, but, on the other hand, it disregards the possible returns after 1951 from the 1937–51 expenditures. On balance, we will probably overestimate the funds spent on the 1940–50 improvement in technology.

I will assume that total private agricultural research expenditures were of about the same magnitude as the public expenditures. This is approximately half the corresponding ratio for hybrid corn but is probably an overestimate for agriculture as a whole.[20]

[19] *The Economic Organization of Agriculture* (New York: McGraw-Hill Book Co., 1953), pp. 114–22.

[20] Mighell has estimated that recent annual expenditures by industry for research on agricultural products and on machinery and materials used in agriculture were in excess of $140 million. At the same time the USDA and state agricultural experi-

Multiplying the 1937–51 public expenditures by 2, deflating them by the Consumers Price Index, and cumulating them at the rate of 5 per cent, I get the figure of 3,180 million 1951 dollars as my estimate of total cumulated research expenditures in agriculture. Comparing this with the two estimated limits of the annual social returns of $1,110 and $5,430 million, I get a lower limit of 35 and an upper limit of 171 per cent as estimates of the annual rate of return per dollar spent on agricultural research. These are substantially lower than the estimated returns from hybrid corn but are comparable to estimates made by Ewell for the economy as a whole (100–200 per cent per year per dollar spent on "research and development") and to figures quoted by major industrial companies on their returns on research.[21]

Of course, these estimates are quite consistent with the estimated returns on hybrid corn, if the probability of success in research on innovations like hybrid corn is on the order of one-tenth or one-twentieth. Nevertheless, all these figures indicate that, in spite of the large growth in research expenditures during this century, the social returns to this activity are still very high.

HYBRID SORGHUM

The approach previously outlined can be used to estimate the probable rate of return on a new development: for example, hybrid sorghum. The development of hybrid sorghum began seriously only after World War II but has gained momentum rapidly since. Hybrid sorghum is now being introduced commercially. Very little was planted in 1956, but substantial amounts were planted in 1957. The experimental data to date suggest that the superiority of hybrid sorghum over previous seed may be even greater than that of hybrid corn.[22]

The 1956 value of the grain sorghum crop was approximately $232 million. Assuming a 15 per cent superiority of hybrids, a demand elasticity of −1.0, and an infinitely elastic supply curve, the estimated gross social returns from sorghum hybrids would be about $37 million annually.[23] Assuming that the extra cost of hybrid seed will be about $7 million annually, the net annual returns would be about $30 million.[24] I have no official data on research expenditures on hybrid sorghum. The head of one of the major seed companies has estimated that to date all public and private expenditures on hybrid sorghum total approximately $1 million and that current expenditures are at an annual rate of $300,000. Doubling his estimate of past expenditures and projecting into the future an annual research expenditure rate of $500,000, I get $10 or $13 million, depending on the rate of interest

ment stations spent about $118 million annually on research. However, only one-third of the industry research was in aid of farm production, mainly in machinery and chemicals: the rest was used in product research, while four-fifths of public expenditures were for research in aid of farm production (see R. Mighell, *American Agriculture* [New York: John Wiley & Sons, 1955], p. 130).

[21] R. H. Ewell, "Role of Research in Economic Growth," *Chemical and Engineering News*, XXXIII (1955), 298–304.

[22] *Sorghum Hybrids* (USDA, ARS, Special Report 22-26 [Washington, May, 1956]).

[23] This figure is derived as follows: $0.15\ (1 + \frac{1}{2} \times 0.15 \times 1.0) \times \$232$ million $= 0.16 \times \$232 =$ $37 million. The figure would have been approximately twice as large if I had used the value of the 1957 crop—$493 million—as my base.

[24] The additional cost of hybrid seed, seven million dollars, is estimated from the following data: 10 million acres sown to sorghum; ten dollars difference between the average prices per hundredweight of hybrid and open-pollinated grain sorghum seed (*Agricultural Prices*, April, 1958, p. 42); and an average seeding rate of seven pounds per acre.

used, as my "estimate" of cumulated hybrid-sorghum research expenditures in 1967. I choose 1967 as the reckoning date on the assumption that it will take ten years for hybrid sorghum to capture most of the sorghum seed market. While the projected rate of development is faster than that of hybrid corn in the United States as a whole, it is equivalent to the rate of acceptance of hybrid corn in Iowa. It is reasonable to assume that hybrid sorghum will spread much faster than hybrid corn did. Sorghum production is more localized than corn production, almost all sorghum is grown for commercial purposes, and hybrid sorghum will probably encounter much less resistance than hybrid corn encountered. I also assume that the use of hybrid sorghum in the United States will follow the same time path that the use of hybrid corn followed in Iowa.[25] This assumption allows me to estimate the social returns during the "transition period," 1957–67.

Table 3 outlines the calculation of the estimated rate of return on research expenditures on hybrid sorghum, which is approximately 400 per cent per annum. While this is somewhat lower than the estimated rate of return on expenditures on hybrid-corn research, it is still very high indeed.

SOME IMPLICATIONS

One might have expected, on a priori grounds, that the rate of return on expenditures on hybrid-sorghum research would be higher than the return on hybrid-corn research. The cost of hybrid sorghum has been and will be lower than the cost of hybrid corn both because of the cumulated experience in hybrid-corn breeding and because sorghum-growing is much more localized than corn-growing. Therefore, adaptable hybrids will have to be developed for a much smaller portion of the United States. Nevertheless, the estimated returns from hybrid sorghum are lower than those from hybrid corn. Why? To some extent this lower rate may be a result of overestimating research costs for hybrid sorghum, but the principal explanation is that the total value of the sorghum crop is substantially smaller than that of corn; sorghum is a relatively unimportant crop. It has recently been suggested that we should redirect our research efforts away from the major commodities that are in "surplus" and away from commodities with low elasticities of demand, where technical improvements result in reduced total returns to farmers in the long run. However, if we assume, in the absence of any other information, that technological change operates somewhat like a percentage increase in yield, and that the cost of achieving a given percentage boost in yield is the same for different crops or at least independent of their price elasticities and relative "importance," then the highest *social* returns

TABLE 3

ESTIMATED COSTS AND RETURNS OF HYBRID-SORGHUM RESEARCH AS OF 1967

(Millions of Dollars)

	$r=0.05$	$r=0.10$
Cumulated social net returns, 1957–66	155	171
Value in 1967 of returns beyond 1967	590	295
Total value of net returns in 1967	745	466
Cumulated research expenditures	9.4	13
Benefit-cost ratio	79	36
"Rate of return" in per cent per annum	395	360

[25] The percentage of all corn planted with hybrid seed in Iowa followed the following time path by years: 0.02, 0.06, 0.14, 0.31, 0.52, 0.73, 0.90, 0.99, and 1.00.

per research dollar are to be found in the important, low-elasticity commodities. For it can be easily shown that the absolute social gain from a given percentage increase in yield will vary proportionately with the total value of the crop and that the impact of different demand-and-supply elasticities is of a second order of magnitude. The latter affect only the "triangle," or the magnitude of the price-change adjustment. Among all the different factors, the total value of the crop is by far the most important determinant of the absolute social gain from a given percentage increase in yield.

No matter how we calculate them, there is little doubt that the over-all social returns on publicly supported technological research have been very high. It is not clear, however, whether or not this fact has any normative implications. I am afraid it has very few. More knowledge than is now at hand is required for presciption.

It is clear that we have not succeeded in equalizing the returns on different kinds of public investments. The returns from technological research have been much higher than the returns from reclamation and watershed projects. But should we have had more technological research? Surely, we have yet to reach the optimal level of expenditures on research, but this can only be a hunch. Should the public support agricultural research? My analysis illustrates and quantifies one of the major arguments for *public* investments in this area—the divergence between the social and private rates of return. Almost none of the calculated social returns from hybrid corn were appropriated by the hybrid-seed industry or by corn producers. They were passed on to consumers in the form of lower prices and higher output. Entry into the hybrid-seed industry was easy, and in the long run no "abnormal" profits were made there. By valuing the extra cost of seed production at the market price, I have counted as a cost whatever profit was made in this area by private producers, and the resulting estimate of returns consists almost entirely of social rather than private returns. These social returns were diffused widely among consumers of corn and corn products. Given the difficulty of patenting most of the valuable ideas in this area, the short life of a patent, and the general precariousness of a monopoly position in the long run, the incentive for private investment was very much smaller than that implied by the social rate of return.

While a divergence between social and private rates of return is a necessary reason for public intervention, it is not, by itself, a sufficient reason. We must ask not only whether social returns are higher than private—this is also true of many private investments—but also whether the private rate of return is too low, relative to returns on alternative private investments, to induce the *right* amount of investment at the *right* time. The social returns from nylon were probably many times higher than DuPont profits, but the latter were high enough to induce the development of nylon without a public subsidy, although, perhaps, not soon enough. To establish a case for public investment one must show that, in an area where social returns are high, private returns, because of the nature of the invention or of the relevant institutions, are not high enough relative to other private alternatives. This was undoubtedly true of hybrid corn, and it is probably true of many other areas of agricultural research and basic research in general. But it is not universally true. Hence a high social rate of return is not an unequivocal signal for public investment.

In this paper I have estimated that the rate of return on public investments in one of the most *successful* ventures of the past has been very high. This may give support to our intuitive feeling that the returns to such ventures in general have been quite high and to our feeling that "research is a good thing." But that does not mean that we should spend any amount of money on anything called "research." The moral is that, though very difficult, some sort of cost-and-returns calculation is possible and should be made. Conceptually, the decisions made by an administrator of research funds are among the most difficult economic decisions to make and to evaluate, but basically they are not very different from any other type of entrepreneurial decision.

[5]

SOCIAL AND PRIVATE RATES OF RETURN FROM INDUSTRIAL INNOVATIONS*

EDWIN MANSFIELD
JOHN RAPOPORT
ANTHONY ROMEO
SAMUEL WAGNER
GEORGE BEARDSLEY

I. INTRODUCTION

For many years, economic analysts and policy-makers have been interested in developing and obtaining better and more complete data concerning social and private rates of return from industrial innovations. It has long been recognized that information of this sort is essential if public policy concerning civilian technology is to be formulated rationally. Yet despite the work of Denison, Fellner, Griliches, Mansfield, Minasian, Peterson, Terleckyj, and others, existing knowledge in this area is far too weak to provide a reasonably adequate foundation for analysis and policy.[1] The purpose of this paper is to report the results of seventeen case studies, each of which estimates the social and private rate of return from the investment in a particular industrial innovation. The results of this paper, although subject to obvious limitations and shortcomings, should provide us with a somewhat better understanding of this topic. The studies described

* The work on which this paper is based was supported by a grant to Mansfield from the Office of National R and D Assessment of the National Science Foundation. The bulk of the basic data on which this study is based was collected by Beardsley, Rapoport, Romeo, and Wagner. A preliminary version of this paper was presented by Mansfield at the meetings of the Eastern Economic Association on October 26, 1974; at the National Bureau of Standards; in a paper at the 1975 meetings of the American Economic Association; and in a paper at the First U.S.–U.S.S.R. Symposium; at M.I.T.; at the RAND Corporation; and at the National Science Foundation. Also, he presented some of this material in his Schmookler Memorial Lecture at the University of Minnesota in 1975 on the Economics of Information (which took place in Leningrad in 1975).

1. For a summary of existing knowledge in this area, see E. Mansfield, "Contribution of R and D to Economic Growth in the United States," *Science*, CLXXV (Feb. 4, 1972), 477–86.

here seem to be the first attempts to make direct measurements of this sort in the industrial sector.

II. The Sample of Innovations

Our first step in carrying out this investigation was to contact a number of business firms in the Northeast and to try to persuade them to provide us with data bearing on the social and private returns from innovations that they had carried out. As would be expected, a substantial percentage of those who were contacted refused to cooperate because, despite our assurance that the data would be held in strictest confidence, they felt that such data were too sensitive to show outsiders. Those firms that were willing to cooperate were asked to pick one or more of their recent innovations more or less at random. Then many manweeks were spent gathering data concerning each innovation and its effects from the innovating firm, from firms using the innovation (if it was used by firms), and from other sources. These innovations occurred in a wide variety of industries (described below), and in firms of quite different sizes. Most of them are of average or routine importance, not major breakthroughs. Although the sample cannot be regarded as randomly selected, there is no obvious indication that it is biased toward very profitable innovations (socially or privately) or relatively unprofitable ones. The sample is described in Table I.

III. Estimation of Social Benefits: Product Innovations Used by Firms

As is evident from Table I, the innovations in our sample can be divided into three classes: product innovations used by firms, product innovations used by households, and process innovations. Based on an intensive study of each of the innovations in each of these classes, it appears that the same general kind of model is applicable to all of the innovations in our sample in a particular class. In this and the following sections we describe the model that is used to measure the social benefits in a particular period from product innovations used by firms.

In each case these new products resulted in a potential saving to users. For example, the product innovation in the primary metals industry resulted in a potential saving to makers of household appliances. Thus, each of these innovations could shift downward the supply curve of the industry using the innovation. How far downward

TABLE I

CHARACTERISTICS OF SAMPLE OF SEVENTEEN INNOVATIONS

Industry producing the innovation	Type of innovation	Nature of innovation	Type of user	Approximate date of innovation
1. Primary metals	New product	New type of metal that reduced cost of appliances	Firms	Late 1950's
2. Machine tools	New product	New computer controls	Firms	Early 1970's
3. Control systems	New product	New type of component	Firms	Late 1960's
4. Construction	New product	New material that cut cost of building	Firms	Mid 1950's
5. Drilling	New product	Reduced cost of drilling wells	Firms	Mid 1960's
6. Industrial equipment	New process	New type of drafting	Firms	Early 1960's
7. Paper	New product	New paper product that cut costs of users	Firms	Mid 1960's
8. Thread	New product	New type of thread that cut costs of garment makers	Firms	Early 1960's
9. Industrial controls	New product	New mechanism for doors	Firms	Early 1970's
10. Electronics	New product	New device that reduced costs of certain video tape operations	Firms	Early 1970's
11. Chemicals[a]	New product	New product that reduced costs of users	Firms	Late 1960's
12. Chemicals[a]	New process	Reduced costs of production	Firms	Mid 1960's
13. Chemicals[a]	New process	Reduced cost of certain aromatic chemicals	Firms	Late 1960's
14. Chemicals[a]	New process	Major process improvement	Firms	Early 1960's
15. Household cleaners	New product	New product that reduced cost of cleaning floors	Households	Early 1960's
16. Stain removers	New product	New stain remover	Households	Mid 1960's
17. Dishwashing liquids	New product	New product that cut costs of operating dishwashers	Households	Early 1960's

a. Chemicals are defined to include petroleum refining.

this supply curve will shift depends, of course, on the pricing policy of the innovator. If the innovator charges a relatively high price for the new product, the supply curve may shift only slightly. Indeed, if the innovator charges a high enough price, the supply curve will not shift downward at all.

Suppose that the innovator decides to set a price for its new product that yields a profit[2] to the innovator equivalent to r dollars per unit of output of the industry using the innovation (for example, r dollars per appliance in the case of the new type of metal). Also, suppose that the industry using the innovation is competitive and that its supply curve is horizontal in the relevant range. In particular, assume that, before the advent of the innovation, this supply curve was S_1 in Figure I, and the price charged by the industry using the innovation was P_1. After the advent of the innovation, this supply curve is S_2, and the price is P_2.

Under these circumstances, the social benefits from the innovation can be measured by the sum of the two shaded areas in Figure I. The top shaded area is the consumers' surplus due to the lower price (P_2 rather than P_1) resulting from the use of the innovation.[3] In addition, there is a resource saving, and a corresponding increase in output elsewhere in the economy, due to the fact that the resource

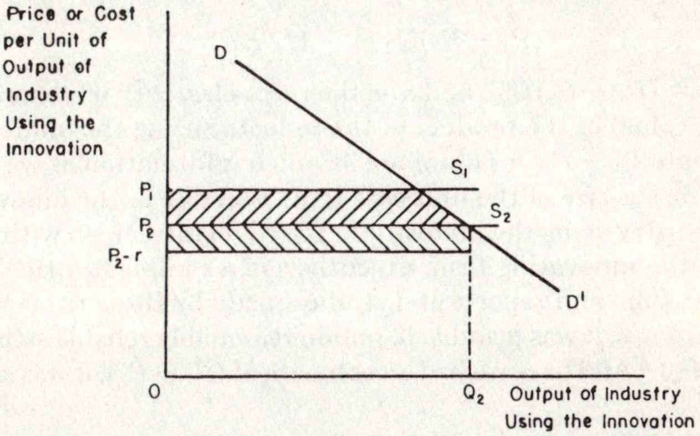

FIGURE I
Social Benefit from Product Innovation That Reduces the Costs of the Industry Using It

2. Profit here is defined to be gross of any depreciation on the investment in the innovation. Basically, it is a cash flow concept.

3. Of course, one problem in any model of this sort is that the concept of consumers' surplus is a controversial one. But as so many writers have pointed out, it is hard to do applied work in this area without adopting some such concept. See, for example, J. Currie, J. Murphy, and A. Schmitz, "The Concept of Economic Surplus and Its Use in Economic Analysis," *Economic Journal*, LXXXI (Dec. 1971), 741–99.

costs of producing the good using the innovation—including the resource costs of producing the innovation—are less than P_2Q_2. Instead, they are P_2Q_2 minus the profits of the innovator from the innovation, the latter being merely a transfer from the producers of the good using the innovation to the innovator. Thus, besides the consumers' surplus arising from the price reduction, there is a resource saving amounting to the profits of the innovator.[4]

However, two adjustments must frequently be made in this estimate corresponding to the lower shaded area in Figure I. First, if the innovation replaces another product, the resource saving cited in the previous paragraph does not equal the profits of the innovator (from the innovation), but these profits less those that would have been made (by the innovator or other firms) if the innovation had not occurred and the displaced product had been used instead. This is the correct measure of the resource saving. Second, if other firms imitate the innovator and begin selling the innovation to the industry that uses it, their profits from the sale of the innovation must be added to those of the innovator to get a full measure of the extent of the resource saving due to the innovation.

Based on this model, an estimate was made of the social benefit in each period from the investment in each of these innovations. For each innovation, the top shaded area in Figure I equals

$$(P_1 - P_2)Q_2\,(1 - 1/2\,Kn), \tag{1}$$

where $K = (P_1 - P_2)/P_2$, and n is the price elasticity of demand (in absolute value) of the product of the industry using the innovation. To estimate $P_1 - P_2$, we obtained as much information as we could concerning the size of the unit cost reduction due to the innovation in the industry using the innovation. Based on interviews with executives of the innovating firm, executives of a sample of firms using the innovation, and reports and studies made by these firms for internal purposes, it was possible to obtain reasonably reliable estimates of $(P_1 - P_2)$.[5] And once we had an estimate of $(P_1 - P_2)$, it was simple

4. To put matters somewhat differently, the sum of the two shaded areas in Figure I can easily be shown to equal the social value of the extra output of the product (as measured by the area under the demand curve) plus the value of the resources saved.

5. We assume that the cost reduction experienced by the industry using the innovation is passed on to consumers. In view of the market structures of the industries in question, this seems to be a reasonable assumption. Also, the available evidence seems to support this assumption. But it is important to note that, if the cost reduction were kept by the industry using the innovation, the social benefit would equal this industry's cost saving due to the innovation plus the innovator's (adjusted) profits. Since the industry's cost saving must be greater than our measure of the upper shaded area in Figure I, our measure of social benefit would be conservative.

to compute K. Also Q_2 was generally available from published records. Rough estimates of n were obtained from published studies and from the firms. Since K is generally very small, the results are not sensitive to errors in n. Indeed, the expression in equation (1) can be approximated quite well in most cases by $(P_1 - P_2)Q_2$, which is the total savings to consumers due to the lower price, if they buy Q_2 units of the product of the industry using the innovation.

To estimate the additional resource saving from the innovation, which equals the bottom shaded area in Figure I (if the adjustments described in the paragraph before the last are unnecessary), the innovator's profit from the new product was obtained from detailed discussions with the firm's executives, as well as study of relevant financial records. For each year, the costs of marketing and producing the innovation, as well as the costs of carrying out the innovation (R and D, plant and equipment, manufacturing start-up, and marketing start-up),[6] were deducted from the innovator's revenues from the innovation. Note that the R and D costs were adjusted to allow for the fact that the innovator invested R and D resources in uncommercialized R and D projects. To make this adjustment, we obtained estimates from each of the innovating firms of the average number of dollars spent on uncommercialized R and D projects per dollar spent on a commercialized R and D project during the relevant period. Then we multiplied the innovator's R and D outlays (in each year) on the innovation by this number in order to get an estimate of the total R and D investment, including a prorated allowance for uncommercialized projects. In cases where the adjustments described in the paragraph before the last were necessary, estimates of the foregone profits from displaced products were deducted, and the profits of imitators were added, to the innovator's profits. These estimates were obtained from the relevant firms.

IV. Parallel Innovative Efforts, Time Horizon, and Rates of Return

In calculating the social rate of return, we must recognize that, besides the innovator, other firms or research organizations in the United States[7] may have invested resources (prior to the introduction

6. For discussions of the cost categories involved, see E. Mansfield and J. Rapoport, "The Costs of Industrial Product Innovations," *Management Science*, XXI (Aug. 1975), 1380–86; and E. Mansfield, J. Rapoport, J. Schnee, S. Wagner, and M. Hamburger, *Research and Innovation in the Modern Corporation* (New York: Norton, 1971).

7. We are concerned here only with the costs and benefits arising within the United States. Thus, we ignore the fact that some of these innovations resulted in significant

of the innovation in question) in R and D and related innovative activities aimed at innovations of essentially the same kind as this one. Clearly, it is not easy to obtain data on the extent of such investments, but fortunately the difficulties seem to be less than they might appear, for two reasons. First, only a limited number of organizations could reasonably have been expected to be doing R and D in the relevant area, and if they had been devoting any substantial amount of resources to such work in the relevant time frame, it is inconceivable that the current executives of these organizations (and all their competitors) would have been unaware of it. Second, since these innovations occurred some time ago, firms generally are quite willing to discuss whether they were carrying out work of the relevant kind at that time. Moreover, and this is particularly fortunate, our results are quite insensitive to errors in the estimated investment in R and D carried out by others. Even if the true social research expenditures were ten times our estimate, and the true social development expenditures were double our estimate, the results change remarkably little.

Based on interviews with executives of the innovating firm, as well as with other firms that could reasonably be expected to do (and be aware of) R and D of the relevant kind, it appears that, in most of the cases in our sample, no other firm or research organization was doing work aimed at roughly the same kind of innovation. Thus, in these cases, the private investment seems to be a good approximation to the social investment. In the remaining cases, other firms or individuals were engaged in R and D aimed at the same kind of innovation. In the cases where this R and D was *unsuccessful,* we obtained as accurate an estimate as possible of the cost of this unsuccessful R and D, and added this figure to the R and D costs described in the previous section to get an estimate of the total social cost of the relevant R and D carried out by the innovator and others.[8]

---

social benefits in other countries, and we do not attempt to include whatever costs may have been incurred in other countries in an attempt to produce innovations of roughly the same kind. Also, each rate of return estimated here is on the investment in the innovation in question, given that previous investments in basic science, education, and so forth had already occurred. Clearly, the rates of return on the investments considered here were dependent on these previous investments.

8. Of course, some of the unsuccessful R and D directed at one innovation may be part of the uncommercialized R and D carried out by the innovator in the case of another innovation. Really, the social cost of an innovation should include a prorated share of the cost of uncommercialized R and D carried out by the innovator, excluding the cost of its uncommercialized R and D aimed at innovations where other firms beat it to the punch. But there is no way to estimate the latter cost, and our procedure clearly biases the social rates of return downward.

In the one case (among product innovations used by firms) where other firms or individuals were engaged (prior to the introduction of the innovation in question[9]) in R and D aimed at roughly the same kind of innovation, and where this R and D was *successful,* we must recognize that the innovator's investment resulted only in the innovation's availability at an earlier point in time, not in all of the social benefits from the innovation date up to the relevant time horizon. In other words, the proper comparison is between what would have occurred if the innovator had not carried out the innovation (but other firms were free to do so) and what in fact occurred. In this case, we obtained as accurate an appraisal as we could of the date when the innovation would have appeared if the innovator had not carried out the innovation, and we calculated the social benefits only during the period between the date when the innovation occurred and the date when it would have appeared if the innovator had done nothing.[10] Of course, it frequently is very difficult to estimate when the innovation would have occurred if the innovator had not carried it out. But in this particular case a realistic (if perhaps somewhat conservative) estimate is that the second firm to produce the innovation would have come up with it when in fact it did so, regardless of whether the innovator preceded it or not.

For most innovations these calculations were carried out for each year from the beginning of work on the innovation until 1973. Thus, our estimates of the social benefits are conservative, since all benefits after 1973 are ignored. But in some cases, this would introduce a serious distortion, since the innovation is relatively new. In these cases, forecasts were made of the consumers' surplus and the innovator's profits (adjusted for imitators' profits and for profits on older products) in each year up to 1980. These forecasts are based on firms' expectations concerning $(P_1 - P_2)$, Q_2, and the relevant profits in the next few years. They are intentionally very conservative, so whatever bias there is in the resulting rates of return should be downward.

Finally, having made the calculations described in this and the previous section for each year, we have an estimate of the net social benefits from the innovation for each year. Then we can compute the

9. Work of this sort that occurs after the innovation takes place is of a different kind. It is directed at *imitating* or *improving on* the innovation, and is not properly regarded as part of the work *producing* the innovation.

10. Note that this procedure assumes that, once the innovator is joined by another firm producing the innovation, they adopt a pricing policy that is just the same as the other firm would have adopted if it were the sole producer of the innovation. Needless to say, this may not be true, since there may be advantages stemming from competition between the two firms. However, we adopt this assumption because it results in a conservative estimate of the social rate of return and because any other assumption would have to be based largely on speculation.

internal social rate of return, the interest rate that makes the present value of the net social benefits equal to zero. In other words, it is the interest rate i that results in the following equality:

$$B(t) + \frac{B(t+1)}{1+i} + \frac{B(t+2)}{(1+i)^2} + \cdots + \frac{B(t+n)}{(1+i)^n} = 0, \tag{2}$$

where $B(t)$ is the net social benefit year in t, t is the first year in which the net social benefit is nonzero, and $(t + n)$ is the last year in which the net social benefit is nonzero.

Finally, we also compute the private rate of return from the innovator's investment in each innovation. To do so, we calculate the cash flow to the innovator from the innovation during each year. This calculation involves the subtraction of all costs incurred by the innovator to carry out, produce, and sell the innovation (including the allowance described in the previous section for R and D on uncommercialized projects) from the innovator's revenues from the innovation. Also, profits that the innovator would have earned on products displaced by the innovation must be subtracted as well.[11] The time period over which these computations were made was generally up to 1973, but in some cases (as in the case of the social rate of return) it extended to 1980. Again, the forecasts in the latter cases are decidedly conservative. The net private benefits in each year, like the B's in equation (2), were deflated. The Consumer Price Index, which generally was used, is not ideal, but it seems very unlikely that the results will be affected in an important way by this choice of a deflator.

V. Product Innovations Used by Households

In this section we turn to the model used to calculate the social benefits for product innovations used by households. Since all of the innovations of this type in our sample are meant to reduce the cost of some particular household activity, basically the same model will apply. Suppose that the demand curve for this activity is DD' in Figure II and that the innovation reduced the price of performing this activity from P_1 to P_2. Suppose too that the innovator makes a profit

11. In all but one case, it is assumed that the profits from the displaced product would have continued (up to the time horizon indicated below) at their previous level if the innovation had not occurred. This seems reasonable, based on the facts. But in the one remaining case, there is reason to believe that, if the innovation had not occurred, another innovation superior to this one would have occurred at a particular point in time afterward. Thus, we assume that the innovator would have earned nothing on the displaced product after this point in time. Of course, it also earned nothing on the innovation after this point.

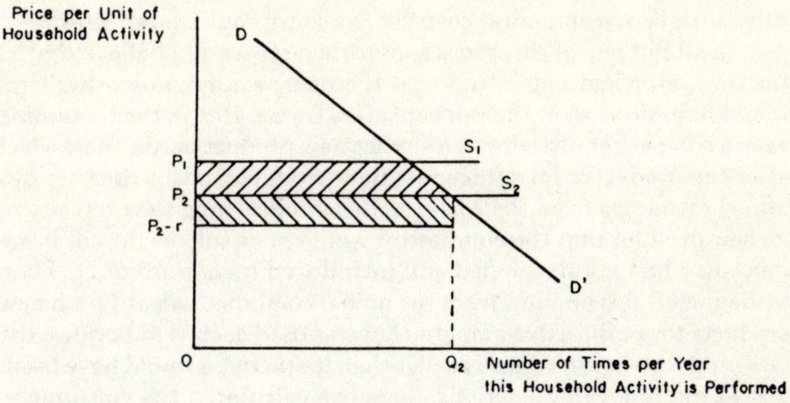

FIGURE II
Social Benefit from Product Innovation That Reduces the Price of a Household Activity

of *r* dollars every time this activity is performed using the product innovation. Then the social benefits from the innovation can be measured by the sum of the two shaded areas in Figure II. These two areas are exactly analogous to those in Figure I, and the methods used to estimate them are similar to those described above, except that no interviews were carried out with users. Instead, the results are based entirely on the estimates and findings of the firms. As in the previous sections, it is necessary to correct the R and D costs for uncommercialized projects, to estimate R and D costs incurred by firms and organizations other than the innovator, to deduct the profit from displaced products from the innovator's profit, and to include the profits of imitators.

However, not everything is the same as in previous sections. For one thing, one of these innovations, the stain remover, resulted in some environmental costs, since it increased the amount of a certain chemical that is put into our water supplies. As a rough measure of these costs, we used an estimate (obtained from an official of the Environmental Protection Agency) of the cost of removing a unit of this chemical from water, given that it occurs in a particular concentration. Then we multiplied this cost times the number of units of the chemical that is deposited in the water, given that consumers use the innovation to perform the relevant household activity Q_2 times per year. Finally, this estimate was deducted from the social benefit to correct, at least roughly, for the environmental costs. Note that this was the only innovation in our entire sample that resulted in any substantial environmental costs, so far as we could determine.[12]

In all but one of these cases, there is no reason to believe that, if the innovator had not introduced the new product, any other firm would have done so in the relevant time frame. But in the remaining case, a competitor introduced a similar new product only a short while after the innovator introduced its new product. In this case, we obtained estimates from the relevant firms concerning the quantity of its new product that the competitor would have sold each year if the innovator had not developed and introduced its new product. Then we deducted this amount from the actual combined sales of both new products to get the extra amount that was used each year because the new product was introduced earlier than it otherwise would have been, due to the innovator's actions. Then we calculated the consumers' surplus based on this extra amount. Also, we calculated the innovator's profit less the profit that would have been earned by the competitor on the share of the innovator's sales that the competitor would have gotten if the innovator had not introduced its product (and less the profit on products displaced in the remainder of cases). Based on these calculations, we computed the social rate of return.

Finally, as in the case of product innovations used by firms, we computed private as well as social rates of return. The calculations are entirely analogous to those described in the previous section. In all cases the calculations end with 1973.

VI. Process Innovations

Next, let us turn to the model underlying our calculation of the social benefits from process innovations. In the case of three of the four process innovations included in our sample, there was no apparent effect on product prices. By lowering the costs of the innovators, these process innovations increased the innovator's profits. Also, since they were imitated (or used at nominal cost) by other firms, they soon increased the profits of other firms as well. The total decrease in costs (which equals the increase in profits) of all of the relevant firms is a measure of the social benefit of each of these innovations in a particular period. It equals the social saving in resources due to the innovation. For each of these innovations, we estimated the total reduction in costs on the basis of interviews with the relevant firms

12. Note that some of these innovations had positive environmental effects. However, since these benefits are hard to measure, we ignore them. Of course, whatever bias this introduces results in our estimates of the social rate of return being conservative.

and of studies and reports concerning the cost reductions due to the innovations and the extent of their utilization.

In the case of the fourth process innovation (the innovation in industrial heating equipment), the situation was different. In this case, the innovator shared the gains with its customers. As shown in Figure III, it reduced the price of its product from P_1 to P_2. Thus, the social

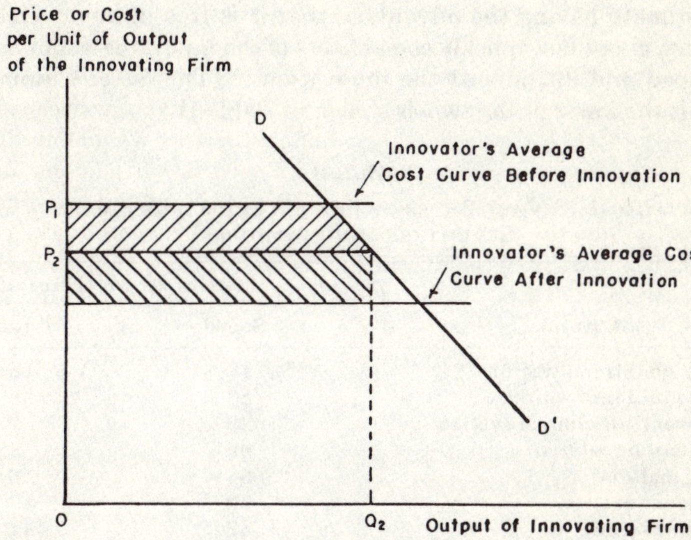

FIGURE III
Social Benefit from Process Innovation That Results in Price Reduction by Innovator

benefit equals the sum of the two shaded areas, the top shaded area being the consumers' surplus resulting from the price reduction, and the bottom shaded area being the profits to the innovator from the innovation. The situation is almost exactly the same as in Figures I and II.

For one of the process innovations, it was necessary to estimate the length of time that the innovator hastened the appearance and introduction of the new process. (In the other cases, there is no reason to believe that if the innovator had not introduced the innovation in the relevant time period, any other firm would have done so.) In this case other firms were doing R and D along somewhat similar lines, and there was some feeling among knowledgeable engineers and managers that the new process might have been developed and introduced about four years later than in fact occurred if the innovator had not done so. In this case we assume that the diffusion of the new process would have begun four years later (but that it would have proceeded at the same rate) if the innovator had not developed and introduced it. Then the social and private savings are computed by comparing the actual extent of use of the innovation with what it would have been under this assumption. In addition, the social rate of return based on the entire industry's investment in this sort of innovation is estimated, this estimate having the advantage that it is free of any such assumption about how quickly competitors of the innovator would have developed and introduced the innovation.[13] (The latter estimate, which is the lower of the two, is shown in Table II.)

TABLE II

SOCIAL AND PRIVATE RATES OF RETURN FROM INVESTMENT IN SEVENTEEN INNOVATIONS

Innovation	Rate of return (percent)	
	Social	Private
Primary metals innovation	17	18
Machine tool innovation	83	35
Component for control system	29	7
Construction material	96	9
Drilling material	54	16
Drafting innovation	92	47
Paper innovation	82	42
Thread innovation	307	27
Door control innovation	27	37
New electronic device	Negative	Negative
Chemical product innovation	71	9
Chemical process innovation	32	25
Chemical process innovation	13	4
Major chemical process innovation	56[a]	31
Household cleaning device	209	214
Stain remover	116	4
Dishwashing liquid	45	46
Median	56	25

a. Based on investment of entire industry. See Section VI.

VII. SOCIAL AND PRIVATE RATES OF RETURN

The results, shown in Table II, seem to have at least three implications. First, they indicate that the social rate of return from the

13. For two of the processes the calculations end with 1972 or 1973; for the other two, forecasts are used up to 1980 (and in one case, beyond).

investments in these seventeen innovations has been very high. The median estimated social rate of return is about 56 percent. Moreover, for a variety of reasons, these estimates are likely to be conservative lower bounds.[14] To put these results in perspective, recall that Griliches found that the internal social rate of return from hybrid corn—a very successful innovation—was 37 percent.[15] Clearly, the investments that have been made in industrial innovations have, on the average, yielded handsome social returns, if these innovations are at all typical.

Second, the results indicate that the private rates of return from the investments in these innovations have been much lower than the social rates of return.[16] The median private rate of return (before taxes) was about 25 percent. In interpreting this number, it is important to recognize the riskiness of this kind of investment (and the fact that these are pretax figures). This riskiness is evidenced by the enormous variation in the private rates of return in Table II. In the case of six innovations the private rate of return (before taxes) was less than 10 percent, while for five innovations, it was more than 40 percent.

14. For example, in many cases, we ignore all post-1973 benefits from the innovations. Throughout this study, we have tried to use estimates and procedures that, if biased at all, would be biased in this direction.

15. See his "Research Costs and Social Returns: Hybrid Corn and Related Innovations," *Journal of Political Economy*, LXVI (Oct. 1958), 419–31. The models used in Sections III and IV are in the same spirit as the one applied by Griliches. The principal differences are that we have extended the analysis to include the pricing policies of the innovator, the effect on displaced products, and the costs of uncommercialized R and D and of R and D done outside the innovating organization. Another noteworthy study of this kind is W. Peterson, "Returns to Poultry Research in the United States," *Journal of Farm Economics*, XLIX (Aug. 1967), 656–69. This study concludes that the average rate of return from the investment in poultry research was 18 percent.

16. Note that there is nothing in our procedures that makes it inevitable that an innovation's social rate of return exceeds its private rate of return. For a variety of reasons (such as investments in R and D by organizations other than the innovator, and the transfer of profits from the producers of displaced products to the innovator), the social rate of return can be less than the private rate of return—and in fact this turns out to be the case for four of the innovations in our sample. For a study of some of the factors associated with a firm's probabilities of success in R and D, see E. Mansfield and S. Wagner, "Organizational and Strategic Factors Associated with Probabilities of Success in Industrial R and D," *Journal of Business*, XLVIII (April 1975), 179–98.

One of the innovations was developed by a nonprofit organization. In this case the private rate of return is defined to be the rate of return to this organization and the licensees that produced and marketed the innovation.

When we compute a confidence interval for the proportions of innovations where the social rate of return exceeds the private rate of return, we find that the probability is 0.90 that this percentage lies between 60 percent and 92 percent. Thus, if we could treat our sample as randomly chosen, we could be quite confident that the social rate of return exceeds the private rate of return in the majority of cases. Also, a t-test indicates that the mean difference between social and private rates of return, which is 47 percent, differs significantly from zero at the 0.01 level. And so does a sign test, which does not assume normality.

Third, the results indicate that, in about 30 percent of the cases, the private rate of return was so low that no firm, with the advantage of hindsight, would have invested in the innovation, but the social rate of return from the innovation was so high that, from society's point of view, the investment was well worthwhile. We would also like to estimate the extent to which innovations with favorable social rates of return have anticipated rates of return that are so low that they are not carried out. Unfortunately, Table II cannot provide much information on this score for the obvious reason that all of these innovations were carried out. Also, expected and actual private rates of return may differ considerably.[17]

VIII. Factors Associated with the Gap Between Social and Private Rates of Return

Economists have long been interested in the sorts of innovative activity where the gap between social and private rates of return, if it exists, is particularly large. There has been considerable speculation on this score, but it seems fair to say that the current state of knowledge in this regard is extremely limited. Economic theorists have generally singled out three factors as being of considerable importance in determining the extent of the gap between the social and private rates of return from an innovation. The first of these factors is the market structure of the innovator's industry. If the innovator is faced with a highly competitive environment, it is less likely that it will be able to appropriate a large proportion of the social benefits than if it has a secure monopoly position or if it is part of a tight oligopoly. Of course, the extent to which the innovator is subjected to competition, and how rapidly, may depend on whether the innovation is patented. Moreover, another consideration of at least equal importance is how expensive it is for potential competitors to "invent around" the innovator's patents, if they exist, and to obtain the equipment needed to begin producing the new product (or using the new process). In some cases, like DuPont's nylon, it would have been extremely diffi-

17. However, a limited amount of evidence can be given on this score. We could obtain data for nine of the innovations concerning the approximate private rate of return expected from the innovation by the innovator when it began the project. In five of the nine cases, the expected private rate of return was less than 15 percent, indicating that these five projects were quite marginal; yet the average social rate of return of these five projects was over 100 percent! Unless there is a marked discontinuity of some sort, it would appear from this that there may be many projects where the expected private rate of return was a bit lower (with the result that they were not carried out), but where the social rate of return would have been quite high. (The rank correlation between the expected private rate of return and the social rate of return does not differ significantly from zero in this sample.) However, the very small sample and the crudeness of the analysis should be borne in mind.

cult to imitate the innovation (legally). In other cases, a potential competitor could obtain and begin producing a "me too" product (or using a "me too" process) at relatively little cost.

A second factor that economists have emphasized as a determinant of the size of the gap between social and private rates of return is whether the innovation is major or minor. (A reasonable measure of an innovation's importance (at least in our sample) might be the annual net social benefits derived from it.) According to R. C. O. Matthews, the "degree of appropriability is likely to be less . . . in major innovations than in minor ones. . . ," [18] since major innovations are more likely, in his view, to be imitated quickly. Similarly, on the basis of a model stressing the indivisibility of information, Kenneth Arrow concludes that "the inventor obtains the entire realized social benefit of moderately cost reducing inventions but not of more radical inventions." [19]

A third factor that is sometimes cited is whether the innovation is a new product or process. Thus, Matthews thinks that the degree of appropriability may be less for process innovations than for product innovations. On the other hand, Nelson, Peck, and Kalachek stress that new processes can often be kept secret and that it frequently is difficult for one firm to find out what processes another firm is using.[20] This, of course, suggests that the gap between social and private rates of return might be greater for products than for processes. Neither Matthews nor Nelson, Peck, and Kalachek present any evidence on this score.

Although most of these hypotheses seem quite plausible, the truth is that none of them have been subjected to any sort of systematic empirical tests. Based on the results presented in the previous section, it is possible for the first time to attempt some empirical tests of these hypotheses. These tests are made in two steps. First, we divide the sample into two groups, processes and products, to see whether the average difference between the social and private rates of return is different in one group than in the other. Second, looking at products alone (since there are too few processes to support such an analysis),

18. R. C. O. Matthews, "The Contribution of Science and Technology to Economic Development," in B. Williams, ed., *Science and Technology in Economic Growth* (London: Macmillan, 1973), p. 14.

19. K. Arrow, "Economic Welfare and the Allocation of Resources for Invention," in National Bureau of Economic Research, *The Rate and Direction of Inventive Activity* (Princeton: Princeton University Press, 1962), p. 622. Arrow's model bears a resemblance to ours in Section III, but he was interested in a somewhat different set of questions, and no attempt was made to use it as a basis for empirical work.

20. R. Nelson, M. Peck, and E. Kalachek, *Technology, Economic Growth, and Public Policy* (Washington: Brookings Institution, 1967), p. 161.

we assume that

$$D_i = \alpha_0 + \alpha_1 S_i + \alpha_2 P_i + \alpha_3 I_i + z_i, \quad (3)$$

where D_i is the difference between the social and private rates of return for the ith innovation, S_i is the annual net social benefit (in thousands of dollars) from the ith innovation three years after its initial introduction (a crude, but reasonable, measure of the innovation's importance), P_i is a dummy variable that equals 1 if the innovation is patented and 0 otherwise, I_i is an estimate of the minimum amount of money (in thousands of dollars) it would have cost any of the innovator's competitors to imitate legally and introduce the ith innovation, and z_i is a random error term. According to the hypotheses stated above, one would expect α_1 to be positive and α_2 and α_3 to be negative. Ordinary regression techniques were used to determine whether this was the case.

The results suggest that the difference between the social and private rates of return tended to be higher for products than for processes. (The average value of D_i was about 57 percent for products and about 21 percent for processes.) But this difference is not statistically significant. Turning to the estimation of the α's, we obtained data concerning I_i for each innovation.[21] Then, combining these data with the information concerning D_i, P_i, and S_i (which stems from the work in previous sections), we used least squares to estimate the α's, the resulting equation being

$$D_i = \underset{(0.46)}{12.4} + \underset{(4.34)}{0.0313 S_i} - \underset{(1.87)}{0.0509 I_i}, \; (R^2 = 0.67), \quad (4)$$

where the t-statistic of each of the regression coefficients is shown in parentheses, and P_i is excluded because its regression coefficient is not even close to being statistically significant. Despite the relatively small sample size, the estimate of α_1 is highly significant, and the estimate of α_3 is significant at the 0.05 level. Both have the expected signs.

Thus, the results seem to support the hypothesis that the gap between social and private rates of return tends to be greater for more important innovations and for innovations that can be imitated relatively cheaply by competitors. Apparently, when the cost of imitating

21. These estimates were obtained from the innovating firms. In each case executives of the innovating firm were asked to estimate the minimum amount that it would have cost a competitor to imitate the innovation (legally). In some cases the estimate of I_i was as high as \$2 million; in other cases, it was as low as \$3,000. Although these estimates are rough, the executives seemed to think that they were not very wide of the mark.

the innovation is held constant, it makes little or no difference whether the innovation is patented—which seems reasonable because whether or not a patent exists is of relevance largely (perhaps only) because of its effects on the costs of imitation. It is important to note that S_i and I_i can explain about two-thirds of the observed variation in this gap among the product innovations in our sample.[22] However, at the same time, it is important to bear in mind the smallness of the sample and its nonrandom character.

IX. Unemployment, Repercussions on Other Markets, and Future Changes in Technology

Before concluding, we must take up several additional factors that have not been mentioned explicitly in previous sections. First, new technology can, of course, result in prolonged and serious unemployment of particular kinds of labor. Recognizing this fact, we tried to determine whether there was any substantial unemployment (or change in job content) resulting from each innovation described in Section II. In no case could we find any evidence of an appreciable change of this sort: although labor requirements sometimes were reduced by the innovation, growth in output offset these reductions. Thus, in this sample at least, this factor does not seem of substantial importance. Obviously, however, it can be of importance in other cases; and in situations where it is of importance, one must adjust the social benefits, as calculated in previous sections, for the fact that (for a certain period of time) some of the resources saved by the innovation do not find employment elsewhere.[23]

Second, new technology can, of course, have an effect on other industries and markets besides the ones directly involved. For example, by reducing the cost and price of the product using an innovation, the innovation affects the outputs of goods that are substitutes or complements of this product. However, so long as the prices of these and other goods (other than the product using the innovation) are not materially affected, the effects of these output changes are already included in the measures specified above.[24] In our sample there is no

22. One innovation had to be excluded from the regression (and the t-test in note 16) because its private rate of return was incalculable, since the private net benefit in each year was negative. Note too that the ratio of I_i to the innovator's cost in carrying out the ith innovation was tried as an additional explanatory variable in equation (4), but it was not close to being significant. Apparently, the absolute, not relative, cost is what matters most.

23. See A. Schmitz and D. Seckler, "Mechanized Agriculture and Social Welfare: The Case of the Tomato Harvester," *American Journal of Agricultural Economics*, LII (Nov. 1970), 569–77.

24. See E. Mishan, *Economics for Social Decisions* (New York: Praeger, 1972), Ch. 6.

evidence that any of these innovations had any appreciable impact on any price other than that of the good using the innovation (in Section IV) or of the household activity in question (in Section V) or of the good based on the one relevant process (in Section VI). Also, one must assume that inputs that were displaced from industries producing goods competing with the product using the innovation readily found employment elsewhere. There is no evidence that this was not the case.

Third, new technology can, of course, play an important role in suggesting and prompting further innovations. Without question, practically all of the innovations in our sample had this effect. Thus, since each of these innovations is instrumental in bringing about future innovations, each results in social benefits above and beyond what we have measured. However, each of these innovations may have been suggested, prompted, or aided by previous innovations, and there is no feasible way to estimate the extent to which there is a net understatement of benefits.

X. Conclusion

As everyone who writes on this subject is quick to point out, very little is known about the social and private returns from industrial research and technological innovation. The need for measurement in this area is acute and widely recognized. For example, the Council of Economic Advisers, in its 1972 Annual Report, stressed the need for further analysis "to show the benefits, costs, and processes associated with R and D." [25] Our purpose in this study has been to measure, as best we can, the social and private rates of return from a sample of innovations.[26] We believe that our results go far beyond anything heretofore attempted in this area. (Also, to check and extend our results, we obtained very rich and detailed data concerning the returns from the innovative activities from 1960 to 1972 of one of the nation's largest firms. The results bear out our present findings.[27])

25. Economic Report of the President, 1972, pp. 127–28.

26. Griliches, Mansfield, Minasian, and Terleckyj have estimated rates of return from nonagricultural R and D based on econometric studies where R and D is inserted into production functions. Where the results of these studies can be compared with this one, the results are quite similar. For example, see N. Terleckyj, *Effects of R and D on the Productivity Growth of Industries: An Exploratory Study* (Washington: National Planning Association, 1974).

27. For each year, this firm has made a careful inventory of the technological innovations arising from its R and D and related activities, and it has made detailed estimates of the effect of each of these innovations on its profit stream. We computed the average rate of return from this firm's total investment in innovative activities during 1960–1972, the result being 19 percent, which is not too different from the

Of course, proper caution should be exercised in interpreting our results. For a host of reasons, they can tell us little about the extent to which there is an underinvestment (or whether an underinvestment exists) in innovative activities of various kinds.[28] Our sample of innovations is not a random sample, the data sometimes are rough, and for a few of the innovations they are based partly on forecasts. Further, the models that we use are simplified in many respects.[29] For these and other reasons, our results should be treated with considerable caution. The measurement of social and private rates of return from investments in new technology is an extremely difficult business, which is one good reason why so few such measurements have been attempted. Nonetheless, it is important that we make the best estimates we can.

UNIVERSITY OF PENNSYLVANIA
MOUNT HOLYOKE COLLEGE
UNIVERSITY OF CONNECTICUT
TEMPLE UNIVERSITY
CALIFORNIA POLYTECHNIC STATE UNIVERSITY

median private rate of return in Table II. Also, we computed lower bounds for the social rate of return from the firm's investment, and found that they were about double its private rates of return. For further discussion of this study, see G. Beardsley's Ph.D. thesis, University of Pennsylvania; E. Mansfield, J. Rapoport, A. Romeo, E. Villani, S. Wagner, and F. Husic, *The Production and Application of New Industrial Technology* (New York: W. W. Norton, 1977); and E. Mansfield, "Comments on Returns to R and D Expenditures in the Private Sector," Conference on Research in Income and Wealth, November 1975.

28. See E. Mansfield, "Federal Support of R and D in the Private Sector," *Priorities and Efficiency in Federal Research and Development*, Joint Economic Committee of Congress, October 29, 1976; his testimony before the House Committee on Science and Technology, published in *Federal Research and Development Expenditures and the National Economy* (Washington: Government Printing Office, 1976); and G. Eads, "U. S. Government Support for Civilian Technology: Economic Theory vs. Political Practice," *Research Policy* (1974).

29. An important point to bear in mind is that the models used here cannot cope with innovations whose effects cannot be regarded as a form of cost reduction. This limitation is likely to be much more important for very major innovations than for lesser ones.

[6]

Interindustry *R&D* Spillovers, Rates of Return, and Production in High-Tech Industries

By JEFFREY I. BERNSTEIN AND M. ISHAQ NADIRI*

Firms undertaking *R&D* investment are unable to completely appropriate all of the benefits from their *R&D* projects. J. Schumpeter (1950), J. Schmookler (1966), R. E. Evenson and Y. Kislev (1973), N. Rosenberg (1979), Z. Griliches (1979), and M. Spence (1984) have pointed out that the degree of appropriability can influence both the causes and consequences of *R&D* investment. The *R&D* investment by a firm reduces its own production cost and, as a result of spillovers, costs of other firms are also reduced. R. C. Levin and P. C. Reiss (1984) estimated that a 1 percent increase in the *R&D* spillover caused average cost to decline by .05 percent. A. Jaffe (1986) estimated that when spillovers increased by 1 percent profit increased by .3 percent. Our paper (forthcoming) focused on intra-industry spillovers of four manufacturing industries. Generally, average cost declined by .2 percent in response to a 1 percent growth in *R&D* spillovers.

A common feature of the previous empirical studies was the measurement of *R&D* spillovers as a single variable. This meant that individual firms or industrial sources of spillovers could not be distinguished as to their relative significance in influencing production cost and factor demand. The first purpose of this paper is to investigate the effects of interindustry *R&D* spillovers in five high-tech industries where each industry is treated as a separate spillover source. This treatment of *R&D* spillovers allows us to estimate a matrix characterizing the sources and beneficiaries of each interindustry spillover.

The *R&D* spillovers create a dichotomy between private and social rates of return to *R&D* capital. The second purpose of this paper is therefore to compute both the social and private rates of return to *R&D* capital. The private return is measured as the variable cost reduction in an industry due to its own *R&D* capital expansion. The social rate of return to an industry's *R&D* capital consists of the private rate plus the interindustry marginal cost reductions due to the spillovers generated by the industry's *R&D* capital. Since each industry has the potential to cause distinct spillovers on each of the other industries, the spillover components of the social rates of return are decomposed by externality-receiving industries.

I. Cost, Factor Demand, and Interindustry Spillover

Production cost and factor demands of an industry are influenced by *R&D* capital accumulated by other industries. The existence of these spillovers and their effects on production processes can be estimated by specifying a variable cost function for each industry,

$$(1)\quad \ln c_v/w_m = \beta_0 + \beta_l \ln \omega_l + \beta_p \ln \omega_p + \beta_y \ln y + \beta_i \ln K_r^i + \beta_{lp} \ln \omega_l \ln \omega_p + \beta_{ly} \ln \omega_l \ln y + \beta_{li} \ln \omega_l \ln K_r^i + \beta_{py} \ln \omega_p \ln y + \beta_{pi} \ln \omega_p \ln K_r^i + \beta_{yi} \ln y \ln K_r^i + \left(\ln K_r^i + \beta_{ls} \ln \omega_l + \beta_{ps} \ln \omega_p\right) \sum_{\substack{j=1 \\ j \neq i}}^{N} \beta_{ij} \ln K_r^j + u_c,$$

*Department of Economics, Carleton University, Ottawa, Canada and Department of Economics, New York University, New York, 10003, respectively. We thank Erwin Diewert, Zvi Griliches, Peter Mohnen, and Peter Reiss for comments and discussions on the research reported in this paper. Financial support was provided by NSF grant PRA-8108635. Technical support from the C.V. Starr Center for Applied Economics is also gratefully acknowledged.

where c_v is variable cost, w_m is the factor price of materials, $\omega_l = w_l/w_m$, where w_l is the wage rate, $\omega_p = w_p/w_m$, where w_p is the rental rate on physical capital, y is output, K_r^i is the industry's own *R&D* capital and K_r^j is another industry's *R&D* capital, N is the number of industries, and u_c is the error term.

The variable cost function is a truncated translog, since there are no own or squared second-order terms in the function.[1] Each industry's *R&D* capital generates a distinct influence on the variable cost and factor demands of every other industry. In this model there are $N-1$ spillovers for each industry and the sum of these spillovers or the pool of knowledge is given by $\sum_{j=1, j\neq i}^{N} \beta_{ij} \ln K_r^j$. The pool of knowledge is determined by the estimation of the model.

The existence of the parameters β_{ls} and β_{ps} enables the spillover to affect the labor, physical capital, and materials cost shares. We can see this by using Shephard's Lemma,

$$(2)\quad s_l = \beta_l + \beta_{lp} \ln \omega_p + \beta_{ly} \ln y + \beta_{li} \ln K_r^i + \beta_{ls} \sum_{\substack{j=1 \\ j\neq i}}^{N} \beta_{ij} \ln K_r^j + u_l,$$

$$(3)\quad s_p = \beta_p + \beta_{lp} \ln \omega_l + \beta_{py} \ln y + \beta_{pi} \ln K_r^i + \beta_{ps} \sum_{\substack{j=1 \\ j\neq i}}^{N} \beta_{ij} \ln K_r^j + u_p,$$

where $s_l = w_l v_l / c_v$ is the labor variable cost share, v_l is the labor demand, $s_p = w_p v_p / c_v$ is the physical capital variable cost share, v_p is the physical capital demand, the material cost share is $s_m = 1 - s_l - s_p$ by definition, and u_l and u_p are the error terms. The two parameters β_{ls} and β_{ps} create the nonlinearity in the model, and in conjunction with the other spillover parameters (β_{ij}) determine the factor biases associated with the *R&D* spillovers. The equilibrium characterized by equations (1), (2), and (3) is short run. The *R&D* capital is assumed to be a quasi-fixed factor because of the development costs which generate lags in the completion of *R&D* projects. Thus short-run cost is not minimized with respect to *R&D* capital.[2]

The variable cost or productivity effects associated with each of the *R&D* spillovers are

$$(4)\quad \partial \ln c_v / \partial \ln K_r^j = \beta_{ij} \left(\ln K_r^i + \beta_{ls} \ln \omega_l + \beta_{ps} \ln \omega_p \right), \quad j \neq i,\ j = 1,\ldots,5.$$

The productivity effects are not constant but depend on the *R&D* capital of the spillover-receiving industry, along with the relative factor prices. The parameter β_{ij} defines the distinct effect that the *R&D* capital from industry j exerts on the ith receiving industry.

The parameters β_{ls}, β_{ps} transform the industry-specific productivity effects of the spillovers into factor bias effects.

$$(5)\quad \partial s_k / \partial \ln K_r^j = \beta_{ks} \beta_{ij}; \quad k = l, p, m,\ i \neq j,\ j = 1,\ldots,5,$$

where $\beta_{ms} = -(\beta_{ls} + \beta_{ps})$. If a variable factor cost share increases (decreases, or does not change), then the interindustry spillover is variable factor using (reducing, or neutral). The effect on each variable factor demand consists of the sum of the productivity and factor bias effects. It is possible for the variable factors to be complementary, substitutable or independent of each of the interindustry *R&D* spillovers.

[1] The translog variable function was estimated but we were unable to find parameter estimates which satisfied the regularity conditions for this function.

[2] Physical capital and certain types of labor could also be quasi fixed. However, the focus of the paper is on *R&D* capital. The model captures the relative inflexibility of *R&D* capital compared to other factors of production.

II. The Effects of *R&D* Spillovers

Five high-technology industries were analyzed in this paper: chemical products (SIC 28), non-electrical machinery (SIC 35), electrical products (SIC 36), transportation equipment (SIC 37), and scientific instruments (SIC 38). The sample period was 1958 to 1981 and most of the data for these industries were obtained from published sources of the Bureau of Economic Analysis (BEA).[3] The data consisted of gross output (y) in 1982 dollars; labor (v_l) was measured as total man-hours. The wage rate (w_l) was defined as the ratio of adjusted total payroll to total man-hours. The physical capital stock (v_p) was measured as net capital stock. The *R&D* capital stock (K_r) was constructed by perpetual inventory method with a depreciation rate of 10 percent. Adjustments were made to avoid double counting with respect to capital used in *R&D* activities. The after-tax rental rates or factor prices of physical and *R&D* capital were derived using the familiar user-cost formula taking account of tax rules applicable to each asset. Real materials was defined as gross output minus real value-added. The materials price was implicitly calculated as the ratio of nominal materials to real materials.

Equations (1), (2,), and (3) were estimated for each of the five industries. The sample period for the estimation was 1958 to 1981. The estimator used was full information maximum likelihood and the convergence criteria was .001. Each industry's own *R&D* capital and each of the spillovers were lagged one period in the estimation since all *R&D* capital stocks were exogenous in the model.[4]

The estimation of the model for each industry proceeded in the following manner. First, each spillover source industry was included in the model individually, next, pairs of spillover source industries were included, then groups of three, and lastly, all four spillover source industries were included in estimating the model. The criteria for accepting an industry as a spillover source were the satisfaction of the regularity conditions pertaining to the variable cost function and the statistical significance of the parameters. The estimation results were quite stable in the sense that when a specific spillover source industry generated by itself either a statistically insignificant effect, or caused the violation of regularity conditions, these problems also occurred when the specific industry was grouped with the other source-industries. Thus the empirical results were robust, both in an economic and statistical sense, as to the acceptance of a particular industry as a source of spillover for each receiving industry.[5]

The effects of the spillovers are presented in Table 1. For example, for the chemical products industry, the *R&D* capital spillover emanated from the scientific instruments industry. Variable cost declined as a result of the spillover over the sample period. A 1 percent increase in the spillover caused variable cost in this industry to decline by .21 percent in 1961 and by .09 percent in 1981. Both labor and material demands declined in response to the spillover suggesting that these two factors were partly substituted by the *R&D* capital spillover. However, the demand for physical capital increased in response to the spillover. Physical capital was a complement to the spillover. It was estimated that the spillover was labor and material reducing, while physical capital using.[6]

Our empirical results point to some general findings. First, variable cost for each industry was reduced by *R&D* capital spill-

[3] For a more detailed discussion of the data, a longer version of the paper is available from the authors.

[4] Details of the estimation results are available upon request.

[5] The variable cost function was assumed to be homogeneous of degree one in variable factor prices. This was a maintained hypothesis as relative variable factor prices affected variable cost. This assumption was reasonable to maintain because it implies that each variable factor demand depends on relative and not nominal prices.

[6] Besides looking at the product of β_{ij} with each of β_{ls}, β_{ps} and $\beta_{ms} = -(\beta_{ls} + \beta_{ps})$, if the factor demand elasticity is positive or if negative smaller in absolute value than the cost elasticity, then the spillover is factor using, otherwise it is factor reducing.

TABLE 1—SPILLOVER EFFECTS ON VARIABLE COST AND FACTOR DEMAND

Receiving Industry	1961	1971	1981
Chemical Products[a]			
Var. Cost	−0.208	−0.133	−0.089
Labor	−0.825	−0.797	−1.088
Phys. Cap.	3.918	2.506	1.420
Materials	−0.382	−0.312	−0.269
Non-Electrical Machinery[b]			
Var. Cost	−0.029	−0.036	−0.058
Labor	0.059	0.056	0.047
Phys. Cap.	−1.740	−1.969	−0.689
Materials	0.013	0.000	−0.017
Electrical Products[a]			
Var. Cost	−0.109	−0.117	−0.119
Labor	−0.109	−0.117	−0.119
Phys. Cap.	−0.109	−0.117	−0.119
Materials	−0.109	−0.117	−0.119
Transportation Equipment[c]			
Var. Cost	−0.113	−0.106	−0.092
Labor	−0.360	−0.346	−0.363
Phys. Cap.	3.637	4.544	1.398
Materials	−0.204	−0.197	−0.187
Scientific Instruments[b]			
Var. Cost	−0.051	−0.059	−0.078
Labor	−0.025	−0.031	−0.045
Phys. Cap.	−1.240	−0.846	−0.391
Materials	−0.012	−0.022	−0.041

[a]Source Industry: Scientific Instruments.
[b]Source Industry: Chemical Products, Electrical Products, Transportation Equipment.
[c]Source Industry: Non-Electrical Machinery.

overs. Second, the spillover for each receiving industry emanated from a very narrow range of industries. There was only a single spillover source for three industries, while for the other two industries, although they were affected by the three industries, the spillover effects were equal across source-industries. Third, in four of the five industries, there were factor bias effects associated with the spillovers. The bias effects for labor and materials were always in the same direction and in the opposite direction to physical capital.

III. Rates of Return *R&D* Capital

The private rate of return is defined by the real value of the variable cost reduction due to an increase in an industry's own *R&D*.

TABLE 2—PRIVATE RATES OF RETURN

Industry	Year	*R&D* Capital	Physical Capital
Chem. Prods.	1961	0.194	0.067
	1971	0.132	0.082
	1981	0.133	0.135
Non-Elec. Mach.	1961	0.160	0.075
	1971	0.267	0.085
	1981	0.240	0.136
Electr. Prods.	1961	0.201	0.075
	1971	0.150	0.084
	1981	0.224	0.139
Transp. Equip.	1961	0.085	0.071
	1971	0.095	0.086
	1981	0.119	0.117
Scient. Instr.	1961	0.168	0.080
	1971	0.173	0.083
	1981	0.161	0.118

Namely, $\rho^i = -(\partial c_v^i/\partial K_r^i)p_r^i$, $i=1,\ldots,N$, where ρ is the gross private rate of return and p_r is the price of *R&D* capital. Since the model is characterized by a short-run equilibrium, then the rate of return on *R&D* capital is endogenous and not assumed to be related to its rental rate. However, for physical capital, since it is part of short-run cost minimization, its gross rate of return is exogenous and given by the ratio of the rental rate to the purchase price for physical capital. In Table 2, we present the estimated private rates of return to *R&D* capital and physical capital. In four of the five industries, the gross private rate of return on *R&D* capital was, on average over the sample period, 1.5 to 2 times greater than the rate on physical capital. In addition, in these four industries the private rates of return on *R&D* were generally the same and ranged from .15 to .20. In the transportation equipment industry, the rates of return on the two types of capital were virtually equal.

The social rate of return to *R&D* differs from the private rate because of the existence of *R&D* spillovers. The social rate is defined inclusive of the spillover and it is equal to the private rate plus the sum of the marginal real interindustry variable cost reductions. Thus, $\gamma^i = \rho^i - \sum_{j=1,j\neq i}^{N}(\partial c_v^j/\partial K_r^i)/p_r^i$, $i=1,\ldots,N$, where γ is the social rate of return to *R&D* capital.

In Table 3 we present the social rates of return and their decomposition. First, the

TABLE 3—SOCIAL RATES OF RETURN

Industry Source and Receiving Industries:	1961	1971	1981
Chemical Products			
Non-El. Machinery	.062	.058	.126
Scient. Instr.	.025	.020	.032
Social *R* of *R*	0.281	0.210	0.291
Non-El. Machinery			
Trans. Equip.	.418	.316	.210
Social *R* of *R*	0.577	0.583	0.450
Electrical Products			
Non-El. Machinery	.024	.024	.062
Scient. Instr.	.010	.008	.016
Social *R* of *R*	0.235	0.182	0.302
Transp. Equipment			
Non-El. Machinery	.014	.013	.035
Scient. Instr.	.006	.004	.009
Social *R* of *R*	0.105	0.112	0.163
Scient. Instruments			
Chemical Products	.926	.505	.657
Electr. Products	.522	.429	.471
Social *R* of *R*	1.615	1.107	1.289

R&D capital from chemical products generates spillovers on non-electrical machinery and scientific instruments. From the third row in Table 3, we note that the social rate of return to *R&D* capital in the chemical products industry was .29 in 1981. Indeed, over the sample period, the social rate was 1.5 to 2 times the private rate of return to *R&D* capital. The non-electrical machinery industry only generated a spillover on the transportation equipment industry. However, the effect was quite large so that the social rate of return on *R&D* capital in the non-electrical machinery industry was 2 to 3 times the private rate of return. The *R&D* capital from electrical products and transportation equipment each affected the same industries as the chemical products industries. In both cases, the social rate was greater than the private rate, but only by about 10 to 20 percent. The last industry is scientific instruments, which affected the chemical products and electrical products industries. The *R&D* capital from scientific instruments generated substantial spillover effects on these two industries. In fact, the social rate of return was around 10 times the private rate.

The results on the social rates of return point out that each industry was a *R&D* capital spillover source. Four of the five industries generated spillovers on two industries. Our results on the differences between social and private rates of return were consistent with the findings of E. Mansfield et al. (1971), Jaffe, and our cited paper. However, the results show clearly that it is important to distinguish among industries as sources of *R&D* spillovers.

IV. Conclusion

In this paper we have estimated a spillover network linking the receiving and sending industries. The findings suggest that there were significant differences among industries as both spillover senders and receivers. Several issues would be of interest for further research. First, we would like to investigate the spillover linkages among other 2-digit manufacturing industries. Another issue of importance is to analyze the relationship between an industry's own *R&D* capital and the spillovers due to the *R&D* activities pursued by other industries. Lastly, the existence of significant intra- and interindustry spillovers has important implications both with respect to tax effects on *R&D* investment and for competition policy relating to joint *R&D* ventures. These issues have yet to be investigated.

REFERENCES

Bernstein, J. I. and Nadiri, M. I., "Research and Development, and Intraindustry Spillovers: An Empirical Application of Dynamic Duality," *Review of Economic Studies*, forthcoming.

Evenson, R. E. and Kislev, Y., "Research and Productivity in Wheat and Maize," *Journal of Political Economy*, November/December 1973, *81*, 1309–29.

Griliches, Z., "Issues in Assessing the Contribution of Research of Development to Productivity Growth," *Bell Journal of Economics*, Spring 1979, *10*, 92–116.

Jaffe, A., "Technological Opportunity and Spillovers of *R&D*," *American Economic Review*, December 1986, *76*, 984–1001.

Levin, R. C. and Reiss, P. C., "Tests of a Schumpeterian Model of *R&D* and Market Structure," in Z. Griliches, ed., *R&D, Patents and Productivity*, NBER, Chicago: University of Chicago Press, 1984.

Mansfield, E. et al., *Research and Innovation in the Modern Corporation*, New York: W. W. Norton, 1971.

Rosenberg, N., "Science, Invention and Economic Growth," *Economic Journal*, March 1974, *84*, 90–108.

Schmookler, J., *Invention and Economic Growth*, Cambridge: Harvard University Press, 1966.

Schumpeter, J., *Capitalism, Socialism and Democracy*, 3rd ed., New York: Harper and Row, 1950.

Spence, M., "Cost Reduction, Competition and Industry Performance," *Econometrica*, January 1984, *52*, 101–21.

[7]

THE SIMPLE ECONOMICS OF BASIC SCIENTIFIC RESEARCH

RICHARD R. NELSON
The RAND Corporation

I. BASIC ECONOMIC FRAMEWORK

RECENTLY, orbiting evidence of un-American technological competition has focused attention on the role played by scientific research in our political economy. Since Sputnik it has become almost trite to argue that we are not spending as much on basic scientific research as we should. But, though dollar figures have been suggested, they have not been based on economic analysis of what is meant by "as much as we should." And, once that question is raised, another immediately comes to mind. Economists often argue that opportunities for private profit draw resources where society most desires them. Why, therefore, does not basic research draw more resources through private-profit opportunity, if, in fact, we are not spending as much on basic scientific research as is "socially desirable"? In order to answer some of these questions, it seems useful to examine the simple economics of basic research. How much are we spending on basic research? How much should we be spending? Under what conditions will these figures tend to be different? Is basic research marked by these conditions? If so, what can we do to eliminate or reduce the discrepancy?

How much are we spending on basic research? In 1953, the latest date for which relatively sophisticated estimates are available, total expenditure on research and development was about $5.4 billion. Of that total, much more than half was for engineering development, much less than half for scientific research. Even less of the total, about $435 million in 1953, was spent on "basic research." All evidence indicates that since 1953 expenditure on research and development has increased markedly; $10 billion seems a reasonable estimate for 1957. Expenditure on basic research has also increased at a rapid rate, perhaps at a faster rate than total research and development expenditure. But basic-research expenditure today is probably under $1 billion, less than one-quarter of 1 per cent of gross national product.[1]

How much should we spend on basic research? Replacing the X_i of the familiar literature on welfare economics with "basic research" provides the theoretical answer. From a given expenditure on science we may expect a given flow, over time, of benefits that would not have been created had none of our resources been directed to basic research. This flow of benefits (properly discounted) may be defined as the social value of a given expenditure on basic research. However, if we allocate a given quantity of resources to science, this implies that we are not allocating these resources to other activities and, hence, that we are depriving ourselves of a flow of future benefits that we could have obtained had we directed these resources elsewhere. The discounted flow of bene-

[1] National Science Foundation, *Basic Research: A National Resource* (Washington, D.C., 1957); *Science and Engineering in American Industry* (Washington, D.C., 1956); *Growth of Scientific Research in Industry—1945–1960* (Washington, D.C., 1957); *Federal Funds for Science—The Federal Research and Development Budget Fiscal Years 1956, 1957, and 1958* (Washington, D.C., 1957).

fits of which we deprive ourselves by allocating resources to basic research and not to other activities may be defined as the social cost of a given expenditure on basic research. The difference between social value and social cost is net social value, or social profit. The quantity of resources that a society should allocate to basic research is that quantity which maximizes social profit.

Under what conditions will private-profit opportunities draw into basic research as great a quantity of resources as is socially desirable? Under what conditions will it not? If all sectors of the economy are perfectly competitive, if every business firm can collect from society through the market mechanism the full value of benefits it produces, and if social costs of each business are exclusively attached to the inputs which it purchases, then the allocation of resources among alternative uses generated by private-profit maximizing will be a socially optimal allocation of resources. But when the marginal value of a "good" to society exceeds the marginal value of the good to the individual who pays for it, the allocation of resources that maximizes private profits will not be optimal. For in these cases private-profit opportunities do not adequately reflect social benefit, and, in the absence of positive public policy, the competitive economy will tend to spend less on that good "than it should." Therefore, it is in the interests of society collectively to support production of that good.[2]

Society does, in fact, collectively support a large share of the economy's basic research. About 60 per cent of our basic-research work is performed by non-profit institutions, predominantly government and university laboratories. And a portion of the basic research performed in industrial laboratories is government financed. (This flow of funds is about equal to the flow of funds from industry to non-profit laboratories in the form of grants and contracts for basic research.)[3] Much, though not all, of the government contribution to basic research is national defense-oriented. But defense-oriented expenditure aside, the American political economy certainly treats basic research as an activity that creates marginal social value in excess of that collectable on the free market.[4] Is this treatment justified? If so, since, in fact, society is collectively supporting much basic research and hence resources directed to basic research do exceed the quantity drawn by private profit opportunity, is present social policy adequate?

II. SCIENTIFIC RESEARCH AND ECONOMIC VALUE

What are the social benefits derived from the activity of science? It is sometimes argued that most of our great social and political problems would simply evaporate if all citizens had a scientific point of view and, hence, that the benefits derived from scientific research are only in small part reflected in the useful inventions generated by science, for science helps to make better citizens. And

[2] Of course, the resources supplied to the industry must be withdrawn from other industries which generate no external economies or less external economies. Although significant external economies are probably rare, they almost certainly exist in education and preventive medicine as well as in basic research. Though the burden of this paper is that more resources should be allocated to basic research, the argument is probably invalid if these resources are taken, for example, from education or preventive medicine.

[3] National Science Foundation, *Basic Research: A National Resource.*

[4] This paper will not consider the vital question of whether the Department of Defense is spending enough on defense-oriented basic research. It probably is not, but the analysis of the paper is independent of this.

many scientists and philosophers take the point of view that the very activity of science—considered as the search for knowledge—is itself the highest social good and that any other benefits society might obtain are just by-products of the activity of science—social gravy. Dissents on both of these points are often sharp. The economist, after the usual perfunctory statement that he is fully aware that his definition does not capture everything, might define the benefits derived from the activity of science as the increase resulting from scientific research in the value of the output flow that the resources of a society can produce. In order to examine the extent to which a private firm can capture through the market the increased value of output resulting from the scientific research, in particular the basic scientific research, that it sponsors, it is necessary to examine the link between scientific research and the creation of something of economic value.

Scientific research may be defined as the human activity directed toward the advancement of knowledge, where knowledge is of two roughly separable sorts: facts or data observed in reproducible experiments (usually, but not always, quantitative data) and theories or relationships between facts (usually, but not always, equations). Of course, no strict line can be drawn between scientific research and all other human activities. Men have always experimented and observed, have always generalized and theorized; thus all men have always been, at least in a limited way, scientists. And knowledge has often (usually?) been acquired in activities in which pursuit of knowledge was of no, or negligible, importance. But even fuzzy definitions often have value. Scientific knowledge rests on reproducible experiments, but science is more than experimentation leading to new observations of facts which are believed observable by any other scientist undertaking the identical experiment. Science is most fruitful when it leads to ability to predict facts about phenomena without, or prior to, experimentation and observation. Scientific knowledge has economic value when the results of research can be used to predict the results of trying one or another alternative solutions to a practical problem.

Scientific research has increasingly been coupled to invention, where invention is defined as the human activity directed toward the creation of new and improved practical products and processes. But though many inventions occur as a result of a reasonably systematic effort to achieve a particular goal, many other inventions do not. They are a by-product of activity directed in a quite different direction, often a scientific research project directed toward solving an unrelated problem. Mauve, the first analine dye, was discovered by W. H. Perkin while he was attempting to synthesize quinine, and calcium carbide and the acetylene gas that it produces were invented by a group attempting to develop a better way to extract aluminum from clay. And though many inventions are made possible by closely preceding advances in scientific knowledge, many others require little knowledge of science or occur long after the relevant scientific knowledge is available: scientific knowledge certainly had little to do with the development of such useful inventions as the safety razor and the zipper; scientists have long known that expanding gases absorb heat, thus cool whatever they contact, but the gas refrigerator is an invention of the twentieth century. But particularly in the institution of the industrial research laboratory, applied science and invention are closely linked, and inventions usually result from a systematic attack on a problem.

In the activity of invention, as in most goal-directed activities, the actor has a number of alternative paths among which he must choose. The greater his knowledge of the relevant fields, the more likely he will be eventually to find a satisfactory path, and the fewer the expected number of tried alternatives before a satisfactory one is found. Thus, the greater the underlying knowledge, the lower the expected cost of making any particular invention.

A rationally planned inventive effort will be undertaken only if the expected revenue of the invention exceeds the expected cost. In many instances the economic utility of a particular invention is so great that an inventive effort is economically rational, even though the underlying scientific knowledge is scanty and hence the expected cost of making the invention is great. Edison's attempt to develop an incandescent lamp and Goodyear's attempt to improve the characteristics of rubber are cases in point. In these cases, since there was little useful underlying scientific knowledge, the invention procedure was trial and error, the next trial being roughly—but only roughly—indicated by a very loose theory formulated as the research proceeded. But though the inventors knew that it would probably prove costly to achieve their objective, they believed that the gains, if they were successful, were sufficiently great to make the effort profitable.

But often, though the inventor believes that there is great demand for a particular invention, it is not rational for him to attempt the invention, given the state of scientific knowledge. Expected cost will exceed expected revenue unless additional scientific knowledge can be obtained. If the expected cost of acquiring the relevant scientific knowledge is low, an organization interested in making a particular invention may undertake an applied scientific research project. A profit-maximizing firm will undertake a research project to solve problems related to a development effort if the expected gains—for example, reduction in development costs, or improvement in the final developed product—exceed expected research costs and if total research and development cost is exceeded by the expected net value of the invention. To the extent that the results of applied research are predictable and relate only to a specific invention desired by a firm, and to the extent that the firm can collect through the market the full value of the invention to society, opportunities for private profit through applied research will just match social benefits of applied research, and the optimum quantity of a society's resources will tend to be thus directed.

However, by no means all scientific research is directed toward practical problem-solving, though the line between basic scientific research and applied scientific research is hard to draw. There is a continuous spectrum of scientific activity. Moving from the applied-science end of the spectrum to the basic-science end, the degree of uncertainty about the results of specific research projects increases, and the goals become less clearly defined and less closely tied to the solution of a specific practical problem or the creation of a practical object. The loose defining of goals at the basic research end of the spectrum is a very rational adaptation to the great uncertainties involved and permits a greater expected payoff from the research dollar than would be possible if goals were more closely defined. For commonly, not just sometimes, in the course of a research project unexpected possibilities not closely related to the original objectives appear, and concurrently it may become clear that the original objectives are unobtainable or will be far more difficult to achieve than originally expected. While the direction of an applied research project must be closely constrained by the practical problem which must be solved, the direction of a basic research project may change markedly, opportunistically, as research proceeds and new possibilities appear. Some of the most striking scientific breakthroughs have resulted from research projects started with quite different ends in mind.

Pasteur's discovery of the value of inoculation with weakened disease strains is one of the more famous cases in point, but what is important is that the case is so similar to many others. While studying chicken cholera, Pasteur accidentally inoculated a group of chickens with a weak culture. The chickens became ill but, instead of dying, recovered. Since Pasteur did not want to waste chickens, he later reinoculated these chickens with fresh culture—one that was strong enough to kill an ordinary chicken—but these chickens remained healthy. At this point Pasteur's attention shifted to this interesting and potentially very (socially) significant phenomenon, and his resulting work, of course, brought about a major medical advance.

Applied research is relatively unlikely to result in significant breakthroughs in scientific knowledge save by accident, for, if significant breakthroughs are needed before a particular practical problem can be solved, the expected costs of achieving this breakthrough by a direct research effort are likely to be extremely high; hence applied research on the problem will not be undertaken, and invention will not be attempted. It is basic research, not applied research, from which significant advances have usually resulted. It is seriously to be doubted whether X-ray analysis would ever have been discovered by any group of scientists who, at the turn of the century, decided to find a means for examining the inner organs of the body or the inner structure of metal castings. Radio communication was impossible prior to the work of Maxwell and Hertz. Maxwell's work was directed toward explaining and elaborating the work of Faraday. Hertz built his equipment to test empirically some implications of Maxwell's equations. Marconi's practical invention was a simple adaptation of the Hertzian equipment. It seems most unlikely that a group of scientists in the mid-nineteenth century, attempting to develop a better method of long-range communication, would have developed Maxwell's equations and radio or anything nearly so good.

The limitations of an applied-research project constrained to the solution of a specific practical problem, and the practical value of many research projects where the goal is simply knowledge, not the solution of a practical problem, is well illustrated by the development of hybrid corn. During the latter half of the nineteenth century several attempts were made to improve corn yields. Many of the researchers directed their attention, at one time or another, to the inbreeding of corn to obtain a predictable and profitable strain. But as corn plants were inbred, though they tended to breed true, they also tended to deteriorate in yield and in quality. For this reason, applied researchers attempting to improve corn dropped this seemingly unpromising approach. But George Harrison Shull, a geneticist working with corn plants and interested in pure breeds not for their economic value but for experiments in genetics, produced several corn strains that bred true and then crossed these strains. His project was motivated by a desire to further the science of genetics, but a result was high-yield, predictable hybrid corn.

III. BASIC RESEARCH AND PRIVATE PROFIT

It is clear that for significant advances in knowledge we must look primarily to basic research; the social gains we may expect from basic research are obvious. But basic research efforts are likely to generate substantial external economies. Private-profit opportunities alone are not likely to draw as large a quantity of resources into basic research as is socially desirable.

Significant advances in scientific knowledge, the types of advances that are likely to result from successful basic-research projects, very often have practical value in many fields. Consider the range of advances resulting from Boyle's gas law or Maxwell's equations. On Gibb's law of phases rests the design of equipment in fields as diverse as petroleum refining, rubber vulcanization, nitrogen fixation, and metal-ore separation. Few firms operate in so wide a field of economic activity that they are able themselves to benefit directly from all the new technological possibilities opened by the results of a successful basic research effort. In order to capture the value of the new knowledge in fields which the firm is unwilling to enter, the firm must patent the practical applications and sell or lease the patents to firms in the industries affected.

But significant advances in scientific knowledge are often not directly and immediately applicable to the solutions of practical problems and hence do not quickly result in patents. Often the new knowledge is of greatest value as a key input of other research projects which, in turn, may yield results of practical and patentable value. For this reason scientists have long argued for free and wide communication of research results, and for this reason natural "laws" and facts are not patentable. Thus it is quite likely that a firm will be unable to capture through patent rights the full economic value created in a basic-research project that it sponsors.

A firm with a narrow technological base is likely to find research profitable only at the applied end of the spectrum, where research can be directed toward solution of problems facing the firm, and where the research results can be quickly and easily translated into patentable products and processes. Such a firm is likely to be able to capture only a small share of the social benefits created by a basic research program it sponsors. On the other hand, a firm producing a wide range of products resting on a broad technological base may well find it profitable to support research toward the basic-science end of the spectrum.

A broad technological base insures that, whatever direction the path of research may take, the results are likely to be of value to the sponsoring firm. It is for this reason that firms which support research toward the basic-science end of the spectrum are firms that have their fingers in many pies. The big chemical companies producing a range of products as wide as the field of chemistry itself, the Bell Telephone Company, General Electric, and Eastman Kodak immediately come to mind. It is not just the size of the companies that makes it worthwhile for them to engage in basic research. Rather it is their broad underlying technological base, the wide range of products they produce or will be willing to produce if their research efforts open possibilities. (Eastman Kodak entered the vitamin business when a research project resulted in a new way to synthesize vitamin B.) Strangely enough, economists have tended to see little economic justification for giant firms not built on economies of scale. Yet it is the many-product giants, not the single-product giants, which have been most technologically dynamic, and, to the extent that we wish the private sector of the economy to support basic research, we must look to these firms.

The importance of a broad technological base as a factor permitting a company to engage profitably in basic research is clearly illustrated by Carothers' famous research project for Du Pont. Carothers' work in linear superpolymers began as an unrestricted foray into the unknown with no particular practical objective in view. But the research was in a new field of chemistry, and Du Pont believed that any new chemical breakthrough would probably be of value to the company. The very lack of a specific objective, the flexibility of the research project, was an important factor behind its success. In the course of research Carothers obtained some superpolymers which at high temperatures became viscous fluids and observed that filaments could be obtained from these materials if a rod were dipped in the molten polymer and then withdrawn. At this discovery the focus of the project shifted to these filaments. Nylon was the result, but at the start of the project Carothers could not possibly have known that his research would lead him to the development of a new fiber.

A wide technological base (usually involving a diversified set of products) does not imply a position of monopoly power in any or all of the product markets, nor does a monopoly position in a market imply a wide technological base. Focusing attention on market position, a business firm operating in a competitive environment will seldom find it profitable to engage in a research project which is not likely to result quickly in something patentable, even if the firm can predict the nature of the research results, unless the firm keeps tight secrecy. For if the results of research cannot be quickly patented and are not kept secret, other firms producing similar products using similar processes will be free to use the results as an input of a development program of their own, designed to achieve a similar patentable objective. If competing firms develop a patentable product first, or develop a competing product, these firms will in effect steal from the research-sponsoring firm, through price and product competition, a large share of the social utility created by research. In fact, many companies engaging in research keep their research findings secret until the new knowledge is put to practical use and the results are patented.

Many industries have attempted to reconcile their need for new knowledge with the lack of incentives to individual private firms to produce that new knowledge by establishing co-operative industry research organizations. To the extent that an industry rests on a field of science that is likely to get little attention in the absence of sponsorship by the firms in the industry, it may be in the interests of all the firms that research in this field be pushed, though each firm would prefer the others to do the financial pushing. An industrial co-operative research laboratory may well develop under these conditions, supported by all or by a large number of the firms in the industry, and undertaking research likely to be applicable to the technology of the industry. The motivation for these co-operative laboratories is only in part the high cost of research. More importantly, these laboratories are motivated by the fact that most of the firms will gain from the results of relatively basic research in certain fields whether or not they pay for it; hence little research will be undertaken in the absence of co-operation.

The preceding argument has been focused on external economies that open a gap between marginal private and marginal social benefit from basic research. Two other factors, working in the same direction, must be mentioned, if not discussed. First, the long lag that very often occurs between the initiation of a basic-research project and the creation of something of marketable value may cause firms much concerned with short-run survival, little concerned with profits many years from now, to place less value on basic-research projects than does society, even in the absence of external economies. This is not to say that all firms have a greater time-discount factor than does society as a whole, but it can be argued that many firms do. Second, the very large variance of the profit probability distribution from a basic research project will tend to cause a risk-avoiding firm, without the economic resources to spread the risk by running a number of basic-research projects at once, to value a basic-research project at significantly less than its expected profitability and hence, even in the absence of external economies, at less than its social value.

IV. IS CURRENT SOCIAL POLICY ADEQUATE?

It seems clear that, were the field of basic research left exclusively to private firms operating independently of each other and selling in competitive markets, profit incentives would not draw so large a quantity of resources to basic research as is socially desirable. But in fact basic research has not been the exclusive domain of private firms. Government and other non-profit institutions (principally universities) together spend more on basic research, and undertake more basic research in their own laboratories, than does industry. Since we are presently supporting collectively such a large share (more than half) of basic research, is it not possible that total basic-research expenditure (the sum of private and public efforts) equals or exceeds the social optimum? This is a tricky theoretical question. However, if basic research can be considered as a homogeneous commodity, like potato chips, and hence the public can be assumed to be indifferent between the research results produced in government or in industry laboratories; if the marginal cost of research output is assumed to be no greater in non-profit laboratories than in profit-oriented laboratories, and if industry laboratories are assumed to operate where marginal revenue equals marginal cost, then the fact that industry laboratories do basic research *at all* is itself evidence that we should increase our expenditure on basic research.

Public support of basic research has primarily been in the form of contracts let with private firms and in the establishment and support of a large number of non-profit laboratories. Save for the effects of tax laws (which apply to all business cost-incurring activities), public policy has not acted to *shift* the marginal cost curve of the basic-research industry. Public policy has resulted in shifts *along* the curve. Nor has public policy acted to drive marginal social utility to marginal private utility. External economies still exist at the margin. Clearly then, if industry laboratories are in profit-maximizing equilibrium, society would benefit from an increase in basic-research expenditure in industry laboratories, holding research efforts elsewhere constant, for the marginal social benefit of basic research in private laboratories exceeds marginal cost to the firm, which under our assumptions still equals alternative cost. But perhaps non-profit laboratories are spending too much on basic research —are operating beyond the point at which marginal cost equals marginal social benefit—and therefore it is desirable to reduce research expenditure in this sector. Given our assumptions, this cannot be. For, if marginal cost is no greater in non-profit laboratories than in industry laboratories, and society cannot distinguish between the fruits of research undertaken in the two kinds of laboratory—that is, if marginal social benefit is the same in the public and the private sector—and if it is socially desirable that expenditure on basic research be increased in industry laboratories, then it is also socially desirable that research expenditure be increased in non-profit laboratories. For if marginal social benefit exceeds marginal cost in industry laboratories, so does it in non-profit laboratories.

The assumptions on which the preceding argument is based rest but shakily on fact. Basic research certainly is not a homogeneous commodity. The types of knowledge generated, say, in an air-force project on high-speed gas flows, a Du Pont project on high polymer chemistry, or a Harvard project on solid-state physics are not perfectly substitutable. The knowledge generated will certainly be different, and in a reasonably predictable way. And, once the non-homogeneity of basic research is admitted, the concept of relative marginal cost becomes fuzzy. Thus one cannot make an airtight statement, based on welfare economics, that we are not spending as much on basic scientific research as we should. But I believe that the evidence certainly points in that direction.

V. SOME POLICY IMPLICATIONS

Though the profit motive may stimulate private industry to spend an amount on applied research reasonably close to the amount that is socially desirable, it is clear from the preceding analysis that under our present economic structure the social benefits of basic research are not adequately reflected in opportunities for private profit. Indeed, there is a basic contradiction between the conditions necessary for efficient basic research—few or no constraints on the direction of research with full and free dissemination of research results—and full appropriation of the gains from sponsoring basic research in a competitive economy.

This is not to say that some firms could not profitably increase their basic-research effort. Some may presently be operating well to the left of their maximum profit point. But to the extent that we want our economy to remain competitive and want efficient use of basic-research funds, the laboratories of colleges, universities, and other non-profit institutions must perform a large share of our basic research if we are to put as much of our resources into basic research as we should. Although several laboratories of private industry have made significant contributions in the field of basic science, these contributions have been few and far between. If we advocate that basic research be increasingly undertaken by business and if we believe that business should be motivated by profit, we must accept the growth of large firms with a wide technological base, with virtual monopolies in several markets. If we do not want such an economic structure, then only to the extent that we think it desirable that private firms look to motives other than profit can we argue that industry laboratories should perform a significantly enlarged share of our basic research. In either case we undermine many of the economic arguments for a free-enterprise economy. If we want to maintain our enterprise economy, basic research must be a matter of conscious social policy.

This is not the place to suggest a menu of policies—the bill of fare offered in the National Science Foundation booklet on basic research lists some of the actions that might be considered. However, it does seem appropriate to suggest that public policy on basic research should recognize the following points:

1. The problem of getting enough resources to flow into basic research is basically the classical external-economy problem.[5] External economies result from two facts: first, that research results often are of little value to the firm that sponsors the research, though of great value to another firm, and, second, that research results often cannot be quickly patented. It therefore seems desirable to encourage the further growth of a "basic-research industry," a group of institutions that benefit from the results of almost any basic-research project they undertake. University laboratories should certainly continue to be a major part of this industry. However, an increasingly important role should probably be played by industry-oriented laboratories not owned by specific industries but doing research on contract for a diversified set of clients. Such laboratories would usually have at least one client who could benefit from almost any research breakthrough.

2. The incentives generated in a profit economy for firms to keep research findings secret produce results that are, in a static sense, economically inefficient. The use of existing knowledge by one firm in no way reduces the ability of another firm to use that same knowledge, though the incentive to do so may be reduced. The marginal social cost of using knowledge that already exists is zero. For maximum static economic efficiency, knowledge should be administered as a common pool, with free access to all who can use the knowledge. But, if scientific knowledge is thus administered, the incentives of private firms to create new knowledge will be reduced. This is another case in which static efficiency and dynamic efficiency may conflict. It is socially desirable that as much of our basic research effort as possible be undertaken in institutions interested in the quick publication of research results if marginal costs are comparable. In the absence of incentives to private firms to publish research results quickly (such incentives might be legislated) a dollar spent on basic research in a university laboratory is worth more to society than a dollar spent in an industry laboratory, again, if productivity is comparable.

3. If society places the brunt of the basic-research burden on universities, funds must be provided for this purpose. The current Department of Defense policies of letting huge applied research projects with universities should either be reconsidered or complemented with other policies designed to prevent the increased applied-research burden from drawing university facilities and scientists away from basic research. This is not to say that universities cannot effectively undertake applied research. Rather it is to say that their comparative advantage lies in basic research.

[5] The external economy aspect of basic research reacts back through the price system to undervalue pure scientists relative to engineers.

[8]

Productivity, *R&D*, and Basic Research at the Firm Level in the 1970's

By ZVI GRILICHES*

This paper reports new results on the relationship of research and development (*R&D*) expenditures, especially expenditures on basic research, to productivity growth in U.S. manufacturing firms during the 1970's. It is based on a unique data set, the National Science Foundation (NSF) *R&D*-Census match, containing information on *R&D* expenditures, sales, employment, and other detail for approximately 1000 largest manufacturing firms from 1957 through 1977. It updates my earlier work (1980) on the precursor of this data set, replicates some of Edwin Mansfield's (1980) work on the contribution of basic research to productivity growth using a larger, more recent, and more representative sample of firms, and complements similar work by myself and J. Mairesse (1983,1984) based on a publicly accessible but more limited data set.

Two topics are explored in some detail: 1) Is there any evidence of a decline in the returns to industrial *R&D* expenditures, a decline in their "fecundity" in the 1970's as compared to earlier time periods? 2) Is there evidence that basic research is a relatively more important component of *R&D* and that there may have been an underinvestment in this component?

A few background facts are worth stressing at this point. In the United States, total *R&D* expenditures in industry peaked (in real terms) around 1968, dropped slightly in the early 1970's and recovered somewhat in the late 1970's. Relative to total sales, *R&D* expenditures in industry declined from 4.2 percent in 1968 to a trough of 2.6 percent in 1979 and then recovered to 3.7 percent by 1982. This pattern masks a strong divergence between the trends in federally and privately supported industrial *R&D*. Federally supported *R&D* fell from 2.1 percent of manufacturing sales in 1967 to 0.7 percent in 1979 and has only recently begun to recover, while company-financed *R&D* stayed essentially constant (relative to industry sales) with almost all of the fluctuation coming from the decline in federal support (NSF, 1983; 1984). During the same period, the economy experienced one of the sharpest and prolonged recessions of the postwar period and a large and pervasive productivity slowdown. Hardest hit were the primary metals, motor vehicles, and other heavy, energy-related industries. On the whole, these were the less *R&D* intensive industries, resulting in a largely accidental correlation between *R&D* intensiveness and the productivity slowdown. (See my 1980 papers and my article with F. Lichtenberg, 1984, for more discussion of these issues.)

The remainder of the paper is organized as follows. First, I describe the data set with its advantages and limitations and present some overall comparative statistics. Second, I outline briefly the framework that underlies the computations to be performed. The results are presented and discussed, and the paper closes with some conclusions, caveats, and suggestions for further research.

I. Previous Work and the Current Data Set

The current project is an extension of work originally begun in the mid-1960's. That work was based on the matching of *R&D* data

*Harvard University, 125 Littauer Center, Cambridge, MA 02138, and the National Bureau of Economic Research. I am indebted to NSF Grant no. SES-82-08006 for the support of this work, to Douglas Dobas and Bronwyn H. Hall for making the data gathering effort possible, and to David Body for research assistance. I have also benefited from comments of seminar participants at the NBER and Yale University.

collected on behalf of the NSF by the Bureau of the Census during 1957–65 with additional company data from the 1958 and 1963 Census of Manufactures and Enterprise Statistics. The universe consisted of large (1000 or more employees) U.S. manufacturing companies performing *R&D*. The final sample of 883 of such companies accounted for over 90 percent of total sales and *R&D* expenditures of all firms in this universe.

The main finding of that work (see my 1980a paper) was a rather consistent and positive relationship between various measures of company productivity and its investments in research and development. The Cobb-Douglas-type production functions, estimated on both levels (1963) and rates of growth (1957–65) yielded an elasticity of output with respect to *R&D* investments of about .07 and an implied average gross excess rate of return of 27 percent (as of 1963), a significantly lower rate of return to federally financed *R&D* expenditures, and no clear evidence of significant scale effects either in *R&D* investment policies or the returns from it.

In trying to extend the earlier study to the more recent time period, it became clear that the earlier work could not be simply updated because much of the earlier data was lost and a new data set had to be created instead. The basic objective was to create a matched body of data on most of the large *R&D* performing corporations in the United States, making it possible to analyze both the determinants and consequences of *R&D* spending *over time*. For this purpose a time-series record has been created for each company consisting of the major variables in the annual *R&D* survey for each of the years 1957–77, supplementary *R&D* information for selected years (1962, 1967, 1972, and 1975), data from the Enterprise Statistics (i.e., company level questionnaires) for 1967, 1972, and 1977, and a few additional items from the Census of Manufactures establishment record summaries for 1967 and 1972. The data set began with all the "certainty" companies in the NSF *R&D* survey as they existed in 1972. There were approximately 1100 such companies, but a "complete" record is available only for a much smaller number.[1]

Table 1 lists the sample size, means, and standard deviation for the major variables as of 1972 and their growth rates from 1966 to 1977. It is intended to describe three aspects of these data: 1) the general characteristics (means and standard deviations) of the sample as of 1972; 2) average rates of growth of the major variables of interest during the 1967–77 period; and 3) how these measures change when the sample is changed to select observations according to the availability of the requisite information.

Turning to the last topic first, note that we tend to lose smaller and more *R&D* intensive firms as the sample gets more restrictive. The first column of Table 1 corresponds to the most liberal criterion: a firm had to exist in 1972 and report positive *R&D*. Column 2 requires both the ability to compute a growth rate for the 1967–77 period (i.e., at least five good time-series observations) and a successful match to the 1972 Census of Enterprise data (NCK–1). In column 3, I add the requirement of a successful match to the 1977 Census data, while in column 4 the subsample is based on a match with the 1967 and 1972 Census data instead. The major differences occur in the transition from column 1 to column 2 where trying to match to the Census we lose a relatively large number of smaller firms for which there are still data in the *R&D* survey files. The firms that can be also found in the 1977 Census are slightly larger and have had a somewhat higher rate of growth in employment, *R&D*, and productivity. The firms that also existed in 1967

[1] The universe of this data match consists of all "certainty" cases in the 1972 *R&D* survey; i.e., the basic definition is the population of companies as they existed in 1972 (as against 1962 in the earlier study) and the requirement of "certainty" assures that the Census Bureau tried to collect consistent data for these firms for more than one year. The "certainty" cases correspond closely to the earlier restriction to companies with 1000 or more employees, though it is a bit more inclusive. See my paper with Bronwyn Hall (1982) and Hall (1984) for more detail on sample definition and variable construction.

TABLE 1—MAJOR VARIABLES IN 1972 AND 1966-77 GROWTH RATES BY SUBSAMPLE: MEANS AND STANDARD DEVIATIONS[a]

	Data Set, Selection Criteria, and Sample Size			
Variable	1972 *R&D* Survey Universe (*N* = 1105) (1)	1966–77 Growth Rates Computable and Matched to 1972 Census (*N* = 652) (2)	(2) and Matched to 1977 Census (*N* = 491) (3)	(2) and Matched to 1967 Census (*N* = 386) (4)
A. Levels in 1972				
Sales in Million Dollars	146	205	223	236
	(1.61)	(1.43)	(1.40)	(1.44)
Total Employment	4038	5570	6212	6698
	(1.48)	(1.27)	(1.30)	(1.31)
R&D Scientists and Engineers	89	74	82	106
	(1.66)	(1.70)	(1.71)	(1.72)
R&D in Million Dollars	2.3	3.0	3.4	4.3
	(1.74)	(1.78)	(1.77)	(1.83)
R&D to Sales Ratio (*RS*)	.051	.033	.032	.035
	(.131)	(.064)	(.051)	(.048)
Company *R&D*/Sales Ratio (*CRS*)	.028	.022	.023	.025
	(.069)	(.026)	(.026)	(.026)
Basic to Total *R&D* Ratio (*BR*)	.025	.026	.026	.027
	(.074)	(.071)	(.075)	(.073)
Value-Added, Million Dollars		100	113	121
		(1.32)	(1.31)	(1.34)
Gross Fixed Assets Million Dollars		115	124	147
		(1.67)	(1.59)	(1.65)
B. Growth Rates 1966–77				
Employment Growth		.012	.015	.006
		(.046)	(.041)	(.040)
Partial Productivity Growth (*BPT*)		.025	0.26	.025
		(.036)	(.034)	(.035)
Total *R&D* Growth, Deflated (*BTRD*)		–.001	.003	–.007
		(.079)	(.074)	(.070)
Scientists and Engineers Growth		.008	.012	.004
		(.087)	(.084)	(.078)
Company *R&D* Growth, Deflated (*BCRD*)		.004	.008	–.000
		(.081)	(.076)	(.071)

Notes: Col. 1: "Certainty" firms in the NSF *R&D* Survey with positive *R&D* in 1972; Col. 2: Growth rates for 1966–77 computable (at least 5 years of good data on sales, employment, and *R&D*) and a successful match to the 1972 Enterprise Census (NCK–1); Col. 3: (2) and a successful match to the 1977 Census (NCK–1); Col. 4: (2) and a successful match to the 1967 Census and growth rates computable for 1957–65; Partial productivity growth = deflated sales growth – (share of labor compensation in total sales) × growth in employment; Sales deflated by NIPA based output price indexes at the 2–3 digit SIC level. *R&D* deflator based on the methodology suggested by Jaffe (NSF), from my 1984 comment.

[a]Geometric means and standard deviations (shown in parenthesis) of the logarithms (approximate coefficient of variation) except for growth rates or ratios.

are even larger but have on average grown somewhat more slowly than those that existed in the 1972–77 period. If we look at two of the major variables of interest, partial productivity growth and the ratio of basic to total *R&D*, there is almost no difference in their means across the relevant columns (2, 3, and 4), and hence it is unlikely that subsequent conclusions will be subject to a serious sample selection bias. I will, therefore, ignore this topic here.

Looking at the levels of the variables in 1972, we see that the average firm in the sample is quite large (5000+ employees), employs close to one hundred $R\&D$ scientists and engineers, and is making only a relatively modest investment of its own money (about 2.5 percent of sales) in $R\&D$, with very little of that, less than 3 percent, being devoted to basic $R\&D$. This picture is somewhat misleading, however. The actual distribution of firms is quite skewed, with a small number of larger firms spending much larger amounts on both total and basic $R\&D$. Looking at growth rates one can observe that on average these firms grew only moderately during this period: about 1 percent per year in total employment, about 2.5 percent per year in partial productivity, and almost zero growth in deflated $R\&D$ expenditures (though a slightly positive rate of growth in the number of $R\&D$ scientists and engineers). Here again, while on average there is little movement, there is a great deal of variability at the individual firm level. The standard deviations of the rates of growth of partial productivity and total $R\&D$ are 3.5 and 8 percent per annum, respectively, with many firms growing much faster (and also much slower) than the average.

Looking at some of the $R\&D$ ratios over time, not reported in Table 1, one cannot see any significant decline in the rate of private investment in $R\&D$. While the total $R\&D$ to sales ratio falls from .042 in 1962 to .035 in 1972 and again from .032 in 1972 to .029 in 1977 for firms in subsamples 4 and 3, respectively, the company-financed $R\&D$ to sales ratios (CRS) are essentially unchanged (.025 in 1962 and 1972 in subsample 4 and .023 in 1972 and 1977 in subsample 3). On the other hand, while the basic research ratio (BR) fell only modestly from .033 to .031 between 1962 and 1972, and from .027 to .023 between 1972 and 1977, coupled with the decline in the overall total $R\&D$ to sales ratio, this implies about a 40 percent reduction in the relative intensity of industrial investment in basic research, relative to industry sales. Almost all of this decline came from the overall decline in federally financed $R\&D$ which declined from about 55 percent of total $R\&D$ in industry in 1965 to about 35 percent in 1982. The federal government financed about 32 percent of all basic research in industry in 1967 but only 19 percent in 1982 (see NSF, 83–302 and 84–311). The reduction was so steep that basic research in industry declined not only relatively (to sales) but also absolutely, from a peak of \$813 million in 1966 (in 1977 dollars) to a trough of \$581 million in 1975 and did not surpass the 1960's levels until the early 1980's. How one interprets the consequences of such a decline depends on one's view of the relative productivity of governmentally financed $R\&D$ expenditures in industry, a topic I will be exploring below.

II. The Analytical Framework and Econometric Results

The work reported here focuses primarily on the analysis of productivity growth for these companies, using a rather simple Cobb-Douglas production function approach:

$$(1) \qquad Q_t = Ae^{\lambda t}K_t^{\alpha}C_t^{\beta}L_t^{1-\beta},$$

where Q is output (sales, or value-added), C and L are measures of capital and labor input, respectively, $K = \sum_i w_i R_{t-i}$ is a measure of the accumulated and still productive research capital ("knowledge"), R_t measures the real (deflated) gross investment in research in period t, and the w_i's connect the levels of past research to the current state of knowledge. In addition, λ measures the rate of disembodied "external" technical change (where t is time in years), A is a constant, and constant returns to scale have been assumed with respect to the conventional inputs (C and L).

A number of serious difficulties arise when one turns to the operational construction of the various variables (see my 1979 article for more detailed discussion). Perhaps the two most important problems are the measurement of output (Q) in a research-intensive industry (where quality changes may be rampant), and the construction of the unobservable research capital measure (K). Turning to the second problem first, note that $K_t = \sum w_i R_{t-i}$ can be thought of as a measure of the distributed lag effect of past research investments on productivity. There are at least three forces at work here: the lag between investment in research and the actual invention of a new technique or product; the lag between invention and the development and complete market acceptance of the new product; and its disappearance from the currently utilized stock of knowledge due to changes in external circumstances and the development of superior techniques or products by competitors (depreciation and obsolescence). There is some scattered evidence, based largely on questionnaire studies that such lags are rather short in industry, where most of research expenditures are spent on development and applied topics, and where the private returns from *R&D* become obsolete much faster due to the erosion of a firm's specific monopoly position (Ariel Pakes and M. Schankerman, 1984).

While my models are written as if the main point of research expenditures is to increase the physical productivity of the firm's production process, most of the actual research in industry is devoted to the development of new products or processes to be sold and used outside the firm in question. Assuming that, on average, the outside world pays for these products what they are worth to it, using sales or value-added as the dependent variable does, in fact, capture the private returns to such research endeavours. However, the observed private returns may underestimate the social returns because, given the competitive structure of the particular industry, the firm is unlikely to appropriate all of these returns. On the other hand, part of the increase in the revenues of a particular firm may come at the expense of other firms, or from changes in the market power induced by the success of its research program. I cannot say much about the net impact of such forces on the basis of the data at hand. This would require a detailed comparison of the individual firm results with estimates based on industry and economy-wide returns to research, a topic beyond the scope of this project. But since expected private returns are a determinant of private investment flows into this activity, they are of some interest even if one cannot answer the social returns question unequivocally.

This framework can be extended to ask whether different types of *R&D* (private vs. federal, or basic vs. applied) are equally "potent" in generating productivity growth. One way of answering this question is to look at the "mix" of *R&D* expenditures and ask if it matters for the question at hand. Let there be two types of *R&D* expenditures, R_1 and R_2, and let us assume that the overall analysis is in terms of the logarithm of total *R&D* expenditures but that we believe that R_2 should have been weighted more, given a δ premium (or discount). That is, the right variable is

$$(2) \quad R^* = R_1 + (1+\delta)R_2 = R(1+\delta s),$$

where $s = R_2/R$ is the "share" of R_2 in total $R = R_1 + R_2$. Then the α log R^* term can be approximated by α log $R^* \simeq \alpha$ log $R + \alpha\delta s$. The sign and significance of the mix term s will give us some clue about the size and magnitude of the δ term.

A similar argument can be made also in the context of a growth-rate formulation. Let lower case letters denote growth rates. Then $r = (1-s)r_1 + sr_2$ while $r^* = (1-s)r_1 + (1+\delta)sr_2$. If, as is mostly the case in our data, the growth rates of r_1 and r_2 are roughly equal, then $r^* = r(1+\delta s)$, and again, the coefficient of the mix term s provides us with some information about the "premium" or "discount" on R_2 since αr^* can be approximated by

$$(3) \quad \alpha r^* \simeq (\alpha + \delta\bar{s})r + (\alpha\bar{r}\delta)s.$$

Given the peculiarities of my data set—its unbalanced nature (many missing observations towards the beginning and end of the period), the availability of capital and value-added only for Census years, the desire to preserve comparability with the earlier study, and the difficulty of doing elaborate programming inside the Census Bureau, I focus primarily on two major dimensions of the data: levels (in 1967, 1972, and 1977) and growth rates, and eschew any attempt at a complete annual data analysis. The annual data are summarized by computing average growth rates for two subperiods 1957–65 (corresponding to the earlier study period) and 1966–77, based on regressions of the logarithms of the relevant variables on time trends (solving thereby the missing years problem within each of these subperiods).

In implementing such a framework of analysis one has to deal with several serious data problems: missing data, erroneous data and possible erroneous matches, and mergers. Except for *R&D* data, no special effort was made to replace missing values by various imputation procedures. It was my notion that the basic data set represents what the Census did collect, what we actually know, and that any imputation procedure should be done only in the context of a particular research project where its implications for the final analysis could be interpreted. As far as the *R&D* data are concerned, the Census used the shuttle nature of the original questionnaires to fill in many of the original blanks. To the extent that there remain missing values which are not due to the fact that the whole company is missing before or after some date, they were interpolated on the basis of the estimated growth rates (which require at least five good data points within each subperiod). For other variables, missing values were not imputed. It was not possible, within the constraints of this project, to develop optimal imputation procedures. This would have required several repeated passes at the original numbers. Instead, the analysis is based either on reduced "clean" samples or on "pairwise present" correlation coefficient matrices.

From an econometric point of view, we have to deal with the problem of firm effects (or firm-specific left-out variables) and the possibility that the relationships being estimated may not stay constant either across firms or across time. The first is handled by analyzing first differences or growth rates, transformations that eliminate any unchanging effects from the data. The second problem, the problem of differences across firms, is handled in part by calculating a measure of "partial" productivity growth $[BPT = y - (1-\hat{\beta})l]$, using individual firm data on the share of labor in total costs. One can also estimate separate and different parameters for the various industry groupings and include some of the other variables available in the record which might distinguish one firm's environment and response pattern from another's (such as its specialization ratio, size, or vertical integration). The main hypothesis under investigation, that the returns to *R&D* investments may have declined over time, is tested both by comparing estimates based on the more recent data with the earlier results, and by allowing and testing for systematic changes in the estimated relationships between the three available cross sections.

Let us look now at the first set of substantive results. Table 2 reports the results of estimating cross-sectional production functions (equation (1)) separately for each Census year, adding to the standard capital and labor variables a measure of total *R&D* capital accumulated by the firm and two *R&D* mix variables: the fraction of total *R&D* that was spent on basic research and the fraction of accumulated *R&D* that had been financed privately. All the reported estimates allow for 18 to 20 (depending on the subsample) separate industry intercepts. Columns 1 and 3 report estimates that are based on the same number of firms and use the same dependent variables, differing only by the year of observation. Column 2 presents additional estimates for 1972 based on different sample and dependent variable definitions with the main intent being to show that the major conclusions are insensitive to such differences. There are three major points to be made about these estimates. The first is that the stock of *R&D* capital contributes significantly to the explanation of cross-sectional differences in productivity and there is little evidence of a decline in its coefficient over time.[2] There is a minor rise in the estimated coefficient from 1967 to 1972 and a somewhat larger but not really significant decline from 1972 to 1977. Given this particular measure of *R&D* capital, based on a

[2] Here and subsequently, all statements about statistical "significance" should not be taken literally. Besides the usual issue of data mining clouding their interpretation, the "samples" analyzed come close to covering completely the relevant population. Tests of significance are used here as a metric for discussing the relative fit of different versions of the model. In each case, the actual magnitude of the estimated coefficients is of more interest than their precise "statistical significance."

TABLE 2—NSF-CENSUS STUDY: CROSS-SECTIONAL PRODUCTION FUNCTIONS, LOG VALUE-ADDED DEPENDENT VARIABLE[a] U.S. FIRMS: 1967; 1972; 1977

	(1)		(2)		(3)[b]	
Variables	1967	1972	1972	1972	1972	1977
ln Employment	.604	.622	.623	.586	.578	.611
	(.045)	(.046)	(.035)	(.038)	(.038)	(.039)
ln Capital Services	.224	.199	.161	.234	.254	.291
	(.041)	(.044)	(.032)	(.036)	(.036)	(.035)
ln *R&D* Stock (*db*)	.113	.135	.165	.126	.115	.089
	(.023)	(.026)	(.019)	(.019)	(.018)	(.017)
Basic Research (*BR*)	.396	.340	.274	.499	.517	.401
	(.240)	(.261)	(.215)	(.191)	(.189)	(.189)
Company-Financed Research (*FP*)	.190	.247	.068	.133	.138	.044
	(.097)	(.106)	(.100)	(.088)	(.088)	(.084)
N	386	386	652	491	491	491
SEE	.312	.336	.390	.312	.309	.290

Notes: ln Employment = log (total employment-employment of scientists and engineers); ln Capital Services = log of (depreciation plus interest on net assets plus machinery and equipment rentals); ln *R&D* Stock (*db*) = log of the "stock" of total *R&D* expenditures based on a 15 percent per year declining balance depreciation assumption; *BR* = basic research as a fraction of total *R&D*; 1972 in the 1977 equation, 1967 in 1967 and 1972. *FP* = fraction of *R&D* stock "private," company-financed *R&D* stock as a ratio to the total *R&D* stock, as of *t*. All equations include also a constant term and industry dummies. The number of industry dummies used depends on the data set and varies between 18 and 20. Standard errors are shown in parentheses.

[a]Value-added and materials used in research in 1967 and 1972.

[b]Value-added only.

15 percent per year declining balance depreciation formula (the results are insensitive to the particular formula used), the implied average (at the geometric mean of the sample) gross rate of return to *R&D* investment rises in a similar fashion from .51 in 1967 to .62 in 1972 (in col. 1) and falls from .39 in 1972 to .33 in 1977 (in col. 3). In either case the estimated rate of return is quite high and there does not appear to be any dramatic fall in it over time.

The second major finding is the significance and rather large size of the basic research coefficient. It seems to be the case that firms that spend a larger fraction of their *R&D* on basic research are more productive, have a higher level of output relative to their other measured inputs, including *R&D* capital, and that this effect has been relatively constant over time. If anything, it has risen rather than fallen over time. Using the formulation of equation (2) implies a very high premium on basic versus the rest, a $\hat{\delta}$ of between 2.5 to 4.5, a several hundred percent premium on basic research. Before I explore the implications of this result, I want to examine other dimensions of these data and see whether similar effects can be observed there too.

The last major result of interest is the significant positive coefficient on the privately vs. federally financed *R&D* mix variable. This variable is of most import for the older more established firms in subsample 4 (Table 1) but its sign is consistent throughout, indicating a positive premium on privately financed *R&D*, or equivalently a discount as far as federally financed expenditures are concerned. Here the implied premium is smaller, between 50 and 180 percent, but still quite large.

All the above results were based on cross-sectional level regressions that are subject to a variety of biases, the main one being the possibility that "rich" successful firms are both more productive and can afford to spend more of their own money on such luxuries as *R&D* and especially the basic variety. One

TABLE 3—GROWTH RATE OF PARTIAL PRODUCTIVITY, 1966–77

Variables	$N = 911$			$N = 652$ (*with industry dummies*)	
Constant	.019	.009	.012	–	–
BTRD 6677	.107		.117	.119	
	(.014)		(.017)	(.016)	
BCRD 6677		.095			.106
		(.014)			(.015)
BR 72	.056	.056	.059	.035	.034
	(.017)	(.017)	(.019)	(.018)	(.018)
FP 72	.011	.019	.017	.022	.030
	(.005)	(.005)	(.006)	(.007)	(.007)
SEE	.0383	.0384	.0337	.0305	.0307

Notes: Dependent variable: *BPT* 6677 = trend growth rate of deflated sales minus the trend growth of total employment multiplied by the share of payroll in total sales. *BTRD* = trend growth of deflated total *R&D* expenditures; *BCRD* = same for company-financed *R&D* expenditures; *BR* = basic research expenditures as of fraction of total research expenditures; *FP* = ratio of company-financed *R&D* stock to total; *SEE* = residual standard error. All equations contain also a term reflecting the variance of *R&D* and terms representing the growth of physical capital: age composition and depreciation as of 1972.

can reduce somewhat the possibility of this type of bias by focusing on firm-growth rates, the changes that occurred, rather than on their levels. To the extent that firms have idiosyncratic productivity coefficients that may be also correlated with their accumulated *R&D* levels, considering growth rates is equivalent to doing a "within" firms analysis, one that eliminates such fixed effects from the analysis. The next two tables present, therefore, the results of analyzing the growth in the partial productivity of these same firms during the whole 1966–77 period.

Table 3 presents the results of estimating partial productivity equations in the largest possible sample for which 1966–77 growth rates were computable ($N = 911$) and in the subsample with a successful 1972 Census match. Here again we find my three main results confirmed: the *R&D* growth term and the two mix variables, the basic research ratio, and the fraction of research financed privately all contribute significantly to the explanation of productivity growth.

On the assumption that the growth rate in the stock of *R&D* is roughly proportional to the growth in deflated *R&D* itself, the coefficient of *BTRD* should be estimating the same number as the coefficient of the *R&D* stock variable in Table 2. The results are in fact surprisingly close: about .12 in Table 3 as against .09 to .17 in Table 2. Moreover, there seems to have been no decline in this coefficient relative to the earlier 1957–65 period. In my previous study (1980a), I estimated the same coefficient to be .073. In the current replication and extension of this sample a similar equation for 1957–65 yields a *BTRD* coefficient of .086. Thus, if anything, the coefficient of *R&D* went up between the early 1960's and the early 1970's.

The second major finding of interest is the positive and significant basic research coefficient. It is hard to interpret its magnitude since the approximation outlined in equation (3) breaks down when the average growth rate of deflated *R&D* and of basic *R&D* is close to zero or negative. Consider, however, the following illustrative calculation. Raising the *BR* ratio by one standard deviation, from .026 to .097 at the mean, would increase the rate of growth of partial productivity by close to half a percent per year ($.071 \times .059 = .0042$). This same increase would raise the growth of total *R&D* by .107 for one year and would contribute a

TABLE 4—GROWTH RATE OF PARTIAL PRODUCTIVITY, BY INDUSTRY, 1966–77 (MATRIX 6, TOTAL $N = 991$)

Coefficients of	Coefficients by the Estimated t-Ratio			
	< -1.5	$-1.5-0$	$0-1.5$	$1.5+$
BTRD		2	7	10: Miscellaneous, Industrial Chemicals, Drugs, Stone & Glass, Machinery, Electronics, Electrical Equipment, Transportation Equipment, Scientific Instruments, Non-Manufacturing
*BR*72		5	8	6: Wood & Paper, Other Chemicals, Oil, Machinery, Aircraft, Non-Manufacturing
*FP*72	2	6	7	4: Oil, Rubber, Electronics, Aircraft

Notes: All equations contain also a term reflecting the variance of *R&D* and terms representing the growth of physical capital: age composition and depreciation as of 1972.

once-and-for-all increase in the level of productivity of .0125. Discounting the more "permanent" effect of basic research by a real interest rate of .05 yields an "equivalent" one-year effect of .084, or a 7 to 1 ratio in favor of basic research! If one allows for industry dummies which in this formulation represent separate industry trend rates of disembodied technical change, the effect of basic research is cut by about 50 percent, implying perhaps that a signficiant fraction of the estimated effect comes from spillovers that diffuse throughout the industry. Note that it is the only coefficient that is affected substantively when separate industry dummies are allowed for. Nevertheless, even a 3.2 to 1 ratio is quite high!

The third finding is the significant positive premium on company-financed *R&D*. Here too the implied premia are quite high, but given that the mix variable is defined in terms of stocks rather than flows, the calculations are more cumbersome. Consider starting from a zero growth position and a .7 ratio of private to total *R&D* stock. To move this fraction from .7 to .75, one would need to raise the private stock by 29 percent and the overall stock by 20 percent (without reducing absolutely the stock of federally financed *R&D* capital). There are different possible investment paths that would achieve this goal and would have somewhat different present value consequences. If one roughly doubled the rate of privately financed *R&D* expenditures, from the previous replacement level of .105 (.7×.15) to .205, one could achieve this target in slightly over two years. Ignoring discounting, this would lead to a once-and-for-all growth in productivity of .024, due to the growth in the total stock of *R&D and* a .0011 permanent increase in the rate of growth due to the shift of the fraction private ratio from .7 to .75. The present value of this second term is about .022, or of the same order of magnitude as the first term. That is, raising the stock of *R&D* by 20 percent but shifting it all into the private component doubles the effect of such dollars.

There are problems, however, with such an interpretation. If private *R&D* expenditures contribute more to productivity growth, one might have thought that when they are substituted for the total *R&D* growth measure, they might fit better and also have a higher coefficient. But that is not the case as can be seen from the results presented in columns 2 and 5 of Table 4. The total *R&D* measure does a little bit better both in terms of fit and in the overall size of its coefficient, implying that the contribution of federal dollars is not zero. That is perhaps what one should expect. Most of the direct output of federal research dollars is "sold" back to the government at "cost plus" and is unlikely to show up as an increase in the firm's own productivity. Thus all that one could expect to measure here are the within-firm spillover effects of such expenditures. What we may be detecting is that such effects are indeed present and positive, but we should not have expected them to be of the same order of magnitude as would be the case for the firm's own investments in improving its productivity or profitability.

There are a number of econometric questions that can be raised about the robustness and sensitivity of such results. I will discuss only a few of these here. The most obvious question arises from the fact that even though I allowed, in the growth rates version, for separate firm intercepts and different industry trends, I am still assuming common *R&D* and the conventional capital coefficients across rather different industries. This is done from necessity rather than as a virtue. Estimating the same models industry by industry reduces the sample sizes drastically and raises greatly the relative noise level, making it rather hard to interpret the resulting estimates. Nevertheless, these estimates, which are summarized in Table 4, are quite consistent with the earlier story: 17 out of the separately estimated 19 coefficients for the *R&D* growth variable are positive and more than half of them are statistically significant at conventional significance levels. Similarly, the coefficients of the basic research ratio variable are positive in 14 of my 19 industries and signficant in over a third of them. The fraction private variable is less robust to the division of the sample into industries, with more than half of the coefficients still positive, but only 4 of them are statistically significant within particular industries. Two of these industries are indeed the ones where one would expect to find such an effect, aircraft and electronics, industries where the bulk of federal monies is spent. Nevertheless, it seems that the effect that is being caught by the fraction private variable has an important industry component, something that had been already noted in my earlier study (1980a), as does also the effect associated with the basic research variable, though to a lesser extent.

A number of other versions were computed using the growth in capital services rather than the depreciation and age composition variables that had been used to keep the results comparable to the earlier study, and the growth in *R&D* "capital" rather than the flow (and also different definitions of such capital). I also estimated versions using the "intensity" form for the *R&D* variable, to make it more comparable to other studies in the literature (my paper with Lichtenberg; Mansfield; and others).[3] By and large the results of these alternatives were somewhat weaker but not substantively different. Perhaps the most interesting alternative estimate is the intensity version using the growth of capital between 1967 and 1972 as its capital measure:

$$(4)\quad BPT6677 = \ldots \underset{(.069)}{.243ACRS} + \underset{(.024)}{.045ABR} + \underset{(.130)}{.180DLCS} \qquad SEE = .0316 \ \text{(Subsample 4)}$$

where *ACRS* is the average company *R&D* to sales ratio, averaged over 1967 and 1972, *ABR* is a similar average basic to total *R&D* ratio, and *DLCS* is the rate of growth in deflated capital services between 1967 and 1972. This version is closest in form to the equation estimated by Mansfield on much smaller samples. The basic results are similar, however. Basic *R&D* is a signficant contributor to productivity growth with an implied basic to company premium of about 5 to 1 (given an average *R&D* to sales ratio of .035).

The final set of results to be presented here, in Table 5, relate to the relative profitability of our firms in 1972 and 1977. The dependent variable, *GRR*, is the ratio of gross profits (value-added minus labor costs and plus *R&D*) to total gross fixed assets. The independent variables include the ratio of *R&D* capital (undepreciated) to total fixed assets and our ubiquitous *R&D* mix variables: the basic research and fraction private

[3]The intensity version uses the fact that $\alpha = (\partial Q/\partial K)K/Q$ and reexpresses $\alpha\dot{K}/K$ as $\rho[R/Q]$, where $\rho = \partial Q/\partial K$ is the marginal product (gross rate of return) of *R&D* capital and it has been assumed that $\dot{K} = R - \delta K \simeq R$, i.e., either $\delta \sim 0$ (no depreciation) and/or initial K very small. This formulation has the advantage that it does not impose the assumption of a constant *elasticity* across different firms, replacing it instead by the, possibly more plausible, assumption of the constancy of rate of return.

TABLE 5—GROSS PROFIT RATE REGRESSIONS
(*GRR* = (Value-Added-Payrolls + *R&D*)/Gross Assets))

Dependent Variable and Sample Size	Coefficients of: Constant	*R&D* Capital to Total Fixed Assets Ratio	Basic *R&D* Ratio	Fraction Private	*SEE*
GRR 72					
N = 652 (a)	.144	.088	.344	.107	.262
	(.049)	(.012)	(.144)	(.048)	
(b)		.060	.187	−.012	.237
		(.013)	(.138)	(.052)	
N = 491 (a)	.117	.080	.514	.154	.264
	(.052)	(.013)	(.139)	(.051)	
(b)		.061	.366	.074	.227
		(.015)	(.138)	(.057)	
GRR 77					
N = 491 (a)	.341	.031	.402	.033	.313
	(.064)	(.019)	(.187)	(.068)	
(b)		.004	.261	−.028	.292
		(.022)	(.187)	(.077)	

Notes: (a) Regressions do not contain industry dummies; (b) do.

ratios. Even though the dependent variable is quite different, the overall results are rather similar to the earlier ones. The *R&D* capital variable is positive and almost always statistically significant through its coefficient is a bit low if it is to be interpreted as a rate of return to it. The basic research variable is both large and significant though possibly too large to be credible. Given that the ratio of total *R&D* capital to total fixed assets is only about .05 on average, the 1972 coefficients imply a δ of about 30 to 60. The fraction private ratio also contributes positively to profitability but its effect largely disappears once industry differences are allowed for. The results for 1977 are weaker than those for 1972, the residual variance is signficantly higher, but they too suggest the importance of basic research even in this context.

A similar analysis was performed using an estimate of the net rate of return as the dependent variable, subtracting depreciation from the numerator of *GRR* and using a net stock concept for the denominator and also in the definition of the *R&D* capital variable. While the fit was significantly worse when using this definition of the dependent variable, the overall results were rather similar. The net return version was also available for 1967 and the results using it indicate a relatively constant and significant coefficient for the basic research ratio while the coefficient of the total *R&D* stock rises from 1967 to 1972 and than falls again in 1977 (from .11 to .16 and down to .06). It is doubtful whether these fluctuations represent real trends or, more likely, reflect the larger noise level in the 1977 data and the changing composition of these samples. In any case, the profitability regressions are consistent with the productivity level and productivity growth rate based results described earlier (Tables 2, 3, and 4).

III. Discussion and Summary

There are three major findings in this paper: *R&D* contributed positively to productivity growth and seems to have earned a relatively high rate of return; basic research appears to be more important as a productivity determinant than other types of *R&D*; and privately financed *R&D* expenditures are more effective, at the firm level, than federally financed ones. These findings are not entirely new. The first finding has been documented in a number of earlier studies (see my 1980a, b papers; my article with Mairesse, 1984; A. N. Link, 1981a; and others). What is new in this paper in this regard is a confirmation of this finding on a much larger and more recent data set. It also presents evidence for the view that this effect has not declined significantly in recent years, in spite of the overall slowdown in productivity growth and the general worry about a possible exhaustion of technological opportunities.[4]

The evidence for a "premium" on basic research is much more scarce. The major previous paper suggesting this type of a result is Mansfield which uses aggregate data for 20 industries for 1948–66 and data for 16 firms during 1960–76, and finds a signficant premium on basic research, on the order of 2 to 1 at the industry level and 16 to 1 at the firm level. (See also Link, 1981b, for similar results for 1973–78 based on data for 55 firms.) In this paper I get similar though somewhat smaller effects at the firm level, using a much larger and more representative sample. I also find that differences in levels of productivity and profitability are related to differences in the basic research intensity of firms.

Such findings are always subject to a variety of econometric and substantive reservations. In this context the two major related issues are simultaneity and the question of how major divergences in private rates of return persist for such long periods. It is possible to argue that it is not *R&D*, or its basic research component, that causes firm "success" as measured by productivity and profitability, but rather that success allows firms to indulge in these types of luxury pursuits. It is difficult to argue about causality on the basis of what are essentially correlational data. It is possible to use simultaneous equation techniques to estimate such models, but then the argument shifts to the validity of the exogeneity assumption for the particular instruments. In the context of my specific data set, it is hard to think of any valid instruments except for possibly lagged values of the same variables, which raises some problems of its own. The best evidence for the notion that these results are not entirely spurious is provided by the growth rates where the individual firm levels are partialed out of the analysis. But, here too, one could argue about the impact of common unanticipated "luck" elements. Unfortunately, it is unlikely that one could use lagged growth rates as instruments, since there is very little correlation in growth rates over time at the firm level. While an attempt will be made in further work with these data to estimate more extended simultaneous equations versions of such models, I am not too optimistic as to what can be accomplished in this regard. The evidence presented here should not be interpreted as "proving" that *R&D*, and especially its basic component, are important for productivity growth but rather as presenting some prima facie evidence in support of such an interpretation. In this sense it is an exercise in economic rhetoric (Donald McCloskey, 1983).

It is even more difficult to respond to the theoretical a priori argument that such results cannot be true since they imply widely differing rates of return to different activities under the control of the same firm. One's response to this depends on one's views as to the prevalence of equilibria in the economy. While it is likely that major divergences in rates of return are eliminated or reduced in the long run, the relevant runs can be quite long. *R&D* as a major component of firm activity was undergoing a diffusion process in the 1950's and 1960's and may not have reached full equilibrium even by the end of our period. This may be especially true of the basic research component where the risks are much greater and the uncertainty introduced

[4]The finding that the coefficients in a logarithmic regression have not declined over time does not dispose of the possibility that there could have been an overall loss in accumulated knowledge capital due to accelerated obsolescence. A proportional decline in the effectiveness of past capital or in the rate that *R&D* is converted into new knowledge capital need not show up as decline in the slope coefficient, it would get absorbed into the shifting constant. Disproportionate shifts should, however, have an impact on the estimated slope coefficient. Also, a pure obsolescence shock to old knowledge capital would have called forth an increase in the rate of *R&D* expenditures, something which has not been observed in the data. I am indebted to M.N. Baily for this point.

by changing government policies and the changing economic environment make it quite difficult to decide what is the right level for it.

A somewhat different version of this argument would claim that the world is indeed in approximate equilibrium but that different firms face different opportunities for doing research, basic or otherwise, are in different ecological niches, and hence have different coefficients in their "production functions." This would explain why different firms are observed to spend different amounts on *R&D* while actually earning about the same rate of return on it. When a constant coefficients production function is fit to such data, it will fit because it is approximating a market equilibrium relation. If the level of *R&D* invested were independent of the coefficient, then such a function would just reproduce its average share and not produce any evidence of excess returns. But if, as is reasonable, *R&D* is invested optimally with firms which have better opportunities, higher coefficients, investing more, this will induce a positive correlation between *R&D* and its individual coefficient and lead to an upward bias in the estimated "average" coefficient.[5] The resulting "larger" coefficient, larger than the observed factor share, will be interpreted, wrongly, as implying a higher rate of return than is actually prevailing at the individual level.

This argument may be recognized as a version of the earlier attacks on the Cobb-Douglas production function combined with a random coefficients interpretation of the same phenomenon. In its extreme form it is testable. Since there are time-series data available for individual firms, one could try to estimate individual firm parameters and check whether they are in fact distributed as is predicted by this particular argument. While individual parameters are unlikely to be well estimated, given the relative shortness of the available time-series, the parameters of the distribution of such coefficients might be estimable with more precision. I intend to pursue this possibility in future work.

To restate again the major points of the paper: a newly available body of data on all the major firms performing *R&D* in the United States has been examined and evidence has been presented for the proposition that *R&D* contributes significantly to productivity growth, that the basic research component of it does so even more strongly, and that privately financed *R&D* expenditures have a significantly larger effect on private productivity and profitability than federally financed *R&D*. These findings are open to a number of reservations. Nevertheless, they do raise the issue that the overall slowdown in the growth of *R&D* and the absolute decline in basic research in industry which occurred in the 1970's may turn out to have been very costly to the economy in terms of foregone growth opportunities.

[5]A positive correlation is not enough, by itself, for a positive bias. The weight of an individual firm slope coefficient in the cross-sectional estimate is proportional to the *square* of the deviation of *R&D* stock from its mean. A positive correlation between levels does not translate itself directly into a positive correlation between the level of one variable and the square of the other, except for certain skewed distributions. Since we do not observe the individual coefficients directly, it is rather difficult to check out this conjecture.

REFERENCES

Bound, John et al., "Who Does *R&D* and Who Patents?," in Z. Griliches, ed., *R&D, Patents, and Productivity*, Chicago: University of Chicago Press, 1984, 21–54.

Griliches, Zvi, "Issues in Assessing the Contribution of *R&D* to Productivity Growth," *Bell Journal of Economics*, Spring 1979, *10*, 92–116.

______, (1980a) "Returns to Research and Development Expenditures in the Private Sector," in J. W. Kendrick and B. Vaccara, eds., *New Developments in Productivity Measurement*, NBER *Studies in Income and Wealth* No. 44, Chicago: University of Chicago Press, 1980, 419–54.

______, (1980b) "*R&D* and The Productivity Slowdown," *American Economic Review, Proceedings*, May 1980, *70*, 343–48.

______, "Comment" on Edwin Mansfield, "R and D and Innovation: Some Empirical Findings," in Z. Griliches, ed., *R & D, Patents and Productivity*, Chicago: University of Chicago Press, 1984.

______ **and Hall, B. H.,** "Census-NSF *R&D* Data Match Project: A Progress Report," in *Development and Use of Longitudinal Establishment Data*, Economic Research Report, ER-4, Bureau of the Census, 1982, 51–68.

______ **and Lichtenberg, F.,** "*R&D* and Productivity Growth at the Industry Level: Is There Still a Relationship?," in Z. Griliches, ed., *R&D, Patents and Productivity*, Chicago: University of Chicago Press, 1984, 465–96.

______ **and Mairesse, J.,** "Comparing Productivity Growth: An Exploration of French and U.S. Industrial and Firm Data," *European Economic Review*, April 1983, *21*, 89–119.

______ **and** ______ ,"Productivity and *R&D* at the Firm Level," in Z. Griliches, ed., *R&D, Patents, and Productivity*, Chicago: University of Chicago Press, 1984, 339–74.

Hall, Bronwyn H., "Historical *R&D* Panel: 1957–77. Public Use Correlation, Matrices Tape Documentation," unpublished, September 1984.

Link, A. N., (1981a) *Research and Development Activity in U.S. Manufacturing*, New York: Proeger, 1981.

______, (1981b) "Basic Research and Productivity Increase in Manufacturing: Additional Evidence," *American Economic Review*, December 1981, *71*, 1111–12.

McCloskey, Donald N., "The Rhetoric of Economics," *Journal of Economic Literature*, June 1983, *22*, 481–517.

Mansfield, Edwin, "Basic Research and Productivity Increase in Manufacturing," *American Economic Review*, December 1980, *70*, 863–73.

Pakes, Ariel and Schankerman, M., "The Rate of Obsolescence of Knowledge, Research Gestation Lags, and the Private Rate of Return to Research Resources," in Z. Griliches, ed., *R&D, Patents, and Productivity*, Chicago: University of Chicago Press, 1984, 209–32.

National Science Foundation (Jaffe, S.A.), "A Price Index for Deflation of Academic *R&D* Expenditures," NSF 72–130, Washington: USGPO, 1972.

______, *Trends to 1982 in Industrial Support of Basic Research*, NSF 83–302, Washington: USGPO, 1983.

______, *National Patterns of Science and Technology Resources*, NSF 84–311, Washington: USGPO, 1984.

Part III
Market Structure and Technical Change

[9]

The Economic Journal, **90** (*June* 1980), 266–293
Printed in Great Britain

INDUSTRIAL STRUCTURE AND THE NATURE OF INNOVATIVE ACTIVITY*

A substantial portion of the increase in the output of advanced industrial nations is widely judged to have been attributable to technical progress. There is also overwhelming evidence that this progress has not occurred merely in a random manner (see, e.g. Schmookler, 1962). Yet there is not much in the way of economic theory to explain either its rate or its direction, and particularly little that has a reasonably precise micro-economic foundation.[1] This lacuna is important, because a recognition of the importance of technical progress raises serious doubts about the adequacy with which traditional micro-economic models allow one to understand the functioning of modern market economies, and to develop policy prescriptions (e.g. with respect to anti-trust policy). The lacuna is also surprising, for it has become a commonplace, at least since the work of Schumpeter (1947), that the pace of inventive and innovative activity is related to market structure. Moreover, there must be many who share the Schumpeterian belief that perfect competition is inimical to inventive activity, and that the gains from such activity more than offset the welfare loss arising from production inefficiency associated with market power.[2]

However, a substantial part of the recent industrial organisation literature would appear to have interpreted tests of the Schumpeterian thesis in a somewhat different manner.[3] It is often argued now that industrial concentration and large size *spur* inventive activity. It is as though concentration is a *cause* of innovations.[4]

* This is a revised and expanded version of the first part of an invited paper (Dasgupta and Stiglitz, 1977) presented at the World Congress of the *International Economic Association* on *Economic Growth and Resources*, held in Tokyo during 27 August–3 September 1977. This version was prepared whilst Dasgupta was a Visiting Professor at the School of International Studies, Jawaharlal Nehru University, and the Delhi School of Economics, Delhi, and Stiglitz was Oskar Morgenstern Distinguished Fellow at Mathematica, and Visiting Professor at the Institute of Advanced Studies, Princeton, during the Autumn of 1978. We have gained much from discussions with Hans Biswinger, Sukhamoy Chakravarty, A. K. Dasgupta, Paul David, Richard Gilbert, Sanford Grossman, Glen Loury, Edwin Mansfield, and most especially, from the comments of Ashok Guha.

[1] For this last, see, for example Arrow (1962), Barzel (1968), Stiglitz (1970), Kamien and Schwartz (1972), Evenson and Kislev (1975), Nelson, Winter and Schuette (1976), Dasgupta and Stiglitz (1977), Loury (1977), Nelson and Winter (1977), Dasgupta and Stiglitz (1978), Levin (1978) and, of course, the pioneering work of Schumpeter (1947).

[2] 'A system . . . that at *every* given point of time fully utilises its possibilities to the best advantage may yet in the long run be inferior to a system that does so at *no* given point of time, because the latter's failure to do so may be a condition for the level or speed of long run performance' (Schumpeter (1947), p. 83). It should be noted that Schumpeter is here comparing industrial capitalism with perfect competition.

[3] The empirical findings suggest that whilst up to a point industrial concentration is positively correlated with innovative activity, it is negatively correlated when an industry is too concentrated. See Scherer (1970), and Kamien and Schwartz (1975) for a survey of the empirical literature.

[4] Thus, for example, after noting in a sentence that technological change has effects on market structure, the body of the chapter devoted to R & D in Scherer (1970), is '. . . concerned with a possible causal flow in the opposite direction; from market structure to technological innovation' and, among others, with the question: 'Is monopoly power, e.g. as manifested in high market concentration, a favourable climate for innovation and technical progress?' Scherer (1970), p. 347.

In this paper we attempt to provide an analytical framework relating market structure to the nature of inventive activity. In doing so we have come to modify this neo-Schumpeterian view in a fundamental way. We shall argue that except in the short run *both* market structure and the nature of inventive activity are endogenous; that the degree of concentration in an industry ought not to be treated as given, as it recently has been in the industrial organisation literature; that they both depend on more basic ingredients, such as the technology of research, demand conditions, the nature of the capital market (i.e. market rates of interest, and the ability of firms to borrow to finance research and development (R & D)), and the legal structure (e.g. patent rights). We shall, to be sure, explore the relationship between the degree of concentration and the nature of innovative activity. But as they are both endogenous, their relationship, unlike the neo-Schumpeterian thesis, ought not to be regarded as a *causal* one.

A major objective of our study is to formulate models within which the efficiency of a market economy can be assessed and where the tradeoff between atemporal production efficiency and dynamic gains can be meaningfully discussed. Not surprisingly these questions, once properly posed, turn out to be far more complicated than they appear at first sight. It is not a case of a single firm making a single decision (e.g. the total volume of R & D expenditure), but rather a case in which several firms make a complex of decisions; and it is the consequences of these with which we are concerned. For example, each firm needs to decide both on how much to spend on R & D and also on which research strategies to pursue. Moreover, research strategies can differ not only with respect to the probability distribution of the dates at which success occurs, but also as regards how much is learnt even if the principal objectives of the research project fail, and to how similar, both in process and in objectives, they are to those being pursued by other firms. Each of these decisions has important consequences not only for the aggregate rate of technological progress, but also for industrial structure and the performance of a market economy. If firms tend to imitate each other's research strategies then much of R & D expenditure may be essentially duplicative, and consequently socially wasteful. If firms engage in excessively risky projects it may lead to too fast a rate of technical progress and high degrees of industrial concentration. This last in turn may imply large losses in production efficiency. What makes the analysis all the more difficult is that each decision on the part of a firm has to be made within an industrial structure which is itself endogenous.

Questions regarding the *consequences* of market structure on R & D are, of course, not new. That there is *underinvestment* in R & D both under competition and monopoly is probably a general presumption. This presumption is based partly on the fact that knowledge – the output of R & D – has the attributes of a public good; but partly also on an argument, due to Arrow (1962), that relies solely on a comparison of the magnitudes of the payoff to the successful firm under alternative market structures.

In Section I, we shall review this argument and see where it is in error. This review will also help set the stage for the formal analysis that follows. In fact a central conclusion of this paper is that on balance there is no reason for supposing that a market economy sustains too low a level of investment in R & D. There may well be over-investment.

Sections II, III and IV contain analyses of various aspects of R & D activity and their relationship to market structure. The discussions are based on models that are natural extensions of a simple model of process innovation presented in Section I. In Section II, we explore the relationship between market structure and expenditure in R & D. The model we shall present, whilst very simple, is at once rich enough to illustrate some points that we have raised above and several that will be raised subsequently. However, it is as well to mention at the outset that a central feature that the model of Section II is designed to display is that while R & D expenditures on the part of firms are rather like fixed-costs in production, the levels of such 'fixed costs' are themselves choice variables for firms (firms can engage in less or more R & D expenditure). This results in a key difference between the analysis of competition in R & D and the more conventional analysis of product competition, as our subsequent discussion will display. Moreover, we shall note formally that there are some basic non-convexities that may be present in the production and use of knowledge, and since knowledge is the output of R & D effort, any analysis of R & D must take into account such possible non-convexities. To illustrate the matter simply, it should be noted that the indirect social benefit function of a commodity is a decreasing but *convex* function of the unit cost of production (see Fig. 4). An immediate implication of this is that as between two research strategies that yield the same mean reduction in production cost, a risk-averse society will prefer the *riskier* one if the expenditure on the two are more or less the same. Another implication of this, as we shall note in Section II, is that net social benefit of R & D expenditure is not necessarily a concave function even if one postulates diminishing returns in cost reduction due to increased R & D effort. The point is that the same piece of knowledge can in principle be applied at any scale of operation. Thus, the cost of information per unit of scale decreases as the scale increases; but the value of information per unit scale need not. Earlier writings, in the main, have concentrated on the failure of the price system to sustain an efficient production and utilisation of knowledge, not only because knowledge has the attributes of a public good, but also because of the impossibility of establishing a complete set of perfectly competitive contingent markets in the face of, say, the phenomenon of moral hazard. A part of our analysis will be directed at noting the natural non-convexities that arise in the production and use of knowledge. Our aim in general is to see how each of these features affects the structure of an industry and the scale and direction of innovative activity.

The example of Section II, being timeless and devoid of uncertainty, suffers from several drawbacks, two crucial ones being that one is unable to study the *degree of risk-taking* in research activity, and that one cannot identify the *speed of research* (the pace of inventive activity). Consequently, it is not possible to analyse how each of these characteristics is related to market structure. It is clear enough in advance, of course, that each is in turn related not only to the structure of the product market but also to the degree of competition in R & D activity.[1] The models that follow will capture this feature sharply. The example discussed in Section III is similar to that of Section II, but it is now supposed that research activities have uncertain payoffs. In this paper we are concerned with *process* innovation (i.e. R & D designed to reduce cost of production). This means that attention is drawn to the distribution of extreme values. This in turn implies, as was noted by Evenson and Kieslev (1975), that even risk-averse firms might wish to engage in randomisation. The example to be discussed in Section III will bear this out. But the desire for randomisation will, in our model, be reinforced by the feature which we have noted earlier.

In Section IV time is introduced into the analysis. It will be supposed that the first firm to invent captures all benefits that are to be had among firms (i.e. we suppose that the winner takes all). No doubt this is a simplification. But it should be transparent in which direction the model needs to be modified, were one to recognise the fact that firms usually are able to invent around patents and that as a result the first firm to make a breakthrough is not necessarily the most advantaged.

A critical assumption of the construction in Section IV is that all firms are obliged to follow the *same* research strategy; i.e. they all face the same decision tree. This has important consequences. In a sequel to this paper we explore the opposite extreme and suppose that firms face uncertainties about the date of success that are independent of one another.

In Section V, we gather together what we regard as the basic morals emerging from our analysis. The Appendices explore the constructions in Sections II and III in detail, verifying some of the claims made in the text. In particular, Appendix 1 contains a theorem regarding the existence of an equilibrium with free entry which may be of wider interest.

I. MARKET BIAS IN PROCESS INNOVATION

In what follows we suppose an absence of income effects. Let Q denote the quantity of a given commodity. The gross social benefit of consuming Q is $u(Q)$ with $u'(Q) > 0$ and $u''(Q) < 0$. Market demand is given by

$$p = p(Q) = u'(Q). \tag{1}$$

Write $R(Q) = p(Q)Q$ as gross revenue to a monopolist. Assume for the moment that marginal revenue is decreasing in output.

Suppose that the current best-practice technique for producing the commodity involves c as the unit cost of production. Suppose also that a particular process innovation reduces the unit cost of production to c^*. In a pioneering paper Arrow (1962) explored the gain to the innovator under three forms of the product market. The first, the socially managed market, is one where the

[1] Among other things it is this last that distinguishes the analyses undertaken in this paper from those undertaken by Arrow (1962), in that Arrow concentrated exclusively on differing structures of the product market and supposed no competition in R & D activity when analysing the incentives to innovate under different market structures. We shall comment on this at a greater length in Section I.

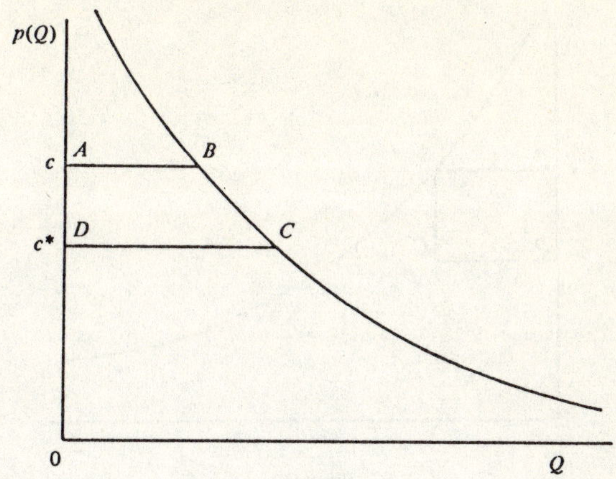

Fig. 1

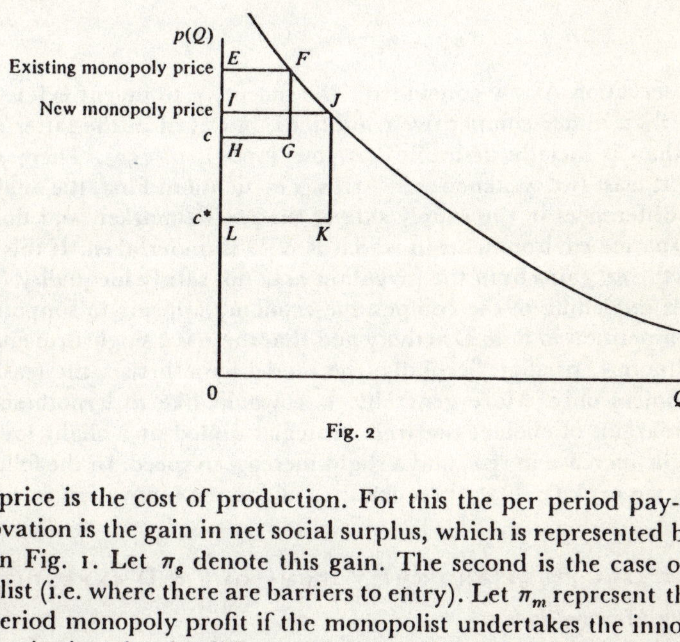

Fig. 2

market price is the cost of production. For this the per period pay-off due to the innovation is the gain in net social surplus, which is represented by the area *ABCD* in Fig. 1. Let π_s denote this gain. The second is the case of the pure monopolist (i.e. where there are barriers to entry). Let π_m represent the increase in per period monopoly profit if the monopolist undertakes the innovation. In Fig. 2, π_m is given by the difference between the areas *IJKL* and *EFGH*. The third is the competitive economy. Assume then that the *c*-technology is competitively exploited and that the innovator acquires a patent on the c^*-technology. Let π_c denote the per-period monopoly profit accruing to the innovator during the life of the patent. There are clearly two cases to be considered. In Fig. 3 the case where the monopoly price exceeds *c* is shown. Thus there is a limit price phenomenon here, and the innovator will supply the entire market and charge *c*. For this case π_c is given by the area *ABC′D*.

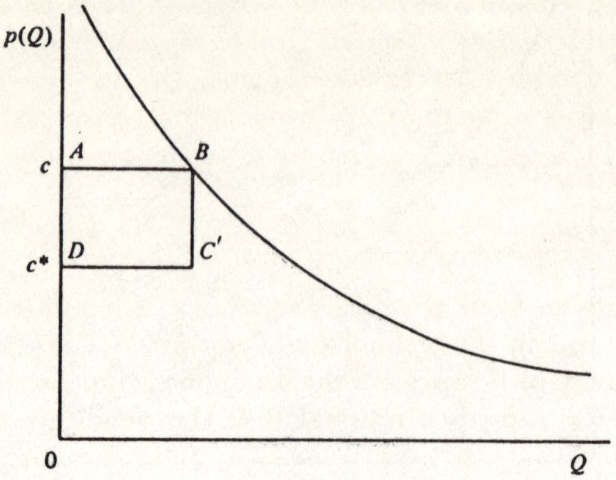

Fig. 3

Now it is possible to show that

$$\pi_s > \pi_c > \pi_m.^1 \tag{2}$$

From this observation Arrow concluded, 'the incentive to invent is less under monopolistic than under competitive conditions, but even in the latter case it will be less than is socially desirable' (Arrow (1962), p. 152). There would appear to be at least two weaknesses in Arrow's contention. First, the analysis is restricted to differences in the supply side of the *product* market, and does not consider alternative environments in which R & D is undertaken. If this latter is considered the *net* gains from the invention may not satisfy inequality (2). In particular, his modelling of the competitive economy appears to suppose that there is no competition in R & D activity and that there is a single firm engaged in it (see footnote 1, p. 269). Secondly, the model hypothesises the feasibility of discrete choices only. More generally, one would like to hypothesise the existence of margins of choice: research strategies aimed at a slight lowering of costs, a slight increase in risk, and a slight increase in speed. In the following three sections we explore these three features of process innovation in turn.

II. MARKET STRUCTURE AND THE VOLUME OF R & D EXPENDITURE

(1) *The Socially Managed Industry*

Imagine first that the market for the commodity in question is socially managed. If c is the (constant) unit cost of production then the net social benefit of consuming Q is $u(Q) - cQ$. Define

$$V(c) = \max_Q [u(Q) - cQ] \tag{3}$$

as the indirect utility function. It is well known that $V(c)$ is a declining convex function of c (see Fig. 4). We suppose that R & D expenditure is designed to

[1] For a demonstration, see Arrow (1962).

reduce the cost of production. Thus if x is expenditure on R & D then $c(x)$ is the unit cost of production. It is then natural to suppose that $c'(x) < 0$.[1]

Assume for the moment that the government can raise the funds for R & D expenditure from general taxation. Its problem then is to

$$\max_{x \geqslant 0} \{V[c(x)] - x\}. \tag{4}$$

Suppose that a solution to (4) exists. The first point to note is that unless further structure is imposed on $c(x)$ we cannot guarantee that $V[c(x)]$ is a concave function of x. This is so even if we were to assume in addition that $c''(x) > 0$ (i.e. decreasing returns in the technology of research). The problems that the possible non-concavity of V raises are the usual ones. Suppose that the solution to (4) dictates that $x > 0$, then optimal R & D expenditure must satisfy the condition

$$\frac{dV}{dc} c'(x) = 1,$$

but

$$\frac{dV}{dc} = -Q,$$

hence

$$-c'(x)Q = 1. \tag{5}$$

The interpretation of equation (5) is clear enough. The LHS is the marginal social benefit of increasing R & D expenditure, which in fact is the reduction in production cost when output is optimal. The RHS is, of course, the marginal cost of increasing R & D expenditure. If V is concave in x there is no need for the government to calculate consumer surplus in locating the optimum. But if V is not concave in x the government needs in general to conduct global cost-benefit analysis to locate the optimum. The need for computing consumer surplus in such circumstances is obvious.

In order to bring these points out more clearly it will prove helpful to specify the functions somewhat. Suppose that

$$u(Q) = \frac{\sigma Q^{1-\epsilon}}{(1-\epsilon)} \quad (\sigma, \epsilon > 0). \tag{6}$$

Thus if we write by $p(Q)$ the market demand curve,

$$p(Q) = \sigma Q^{-\epsilon}, \tag{7}$$

where ϵ^{-1} is the elasticity of demand. Therefore, (7) supposes that market demand is iso-elastic.

It is then simple to confirm that

$$V[c(x)] = \left(\frac{\epsilon}{1-\epsilon}\right) (\sigma)^{1/\epsilon}/[c(x)]^{(1-\epsilon)/\epsilon}. \tag{8}$$

Suppose in addition that it is a new product which is under consideration.

[1] It is as though Mother Nature has a patent on all techniques of production with unit cost $c(x)$, $(x > 0)$ and that society has to pay x to purchase the right to use the technique of production with unit cost $c(x)$.

In particular assume

$$c(x) = \beta x^{-\alpha} \quad (\alpha, \beta > 0). \tag{9}$$

In this case

$$V[c(x)] = \left(\frac{\epsilon}{1-\epsilon}\right) \frac{\sigma^{1/\epsilon}}{\beta^{(1-\epsilon)/\epsilon}} x^{\alpha(1-\epsilon)/\epsilon}.$$

We may now note that $V[c(x)]$ is strictly concave in x if $\epsilon > \alpha(1-\epsilon)$. If the reverse inequality were to hold, $V[c(x)]$ would be strictly convex in x, and in particular, (4) will not possess a finite solution. Social benefits will continually increase with increasing R & D expenditure.[1]

The foregoing condition for ensuring that $V[c(x)]$ is strictly concave is really rather transparent. It says that $1 + 1/\alpha > 1/\epsilon$. Since $1/\epsilon$ is the elasticity of demand, and since $1/\alpha$ is the production cost elasticity of R & D investment, the condition says that this latter elasticity must be sufficiently large to entertain continual diminishing (utility) returns on R & D expenditure. It should be noted that the condition is readily satisfied if demand is inelastic (i.e. $\epsilon > 1$). It is satisfied even if demand is elastic (i.e. $\epsilon < 1$) so long, of course, it is not too elastic. The condition makes precise the range on ϵ that yields a well defined planning problem on the assumption that income effects are nil.[2]

Suppose then that $\epsilon > \alpha(1-\epsilon)$. Let x_s and Q_s denote the socially optimal levels of R & D expenditure and total output, respectively. If we now use (7) and (9) in the optimality conditions it is simple to verify that

$$x_s = (\alpha^{\epsilon}\sigma\, \beta^{\epsilon-1})^{1/[\epsilon-\alpha(1-\epsilon)]} \tag{10}$$

and

$$Q_s = (\alpha\beta)^{-1}\, (\sigma\alpha^{\epsilon}\beta^{\epsilon-1})^{(1+\alpha)/[\epsilon-\alpha(1-\epsilon)]}. \tag{11}$$

It follows immediately from equations (10) and (11) that both optimum output and optimum R & D expenditure are greater in industries that face larger markets (i.e. larger value of σ). Thus the larger is the size of the market the greater ought (unit) cost reduction be. It is also worth noting that total output ought to be less in industries characterised by more costly R & D technology (i.e. a larger value of β). One expects these features of the optimum. However, equation (10) implies that optimum R & D expenditure is *less* in industries characterised by more costly R & D technology if demand is *elastic* ($\epsilon < 1$), and is *greater* if demand is *inelastic* ($\epsilon > 1$). On reflection, this too is possibly in line with intuition.

[1] This is, of course, absurd since we would not be justified in supposing the marginal utility of income to remain constant.

[2] Our purpose so far has been solely to draw attention to the *existence* of non-concavities in the payoff function in the context of process innovation under certain circumstances; nothing more. As we noted in footnote 1 it is rather silly to explore non-concavities while assuming away income effects. The obvious question to ask is how to reformulate the model if in fact $\epsilon < \alpha(1-\epsilon)$. One obviously needs to assume that the marginal utility of income is not constant. Suppose, for example, that in the obvious notation $u(Q, x) = \sigma[Q^{(1-\epsilon)}/(1-\epsilon)]\,[M-x-c(x)\,Q]^{\delta}$, where $1 > \epsilon, \delta > 0$, and M is total income. Let $W(x) = \max_{Q} u(Q, x)$. It is simple to verify that if $c(x)$ satisfies (9) then $W(x)$ is strictly convex for small x if $1 < \alpha(1-\epsilon)$. Nevertheless, $W(x)$ has a unique maximum.

(2) *Oligopoly with Free-Entry*

It remains to explore a market economy. Suppose now that R & D activity is undertaken exclusively in the private sector. For simplicity of exposition we consider a market for a new product and suppose, also for simplicity, that all firms face the same R & D technology, $c(x)$. Thus if $x_i(\geqslant 0)$ is the R & D expenditure of firm i, $c(x_i)$ is the unit cost of production that it faces for the product. In pursuing this formulation we are supposing that knowledge is monopolised by a firm when it pays for it.[1] Firms are profit maximising, behave non-cooperatively, and our task is to describe an equilibrium for this market. Towards this we assume that firms behave in a Cournot fashion (i.e. each firm chooses its own R & D expenditure level and its own output level) and that they all entertain Cournot conjectures (i.e. each firm supposes that none of the remaining firms will deviate from its course of action if the given firm deviates).

Begin by supposing that there is free entry into the industry; i.e. assume that market structure is endogenous. Let n be the number of firms in the industry and let Q_i ($\geqslant 0$) denote the output of the ith firm ($i = 1, \ldots, n$). It follows from our assumptions that

$$[n^*, (Q_1^*, x_1^*), (Q_2^*, x_2^*), \ldots, (Q_i^*, x_i^*), \ldots, (Q_{n^*}^*, x_{n^*}^*)]$$

is an equilibrium with free entry if, for $i = 1, 2, \ldots, n^*$,

$$[p(\sum_{j \neq i} Q_j^* + Q_i^*) - c(x_i^*)]\, Q_i^* - x_i^* \geqslant [p(\sum_{j \neq i} Q_j^* + Q) - c(x_i)]Q_i - x_i, \quad \text{for all} \quad x_i, Q_i \geqslant 0, \qquad (12)$$

and

$$\left[p\left(\sum_{i=1}^{n^*} Q_i^* + Q\right) - c(x)\right]Q - x \leqslant 0 \quad \text{for all} \quad x, Q \geqslant 0. \qquad (13)$$

Condition (12) is clear enough. It says that (Q_i^*, x_i^*) is profit maximising for firm i ($i = 1, \ldots, n^*$) if it assumes that each of the remaining firms in the industry, j ($j \neq i$), chooses output level Q_j^*, and that those firms outside the industry remain outside. Condition (13) must be satisfied if these are no barriers to entry. It says that there is no way for a firm not in the industry to enter and make positive profits if the ith firm in the industry ($i = 1, \ldots, n^*$) chooses to produce at the level Q_i^*. We have supposed, therefore, that each firm, whether in the industry or not, entertains Cournot conjectures regarding all other firms.

We are here concerned not only with the question of whether a free-entry market equilibrium exists but also with the characteristics of such equilibria in those circumstances where they do exist. This latter task is greatly simplified if we restrict our attention to *symmetric* equilibria; i.e. equilibria in which firms in the industry behave identically. In Appendix 1, we shall locate circumstances in which symmetric equilibria can be shown to exist. For the moment we suppose that such equilibria exist and we proceed to characterise them.

[1] Since the product is new, $c_i(0) = \infty$, by assumption. We leave it to the reader to analyse the case where the existing competitive product price is c and firms spend resources to cut costs (i.e. $c_i(0) = c$). An alternative interpretation of this model would be to suppose that preferences are defined over the *characteristics* of commodities, and that firms in the industry compete to produce *different* commodities but which possess the *same* characteristic. Hence the commodities are perfect substitutes in consumption.

Let n^* be the equilibrium number of firms in the industry (to be determined, of course). Let $\hat{Q}_i$ denote the total output of all firms other than the ith. Thus $Q = Q_i + \hat{Q}_i$. Firm i chooses x_i and Q_i with a view to

$$\text{maximising } \{[p(Q_i+\hat{Q}_i) - c(x_i)]Q_i - x_i\}. \tag{14}$$

On the assumption that profit maximising Q_i and x_i are both positive, the first-order conditions are

$$p(Q)[1-\epsilon(Q)Q_i/Q] = c(x_i) \tag{15}$$

and

$$-Q_i\,c'(x_i) = 1, \tag{16}$$

where $\epsilon(Q) \equiv -Qp'(Q)/p(Q)$, is the inverse of the elasticity of demand. Since we are exploring symmetric equilibria, Q_i and x_i must be independent of i for $i = 1, \ldots, n^*$. Therefore, if n^*, Q^*, x^* characterise a symmetric equilibrium with free entry they must satisfy conditions (15) and (16), which reduce to

$$p(Q^*)[1-\epsilon(Q^*)/n^*] = c(x^*) \tag{17}$$

and

$$-c'(x^*)\,Q^*/n^* = 1. \tag{18}$$

(Note that Q^* denotes total output in equilibrium and therefore Q^*/n^* is output per firm in the industry.)

Moreover, the free entry condition (13) reduces to

$$[p(Q^*+Q) - c(x)]Q - x \leqslant 0 \quad \text{for all} \quad x, Q \geqslant 0. \tag{19}$$

Finally, note that since $(x^*, Q^*/n^*)$ is the profit maximising pair of choices for the representative firm in the industry in equilibrium, it must yield non-negative profits. Thus

$$[p(Q^*) - c(x^*)]\,Q^* \geqslant n^*x^*. \tag{20}$$

It is clear that the analysis would be greatly eased if free-entry were to result in firms earning negligible profits. Suppose then that $\{[p(Q^*)-c(x^*)]Q^* - n^*x^*\}/n^*x^*$ is small enough to be neglected. We can then, instead of (20), work with the zero-profit condition.

$$[p(Q^*) - c(x^*)]Q^* = n^*x^*. \tag{21}$$

If (21) is satisfied, condition (19) is most certainly satisfied. But one can locate conditions in which (19) is satisfied even though (20) is a strict inequality. We shall explore these issues in detail in Appendix 1, where we shall also enquire into the circumstances in which (21) is a good approximation. In fact one expects that (21) is a reasonable approximation in those circumstances where n^* is 'large' and that one can identify parametric conditions for which the number of firms in equilibrium *is* 'large'. In Appendix 1 we shall note by way of an important class of cases that this is so, but that n^* need not *necessarily* be large in those circumstances in which the use of (21) as an equilibrium condition is justifiable. For the moment we assume that it is a good approximation. It follows that n^*, Q^*, x^* are obtained from equations (17), (18) and (21).

A glance at these conditions shows that a market equilibrium is not characterised by an optimal allocation of resources. For one thing, producers exercise a degree of monopoly power at a market equilibrium which sustains the level of R & D expenditure they incur. For another the private firm's marginal benefit of its R & D expenditure is calculated on the basis of its contribution to *its* scale of output (condition (18)) and not for the entire market (condition (5)).

We now proceed to analyse the market equilibrium conditions in detail. On using equations (17) and (21) we note that

$$1/n^* = Z^*/\epsilon(Q^*) \quad (\text{where } Z^* = n^*x^*/p(Q^*)Q^*), \tag{22}$$

(22) is the fundamental equation of this section. As we are analysing symmetric equilibria we cannot compare the degrees of concentration in different markets by indices such as the concentration ratio. For our model it would seem natural to regard $1/n^*$ as the index of the *degree of concentration.* Z^*, being the fraction of industry sales that is spent on R & D, is an obvious index for research intensity in the industry. Equation (22) says that if ϵ is constant they are proportional to each other. Thus, in a cross-section study of different industries with the same demand elasticity in equilibrium, but varying by way of the size of the market and R & D technologies they face, one would observe a linear relationship between research intensity and concentration. *But there is no causality to be imputed to this relationship: industrial concentration and research intensity are simultaneously determined.*[1]

It will have been noted that in arriving at equation (22) no use was made of equation (18). Thus (22) obtains under more general circumstances than we have allowed, in that it does not depend on firms choosing their R & D strategies with a view to profit maximisation. We therefore proceed to obtain a somewhat sharper characterisation. Let $\alpha(x) \equiv -xc'(x)/c(x)$ denote the elasticity of unit cost of production with respect to R & D expenditure. It then follows from equations (18) and (22) that

$$Z^* \equiv n^*x^*/p(Q^*)Q^* = \alpha(x^*)/[1+\alpha(x^*)]. \tag{23}$$

Therefore, in a cross-section study of industries facing different demand conditions but the same elasticity of the unit cost function (α) in equilibrium, one would observe that the index of research intensity, Z^*, is the same. But these industries would be characterised by different degrees of concentration; for on using (23) in (22) one obtains the equilibrium number of firms as

$$n^* = \epsilon(Q^*)\,[1+\alpha(x^*)]/\alpha(x^*). \tag{24}$$

Therefore, the greater is the elasticity of demand $(1/\epsilon(Q^*))$ the smaller is the number of firms that one will observe in such a cross-section study.

[1] The relationship between the degree of concentration and R & D expenditure established here is not inconsistent with the empirical findings mentioned in footnote 3, p. 266 above, for note that equation (22) has been obtained on the assumption that (21) is a good approximation. In Appendix 1 we shall note that (21) is indeed a good approximation for certain ranges of the parameters underlying the model and, in particular, for those ranges for which n^* is large (i.e. the degree of concentration is small). To get a feel for orders of magnitude it may be noted that for firms in the United States in 1961 employing more than 5,000 persons, Z was on average 5·2 %, for those employing 1,000–5,000 persons it was 2·2 % and for those employing less than 1,000 persons it was 2 % (see Nelson *et al.* (1967), p. 67).

Notice, however, that equations (22) and (24) tell us that industries with smaller demand elasticities will, with any given number of firms, be characterised by a higher value of the index of research intensity. The lower demand elasticity leads, with a given degree of concentration, to higher markups; equilibrium is maintained not by entry, but by firms spending enough on R & D to forestall entry (essentially by spending all of their profits on R & D).

These foregoing propositions have an unmistakable Schumpeterian flavour to them. Since R & D involves fixed costs we cannot expect an industry that engages in it to be characterised by perfect competition. Nevertheless, *effective* competition is maintained by firms entering the market. Restrictive practices in our model, such as market price exceeding the unit cost of production (equation (17)), must be understood in the context of what Schumpeter called a 'perennial gale of creative destruction'. Our analysis has shown that the number of firms in an industry is no measure of the extent of this effective competition. Nor does the size of the market directly influence the number of firms in an industry for, as equation (24) makes clear, the equilibrium number of firms depends solely on the elasticities of the demand and innovation functions. If these elasticities are constant, as in (7) and (9), the equilibrium number of firms can be calculated directly, and is independent of the size of the market. However, the size of the market does influence the extent to which process innovation occurs in a market economy. To see this sharply suppose that market demand and innovation functions satisfy (7) and (9). If we now use these functional forms in the equilibrium conditions (17), (18) and (21), routine calculations yield their solution as:

$$n^* = \epsilon(1+\alpha)/\alpha, \tag{25}$$

$$Q^* = \frac{\epsilon(1+\alpha)}{\alpha^2\beta}\left[\sigma\alpha^{2\epsilon}\beta^{\epsilon-1}\epsilon^{-\epsilon}(1+\alpha)^{-(1+\epsilon)}\right]^{(1+\alpha)/[\epsilon-\alpha(1-\epsilon)]} \tag{26}$$

and

$$x^* = \left[\sigma\alpha^{2\epsilon}\beta^{\epsilon-1}\epsilon^{-\epsilon}(1+\alpha)^{-(1+\epsilon)}\right]^{1/[\epsilon-\alpha(1-\epsilon)]}. \tag{27}$$

Now it will be recalled that (17) and (18) are merely the first-order conditions for each firm's profit maximising exercise. In Appendix 1 we shall note that for (25)–(27) to represent a market equilibrium we shall need to suppose that $\epsilon > \alpha(1-\epsilon)$. Otherwise an equilibrium does not exist. Assume then that $\epsilon > \alpha(1-\epsilon)$. (This, as we noted earlier, is trivially true if market demand is inelastic (i.e. $\epsilon > 1$).) It is then immediate from (27) that the greater is the size of the market (i.e. the greater is σ), the greater is R & D expenditure per firm and, therefore, the greater is (unit) cost reduction. And so is industry output in equilibrium the greater.[1] Likewise, the costlier is R & D technology (i.e. the greater is β), the smaller is industry output. This is precisely what intuition suggests. However, note that the costlier is R & D technology the *smaller* is equilibrium R & D expenditure per firm (and hence industry-wide R & D expenditure) if demand is *elastic*; but it is *greater* if demand is *inelastic*. It is this last which is not intuitively immediate.

[1] The late Jacob Schmookler, in a series of writings, stressed the importance of the growth in demand for a product in stimulating R & D activity designed towards cost reduction and quality improvement. See, for example Schmookler (1962).

It will be recalled that equations (25)–(27) have been arrived at on the supposition that firms in equilibrium earn negligible profits. It is then clear that if $\epsilon(1+\alpha)/\alpha$ is an integer, a free-entry symmetric equilibrium exists, where firms in equilibrium earn precisely zero profits and where the number of firms in equilibrium is given by (25).[1] Now, while the number of firms must be an integer, $\epsilon(1+\alpha)/\alpha$ will not be, except by fluke. However, if ϵ/α is 'large', the largest integer less than $\epsilon(1+\alpha)/\alpha$ is 'large'. Then condition (25) suggests that if we set $n^* = [\epsilon(1+\alpha)/\alpha]$ (i.e. the largest integer less than $\epsilon(1+\alpha)\alpha$), then such a number of firms can sustain an equilibrium, in the sense that if each chooses the pair $(x^*, Q^*/n^*)$ which satisfies conditions (26) and (27), each will have maximised its profits given the choice of others, and its maximised profit level will be so low that condition (19) will also be satisfied, thus deterring further entry into the industry. In Appendix 1, we shall note that this is in fact the case. By construction we shall note that a free-entry symmetric equilibrium exists if ϵ/α is 'large', and that by choosing either α to be sufficiently small or ϵ to be sufficiently large we can force the equilibrium level of profit per firm to be as small as we care to make, so as to allow the zero profit condition (21) to be as good an approximation as we want.[2] However, it should be noted that the two limiting values of α and ϵ lead to different characteristics of industry equilibrium. To see this use equation (24) in equation (17) to obtain

$$p(Q^*)/c(x^*) = 1+\alpha(x^*). \tag{28}$$

Assume that α and ϵ are both constants. *Notice now that $\epsilon/n^* \to 0$ as $\alpha \to 0$, and from* (28) *we conclude that $p(Q^*) \to c(x^*) = \beta$ as $\alpha \to 0$. In the limit, as $\alpha \to 0$, the present model reduces to the conventional model of a perfectly competitive industry.* However, $\epsilon/n^* \to \alpha/(1+\alpha)$ as $\epsilon \to \infty$ and in particular, equation (28) implies that *if industry demand is highly inelastic market equilibrium sustains a great many firms. Nevertheless the ratio of market price to marginal cost of production can be much greater than unity.*

Earlier we noted that it was natural to regard n^{-1} as the index of industrial concentration for the model at hand. Indeed, unguided intuition might suggest that if the equilibrium number of firms is large the industry resembles the conventional competitive model. The foregoing result suggests that this is wrong. In fact several authors (e.g. Kalecki (1954)) have used instead the ratio of product price to 'prime cost' (i.e. $p(Q)/c(x)$ here) as a measure of the *degree of monopoly* in an industry. Now, a glance at equation (28) tells us that an industry could at the same time be characterised by a low degree of concentration (large n^*) and a high degree of monopoly (large $p(Q^*)/c(x^*)$). This would be so if both ϵ and α are 'large', so that $n^* \equiv [\epsilon(1+\alpha)/\alpha] \simeq \epsilon$, and therefore, n^* is 'large'. Indeed, presently we shall note that for the model at hand the degree of monopoly is a much better index of market imperfection than the degree of concentration.

Now, in a cross-section study of industries differing from one another in terms of the elasticity of the unit cost function, we would note that those facing a

[1] Since we have already supposed that $\epsilon > \alpha(1-\epsilon)$, it follows that $\epsilon(1+\alpha)/\alpha > 1$.

[2] For equation (17) to make sense n^* must exceed $\epsilon(Q^*)$. But if ϵ and α are both constants, then if either α is 'small' or ϵ is 'large', $n^* = [\epsilon(1+\alpha)/\alpha] > \epsilon$. Here, and in what follows, bold square brackets round a number will denote the largest integer not exceeding the number.

larger elasticity are characterised by a greater degree of monopoly. If these industries face the same demand elasticity, equation (17) tells us that those characterised by a greater degree of monopoly contain a smaller number of firms. But then equation (22) tells us that these in turn are associated with a higher index of research intensity, Z^*. The question arises whether R & D expenditure per firm is higher. This in general is hard to tell. However, suppose α and ϵ are both constant, so that equilibrium is characterised by (25)–(27). It should now be noted from (27) that if α is 'small' and demand is inelastic (i.e. $\epsilon > 1$), $\partial x^*/\partial\alpha > 0$, so that the greater is α the greater is R & D expenditure per firm in oligopoly equilibrium. But this in turn means that cost reduction is greater in such industries. *Somewhat paradoxically, then, we would observe greater cost reduction in those industries that are characterised by a higher degree of monopoly power.*

It remains for us to compare the performance of the free entry oligopoly equilibrium with that of the socially managed industry. Continue to assume that $p(Q)$ and $c(x)$ satisfy (7) and (9) respectively. We then note from (11) and (26) that $Q_s > Q^*$. Market equilibrium output is less than the socially optimal output, a result which is not immediately obvious since one may have thought that the pressure of competition (free-entry) would drive the market price down to a level below what is socially optimal, by forcing each firm to invest more in R & D than is desirable for society. The question then arises whether cost reduction is greater or less in the oligopoly market than in the socially optimal one. On comparing (10) and (27) on notes that $x^* \gtrless x_s$ as $(n^*)^{-\epsilon} \gtrless (1+\alpha)$. It follows that $x^* < x_s$, and hence that there is insufficient cost reduction in the oligopoly industry. But industry-wide R & D expenditure in the market economy is n^*x^* and the question arises how this compares with x_s. On using (10), (25) and (27) it is now easy to confirm that $n^*x^* \gtrless x_s$ as $(n^*)^{\alpha(\epsilon-1)} \gtrless (1+\alpha)$. This implies that if ϵ is 'large' (i.e. demand is highly inelastic), then $n^*x^* > x_s$ and, therefore, that total R & D expenditure in the market economy exceeds the socially optimal level. *Thus the market economy may be characterised both by excessive expenditure on R & D* (n^*x^*) *and too low a rate of technical progress* (x^*) *as compared to the socially managed one.* The point of course is that the market encourages too much duplication here. To see this sharply we may note from (10) that $x_s \to (\alpha\beta)^{1/(1+\alpha)}$ as $\epsilon \to \infty$. However, from equation (27) it is immediate that $x^* \to 0$ as $\epsilon \to \infty$, but $n^*x^* \to \infty$. *Thus welfare loss due to competition is unbounded as* $\epsilon \to \infty$, *even though each firm serves an infinitesimal fraction of a finite demand as* $\epsilon \to \infty$.

Matters are quite different though for the other limit economy. On comparing equations (10) and (27) it is immediate that $x^* \to 0$ and $x_s \to 0$ as $\alpha \to 0$. Note as well that as $\alpha \to 0$ we have $n^*x^* \to 0$, even though $n^* \to \infty$. Therefore, in the limit, as $\alpha \to 0$, welfare loss due to the industry being privately managed is nil. As we noted earlier, the industry in this case is the conventional competitive one, characterised by a lack of distortion. Welfare loss due to competition in this limit economy is nil.

(3) *Oligopoly with Barriers to Entry*

It has been our purpose so far to explore the implications of an endogenous market structure on the amount of innovative activity. We have captured this in our model via the free-entry condition (13). Entry sustains effective competition. This, as we noted in the introduction, implies that the pace of innovative activity must be traced neither to the degree of concentration, nor to the degree of monopoly in the industry in question, but to more basic ingredients such as demand conditions, R & D technology, the nature of the capital market, etc. Nevertheless, it is the case that several authors have claimed a slowing down in the intensity of innovations in advanced capitalist economies, and have traced it to an increasingly monopolistic character of capitalism.[1]

Now, we have already noted on several occasions that the degree of monopoly, $p(Q)/c(x)$, is not an explanatory variable. Nor, if there is free-entry, is the number of active firms an explanatory variable. Indeed, we have noted that with free entry into the industry it can readily happen that industries characterised by a greater degree of monopoly are *more* innovative, *not less*. It is then worthwhile inquiring into the relationship between the degree of monopoly and the reduction in the unit cost of production in an environment where there are *barriers to entry*. We can then see whether allowing new firms to enter would result in greater cost reduction in equilibrium. Thus let the number of firms be exogenously given, say n. Formally, $[(Q_1^*, x_1^*), \ldots, (Q_i^*, x_i^*), \ldots, (Q_n^*, x_n^*)]$ is an equilibrium if for $i = 1, \ldots, n$

$$\left.\begin{aligned} &[p(\sum_{j\neq i} Q_j^* + Q_i^*) - c(x_i^*)]\, Q_i^* - x_i^* \\ &\geqslant [p(\sum_{j\neq i} Q_j^* + Q_i) - c(x_i)]\, Q_i - x_i \quad \text{for all} \quad x_i, Q_i \geqslant 0. \end{aligned}\right\} \tag{29}$$

Since we are interested only in symmetric equilibria, the foregoing definition of an equilibrium implies that conditions (17) and (18) must be satisfied, which we re-write here as

$$p(Q^*)[1 - \epsilon(Q^*)/n] = c(x^*) \tag{30}$$

and

$$-c'(x^*)\, Q^*/n = 1. \tag{31}$$

Since n is given, there are only two unknowns, Q^* and x^*, to be determined from (30) and (31). Suppose, to get explicit solutions, that $p(Q)$ and $c(x)$ satisfy (7) and (9) respectively. From (30) it is clear that since n is exogenously given in the present analysis, we must assume $n > \epsilon$ (otherwise (30) and (31) will not possess a solution). It must also be supposed that $n \leqslant \epsilon(1 + \alpha)/\alpha$, since otherwise the solution of equation (30) and (31) will result in firms making negative profits and so they would not represent equilibrium conditions. Finally, we shall need to suppose that $\epsilon > \alpha(1 - \epsilon)$, a condition we required earlier (for details see Appendix 1). Routine calculations now enable one to solve equations (30) and (31), and they yield

$$x^* = [\sigma(\alpha/n)^{\epsilon}\, \beta^{\epsilon-1}(1 - \epsilon/n)]^{1/[\epsilon - \alpha(1-\epsilon)]} \tag{32}$$

and

$$Q^* = (n/\alpha\beta)\, [\sigma(\alpha/n)^{\epsilon}\, \beta^{\epsilon-1}(1 - \epsilon/n)]^{(1+\alpha)/[\epsilon - \alpha(1-\epsilon)]}. \tag{33}$$

[1] 'Another (reason for the slowing down of the growth of advanced capitalist economies) is the hampering of application of new inventions which results from the increasingly monopolistic character of capitalism' (Kalecki (1954), p. 159).

It will be noticed immediately from equation (33) that in the admissible range $(\epsilon, \epsilon(1+\alpha)/\alpha]$, Q^* is an increasing function of n. *Industry output therefore increases with the number of firms.* Consequently product price decreases. This is the advantage of greater competition. Moreover, from equation (30) it is clear that the degree of monopoly declines as the number of firms in the industry increases. The question that we began with is whether an increase in the number of firms results in greater innovation (i.e. greater cost reduction). From equation (32) it is immediate that the answer is 'no'. *If the number of firms is increased, each firm in equilibrium spends less on R & D, and so unit cost of production in equilibrium is higher.* However, it is easy to confirm from equation (32) that total R & D expenditure, nx^*, increases with the number of firms. The point is that while the industry spends more on R & D as a consequence of increased competition, each firm spends less. The extra expenditure is essentially wasted in duplication. In fact a comparison of equations (10) and (32) shows that $x^* < x_s$. We conclude that for the model at hand (unit) cost reduction is insufficient in a market economy whether or not there are barriers to entry; and consequently, market price for the product is higher than is socially desirable.

This is brought out most forcefully if we consider a monopolist protected by entry barriers. In this case there is, of course, no duplication in R & D. Nevertheless, as we have seen, the monopolist engages in less R & D activity than is socially desirable. This remains true even if the government in a socially managed industry is forced to raise its revenue for R & D expenditure through benefit taxation.

However, the *speed* with which firms carry out their R & D work and, consequently, the *rapidity* with which technological innovations take place in a market economy would appear to be greater if there is free-entry into R & D activity, than if active firms were protected by entry barriers into the research sector. In Section IV we shall study the implication of free-entry into the research sector on the speed with which R & D is undertaken. For a more complete discussion of the issues see Dasgupta and Stiglitz (1980).

III. NON-CONCAVITIES IN THE VALUE OF PROCESS INNOVATION AND THE DEGREE OF RISK-TAKING

It has already been noted that the indirect social utility function $V(c)$ in equation (3) is a declining convex function (see Fig. 4). We revert to the basic model of Section I and suppose that c is the unit cost of production associated with the existing best-practice technique. Our aim is to explore the implications of the convexity of $V(c)$ on the choice among risky research projects. To begin with, suppose there are two research strategies to choose from, one of which reduces the cost of production to c^* with certainty, whilst the other, if successful, reduces it to $\tilde{c}$ (with $\tilde{c} < c^* < c$). If unsuccessful, the cost of production remains at c. But suppose the *expected* cost reduction associated with the risky research project is $c - c^*$. If society's welfare criteria are derived with a view to maximising expected social utility then if the costs of the two research projects are the same, it would prefer the risky project to the riskless one. This is depicted in Fig. 4.[1] Let us now generalise this example.

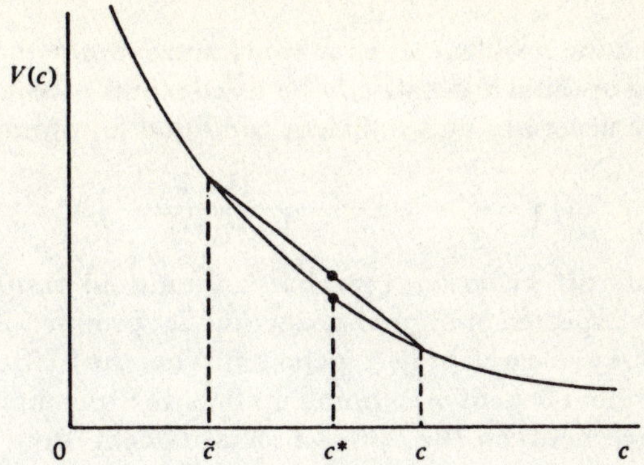

Fig. 4

Suppose that research projects are indexed by α ($\bar{\alpha} \geqslant \alpha \geqslant 0$). Project-$\alpha$ has probability $h(\alpha)$ of success. If successful, the unit cost of production reduces to $c(\alpha)$. If unsuccessful, the cost of production remains at c. The projects are so labelled that we suppose, without loss of generality, that $c'(\alpha) < 0$. For simplicity, continue to assume that this family of research projects has the same mean outcome; i.e.

$$\left.\begin{aligned} E[c(\alpha)] &= c^* = h(\alpha)\,c(\alpha) + [1-h(\alpha)]\,c, \\ \bar{\alpha} &\geqslant \alpha \geqslant 0 \quad \text{and} \quad h(0) = 1. \end{aligned}\right\} \tag{34}$$

(34) implies that $c(0) = c^*$ and hence $\alpha = 0$ is the riskless project. If we differentiate (34) we obtain

$$c'(\alpha)\,h(\alpha) = [c - c(\alpha)]\,h'(\alpha). \tag{35}$$

Consequently $h'(\alpha) < 0$. Therefore a higher value of α is associated with a more risky project.

R & D expenditure associated with project-α is $R(\alpha)$. Quite obviously $R(\alpha) > 0$, $\alpha \geqslant 0$. If $R'(\alpha) \leqslant 0$, nothing of interest remains to be said. Since $\bar{\alpha}$ is the riskiest project available (the highest value of α in the given family of available research projects) the planner ought to choose $\bar{\alpha}$ (or undertake no research at all). Consequently we consider the other special case; where $R'(\alpha) > 0$. For obvious reasons, suppose in addition that $R''(\alpha) > 0$.

In what follows, we take it that $R(0)$ is 'small', so that the optimum involves some R & D activity. Now if the planner were to choose project-α, expected net social benefits, $E(\alpha)$, would be

$$E(\alpha) \equiv h(\alpha)\,V[c(\alpha)] + [1-h(\alpha)]\,V(c) - R(\alpha). \tag{36}$$

[1] If R & D cost is too high society would, presumably, not wish to engage in any research at all, and continue to rely on the c-technology.

Notice that unless we impose further structure on the functions $h(.)$ and $V(.)$, one cannot ensure that $E(.)$ is concave. If it is not, then in general the government will need to conduct global cost-benefit analysis of the 'random' research projects.

Having indicated the problem let us simplify and suppose that (36) is strictly concave. Assuming optimal α lies strictly between 0 and $\bar{\alpha}$ we note that on using condition (35), the necessary and sufficient condition for optimality is:

$$-h(\alpha)\,c'(\alpha)\left(-V'[c(\alpha)]-\frac{\{V[c(\alpha)]-V(c)\}}{[c-c(\alpha)]}\right)=R'(\alpha). \qquad (37)$$

The social benefit–cost criterion (37) has an unusual simplicity about it. $-h(\alpha)\,c'(\alpha)$ is the expected marginal reduction in production cost – when a marginally more risky research project is chosen. What the LHS of condition (37) says is that the marginal benefit of stepping up R & D expenditure is $-h(\alpha)\,c'(\alpha)$ times the difference between the *marginal* social benefit from cost reduction $\{-V'[c(\alpha)]\}$ and the *average* social benefit from cost reduction, $\{V[c(\alpha)]-V(c)\}/[c-c(\alpha)]$.

To look at the matter another way, we know in advance that

$$-V'[c(\alpha)] > \{V[c(\alpha)]-V(c)\}/[c-c(\alpha)]$$

(since $V(c)$ is convex and declining). What (37) says is that at an optimum the difference between marginal and average returns ought to be $-R'(\alpha)/h(\alpha)\,c'(\alpha)$.

We have established that a social planner will wish to engage in risky research projects. But then so will a pure monopolist. To see this suppose that marginal revenue from sales is a declining function of sales. Then it is simple to confirm that the monopolist's maximised profit from sales is a convex and declining function of his unit cost of production. Consequently, if the monopolist is concerned with maximising expected profits, an argument identical to the one we have presented above comes into play. The monopolist too favours risk.

The natural question to ask is whether the monopolist is inclined to engage in the optimum degree of risk and, if not, whether one can establish a bias in his behaviour towards risk. In Appendix 2, we demonstrate that if the family of research projects is restricted to the class we have discussed above, the monopolist undertakes insufficient risk in his R & D effort and invests too little in research and development. But then clearly this would be true even if the family of available research projects were characterised by a slight lowering of expected cost reduction, $(c-c^*)$, with increasing risk (i.e. increase in α). In this case one would observe greater cost reduction *on average* if the industry is controlled by a private monopolist, than if it were socially controlled. The point about this observation is that while cost reduction is a 'good thing' the monopolist, paradoxically, would on average be reducing costs too much, even though his R & D expenditure would be less than that which is socially desirable.

These pure economic environments are the easiest to analyse. With free entry the analysis appears to be unusually complex. But in an oligopolistic environment the market power of a producer increases as his cost advantage over his rivals increases. There is then an *a-priori* presumption that competition may encourage excessive risk-taking on the part of producers engaged in process innovation. We have been unable to settle the issue when there is free entry into an industry.

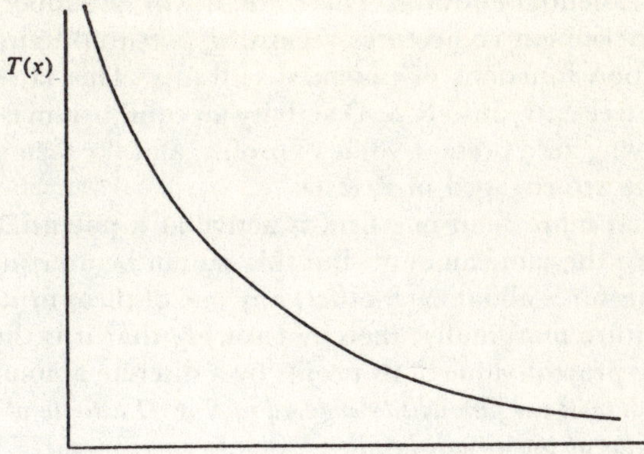

Fig. 5

IV. MARKET STRUCTURE AND THE SPEED OF RESEARCH

We are here concerned with the speed of research, and we continue to illustrate matters in the context of process innovation. Revert to the basic model of Section I and suppose c to be the unit cost of production associated with the existing best practice technique. As in Sections I and II, we simplify and hypothesise highly goal orientated research. Research now is taken to consist of solving a sequence of problems that will enable the commodity to be produced at a unit cost c^* $(c^* < c)$. There is no uncertainty.[1] If a research unit invests x at $t = 0$ it solves the entire set of problems at date $T(x)$, where $T'(x) < 0$ and $T''(x) > 0$ with $T(x) \to 0$ as $x \to \infty$ and $T(x) \to \infty$ at $x \to 0$ (see Fig. 5).[2]

Consider first competitive conditions. We suppose that currently the market price for the commodity is c. It is understood that the first firm to make the invention is awarded a patent of fixed duration. If there are several winners they share the patent. We need not specify whether in this latter case the winners propose to collude by forming a cartel, or whether they play a Cournot game, as in Section II. All that we need to postulate is that each winner enjoys a positive flow of profits while the patent lasts. It follows that firms may well have an incentive to engage in R & D activity. We take it that each firm knows the R & D strategy of the other firms.

Firms are assumed to be able to borrow freely to finance their R & D activity

[1] Models incorporating uncertainty in the date of invention have been discussed in Stiglitz (1970), Dasgupta, Heal and Majumdar (1977), Kamien and Schwartz (1978), Loury (1977) and Dasgupta and Stiglitz (1978).

[2] I.e. all potential research units are of identical ability. This assumption is made solely for expositional ease, and can be relaxed in the obvious manner.

at a competitive rate of interest r (> 0). Since by hypothesis $T'(x) < 0$, each firm engaged in R & D will establish only one research unit. In order to explore game equilibria with free entry we shall, as earlier, suppose that firms entertain Cournot conjectures regarding those that are engaged in R & D activity. However, we shall wish to capture the fact that firms engaged in R & D fear the threat of entry by potential entrants. Therefore, it will be supposed that active firms entertain Stackelberg conjectures regarding potential entrants; i.e. they work on the reaction functions of potential entrants. This latter assumption implies that under free entry into R & D activity an equilibrium is characterised by each firm receiving zero present-value of profits, and the former implies that only future winners are engaged in R & D.

Now suppose that more than one firm is active at a potential equilibrium. They all then invest the same amount. But this cannot be an equilibrium. For, with Cournot conjectures about each other, any one of these firms can increase its R & D expenditure marginally, thereby ensuring that it is the sole winner, and so increase the present value of its profits by a discrete amount. *We conclude that with free entry, at most one firm will be engaged in R & D activity at an equilibrium, and its net present value of profits will be nil.*

What this result is telling one is that the fact that a single firm is engaged in R & D activity is not in itself an evidence that there is little competition in this field. For the model at hand competition is intense. This single firm raises its R & D expenditure and so speeds up its research to a level high enough to forestall entry.

It remains to calculate the equilibrium level of expenditure. Let T^* (> 0) be the length of the patent. We now recall the notation introduced in Section I. If x^* is the equilibrium level of investment undertaken by the single firm, it satisfies the zero-profit condition

$$(\pi_c/r)\,(1 - e^{-rT^*})\,e^{-rT(x)} = x. \tag{38}$$

The date of invention is $T(x^*)$. One notes first that $x = 0$ is a solution of equation (38). One notes as well that the LHS of equation (38) is not necessarily a concave function of x. In Fig. 6 we have drawn both the LHS and the RHS of equation (38). As the figure makes clear, the largest solution of (38) is the equilibrium level of R & D expenditure, x^*. An equilibrium therefore exists and it is unique. A glance at Fig. 6 also shows that in a cross-section study of industries with the same R & D technology those characterised by greater demand (i.e. greater π_c) will sustain a greater speed of research.[1]

We now compare the competitive outcome with the socially optimal speed of research. We take it that r is regarded as appropriate to use for discounting

[1] To obtain a feel for orders of magnitude it may be noted that in large corporations R & D projects are often expected to be completed in 4 to 5 years and that many such projects involve not undue risk, in many cases the estimated probability of success exceeding 0·8 (see Mansfield (1967)). Notable examples of risky R & D projects would appear to be in the pharmaceutical industry (see Schwartzman (1977)). For a theoretical exploration of the relationship between market structure, risk-taking in R & D, and the speed of research, see Dasgupta and Stiglitz (1977, 1980).

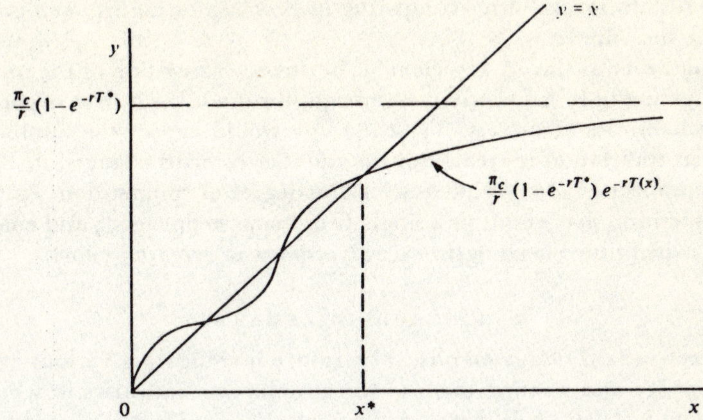

Fig. 6

social benefits and costs. The social planner is then concerned with choosing x so as to maximise

$$\left[\frac{\pi_s}{r}e^{-rT(x)} - x\right]. \tag{39}$$

Notice again that (39) is not in general a concave function of x. However, given the assumptions that we have made about $T(x)$ it is immediate that (39) has a maximum. Let x_s be the optimum R & D expenditure. Then provided it is positive it satisfies the cost-benefit rule

$$-T'(x)\,\pi_s e^{-rT(x)} = 1. \tag{40}$$

It will be recalled that $\pi_s > \pi_c$ (inequality (2)). But a glance at (38) and (40) tells us that one can easily have $x^* > x_s$. When this is so the single firm engaged in R & D activity is forced to incur excessive R & D expenditure because of the *threat* of competition. Market forces encourage too high a speed of research.

The central result of this Section, that under free-entry, a Cournot–Nash equilibrium for the model considered here sustains at most one firm, survives if we introduce uncertainty in the special form where all firms are obliged to follow the *same* steps in solving the problems that are necessary to solve in order to make the c^*-technology viable (i.e. if all research units face the same decision tree). This means that given the pace of R & D activity of the remaining firms a given firm can guarantee that it is the first to invent by choosing a sufficiently high pace of research, even though it is still unable to say at which date the sequence of tasks will be completed by its research unit. With this form of uncertainty it is immediate that at most one firm will incur R & D expenditure at an equilibrium.

It will be granted that the foregoing postulates undue interdependence of research strategies amongst firms. In the sequel to this paper (Dasgupta and Stiglitz, 1978) we move to the opposite extreme and suppose complete independence. In such a case it is clear enough in advance that an equilibrium may well sustain several firms competing in R & D; for no firm can guarantee itself to be the winner.

The point then is this. A key element in the determination of the number of firms competing in R & D activity at an equilibrium is the degree of correlation in the probabilities of success. Typically one would expect the number to be small when correlation is great. But the number of firms engaged in R & D is not an appropriate index for measuring the degree of competition. As we have seen, competition may result in a single firm being so engaged; and engaged in excessive expenditure, leading to excessive speed in research effort.

V. CONCLUDING REMARKS

There is now a fairly large empirical literature investigating various aspects of R & D activity and relating them to the structure of industries in which such activity is undertaken. It is therefore particularly surprising to note the paucity of theoretical explorations in this area. Moreover, so far as we are aware the majority of such explorations have been directed at the behaviour of a single firm engaged in R & D in the midst of an exogenously given environment. In order to make contact with some of the empirical findings and to obtain a less incomplete understanding of these matters it is necessary to move beyond the analysis of a single firm and to consider a set of interacting firms and, if one is interested in the long run as well, to postulate an endogenous market structure. What results is a game environment, and in this paper we have attempted to come to grips with a few of the traditional issues in the theory of industrial organisation in the context of a set of simple constructions. In investigating R & D activity in a market economy we have throughout made use of the concept of a game equilibrium. It is, of course, well known that both the existence and characteristics of such equilibria depend on what types of action agents are postulated to choose and on the conjectures that are entertained by agents regarding the responses by other agents to their choice. In this paper we have supposed that firms choose quantities (viz. output and R & D expenditure levels). As regards conjectures the simplest by far to analyse are Cournot ones, namely that other agents do not respond to the given agent's choice. We have, for the most part, supposed such conjectures. However, while plausible in certain circumstances (e.g. in the model of Section II), such conjectures are not plausible in many others. For example, the reader can readily check that an equilibrium does not exist in the model of Section IV, if all firms entertain Cournot conjectures about *all* other firms. We have consequently resorted to a set of hybrid conjectures on the part of firms in the analysis of Section IV; conjectures that are *a-priori*, not implausible. There is a clear need, though, of a study of the characteristics of oligopoly equilibria under more sophisticated conjectures on the part of firms and to see how these characteristics vary as conjectures are varied.[1]

One of the most oft discussed issues in the industrial organisation literature

[1] For an important analysis of oligopoly equilibrium with barriers to entry, where the focus of attention is on alternative specifications of the conjectures entertained by firms, see Marschak and Selton (1974).

has been the relationship between the size of firms and the pace of R & D activity. We have discussed some aspects of this issue in our introduction and in Section II. Now if firms are forced to finance their R & D expenditure from internal funds there is a clear presumption that industrial concentration is positively correlated with R & D activity. The model of Section II, being timeless, is consistent with the internal finance hypothesis. However, in Section IV, we supposed a perfect capital market with a view to demonstrating that one does not need to resort to the internal finance hypothesis if one wants to argue that only a limited number of firms will typically be engaged in competing R & D activities. Nor in fact, as we have shown, does one need to postulate non-classical goals on the part of firms in order to make contact with some of the empirical findings. We have throughout formalised the market economy as one in which firms are engaged in games with complete information. In such contexts it hardly makes sense to postulate anything other than profits as a firm's goals. Matters are different if incomplete information is postulated for firms; and a particularly interesting avenue that has recently been explored (see, e.g. Winter (1971), and Nelson, Winter and Schuette (1976)) consists in supposing in addition that agents have only limited ability to solve complicated maximisation problems. Endowed only with 'bounded rationality', firms are postulated to follow some 'satisficing' course of action, and to search locally for improvements if existing courses of action cease to produce satisfactory results.

Quite apart from any details that may be found interesting, the central conclusions of our analysis would appear to be:

(1) Even when one regards industrial structure to be endogenous, theoretical considerations are consistent with the empirical finding that when the degree of concentration in industries is small, industry-wide R & D effort is positively correlated with concentration (Section II).

(2) High degrees of concentration are by themselves not an evidence of lack of effective competition (Section II).

(3) When the degree of concentration in industries with free-entry is small, R & D effort per firm (and therefore cost reduction) is often positively correlated with concentration. Moreover, the degree of concentration is positively correlated with the degree of monopoly (Section II).

(4) Both optimal R & D expenditure and R & D expenditure per firm in a market economy increase with the size of the market. They decrease with increasing costs associated with R & D technology if demand is elastic and increase with increasing costs if demand is inelastic (Section II).

(5) If there are barriers to entry an increase in the number of firms would result in a decrease in R & D expenditure per firm in a market economy, although industry output would increase, and therefore, the degree of monopoly would decrease (Section II).

(6) There is some presumption that cost-reducing in an industry in a market economy, even when there is free entry, is less than the socially optimal level (Section II). However,

(7) If demand is highly inelastic, *total* R & D expenditure in an industry with free entry exceeds the socially optimal level (Section II). (6) and (7) imply

(8) There may be excessive duplication of research effort in a market economy in the sense that industry-wide R & D expenditure exceeds the socially optimal level even though cost-reduction is lower. In particular, an industry may be characterised by a very low degree of concentration (i.e. a large number of firms) and at the same time engage in a great deal of social waste.

(9) In the case of process innovation a good case can be made for encouraging investment in risky research projects, even if society is risk-averse (Section III).

(10) A pure monopolist (i.e. one protected by entry barriers) appears to have insufficient incentive (a) to undertake R & D expenditure (Section II) and (b) to engage in risky research ventures (Section III and Appendix 2).

(11) Since the market power of a firm increases as its cost advantage over its rivals increases there is a presumption that competitive markets encourage firms to engage in overly risky research projects (Section III).

(12) If the first firm to succeed is awarded most of the reward for invention, then to the extent the risks that firms undertake are positively correlated, pressure of competition will ensure that only a few firms engage in R & D activity; in extreme cases at most one firm will be engaged in research (Section IV).

(13) But the observation that only a few firms are engaged in R & D is not in itself an evidence that a market economy sustains too little R & D activity (Section IV). In particular

(14) Pressure of competition may result in excessive speed in research (Section IV), and in general

(15) There is no presumption that a market economy has a tendency to generate insufficient information.

London School of Economics PARTHA DASGUPTA
Princeton University JOSEPH STIGLITZ

Date of receipt of final typescript: September 1979

APPENDIX I

We shall by construction locate conditions under which a free-entry symmetric equilibrium exists for the model of section II, when $p(Q)$ and $c(x)$ satisfy (7) and (9) in the text. In doing this we reverse the order followed in the text and suppose first that there are barriers to entry, so that n is exogenously given.

Consider the solution (32) and (33) of the two first-order conditions (30) and (31). For it to be real-valued one must suppose that $n > \epsilon$. Let us do so. It will have been noted that the pair $(x^*, Q^*/n)$ in (32) and (33) is the unique solution of the (interior) first-order conditions for the representative profit maximising firm when it assumes that each of the remaining firms chooses Q^*/n as its output level. We must now locate conditions under which the representative firm earns non-negative profits when it chooses $(x^*, Q^*/n)$.

Let $\pi(x^*, Q^*/n)$ denote the level of its profit. Then $\pi(x^*, Q^*/n) = [p(Q^*) - c(x^*)]\, Q^*/n - x^*$. For this to be non-negative it is necessary and sufficient that

$$[p(Q^*) - c(x^*)]\, Q^*/nx^* \geqslant 1. \qquad \text{(A 1)}$$

If we now appeal to the functional forms (7) and (9) and use (32) and (33), routine calculations yield

$$[p(Q^*)-c(x^*)]\, Q^*/nx^* = \epsilon/\alpha(n-\epsilon). \qquad \text{(A 2)}$$

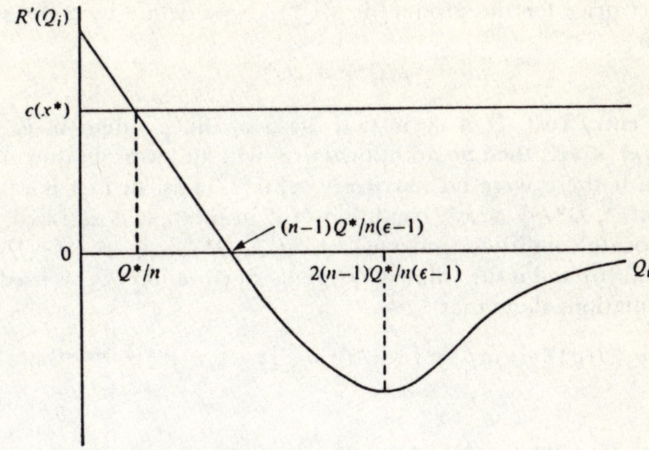

Fig. 7. Marginal revenue schedule, $R'(Q_i)$, of firm i, when gross revenue,

$$R(Q_i) = \sigma[Q_i+(n-1)\,Q^*/n]^{-\epsilon}Q_i,$$

where Q^* is equilibrium industry output, $\epsilon > 1$ and $n > \epsilon$. x^* is equilibrium R & D expenditure per firm.

It will be noted that for $n > \epsilon$, the RHS of (A 2) is a decreasing function of n and is equal to unity at $n = \epsilon(1+\alpha)/\alpha$. Combining (A 1) and (A 2) we therefore conclude that

$$\pi(x^*, Q^*/n) \geqslant 0 \quad \text{if } \epsilon < n \leqslant \epsilon(1+\alpha)/\alpha. \qquad \text{(A 3)}$$

(A 3) therefore instructs us to suppose that $\epsilon < n \leqslant \epsilon(1+\alpha)/\alpha$. Let us do so.

It remains to find conditions under which the profit function, $\{p[(n-1)Q^*/n+Q_i]-c(x_i)\}\, Q_i - x_i$, for firm i is concave in the neighbourhood of the point $x_i = x^*$, $Q_i = Q^*/n$. In fact it can be readily checked that if $p(Q)$ and $c(x)$ have the functional forms (7) and (9) and x^* and Q^*/n are given by (32) and (33), the profit function of the representative firm satisfies the second-order conditions at x^* and Q^*/n if $\epsilon > \alpha(1-\epsilon)$ (i.e. if $\epsilon(1+\alpha)/\alpha > 1$). Thus, $(x^*, Q^*/n)$ is the global profit maximising choice for the representative firm if each of the other firms chooses $(x^*, Q^*/n)$ (see Fig. 7). We have therefore proved:

Theorem 1. If $p(Q)$ and $c(x)$ satisfy (7) and (9) and if n is a positive integer, then a symmetric Cournot equilibrium amongst n firms exists if

(i) $\epsilon < n \leqslant \epsilon(1+\alpha)/\alpha$, and (ii) $\epsilon(1+\alpha)/\alpha > 1$.

Now clearly $\epsilon(1+\alpha)/\alpha > 1$ if $\epsilon > 1$. What we have therefore shown is that if $\epsilon < n \leqslant \epsilon(1+\alpha)/\alpha$, then a symmetric Cournot equilibrium with n firms exists even if market demand is throughout *inelastic*.

We turn now to locating conditions under which a *free-entry* symmetric Cournot equilibrium exists. Suppose there are n firms in the industry, and suppose for the moment barriers to entry. Then let $(x^*, Q^*/n)$ be the chosen course of action for the representative firm at a symmetric Cournot equilibrium. Then market price for the product is $p(Q^*)$. Now define by $\hat{x}$ the solution of the equation

$$p(Q^*) = c(x). \tag{A 4}$$

$\hat{x}$ is like an entry cost. It is clear that $\exists m > 0$, independent of n, such that *if* $\pi(x^*, Q^*/n) \leqslant m\hat{x}$, then no additional firm will find it profitable to enter the market even if there were no barriers to entry. Thus, in fact if n has such a value that $\pi(x^*, Q^*/n) \leqslant m\hat{x}$, condition (19) in the text is satisfied. We now proceed to locate conditions under which $\pi(x^*, Q^*/n) \leqslant m\hat{x}$. If $p(Q)$ and $c(x)$ satisfy (7) and (9) and if the value of Q^* given in equation (33) is used in (A 4), routine calculations show that

$$\hat{x} = (\beta/\sigma)^{1/\alpha}(n/\alpha\beta)^{\epsilon/\alpha}[\sigma(\alpha/n)^{\epsilon}\beta^{\epsilon-1}(1-\epsilon/n)]^{\epsilon(1+\alpha)/\alpha[\epsilon-\alpha(1-\epsilon)]}. \tag{A 5}$$

Moreover,

$$\begin{aligned}\pi(x^*, Q^*/n) &= [p(Q^*)-c(x^*)]Q^*/n-x^* \\ &= (n-\epsilon)^{-1}[\sigma(\alpha/n)^{\epsilon}\beta^{\epsilon-1}(1-\epsilon/n)]^{1/[\epsilon-\alpha(1-\epsilon)]}[\epsilon(1+\alpha)/\alpha-n]. \end{aligned} \tag{A 6}$$

Using (A 5) and (A 6) one notes that

$$\pi(x^*, Q^*/n) \leqslant \hat{x} \quad \text{if and only if} \quad \epsilon(1+\alpha)/\alpha-n \leqslant mn(1-\epsilon/n)^{(1+\alpha)/\alpha}. \tag{A 7}$$

As we are concerned with locating circumstances under which a free entry symmetric Cournot equilibrium *exists*, we are entitled to choose n. In what follows we shall always choose n to be the largest positive integer not exceeding $\epsilon(1+\alpha)/\alpha$. Call this n^*. Then $n^* = [\epsilon(1+\alpha)/\alpha]$. On appealing to Theorem 1, we may now note:

Theorem 2.[1] If $p(Q)$ and $c(x)$ satisfy (7) and (9) respectively, a free-entry symmetric Cournot equilibrium exists if (i) $\epsilon(1+\alpha)/\alpha > 1$, (ii) $n^* = [\epsilon(1+\alpha)/\alpha] > \epsilon$, and (iii) $\epsilon(1+\alpha)/\alpha - n^* \leqslant mn^*(1-\epsilon/n^*)^{(1+\alpha)/\alpha}$.

The foregoing theorem makes explicit the relationship between the parameters of the economy that are sufficient to guarantee the existence of a free-entry symmetric Cournot equilibrium. It should be noted in particular that the conditions in *Theorem 2* are satisfied if for any given value of ϵ, α is chosen small enough and also, for any given value of α, if ϵ is chosen large enough. That is, an equilibrium exists if ϵ/α is sufficiently large. This confirms the claims made in the text.

It remains to find conditions under which the use of the zero-profit condition (21) is a good approximation. In fact we have already located such conditions. What we are looking for are conditions under which $\pi(x^*, Q^*/n^*)/x^*$ is 'small'. (A 2) then tells us that for the specification given by (7) and (9) in the text

[1] Novshek (1977) has independently proved a result similar to this in a somewhat different context. We are grateful to Morton Kamien for drawing our attention to Novshek's work.

this is the same as finding conditions under which $\epsilon/\alpha(n^* - \epsilon) - 1$ is negligible. Since by construction $n^* = [\epsilon(1+\alpha)/\alpha]$, one can locate ranges for the pair (α, ϵ) for which the zero profit condition is a good approximation. In particular, one notes that it is a good approximation if ϵ/α is 'large'. In this case, n^* is. of course 'large'.

APPENDIX 2

In this appendix we compare the degree of risk-taking on the part of the pure monopolist with the socially optimum degree of risk-bearing. Towards this suppose the family of research projects is characterised by (34) and suppose $R(\alpha) > 0$ for $\alpha \geqslant 0$, $R'(\alpha) > 0$ and $R''(\alpha) > 0$. Let the market demand function be generated by the social utility function (6). Thus the demand function is

$$p(Q) = \sigma Q^{-\epsilon} \quad (1 > \epsilon > 0). \tag{A 8}$$

Let $V_m(c)$ denote the monopolist's maximum profit from sales when c is the unit cost of production. It is simple to confirm that

$$V_m(c) = \epsilon\sigma^{1/\epsilon}(1-\epsilon)^{(1-\epsilon)/\epsilon}c^{-(1-\epsilon)/\epsilon}. \tag{A 9}$$

If the monopolist selects project-α his net expected profit is

$$h(\alpha)V_m[c(\alpha)] + [1-h(\alpha)]V_m(c) - R(\alpha), \tag{A 10}$$

where $V_m(c)$ is given by (A 9). Write

$$A \equiv \epsilon^{1/\epsilon}(1-\epsilon)^{(1-\epsilon)/\epsilon}, \quad g(\alpha) \equiv \{h(\alpha)c(\alpha)^{-(1-\epsilon)/\epsilon} + [1-h(\alpha)]c^{-(1-\epsilon)/\epsilon}\}.$$

Then (A 10) can be expressed as

$$Ag(\alpha) - R(\alpha). \tag{A 11}$$

We now suppose for analytical simplicity that the family of research projects is such that $g(\alpha)$ is concave in α. Then provided the (expected) profit maximising R & D project lies strictly between $\alpha = 0$ and $\alpha = \bar{\alpha}$, the monopolist will choose that α which is the solution of

$$Ag'(\alpha) = R'(\alpha). \tag{A 12}$$

Let α_m be the (unique) solution of (A 12).

We turn now to the social planner. He is concerned with social benefit (36). Since social utility has the form (6), $V(c)$ satisfies (8). Write $B \equiv \epsilon\sigma^{1/\epsilon}(1-\epsilon)^{-1}$. Then (36) can be expressed as

$$Bg(\alpha) - R(\alpha). \tag{A 13}$$

Since by hypothesis (A 13) is strictly concave in α the planner will also wish to engage in an elementary research project. Let α_s $(0 < \alpha_s < \bar{\alpha})$ be the socially optimal research project. Then α_s must be the solution of

$$Bg'(\alpha) = R'(\alpha). \tag{A 14}$$

But $B > A$. On comparing (A 12) and (A 14) it follows that $\alpha_s > \alpha_m$. This establishes the claim made in the text.

References

Arrow, K. J. (1962). 'Economic welfare and the allocation of resources for invention.' In *The Rate and Direction of Inventive Activity: Economic and Social Factors* (ed. R. Nelson). (NBER) Princeton University Press.

Barzel, Y. (1968). 'Optimal timing of innovation.' *Review of Economics and Statistics* (August), pp. 348–55.

Dasgupta, P., Heal, G. M. and Majumdar, M. (1977). 'Resource depletion and research and development.' In *Frontiers of Quantitative Economics*, vol. IIIB (ed M. Intrilligator). Amsterdam: North Holland.

—— and Stiglitz, J. E. (1977). 'Market structure and research and development.' London School of Economics (mimeo).

—— —— (1980). 'Uncertainty, market structure, and the speed of research.' *Bell Journal of Economics* Spring.

Evenson, R. and Kieslev, Y. (1975). 'A model of technological research.' *Journal of Political Economy*.

Kalecki, M. (1954). *Theory of Economic Dynamics*. London: George Allen and Unwin.

Kamien, M. and Schwartz, N. (1972). 'Timing of innovations under rivalry.' *Econometrica*, 40, no. 1, pp. 43–60.

—— —— (1975). 'Market structure and innovation: a survey.' *Journal of Economic Literature*, vol. 13, no. 1, pp. 1–37.

—— —— (1978). 'Optimal exhaustible resource depletion with endogenous technical change.' *Review of Economic Studies*, vol. 45, pp. 179–96.

Levin, R. C. (1978). 'Technical change, barriers to entry and market structure.' *Economica*, vol. 45, no. 180, pp. 347–62.

Loury, G. (1977). 'Market structure and innovation.' NorthWestern University, Center for Mathematical Studies in Economics and Management Science, Discussion Paper, no. 256.

Mansfield, E. (1967). *Econometric Studies of Industrial Research and Technological Innovation*. New York: W. W. Norton & Co.

Marschak, T. and Selton, R. (1974). *General Equilibrium with Price-Making Firms*. Berlin: Springer-Verlag.

Nelson, R. R., Peck, M. J. and Kalachek, E. D. (1967). *Technology, Economic Growth and Public Policy*. Washington, D.C.: Brookings Institution.

—— and Winter, S. (1977), 'Forces generating and limiting concentration under Schumpeterian competition.' *Bell Journal of Economics*, vol. 9, no. 2 (Autumn), pp. 524–48.

—— —— and Schuette, H. L. (1976). 'Technical change in an evolutionary model.' *Quarterly Journal of Economics*, vol. 90, no. 1, pp. 90–118.

Novshek, W. (1977). 'Nash–Cournot equilibrium with entry.' Mimeo. Princeton University.

Scherer, F. M. (1970). *Industrial Market Structure and Economic Performance*. Chicago: Rand McNally.

Schmookler, J. (1962). 'Economic sources of inventive activity.' *Journal of Economic History*, vol. 22 (March).

Schumpeter, J. (1947). *Capitalism, Socialism and Democracy* (2nd edn.). London: Allen and Unwin.

Schwartzman, D. (1977). *Innovation in the Pharmaceutical Industry*. Baltimore: John Hopkins University Press.

Stiglitz, J. E. (1970). 'Perfect and imperfect capital markets.' Yale University, mimeo. Paper presented to the Winter meeting of the Econometric Society (1970).

Winter, S. (1971). 'Satisficing, selection, and the innovating remnant.' *Quarterly Journal of Economics*.

[10]

Oxford Economic Papers 42 (1990), 586–602

INNOVATION, TECHNOLOGICAL OPPORTUNITY, AND MARKET STRUCTURE

By P. A. GEROSKI

1. Introduction

DISCUSSIONS of the social cost of monopoly invariably end with a long list of caveats, of which the Schumpeterian hypothesis is perhaps the best known and most frequently invoked. The notion that monopoly deadweight loss is the price that must be paid for high levels of innovative activity has attracted less support than attention, but still has enough adherents to insure it remains in the forefront of thinking about competition policy. The importance of the question demands that one examine the evidence available on the issue, and this paper does so in the context of inter-industry comparisons using cross-section data. In what follows, we propose to explore the correlation between innovativeness and monopoly power by examining the effect of rivalry using more information on market structure than just concentration ratios, by correcting for inter-industry variations in technological opportunity, and by distinguishing the effect that rivalry has on innovativeness for a given level of post-innovation returns from the effect that rivalry has on innovativeness through its effects on post-innovation returns. Using data on major innovations introduced in the UK during the 1970s, we find fairly strong evidence against the hypothesis that increases in competitive rivalry decrease innovativeness. That is, our data suggest that the price which has to be paid for high levels of innovation may not include tolerating the growth of highly concentrated, imperfectly competitive market structures.

The plan of the paper is as follows. In Section 2, we briefly restate the Schumpeterian hypothesis, distinguishing three senses in which "monopoly" may be thought to affect innovative activity. With this distinction in mind, Section 3 sets out a model for testing the Schumpeterian hypothesis and discusses its empirical implementation. Section 4 presents some empirical results for the UK, looking at the incidence of significant innovations during the period 1970–9 in 73 three-digit industries, and Section 5 contains a summary of the results.

I am obliged to D. Mueller, G. Stewart, J. Black, R. J. O'Brien, P. Stoneman, R. Pomroy, several referees, and to seminar participants at the Universities of Southampton, Exeter, Harvard, Cornell, Penn State, Bristol, Nottingham, the Economic Council of Canada, the Ministry of Finance in Paris, and Middlesex Polytechnic for useful comments. Jim Fairburn provided excellent research assistance. This research was supported by the ESRC, and parts of the data were obtained from the ESRC Data Archive in Essex and from the Business Statistics Office in Newport. Despite this debt, the usual disclaimer applies.

2. The Schumpeterian hypothesis

Popular versions of the Schumpeterian hypothesis contain several different assertions, but the one which we wish to focus on here is the argument that monopoly power stimulates innovative activity. To put this into a testable form, it is necessary to distinguish between *anticipated* and *actual* monopoly power (e.g. Kamien and Schwartz, 1982, p. 27). Anticipated monopoly power refers to an innovator's ability to enjoy the full benefits of its research by preventing imitation. Were the Schumpeterian hypothesis to be expressed solely in terms of the assertion that innovation occurs only when some degree of expected post-innovation monopoly insures that the innovator will at least cover its costs, it would be relatively uncontroversial. Much more disputable are assertions about the effect of actual monopoly power. In principle, actual monopoly can have one or both of two effects on innovative activity: a *direct* effect for any given level of post-invention reward, and an *indirect* effect via the effect that actual monopoly has on the size of the post-innovation reward (Scherer, 1980, pp. 428–9 calls these the "stimulus" and "market room" factors).

An indirect effect of actual monopoly on innovative behaviour arises whenever current monopoly power affects the likelihood of achieving a given degree of post-innovation monopoly power. There are several reasons to suspect that this effect may be positive. Chief amongst these are the notions that a current period monopolist will generally be well placed to erect barriers to future entry (e.g. Levin, 1978), and that the entry barriers which sustain a current period monopoly position may be durable enough to protect the monopolist into the future. Further, whenever the results of future innovations complement innovations already made by the monopolist, then it will gain more than rivals will from their introduction and, if necessary, will pre-empt rivals (e.g. Gilbert and Newberry, 1982). Since it is difficult to imagine how actual monopoly might reduce expected post-innovation returns, the smallest indirect effect that one might plausibly expect is no effect at all.

The direct effect of actual monopoly on innovative behaviour is reflected in the response to a post-innovation return of any given size. Here it is possible to find plausible arguments that work in both directions (e.g. Scherer, 1980, pp. 423–38). A positive direct effect is expected by those who see monopolists as possessors of various types of material advantages. In particular, high current period profits generated by the exercise of actual monopoly power may enable a monopolist to hire more highly qualified personnel, and may provide internal finance which facilitates a rapid response to events and weakens the firm's reliance on costly external finance. As against this, there are at least three reasons to expect a negative direct effect. First, the absence of active competitive forces may allow certain behavioural disadvantages of monopoly to become manifest. Managers may exhibit a preference for leisure or become sleepy, and bureaucratic inertia and control loss may generate substantial x-inefficiency. Secondly, increasing the number of firms searching for an innovation may raise the probability of getting it by some time t, either because the more firms there are undertaking research, the more intense is each firms search activity, or because the more firms there are searching, the more likely it is that one will stumble upon the important clues by date t.[1] Finally, incumbent monopolists whose positions are based on previous innovations will enjoy a lower net return from introducing a new innovation which displaces part of the activities of the old one than entrants would (Arrow, 1962; *see also* Fellner, 1951). This opportunity cost of innovation compounds that which may arise if the incumbents capital stock is locked into a particular technology, and slows the response of a monopolist to a new innovation promising a positive return of any given size.

Thus, there are two effects of actual monopoly on innovative activity. While the indirect effect is likely to be positive, the direct effect may not be. To cast light on the size and direction of the overall effect of actual monopoly on innovative activity, it is necessary to try to isolate and measure the strength of each of these two effects taken separately. The empirical model to be discussed below is designed with precisely this goal in mind.

3. Model specification

When examining the Schumpeterian hypothesis, it is natural to focus in the first instance on the relationship between actual market structure and research activity. There are a number of ways in which this relationship can be expressed, but most theoretical models suggest that research activity, S_i, will not occur if expected post-innovation returns, π_i^e, are zero, and that research activity will increase as the level of such returns increases. Thus, we posit that

$$S_i = \theta_i(\pi_i^e)^{\alpha_1}, \tag{1}$$

for industries indexed by i. Equation (1) has, in fact, appeared in the literature in at least two different guises. The first is in decision theoretic models of where (1) is derived from the first order conditions describing the choice of research intensity by each firm j and then summed over all j in i (e.g. Levin and Reiss, 1984, Stoneman, 1983, Chapter 3, and others). (1) has also appeared as a zero profit condition in several equilibrium models of rivalry (e.g. Loury, 1979 and others). In both cases, S_i is interpreted as research expenditure relative to sales, π_i^e as post-innovation price-cost margins, and θ_i generally depends on the degree of monopoly, M_i, and other factors, Z_i. As $\theta_i \geqslant 0$, it is natural to write

$$\log \theta_i = \alpha_0 + \alpha_2 M_i + \alpha_3 Z_i, \tag{2}$$

in which case,

$$\log S_i = \alpha_0 + \alpha_1 \log \pi_i^e + \alpha_2 M_i + \alpha_3 Z_i. \tag{3}$$

[1] The first of these propositions is a little controversial: compare Loury, 1979, with Lee and Wilde, 1980. On the second, *see also* Scherer, 1967, Dasgupta and Stiglitz, 1980a and 1980b, Reinganum, 1982, Futia, 1980; Dasgupta, 1986, and Kamien and Schwartz, 1982, survey this material.

In terms of the discussion in Section 2, α_1 is the effect of anticipated monopoly on innovations, α_2 is the direct effect of actual monopoly, and $\alpha_1 \cdot \partial \log \pi_i^e / \partial M_i$ is the indirect effect.

Regressions similar to (3) have appeared frequently in the literature, generally producing positive (but not always significant) correlations between R&D expenditures divided by sales (or, the proportion of scientists and engineers in the workforce) and industry concentration (e.g. *see* Scherer, 1980, Kamien and Schwartz, 1982 and Baldwin and Scott, 1987, for surveys). The problem with accepting these results as evidence consistent with the Schumpeterian hypothesis is that R&D expenditures are not only likely to be a poor measure of total research activity, but they may actually mismeasure true research activity with an error which depends on the degree of monopoly, M_i. If, as seems likely, R&D expenditures systematically understate the research activity of small firms and of putative entrants in competitive (i.e. "low" M_i) industries, then estimates of α_1 will be biased upwards, and, indeed, estimated values of α_1 may be positive when the true value of α_1 is negative.[2] Under these circumstances, it seems prudent to treat S_i as unobservable and look for an alternative method of measuring the direct and indirect effects of actual monopoly.

Since research activity will, sooner or later, produce an output in the form of one or more innovations, I_i, then one can still examine the Schumpeterian hypothesis despite the problems with (3) if one knows the relationship between research inputs and output, S_i and I_i. Any systematic association between the two can be exploited to replace the unobservable S_i by an indicator like I_i. In this context, it is natural to think (roughly) in terms of a production function $I_i = F(S_i)$, which exhibits diminishing returns (if only because of duplication in research across firms). Choosing a functional form largely on grounds of convenience, we assume that

$$I_i = \beta \log S_i + \mu_i, \tag{4}$$

where μ_i, the "propensity to innovate" (adapting a term introduced by Scherer, 1965; *see also* Pakes, 1985), reflects other exogenous factors above and beyond research inputs, S_i, which generate innovations (including simple luck, good or bad). Using (4) to eliminate S_i in (3) yields

$$I_i = \gamma_0 + \gamma_1 \log \pi_i^e + \gamma_2 M_i + \gamma_3 Z_i + \mu_i, \tag{5}$$

[2] In partial support of this assertion, note that R&D expenditures are incurred almost exclusively by large firms whereas innovations are produced by large and small firms (e.g. Pavitt *et al.*, 1987, or Geroski and Stewart, 1986), suggesting that the research inputs of small firms are under-recorded (*see also* Kleinknecht, 1987, for some direct evidence). This observation is also consistent with a range of studies which suggest that apparent research efficiency declines with firm size (e.g. Kamien and Schwartz, 1982, pp. 66–70). Part of the problem seems to be that small firms generally do different types of research than large ones, and much of it is less formal and less tied to R&D labs than is true in large firms; Pavitt *et al.*, 1987 also observe that the research of small firms is often supported by the users of innovations in other industries. Putative entrants who do research but fail to enter are also not recorded in industry R&D totals.

where the γ_k, $k = 0, 1, 2, 3$ and proportional to the α_k, being scaled up by a positive constant β. Hence, $\gamma_2 < 0$ points to a negative direct effect of actual monopoly on innovation, and, if $\partial \log \pi_i^e / \partial M_i > 0$, then $\gamma_1 > 0$ points to a positive indirect effect; the total effect is proportional to the quantity $(\gamma_2 + \gamma_1 \, \partial \log \pi_i^e / \partial M_i)$.

Thus, (5) is a model which enables one to chart the direction and measure relative size of the two effects of actual monopoly on innovation without needing to measure research inputs accurately.[3] Given (5) as an empirical vehicle, there are four specification decisions to consider. These translate the variables on the right-hand side of (5) into observables, and we consider each in turn.

The first specification decision concerns $\log \pi_i^e$, the level of price-cost margins expected post-innovation. Being an expectation, $\log \pi_i^e$ is not directly observable, and must be proxied. Two courses of action suggest themselves. The simplest path to follow is to assume that agents have complete foresight, i.e. to take expected post-innovation margins to equal actual, observed post-innovation margins. Alternately, one might assume that agents do not possess complete foresight, but do form rational expectations derived from a model of how innovations affect industry profitability. If the log of expected price-cost margins depends linearly on innovative output, I_i, monopoly M_i, and other factors, W_i, then the assumption of rational expectations means that the log of observed price-cost margins, π_i, is given by

$$\log \pi_i = \phi_0 + \phi_1 I_i + \phi_2 M_i + \phi_3 W_i + \varepsilon_i, \tag{6}$$

where $\varepsilon_i \equiv \log \pi_i - \log \pi_i^e$ is a classical regression error. Following standard procedures and using the predictions from a reduced form regression of $\log \pi_i$ on M_i, W_i, and Z_i to proxy $\log \pi_i^e$ in (5) enables one to generate consistent, if inefficient, estimates of the γ_k.[4]

The second specification decision concerns measuring the degree of monopoly, M_i. The usual practice in the literature is to proxy this using an index of

[3] Measuring I_i inaccurately does not have the same impact on Schumpeterian tests as errors in measuring S_i do. To see this, suppose that the total output of research is $I_i + J_i$, of which only I_i is observed. If $I_i = \beta_1 \log S_i + \mu_i^1$ and $J_j = \beta_2 \log S_i + \mu_i^2$, then $(I_i + J_i) = \beta_3 \log S_i + \mu_i^3$, where $\beta_3 \equiv \beta_1 + \beta_2$ and $\mu_i^3 \equiv \mu_i^1 + \mu_i^2$. It is clear that regressing I_i on S_i gives a biased estimate of β_3 (although estimates of β_1 are unbiased), but neither β_1 nor β_3 are associated with the Schumpeterian hypothesis. That is, as α_1 and α_3 are observed up to the same factor of proportionality in (5), testing for their relative size is unaffected by any bias involved in estimating that factor of proportionality.

[4] On estimating rational expectations models, *see* Wallis, 1980, and Wickens, 1982. One might insist that the relevant post-entry return be that increment in returns attributable to innovations (as stressed, for example, by Futia, 1980). Denoting the expected number of innovations as I_i^e, then replacing $\log \pi_i^e$ by $\phi_1 I_i^e$ changes (5) to $I_i = \gamma_0 + (\gamma_1 \phi_1) I_i^e + \gamma_2 M_i + \gamma_3 Z_i + \gamma \mu_i$. Assuming for the moment that μ_i is purely random and has zero mean, then, taking expectations and solving, $I_i^e = (1 - \gamma_1 \phi_1)^{-1} (\gamma_0 + \gamma_2 M_i + \gamma_3 Z_i)$, and so $I_i = \eta_0 + \eta_1 M_i + \eta_2 Z_i + \mu_i$, say. Thus, interpreting $\log \pi_i^e$ as not price-cost margins but as the increment of returns attributable to expected innovations reduces (5) to a linear relationship between I_i, M_i, Z_i and μ_i, and changes slightly the interpretation of the γ_k. The main problem with this is that it is no longer possible to separately identify direct and indirect effects. None of the remarks to be made below about estimates of γ_2 calculated from our data are affected by omitting proxies for $\log \pi_i^e$ from the estimating equation.

concentration, effectively hanging the test for direct effects on a single coefficient. This seems somewhat restrictive. The degree of rivalry in a market is difficult to determine with any precision, and probably cannot be completely captured by just one variable. To avoid the problems that trying to do this might create (and choosing to err (if at all) on the side of excessive inclusion), we have used six measures of rivalry: the extent of market penetration by entrants (ENTRY), the market share of imports (IMPORT), the relative number of small (less than 99 employees) firms (SFIRM), the within period percentage change in concentration (DCON), the market share of exiting firms (EXIT), and the five firm market concentration ratio (CON).[5] An industry with extensive entry, an active small firm sector, a low and falling concentration ratio, and low exit has many of the structural features that one normally associates with "competitive" markets. By contrast, low entry, a small and contracting fringe sector, and a high and rising level of market concentration are all likely to be features of monopolistic markets in which "rivalry" is weak. Hence, we shall interpret positive coefficients on the first three variables (ENTRY, IMPORT, SFIRM) and negative ones on the last three (DCON, EXIT, CON) as consistent with a negative direct effect of actual monopoly on innovation.

In addition to rivalry, there are a number of further factors which may affect the speed with which industries with a given degree of monopoly respond to changes in post-innovation returns. The third specification decision is concerned with these added factors, the Z_i. None of these factors is of interest *per se* in the present exercise, their inclusion being determined by trade-offs between inefficiency (inclusion of irrelevant variables) or bias (exclusion of relevant variables correlated to M_i) in the estimation of γ_1 and γ_2. We have included five such conditioning variables: industry size and growth (SIZE, GROW), capital intensity (KAYO), export intensity (EXPORT), and a proxy for industry unionization (UNION). Of these, UNION, SIZE and KAYO are known to be highly correlated to M_i, while GROW and EXPORT are obvious inclusions given previous empirical work in this area. While it has not proved possible to collect enough variables to encompass all the Schumpeterian regressions reported in the literature, this group of variables takes us most of the way.

Last but by no means least in (5) is the propensity to innovate, μ_i, which summarizes the numerous exogenous factors which affect the average productivity of research, I_i/S_i.[6] Chief amongst these are the conditions of supply

[5] In support of including the entry and small firm variables, *see* Beesley and Hamilton, 1984, who stress the "seed bed" role of new firms, associating industry innovative activity with high "turbulence" or turnover. The role of small firms and entrants in stimulating innovativeness and diffusion, particularly at various stages of the industry life cycle, is discussed in empirical work by Gort and Klepper, 1982, Abernathy and Utterback, 1975, and Kaplinsky, 1983, and is attested to in numerous case studies; *see also* Comanor, 1967, on the role of entry barriers. The variable DCON is intended to reflect the activities of established fringe firms challenging the leading five.

[6] One might consider allowing the propensity to innovate to affect the marginal productivity of research by writing (4) as $\log I_1 = \beta \log S_i + \mu_i$, say. (5) then would have $\log I_i$ as its dependent variable, but all else would remain the same. Testing whether the propensity to innovate affects the marginal product would, in these circumstances, amount to testing whether I_i or $\log I_i$ is the appropriate dependent variable in (5). As it happens, our data are poorly described by a model with $\log I_i$ as the dependent variable.

which facilitate or inhibit the generation of innovations, conditions sometimes referred to as "technological opportunity" (e.g. Scherer, 1965, or Levin *et al.*, 1985; Rosenberg, 1974, is a classic discussion of the importance of these factors). Previous work in this area has followed a variety of different routes in modelling differences in technological opportunity across industries in terms of observables, none of them completely satisfactory.[7] We differ from much of this literature in not trying to model technological opportunity in terms of observables, a procedure which, however desirable in principle, is, in practice, plagued by problems arising from the omission of relevant factors which are difficult to observe. Instead, we take our cue from the observation that technological opportunity is not only exogenous to current period research decisions, but is also likely to be relatively constant in the short to medium run. Hence, while it takes a specific value for each industry, it is constant over time industry by industry. Thus, we write μ_i as containing two components: "technological opportunity", τ_i, an industry specific constant, and ζ_i, a zero mean normally distributed random variable capturing "all other factors".

The consequence of all the decisions discussed in this section is the regression model

$$\begin{aligned} I_i = f_i + \lambda_1 \log \pi_i^e &+ \lambda_2 \text{CON}_i + \\ &+ \lambda_3 \text{ENTRY}_i + \lambda_4 \text{IMPORT}_i + \lambda_5 \text{SFIRM}_i \\ &+ \lambda_6 \text{EXIT}_i + \lambda_7 \text{DCON}_i + \lambda_8 \text{SIZE}_i \\ &+ \lambda_9 \text{GROW}_i + \lambda_{10} \text{KATO}_i + \lambda_{11} \text{EXPORT}_i \\ &+ \lambda_{12} \text{UNION}_i + \zeta, \end{aligned} \tag{7}$$

where $\log \pi_i^e$ can be proxied either by $\log \pi_i$ or by an instrument derived from (6), and $f_i \equiv \tau_i \gamma_0$. The sign pattern: $\lambda_3, \lambda_4, \lambda_5 > 0$ and $\lambda_2, \lambda_6, \lambda_7 < 0$ will be interpreted as a negative direct effect of monopoly on innovation, and the pattern $\lambda_3, \lambda_4, \lambda_5 < 0$ and $\lambda_2, \lambda_6, \lambda_7 > 0$ as a positive effect. λ_1 is the effect of anticipated monopoly on innovation. Each of the six rivalry variables have an effect on $\log \pi_i^e$ given by (6), say $\phi_{21}, \ldots \phi_{26}$. Thus the indirect effects on innovation associated with them are $\lambda_1 \phi_{11}, \ldots \lambda_1 \phi_{16}$, with total effects $(\lambda_2 + \lambda_1 \phi_{11}), \ldots, (\lambda_7 + \lambda_1 \phi_{16})$. These will be the parameters of interest in the estimation which follows.

4. Results

The basic data that we have used are information on 4378 significant innovations introduced in the UK, 1945–83. Given the need to estimate fixed

[7] Thus, for example, Lunn and Martin, 1983, and Scherer, 1965, use a few subjectively selected industry dummies; Waterson and Lopez, 1983, use a proxy for capital intensity and changes in productivity; Schrieves, 1978, uses information on types of scientists and engineers in firms; Hughes, 1984, uses foreign R&D intensity; and Levin *et al.*, 1985, use a variety of measures which proxy closeness to science, the sources of technical progress, industry maturity, conditions of appropriability, and so on.

effects, we focused on two 73 MLH (or, three digit) industry cross-section panels covering the periods 1970–4 and 1975–9, using average values over the five year period for the independent variables (for further details, *see* the Appendix).[8] The dependent variable, I_i, is non-negative and so a Tobit estimator was used, producing the estimates shown on Table 1.

Column (i) on Table 1 shows our preferred estimates, using a rational expectations proxy for $\log \pi_i^e$. Column (ii) uses the log of actual post innovation margins, and a comparison between the two suggests that the precise specification of $\log \pi_i^e$ has very little qualitative effect on the results.[9] Both columns (i) and (ii) assume that the eleven remaining variables are exogenous, and we have adopted this as a maintained hypothesis for SIZE_i, GROW_i, KAYO_i, EXPORT_i and UNION_i. However, since the endogeneity of market structure is an integral element of Schumpeterian notions of "creative destruction" (Levin and Reiss, 1984, and Dasgupta, 1986, each make this case in slightly different ways), it is important to test the exogeneity assumptions made on the six rivalry variables. Tests conducted on the six considered jointly, singly, and on numerous subsets of the six all uniformly indicated a failure to reject the null of exogeneity, although in the case of CON_i the decision was close to marginal.[10] Column (iii) shows 2SLS estimates of the same model as column (i) when CON_i is assumed to be endogenous. Although the indications are that the value of λ_2 in column (i) may be biased upwards somewhat, there seems to be little doubt that the true value of $\lambda_2 < 0$. Finally, it is worth remarking that the standard errors reported in all three columns have probably been inflated by the perhaps excessive inclusion of variables given the extent of colinearity that exists between them. IMPORT_i and EXPORT_i, CON_i and UNION_i, KAYO_i and SIZE_i, CON_i and SIZE_i, and ENTRY_i and EXIT_i are all noticeably positively correlated, while CON_i and SFIRM_i, CON_i and ENTRY_i, and CON_i and EXIT_i are negatively correlated. It is, of course, not terribly surprising to discover that these variables contain similar information, and it is not surprising to discover that columns (i)–(iii) can be simplified in numerous

[8] The innovations data base only provides 80 possible MLH industry classifications, and 7 further industries were eliminated because of holes in our capital stock series. The number of innovations in each period were 538 and 665 respectively. 27 and 16 industries in each sample recorded no innovations; the maximum number of innovations per sector was 88 and 134, both in sector 354 (Industrial Engines). Five year intervals were chosen to minimize the effect of any inaccuracies in recording the precise date of innovation.

[9] The rational expectations proxy for $\log \pi_i^e$ was generated as the predictions of a reduced form regression of margins on lagged innovations, lagged innovations squared, growth, exports, unionization, risk (the within period standard deviation of price-cost margins), capital output ratio, capital expenditures, industry size squared, and a full set of fixed effects. Including the six rivalry variables in the set of regressors had very little substantive effect, and, indeed, numerous experiments with this set of exogenous variables suggested that the qualitative features of the estimates of λ_1 on Table 1 are reasonably robust. Much of the same applies for a model involving the level rather than the log of π_i^e, and for a model in which π_i is profits on assets rather than price-cost margins.

[10] *See* Smith and Blundell, 1986, for an extension of the Wu test (Wu, 1973) to the Tobit model. The instruments used were those used for $\log \pi_i^e$ (*see* footnote 9 above). It is noticeable that lagged (i.e. pre-1970) innovations had a clear negative influence on current leads of concentration, a correlation that is explored more fully in Geroski and Pomroy, 1987.

TABLE 1
Regression Results for Equation (7)

Independent variables (means)	(i)	(ii)	(iii)	(iv)	(v)
$\log \pi_i$ (– 0.588)	2.27* (0.146)	1.16 (0.176)	11.54* (0.72)	4.20 (0.897)	21.96* (3.70)
CON_i (0.517)	– 58.42 (2.34)	– 57.57 (2.27)	– 439.2* (2.42)	– 50.87 (3.05)	7.12 (0.695)
$ENTRY_i$ (0.05)	18.40 (1.05)	18.51 (1.05)	25.28 (1.44)	31.864 (1.94)	41.03 (0.780)
$IMPORT_i$ (0.213)	– 0.401 (0.129)	– 0.380 (0.122)	– 0.102 (0.033)	– 2.14 (0.239)	10.43 (0.941)
$SFIRM_i$ (0.738)	3.63 (1.09)	3.61 (1.09)	4.94 (1.49)	12.46 (2.13)	– 15.73 (1.09)
$EXIT_i$ (0.049)	– 25.94 (1.36)	– 26.17 (1.38)	– 25.28 (1.34)	– 18.03 (2.207)	59.45 (1.28)
$DCON_i$ (0.013)	– 10.01 (2.22)	– 0.73 (2.02)	– 10.75 (2.35)	– 7.68 (2.24)	– 20.08 (1.42)
$SIZE_i$ (5.76)	4.24 (0.614)	4.27 (0.625)	– 6.16 (0.724)	0.071 (0.016)	8.51 (3.45)
$GROW_i$ (0.742)	3.46 (1.91)	3.44 (1.91)	5.36 (2.58)	2.30 (2.06)	1.14 (0.246)
$KAYO_i$ (1.66)	0.854 (1.26)	0.835 (1.25)	2.19 (2.37)	0.906 (1.75)	– 4.36 (3.66)
$EXPORT_i$ (0.295)	6.84 (0.857)	6.98 (0.883)	33.5 (2.26)	4.01 (0.916)	16.56 (2.77)
$UNION_i$ (0.778)	– 2.618 (0.489)	– 2.615 (0.911)	– 4.922 (1.00)	– 7.77 (1.77)	– 16.417 (1.02)
LogL	– 309.82	– 309.82	– 309.63	– 392.58	– 462.90

Equations (i)–(iii) are Tobit estimates that include fixed effects; values of δ from the Tobit were: 4.38 (13.9), 4.38 (13.9), 4.38 (13.9), respectively (t-values in brackets); equation (iv) duplicates (ii) using OLS, while (v) duplicates (i) suppressing the fixed effects into a single constant common to all industries; * = instrumented (for details, see footnotes 9 and 10); t-values (in absolute value) are in brackets below estimated coefficients. The definition of the variables is: I_i = number of innovations; CON_i = 5 firm concentration ratio; $GROW_i$ = % change in domestic production over the period; $SIZE_i$ = log of industry capital stock; $DCON_i$ = % change in industry concentration; $KAYO_i$ = capital-output ratio; $IMPORT_i$ = imports as a % of sales; $ENTRY_i$ = market share of entrants in year of entry; $EXPORT_i$ = exports as a % of sales; $SFIRM_i$ = No. firms $\leq$ 99 employees as a % of total number of firms; $EXIT_i$ = market share of exiting firms in the year of exit; $UNION_i$ = % workforce covered by colective agreements; and $\log \pi_i^e$ = expected post-innovation price-cost margins.

ways by excluding subsets of the twelve independent variables. However, with the exception of simplifications which exclude $KAYO_i$ and/or $SIZE_i$, parameter estimates were extremely robust to such simplifications; exclusion of $KAYO_i$ and $SIZE_i$ did have noticeable effects on some coefficient sizes, but not on any signs. Needless to say, *t*-statistics were generally higher in more parsimonious representations.

The estimates reported in columns (i)–(iii) all provide fairly strong evidence inconsistent with the notion that there is a positive direct effect of actual monopoly on innovativeness. Only λ_4 violates the sign pattern identified earlier as corresponding to a negative direct effect, and this coefficient is insignificant. λ_2 and λ_7 are both negative and significant, λ_6 is negative, and λ_3 and λ_5 are positive. Thus, it seems to be the case that highly concentrated industries are those in the process of becoming more concentrated are less innovative than more competitive looking ones (given the level of post-innovation returns, $\log \pi_i^e$), and there is a noticeable tendency for this to be true for industries subject to high net entry rates,[11] and for industries which have a large small firm sector. Anticipated monopoly, as captured by $\log \pi_i^e$, has a positive effect on innovativeness, but it is an effect that is extremely difficult to estimate with precision. Industry size, export intensity and unionization all appear to be relatively unrelated to innovativeness (although with exports, this is not completely clear), while growth and capital intensity are positively associated with innovativeness (albeit weakly). Both Tobit and OLS predictions from (i)–(iii) were quite accurate, and OLS regressions produced R^2's in the range of 90–95%. One concludes from (i)–(iii) that increases in rivalry and reductions in monopoly market power increase innovativeness for a given level of post-innovation return, π_i^e.

These results are also robust to the method of estimation used. Comparisons between the estimates in column (i) and those generated using a linear method revealed that, while some bias exists in the latter because of the censored dependent variable, that bias does not substantively affect the features of (i)–(iii) just remarked upon.[12] Tobit estimates are, however, not always robust to apparently minor misspecifications, and the choice between Tobit and least squares or instrumental variable estimates is not necessarily straightforward. Column (iv) on Table 1 reproduces column (ii) using OLS and shows that the choice is, in this case, not controversial. Re-estimating (i) and (iii) by instrumental variables (not shown) leads to much the same conclusion. The Tobit model is of course, based on the rather strong assumption that increases in the independent variables have the same effect on the probability of limit observations as on the density of non-limit ones. So-called "Double Hurdle" models relax this assumption (e.g. *see* Blundell and Meghir, 1987), and, when

[11] The restriction $\lambda_3 = -\lambda_6$ is statistically acceptable, suggesting that it is net entry (rather than gross entry) which has an impact on innovativeness.

[12] I_i is, of course, an integer, but we have not explored models of count data (e.g. Hausman *et al.*, 1984, and Cameron and Trivedi, 1966) since more than 20% of the sample had values of $I_i \geqslant 10$, and 10% had $I_i \geqslant 20$.

applied to our data, suggest that CON_i has little effect on the probability than on innovation will be produced and/or adopted in industry *i*, but has a strong negative effect on the number of innovations produced and/or adopted (given that at least one innovation was adopted or generated in *i*), and a weak positive effect on the probability that an innovation generated elsewhere and adopted in *i* will be adopted in *i* by a new entrant (who presumably uses it as a vehicle of entry) and not an established incumbent. Thus, the result that $\lambda_2 < 0$ seems to reflect not so much the reluctance of monopolists to innovate as to a reluctance to extensively innovate when they do innovate, and a slowness in generating innovation which is more than matched by an unwillingness to adopt innovations generated elsewhere. The consequence of all of this is that entry by outsiders bearing new innovations is more likely in monopolistic industries, and plays an important role in keeping them technically progressive.

The principal problem with columns (i)–(iii) is not so much interpreting the pattern of signs observed as reconciling them with results reported in previous empirical studies. Most of these have focused on the role of CON_i alone among the rivalry variables, and almost all of them have suggested that $\lambda_2 > 0$. In fact, the resolution of this particular conundrum turns out to depend on the treatment of technological opportunity. If, as is generally believed, technological opportunity is higher in more concentrated sectors, then failing to completely capture its inter-industry variation will lead estimates of λ_2 to be biased upward.[13] If this is so, then the relatively complete correction that a fixed effects model provides ought to yield an estimate of λ_2 close to its true value. Further, if other types of correction for technological opportunity are incomplete, then it is possible that positive values of λ_2 will be observed when, in fact, the true value of λ_2 is negative, as reported on Table 1. This is an assertion which is easily tested, since the simplification of (7) which embodies the hypothesis that technological opportunity is the same in all sectors is an equation without fixed effects but with a constant whose value is common to all sectors. Column (v) on Table 1 reports estimates of (i) which embody this (statistically unacceptable) simplification (estimates of (ii) and (iii) suppressing the fixed effects lead to qualitatively similar conclusions). It is clear at a glance that neglecting technological opportunity produces estimates of λ_2 that are not only badly biased, but are so biased as to suggest that this one factor alone may account for the difference between columns (i)–(iii) and previous results. That is, as (v) replicates the major feature in question of previous work, then one is entitled to hazard the conjecture that our results are not sample or data

[13] Most scholars have not questioned the existence of a positive correlation between monopoly and technological opportunity (e.g. Scherer, 1980, pp. 435), although the reasons vary according to the particular interpretation of technological opportunity favoured by particular authors. Two examples are: (1) technological opportunity may reflect government supported, defence related research which, in the UK, is highly concentrated in the hands of a few large firms in highly concentrated industries; and (2) it may also reflect ease of appropriability and so be correlated to π_i^e and thus to M_i through the correlation between these two.

specific. Rather, it seems to be the case that the usual methodology of testing the Schumpeterian hypothesis contains a flaw which imparts a distinctly "pro-Schumpeterian" bias to the results.[14]

The comparison between columns (i)–(iii) and (v) casts some interesting light on the role played by technological opportunity in accounting for inter-industry variations in innovativeness. It is clear that the conditions of technological opportunity are highly correlated with many of the twelve independent variables in (7) and, indeed, a regression of estimates of the fixed effects on these twelve for either subperiod produces an R^2 slightly in excess of 50%. Furthermore, industries with high technological opportunity appear not only to be highly concentrated (and relatively unfavourable to small firms), but also to be large, not very capital intensive, and rather more profitable than the rest. These sectors also seem to experience higher domestic and foreign entry. As a consequence, failing to correct for technological opportunity leads one to substantially overstate the effect of anticipated monopoly on innovativeness (λ_1), incorrectly sign the effects of concentration, imports, small firms and exit (λ_2, λ_4, λ_5, λ_6), exaggerate the effect of market size and unionization (λ_8, λ_{12}), understate that of growth (λ_9), exaggerate the significance of exports (λ_{11}) and mistakenly attribute a retarding influence to capital intensity on innovation. Finally, it appears to be the case that variations in technological opportunity taken alone account for at least 60% of the variation in innovations, while the twelve observables in (7) account, at best, for about 30%.[15] Technological opportunity not only appears to be a dominant influence in innovativeness, it also colours the role that more conventional observable factors appear to play.

Thus, it appears that actual monopoly has an inhibiting direct effect on innovativeness. However, while rivalry may increase innovativeness by increasing the response to a given level of $\log \pi_i^e$, increases in rivalry may also indirectly lower innovativeness by reducing $\log \pi_i^e$. To assess the size of this indirect and potentially offsetting effect, it is necessary to estimate (6). This type

[14]This result is not new, although it is more dramatic than most reported in the literature. Most previous studies report that the introduction of variables intended to correct for variations in technological opportunity reduces the sign and significance of λ_2 (e.g. Scherer, 1965). The Cohen *et al* (1987) exploration of the firm size—R&D link reports results with very similar features to those discussed above; that is, technological opportunity seems to play a central role and failing to take this into account appears to exaggerate the pro-Schumpeterian features of the data. As in Levin *et al.* (1985), this study also reports much progress in modelling technological opportunity, using observables accounting for about 50% of its apparent inter-industry variation. Although mitigating the bias created by ignoring technological opportunity, this only partially successful corection does not eliminate it.

[15]The problem with decomposing the variation in innovations into that associated with technological opportunity and that associated with the rest of (7) is the high correlation between these two. OLS variants of (*v*) yielded R^2's no higher than 30%, while OLS variants of (i)–(iii) yielded R^2's no lower than 90%, leaving a minimum contribution by technological opportunity of 60% or so. It is possible that the true figure may be as high as 75–80%. Results emphasizing the major role played by technological opportunity imply that inter-industry variations in innovativeness will be rather stable over time, an observation which is consistent with failure to reject the null of no parameter variation between the two cross-section panels estimated separately.

of equation is familiar from the literature, (e.g. *see* the survey by Schmalensee, 1987), and, as a series of experiments suggested that the results on indirect effects are not sensitive to the precise details of its specification, we can be brief. Column (i) on Table 2 shows an equation in which the six rivalry variables, the five additional variables in (7), two additional exogenous variables and fixed

TABLE 2
Regression Results for Equation (6)
and the Calculation of Direct and Indirect Effects

Variables	(i)	(ii)	(iii)	(iv)
CON_i	0.323 (0.270)	− 58.42	− 0.499	− 58.919
$ENTRY_i$	− 0.838 (2.56)	18.40	− 1.890	16.51
$IMPORT_i$	− 0.0049 (0.017)	− 0.401	− 0.011	− 0.412
$SFIRM_i$	− 0.260 (1.87)	3.63	− 0.5902	3.37
$EXIT_i$	0.445 (1.48)	− 25.94	1.010	− 24.93
$DCON_i$	− 0.102 (1.20)	− 10.01	− 0.231	− 10.241
$SIZE_i$	− 0.239 (1.71)	4.24	− 0.542	3.698
$GROW_i$	− 0.056 (1.81)	3.46	− 0.127	3.333
$KAYO_i$	− 0.017 (1.30)	0.854	− 0.038	0.816
$EXPORT_i$	0.020 (0.139)	6.84	0.045	6.885
UNION	0.040 (0.234)	− 2.61	0.09	− 2.52

The dependent variable in (i) is $\log \pi_i$, and (i) also includes fixed effects, $Risk_i = 1.17$ (1.32), an instrument for $I_i = 0.0159$ (3.59), and CON_i squared $= -0.806$ (0.704). $Risk_i$ is measured as the standard deviation of price-cost margins within the relevant five year period. The R^2 for this equation = 0.973, $F(85, 59) = 25.0363$, the number of observations = 146, and SSR = 0.4044. The equation was estimated by OLS and the estimates are heteroscedastic-consistent; the absolute value of t-statistics is given in brackets. Column (ii) gives the direct effects taken from (i) on Table 1, column (iii) gives the indirect effects calculated from the coefficient in $\log \pi_i$ in (i) on Table 1 (the indirect effect for CON_i is evaluated at the mean, $\overline{CON_i} = 0.51$) and the numbers in (ii) on 2, and, finally, column (iv) shows the total effects of actual monopoly on innovation.

effects are added to an instrument for I_i to explain $\log \pi_i$.[16] Although colinearity has inflated estimates of standard errors, it seems fairly plain that, in broad terms, monopoly increases and rivalry reduces price-cost margins. The effect of concentration is positive but declining; entry, imports, changes in concentration and small firms all have negative effects, and exit has a positive effect on margins. In short, since rivalry appears to reduce margins, the potential exists for the indirect effects of actual monopoly on innovation to offset its direct effect. The rest of the estimates of (6) are, on the whole, not inconsistent with what one might expect. Risk and innovations have a positive effect on margins, size, growth and capital intensity negative ones, and exports and unionization have extremely weak positive effects that, like the imports effect, are almost certainly insignificant. Virtually identical qualitative results were observed in an equation whose dependent variable was π_i rather than $\log \pi_i$.

Column (ii) on Table 2 shows the direct effects taken from column (i) on Table 1, although, as will be plain in a moment, one could use the estimates from columns (ii)–(iii) on Table 1 and find virtually the same results. Column (iii) on Table 2 shows estimates of the indirect effects using $\lambda_1 = 2.27$ from (i) on Table 1 and evaluating the indirect effect of concentration at its mean, $\text{CON}_i = 0.51$. Even if one used $\lambda_1 = 11.54$ from (iii) on Table 1, it is still easy to see that the direct effects dwarf the indirect effects for all eleven variables on Table 2. The indirect effects of entry, small firms, exit, industry size, growth, capital intensity and unionization mildly offset the direct effects on innovation associated with these variables, and those of concentration (at the mean), imports, changes in concentration, and exports weakly reinforce their direct effects. Column (iv) on Table 2 shows the total effects of each of the eleven variables on innovativeness, and it is apparent that the total effects are not only of the same sign, but virtually of the same size as the direct effects. Under these circumstances, it seems reasonable to conclude that actual monopoly has an unambiguously inhibiting effect, and that rivalry has an unambiguously stimulating effect on innovativeness.

5. Conclusions

Our goal in this paper has been to cast light on the hypothesis that rivalry is inimical and monopoly conducive to innovativeness. The test itself is not completely straightforward since monopoly can, in principle, have both an indirect effect via post-innovation returns and a direct effect given the level of post-innovation returns. Our examination of these effects used conventional interpretations of six rather imperfect measures of rivalry, and sought to discern

[16]The instrument is derived from an OLS regression of I_i on fixed effects, lagged I_i, lagged I_i squared, growth, exports, the six rivalry variables, unionization, risk (the within period standard deviation of price-cost margins), capital intensity, capital expenditures, size and size squared. The results reported in Table 2 are robust to numerous alterations in this specification. The system defined by this equation (which is a reduced form) and (6) as it appears in column (i) of Table 2 is over identified.

a specific pattern of signs in the data. The calculations revealed that monopoly appears to inhibit the response to a given level of post-innovation returns, and that the indirect effects on innovativeness are relatively small. There is, in short, almost no support in the data for popular Schumpeterian assertions about the role of actual monopoly in stimulating progressiveness.

This set of results carries strong implications for policy as well as for future empirical research. On the policy front, one is entitled to be sceptical about the wisdom of Industrial Policies which seek to promote progressiveness by creating "National Champions". Monopolies, whether publicly or privately created, do not seem to be particularly progressive in general, and it is hard to believe that restraining the hand of anti-trust authorities will improve this picture. Small firms and new entrants play a role in stimulating innovativeness which is clearly discernible in the data, and probably considerably understates their total effect on innovation. Competition—disirable on other grounds—certaily seems desirable in this area too. The next step for research also seems obvious. While we have been able to correct for variations in technological opportunity, its importance demands that much more attention be paid to modelling it, and to identifying the factors which seem to be important components of it. The role of rivalry in stimulating innovation is interesting, but it is apparently nowhere near as important as that of technological opportunity.

London Business School and University of Southampton

DATA APPENDIX

I_i = number of significant technically and commercially successful innovations used in industry i. These data are based on a major study (involving 400 experts) by SPRU, University of Sussex, of 4378 major innovations in the UK, 1945–83; for details *see* Townsend, 1981, and Robson and Townsend, 1984. Explicit attempts were made to overcome a natural tendency to concentrate on large firms and, although it is extremely unlikely, this may have imported a slight tendency to exaggerate the contribution of smaller firms. It is not clear that such a bias, if present, would have much affect on our conclusions about monopoly power, since highly concentrated markets may be composed of firms small in absolute size and large firms that operate in very large markets may have only modest market shares in these markets. The data was obtained from the ESRC Data Archive at Essex.

π_i = price-cost margins, defined as net output less the wage bill less net capital expenditure divided by gross output. Data were obtained from the Census of Production.

DCON and CON_i = percentage change and level of the 5-firm concentration ratio, from the Census of Production.

$ENTRY_i$ and $EXIT_i$ = market share of all new firms and of existing firms. The data were drawn from a special compilation made by the Business Statistics Office. ENTRY and EXIT for the 1980–74 period were measured by 1974 values of the variables since data for 1970–3 were not available.

$IMPORT_i$ and $EXPORT_i$ = import and export intensity, obtained from the DTI via the MICRODATA data base compiled at the OECD. These data are virtually the same as is available in the Business Monitor.

$SFIRM_i$ = relative number of firms with no more than 99 employees, from the Census of Production. One or two missing observations were filled using Order averages.

$SIZE_i$ and $KAYO_i$ = log of industry assets and the capital stock–output ratio, obtained from calculations made by D. Allard for the Office of Fair Trading.

$UNION_i$ = % of workers covered by a collective agreement, from the New Earnings Survey. One or two missing observations were filled using Order averages.

All other variables used were derived from these or directly obtained from the Census of Production; all market shares have been adjusted for imports and exports.

REFERENCES

ABERNATHY, W. and UTTERBACK, J. (1975). "Dynamic Model of Process and Product Innovation", *OMEGA*, 3, 639–56.

ARROW, K. (1962). "Economic Welfare and the Allocation of Resources for Inventions", in Nelson, R. (ed.), *The Rate and Direction of Inventive Activity*, Princeton University Press, Princeton.

BALDWIN, W. and SCOTT, J. (1987). *Market Structure and Technological Change*. Harwood Academic Publishers, London.

BEESLEY, M. and HAMILTON, R. (1984). "Small Firms Seedbed Role and the Concept of Turbulence", *Journal of Industrial Economics*, 33, 217–29.

BLUNDELL, R. and MEGHIR, C. (1987). "Bivariate Alternatives to the Tobit Model", *Journal of Econometrics* (forthcoming).

CAMERON, A. and TRIVEDI, P. (1986). "Econometric Models based on Count Data: Comparisons and Applications of Some Estimators and Tests", *Journal of Applied Econometrics*, 1, 29–54.

COHEN, W. M., LEVIN, R. and MOWERY, D. (1987). "Firm Size and R&D Intensity: A Re-Examination", *Journal of Industrial Economics*, 35, 543–66.

COMANOR, W. (1964). "Market Structure, Product Differentiation and Industrial Research", *Quarterly Journal of Economics*, 81, 639–57.

DASGUPTA, P. and STIGLITZ, J. (1980a). "Uncertainty, Industrial Structure and the Speed of R&D", *Bell Journal of Economics*, 11, 1–28.

DASGUPTA, P. and STIGLITZ, J. (1980b). "Industrial Structure and the Nature of Innovative Activity", *Economic Journal*, 90, 266–93.

DASGUPTA, P. (1986). "The Theory of Technological Competition", in F. Mathewson and J. Stiglitz (eds.), *New Developments in the Analysis of Market Structure*. Macmillan, London and Basingstoke.

FELLNER, W. (1951). "The Influence of Market Structure on Technological Progress", *Quarterly Journal of Economics*, 65, 556–77.

FUTIA, C. (1980). "Schumpeterian Competition", *Quarterly Journal of Economics*, 94, 675–95.

GEROSKI, P. and POMROY, R. (1987). "Innovation and the Evolution of Market Structure", forthcoming, *Journal of Industrial Economics*.

GEROSKI, P. and STEWART, G. (1986). "Competitive Rivalry and the Response of Markets to Innovative Opportunities". Forthcoming in S. Arndt and G. McKenzie (eds.), *The Flexibility of the UK Economy*. Macmillan, London.

GILBERT, R. and NEWBERRY, D. (1982). "Pre-emptive Patenting and the Persistence of Monopoly", *American Economic Review*, 72, 514–26.

GORT, M. and KLEPPER, S. (1982). "Time Paths in the Diffusion of Product Innovations", *Economic Journal*, 92, 630–53.

HAUSMAN, J. and TAYLOR, W. (1981). "Panel Data and Unobservable Individual Effects", *Econometrica*, 49, 1377–98.

HAUSMAN, J., HALL, B. and GRILICHES, Z. (1984). "Econometric Models for Count Data with an Application to the Patents—R&D Relationship", *Econometrica*, 52, 909–38.

HUGHES, K. (1984). "Determinants of R&D Expenditure in UK Manufacturing Industry", *Journal of Industrial Economics* (forthcoming).

KAMIEN, M. and SCHWARTZ, N. (1982). *Market Structure and Innovation*. Cambridge University Press, Cambridge.

KLAPINSKY, R. (1983). "Firm Size and Technical Change in a Dynamic Context", *Journal of Industrial Economics*, 32, 39–60.

KLEINKNECHT, A. (1987). "Measuring R&D in Small Firms: How much Are We Missing?", *Journal of Industrial Economics*, 36, 253–56.

LEE, T. and WILDE, L. (1980). "Market Structure and Innovation: A Reformulation", *Quarterly Journal of Economics*, 94, 429–36.

LEVIN, R. (1978). "Technical Change, Barriers to Entry and Market Structure", *Economica*, 45, 347–61.

LEVIN, R. and REISS, P. (1984). "Tests of a Schumpeterian Model of R&D and Market Structure", in Z. Griliches (ed.), *R&D, Patents and Productivity*, University of Chicago Press, Chicago.

LEVIN, R., COHEN, W. and MOWERY, D. (1985). "R&D Appropriability, Opportunity, and Market Structure: New Evidence on the Schumpeterian Hypothesis", *American Economic Review*, 75, 20–24.

LOURY, G. (1979). "Market Structure and Innovation", *Quarterly Journal of Economics*, 93, 395–410.

LUNN, J. and MARTIN, S. (1983). "Market Structure, Firm Structure and R&D", forthcoming, *Quarterly Review of Economics and Business.*

PAKES, A. (1985). "On Patents, R&D, and the Stock Market Rate of Return", *Journal of Political Economy*, 93, 390–409.

PAVITT, K., ROBSON, M. and TOWNSEND, J. (1987). "The Size Distribution of Innovating Firms in the UK: 1945–1983", *Journal of Industrial Economics*, 35, 297–316.

REINGANUM, J. (1982). "A Dynamic Game of R&D: Patent Protection and Competitive Behavior", *Econometrica*, 50, 671–88.

ROBSON, M. and TOWNSEND, J. (1984). "Users Manual for ESRC Archive File on Innovations in Britain since 1945", mimeo, SPRU, University of Sussex.

ROSENBERG, N. (1974). "Science, Invention and Economic Growth", *Economic Journal*, 84, 90–108.

SCHMALENSEE, R. (1987). "Inter-industry Studies of Structure and Performance". Forthcoming in R. Schmalensee and R. Willig (eds.), *Handbook of Industrial Economics*, North-Holland, Amsterdam.

SCHERER, F. M. (1965). "Firm Size, Market Structure, Opportunity and the Output of Patented Inventions", *American Economic Review*, 55, 1097–125.

SCHERER, F. M. (1967). "Research and Development Allocation Under Rivalry", *Quarterly Journal of Economics*, 81, 359–94.

SCHERER, F. M. (1980). *Industrial Market Structure and Economic Performance*, 2nd edn., Rand McNally, Chicago.

SCHRIEVES, R. (1978). "Market Structure and Innovation: A New Perspective", *Journal of Industrial Economics*, 26, 329–47.

SMITH, R. and BLUNDELL, R. (1986). "An Exogeneity Test for a Simultaneous Equation Tobit Model with an Application to Labour Supply", *Econometrica*, 54, 679–85.

STONEMAN, P. (1983). *The Economic Analysis of Technological Change*. Oxford University Press, Oxford.

TOWNSEND, J., HENWOOD, F., THOMAS, G., PAVITT, K. and WYATT, S. (1981). "Innovations in Britain since 1945", Occasional Paper No. 16, SPRU. University of Sussex.

WALLIS, K. (1980). "Econometric Implications of the Rational Expectations Hypothesis", *Econometrica*, 48, 49–73.

WATERSON, M. and LOPEZ, A. (1983). "The Determinants of R&D Intensity in the UK", *Applied Economics*, 15, 379–91.

WICKENS, M. (1982). "The Efficient Estimation of Econometric Models with Rational Expectations", *Review of Economic Studies*, 49, 55–67.

WU, D. (1973). "Alternative Tests of Independence Between Stochastic Regressors and Disturbances", *Econometrica*, 41, 733–50.

[11]

Journal of Economic Literature
Vol. XXX (September 1992), pp. 1416–1433

Schumpeter and Plausible Capitalism

By F. M. Scherer
Harvard University

I am grateful to John Scott and an anonymous referee for helpful comments.

I. *Introduction*

In the fall of 1942, as the Allied and Axis nations marshalled their forces for decisive battles at Guadalcanal, El Alamein, and Stalingrad, Joseph A. Schumpeter's *Capitalism, Socialism, and Democracy* (*CSD*) was published. Perceived by Schumpeter at the time as a "little book of essays" somewhat in the nature of a "potboiler,"[1] it distilled for the intelligent lay reader "almost forty years' thought, observation, and research" (1942, p. ix) by the Austrian economist-statesman-banker turned Harvard don.

Fifty years later, it remains relevant. In particular, it brought into the main stream of economic discourse the question of what market structures were most favorable to technological change and hence economic growth. Both in the United States and abroad, policy debates over that issue persist. Schumpeter's conjectures on market structure, the work by economists to extend them, and the continuing policy puzzles are the main focus of this anniversay essay.

[1] Allen (1991, vol. 2, pp. 99, 133). Allen's two-volume biography is the definitive work on Schumpeter's strikingly contrasting professional and private lives.

II. *The Challenge*

CSD had three main themes: the spectacular success of capitalism in generating economic progress, an analysis of how that success came about, and the perceived "march" of capitalist economies into socialism.

The book is best known for arguing that by virtue of its success in cranking out goods and services, capitalism would undermine its own social, organizational, and moral foundations, setting the stage for the ascendance of socialism. Today, as the tumultous changes in Eastern Europe unfold, that warning appears wildly off the mark.[2] (A mere decade ago, scholars gave it more weight. See the review essays in Arnold Heertje 1981.)

A second thrust of *CSD* has proved to be much hardier. Chapters 5–11 embodied Schumpeter's most complete characterization of "plausible capitalism," contrasting the dynamics of the system, as he viewed them, against the essentially static conception emphasized

[2] Deeply depressed over the events that precipitated it, Schumpeter never viewed it as a prediction. See his address to the American Economic Association (1950, p. 447): "I do not 'prophesy' or predict it. . . . [F]actors external to the chosen range of observation may intervene to prevent . . . consummation."

in contemporary neoclassical economic analyses, both at the time he wrote and (with only modest amendment) fifty years later.

In some respects this theme of *CSD* was a direct extension of the vision first articulated in the most influential work of Schumpeter's youth, his *Theory of Economic Development* (1912). The hallmark of capitalism, Schumpeter argued both in 1912 and 1942, was not the equalibration of supplies with demands for an unchanging array of goods and services, but dynamic evolutionary growth, which permanently increased the output available for consumption and further growth. In 1912, the importance of economic growth was shown by appealing to readers' commonsense observation of the changes swirling about them. By 1942, Schumpeter could estimate (p. 65) that output in the United States had been growing at a long-run average rate sufficient to double real income per capita in what at the time appeared to be the relatively short period of a half century. Real economic growth depended primarily upon innovations: new consumer goods, new methods of production and transportation, new markets, and new forms of industrial organization. Schumpeter used the term "technological progress" to characterize these diverse changes. Innovation in turn was driven by the entrepreneurial activity to which the capitalist system's rewards and opportunities were uniquely conducive. Thus, according to the 1942 Schumpeter (p. 110):

> It is therefore quite wrong . . . to say . . . that capitalist enterprise was one, and technological progress a second, distinct factor in the observed development of output; they were essentially one and the same thing or, as we may also put it, the former was the propelling force of the latter.

All this could be found, albeit less eloquently stated, in *The Theory of Economic Development*. Where *CSD* departed most radically from Schumpeter's earlier vision was in its view of the market structures that best sustained technological progress. In his 1912 book (1934, p. 66), Schumpeter insisted that innovations typically originated in new, characteristically small, firms commencing operation outside the "circular flow" of existing production activities. To be sure, the small innovating firms that succeeded would grow large, and their leaders would amass great fortunes. They started, however, as outsiders. But by 1942, large established business enterprises, frequently enjoying monopoly power over old products as well as new ones, replaced small outsiders in Schumpeter's pantheon of innovative leaders.[3]

A modern economist accustomed to analyzing appropriability conditions, spillovers, scale economies, and innovation races will find Schumpeter's description of large firms' advantages primitive. Important in Schumpeter's schema was the ability of monopolists to attract superior "brains" and secure a high financial standing (p. 101). Even more crucial, he perceived technical innovation to be risky, and risk-bearing was more likely when firms could deploy an array of restrictive practices to protect their investments. "There is no more of paradox in

[3] A transitional stage in Schumpeter's thinking is revealed by a footnote in his book, *Business Cycles* (1939, p. 404):

> It is, of course, true that mere size is not necessarily an advantage and may well be a disadvantage. Judgment must turn on the merits of each case. But statistical evidence to the effect that smaller concerns often do better than the giants should not be uncritically accepted. The smaller concerns may *now* often be in the position of the new, and the giants in the position of the old firms in our model. It is held . . . that the big concerns . . . implied technological and organizational improvement when they were founded. It is not held that they retained their advantages until the present day. Our theory would in fact lead us to expect the contrary.

this than there is in saying that motorcars are traveling faster than they otherwise would *because* they are provided with brakes" (1942, p. 88). Schumpeter went far beyond economists' long-accepted view that the *expectation* of a monopoly position (e.g., through patent protection on inventions) was necessary to make the venture worth while. Monopoly power *already held* also supported investment in technological progress. Here, Schumpeter argued (1942, p. 106), both economists and trust-busters had their priorities wrong:

> What we have got to accept is that [the large-scale establishment or unit of control] has come to be the most powerful engine of . . . progress and in particular of the long-run expansion of output not only in spite of, but to a considerable extent through, this strategy which looks so restrictive . . . In this respect, perfect competition is not only impossible but inferior, and has no title to being set up as a model of ideal efficiency.

Most of the microeconomic theory of the firm focused then (and still does) on pricing behavior, and, in the case of monopolies, on the possibility of pricing that distorts resource allocation. Schumpeter challenged this notion in an extended argument centered on *CSD*'s most enduring catch phrase: "creative destruction." Innovation led not only to superior new goods and services; it simultaneously undermined the market positions of firms committed to old ways of doing business. It destroyed old monopolies while creating new economic value. And to the extent that monopoly power accompanied the value-creating innovations, its possessors had to exercise their power cautiously both in pricing and product policy lest they stimulate another wave of monopoly-eroding changes. Summing up, he wrote (1942, pp. 84–85):

> But in capitalist reality as distinguished from its textbook picture, it is not [price] competition which counts but the competition from the new commodity, the new technology, the new source of supply, the new type of organization (the largest-scale unit of control for instance)—competition which commands a decisive cost or quality advantage and which strikes not at the margins of the profits and the outputs of the existing firms but at their foundations and their very lives. This kind of competition is as much more effective than the other as a bombardment is in comparison with forcing a door, and so much more important that it becomes a matter of comparative indifference whether competition in the ordinary sense functions more or less promptly; the powerful lever that in the long run expands output and brings down prices is in any case made of other stuff.

Thus, the gauntlet was thrown down for an economics profession preoccupied with static "circular flow" theories.

III. *The Reaction*

Schumpeter's hypothesized links between monopoly, competition, and innovation were mere fragments, lacking both conceptual precision and empirical support. Among other things, terms were left unclear. Despite a careful discussion (1942, pp. 79 and 98–100) of single-seller markets (which he considered rare) and monopolistic competition, Schumpeter repeatedly used terms such as "large-scale unit of control" and "big business" to characterize the industrial climate in which innovation thrived. Only three years earlier (1939, p. 1044), in a more scholarly work, he had observed that "'monopoly' really means any large-scale business." At first slowly but at an accelerating pace, economists sought to clarify and extend the Schumpeterian conjectures. Their research proceeded along three main lines: theory building, case studies,[4] and statistical analyses.

[4] The case study literature is huge and cannot be surveyed adequately here. Leading early examples were W. Rupert Maclaurin (1949), Jesse Markham (1952), and John Jewkes, David Sawers, and Richard Stillerman (1958).

A. *Theoretical Advances*

The first significant contributions using modern theoretical methods were by William Fellner (1951) and Kenneth Arrow (1962). Both contrasted polar cases, pure monopoly and pure competition, finding that the incentive to carry out cost-reducing innovations is greater when product market pricing is competitive. This is so because (in both models) post-innovation monopolistic output restrictions engender a deadweight loss that could enhance quasi-rents for the innovator under competition, assuming that the competitor can fully appropriate them (e.g., under a perfect patent). Also, in the Arrow model, some of the rents secured by the monopolist after innovation would have been enjoyed even without innovation, whereas for the competitor, all rents are incremental and attributable to innovation. Arrow's contribution precipitated a series of papers exploring variants on his basic model. See the survey by William L. Baldwin and John Scott (1987, pp. 14–17).

Early qualitative discussions, most prominently by John Kenneth Galbraith (1952, ch. 7), suggested that oligopolists were superior as innovators over both pure monopolists and pure competitors. At first, progress in building models of innovative rivalry under oligopoly was slow, but in the second half of the 1960s, theoretical research flourished. Surveying one main branch of the ensuing literature two decades later, Jennifer Reinganum (1989) listed 75 theoretical contributions. It is neither feasible nor (in view of Reinganum's survey and the one by Baldwin and Scott 1987) useful to review the literature comprehensively here. A few highlights nonetheless warrant extraction.

As in other domains where game theory is applied, the results of this theoretical research are sensitive to the assumptions made, and with the appropriate constellation of assumptions, virtually anything can be shown to happen. Particularly crucial are assumptions about the strength of first mover advantages (e.g., whether a successful innovator's market control remains total, or is easily eroded), the degree of uncertainty associated with the outcome of research and development investments, the rate at which innovation opportunities arise through more or less exogenous advances of knowledge and/or demand, how rapidly firms react to changes in rival behavior, the extent to which new knowledge "spills over" from one firm to another, and whether competitive entry into a field of innovative rivalry drives economic profits to an "equilibrium" expected value of zero.

Early formulations by F. M. Scherer (1967) and Morton Kamien and Nancy Schwartz (1972, 1976) assumed the structure of the market to be set exogenously. They investigated how R&D spending, and hence the speed at which innovations are completed, varies with structure. For most appropriability regimes (and assuming no technological knowledge spillovers), incentives were found to depend nonlinearly upon structure. Within certain bounds, increased rivalry, measured by a larger number of rivals or higher R&D spending by any representative rival, stimulates other participants to greater effort. But this is true only as long as the pool of appropriable quasi-rents remains sufficiently large that each firm tapping the pool can anticipate a reward covering its front-end R&D costs. As the number of rivals grows and/or the pool itself is shrunk through post-innovation price competition, there arrives a point at which appropriable benefits fall short of expected costs, causing participants to cut back their efforts or drop out altogether. As a first approximation, then, innovative vigor rises initially with increased rivalry but then falls at still higher levels of rivalry. This has come to be called the "inverted U" relationship. Where the peak of the inverted U lies depends in turn upon the rate at which advances in knowledge and demand open up new innovation possibilities and the speed at which rivals recognize and act upon those opportunities. If the opportunity set shifts slowly and/or recognition is fast, a monopolist can profitably innovate before fragmented rivals can appropriate sufficient quasi-rents to cover their costs. If opportunities proliferate rapidly, however, innovation is likely to be fastest in fragmented (but not too fragmented) market structures.

Structural and informational asymmetries are also important. Although *able* to innovate at an early date, firms with dominant (Schumpeterian?) market positions maximize secure profits by choosing a leisurely, inexpensive R&D pace. But if smaller rivals or new entrants threaten existing or potential profits through innovation, dominant firms have incentives to accelerate their efforts and perhaps preempt. Scherer's original (1967) paper revealed a tendency for dominant incumbents to be slow starters in a product innovation rivalry, but to react strongly and preempt when their market positions were threatened. Adding technological spillovers from first to later movers and assuming that the market is shared in inverse proportion to product launching delays by first and later arrivals, William L. Baldwin and Gerald L. Childs (1969) originated the theory of "the fast second," in which incumbent dominant firms let smaller rivals (including new entrants) be the technological pioneers. Introducing uncertainty to a simpler preemption model proposed by Richard Gilbert and David Newbery (1982), Reinganum (1983) showed that the dominant incumbent maximizes profits by investing less than outsiders in R&D and leaving some finite probability of being beaten to an important cost reducing innovation. In this way the incumbent increases the expected value of the time period during which it enjoys rents threatened by the innovation. Other models have embroidered rich variations on these themes.

A third major line of attack was initiated by Yoram Barzel (1968) and extended by Glenn Loury (1979) and Partha Dasgupta and Joseph Stiglitz (1980), among others. Their central premise is that market structure is endogenous, not given. Firms will enter the pursuit of innovation prospects until profit expectations are driven to zero. For given costs of carrying out an R&D project, the larger the pool of attainable quasi-rents, the more firms there will be competing to tap that pool. Depending upon the precise model structure, more rivalry may lead each individual firm to cut back its own R&D spending, although in the most general case, *combined* spending rises with more rivalry, and so technology advances more rapidly. But this is not always a blessing. Multifirm rivalry that yields only one winner may entail extensive and wasteful duplication of R&D expenditures. Parallel competing projects are more desirable: (1) the harder it is to identify the "right" approach because of technological uncertainty; (2) the larger is the wedge between an innovation's ultimate social benefits and appropriable private benefits; and (3) the more the products emerging from rivals' parallel efforts are differentiated, satisfying heterogeneous consumer preferences.

A fourth line of theorizing treats market structure as endogenous but determined more in the spirit of Schumpeter's time-lagged "creative destruction" than as a set of zero-profit conditions to be satisfied simultaneously. The boldest departure came from Richard Nelson and

Sidney Winter (1982). They waive the profit maximization assumption of most Schumpeterian models, assuming instead that firms follow rules of thumb and grope experimentally for superior production processes, whose availability may come from the exogenous progress of science or through past cumulative industrial technology improvements. Firms experiencing lucky "draws" reap added profits, invest, and grow; others decline or survive by capturing spillovers from the leaders. Simulating their evolutionary model over a wide range of market structure, opportunity, and appropriability conditions, Nelson and Winter generate productivity growth histories and (characteristically skewed) firm size distributions that correspond plausibly to the outcomes observed in real-world industries.

The only simple conclusion stemming from this and much other theoretical research stimulated by Schumpeter's original conjectures is that the links between market structure, innovation, and economic welfare are extremely complex. Schumpeter was correct in asserting that the incentives for innovation are almost surely inadequate in a world of atomistically competitive markets.[5] But where technological progress is most vigorous—with loose oligopoly, tight oligopoly, or pure monopoly, with or without impeded entry—remains unclear a priori, varying from case to case.

[5] The "world of . . . competitive markets" phrase here is crucial, for deficient structure at one vertical stage (e.g., agriculture) may be remedied if the structure of supplying industries is favorable. See Scherer (1984, pp. 32–58 and 248–55); Eric von Hippel (1988); and Dietmar Harhoff (1991). Even when new products and processes are provided by upstream industries, however, market structure may affect the speed at which the technology is adopted and diffuses through the industry. For a survey of the voluminous literature on this topic, see Baldwin and Scott (1987, pp. 128–44). The weight of evidence supports a conclusion that diffusion is more rapid in less monopolistic markets.

B. *Empirical Research*

Although he presided at the founding meeting of the Econometric Society, Schumpeter did not attempt to formulate his hypotheses in forms amenable to quantitative testing. Nevertheless, since theoretical research indicates that almost anything can happen, empirical research must illuminate what behavioral patterns are most likely. Along with several dozen case studies, there is a sizable statistical literature in the Schumpeterian tradition. Because excellent surveys have been published by Baldwin and Scott (1987) and Wesley Cohen and Richard Levin (1989), the discussion here is confined to a broad overview of the issues and insights.

Statistical research has moved forward with the increasing richness of its motivating theoretical insights, and also with improvements in the available data (often stemming from ambitious new data collection efforts[6]) and econometric techniques. The data used to measure Schumpeterian innovation include such input indices as counts of scientists and engineers employed by companies and outlays for research and development (R&D), intermediate indicators such as patent counts, and output indices such as counts of significant technological innovations or the sales associated with such innovations.

Most of the theorizing on Schumpeterian "plausible capitalism" has emphasized market structure, from competition to monopoly and with open or closed entry. Schumpeter's conjectures, however, were both broader and less clearly articulated. In *CSD* he wrote not only of monopoly but also about the advantages of "big business," which is a quite different

[6] Edwin Mansfield (1963, 1971) pioneered the development of completely new data sets for testing Schumpeterian and many other hypotheses.

structural phenomenon. Favoring large firms is the possibility that economies of scale exist in staffing and equipping R&D laboratories, that risks can be spread more widely, and that financial resources can be tapped at lower cost. Working against size is the possibility of hierarchical incentive failures, coordination breakdowns, and bureaucratic sluggishness. How these advantages and handicaps balance out can only be ascertained empirically.

Many early studies related some measure of company innovative input or output cross-sectionally to a measure of company size—most frequently, sales, although employment and asset measures have also been used. An assumption implicit or explicit in such analyses is that sales revenue provides an unbiased and exogenous measure of firm size.

On one aspect of bias, a sales variable is more neutral than assets, whose intensity of use tends to rise with firm size, and employment, which has the opposite pattern. Censoring biases must also be overcome, e.g., since Financial Accounting Standards Board rules require that company R&D expenditures be reported only when they exceed a certain absolute dollar threshold, which tends to be satisfied more readily for large firms than for small firms.

At least as vexsome is the exogeneity assumption. The Schumpeterian hypothesis implies that large firms innovate more intensively than small firms. It is usually tested by determining whether an index of innovation rises more than proportionately with the size index. Clearly, however, successful innovation raises sales, so there is a danger that sales in year t are influenced by innovative activity in year $t - i$, which in turn, because of habit persistence or other dynamic links, is positively correlated with innovative activity in year t. Exogeneity is usually assumed by arguing that there are multiyear lags between the time when innovative activity occurs and significant sales gains follow. For support, see Ariel Pakes and Mark Schankerman (1984).

To the extent that these and other methodological problems have been overcome, several conclusions emerge fairly consistently from studies of innovative activity and firm size covering both the United States and other nations. First, large firms are much more likely to support formal R&D programs and receive patents than very small firms, but the probability of *some* innovative activity approaches unity at modest size thresholds in most manufacturing industries. See Scherer (1983) and John Bound et al. (1984). Second, formal statistics on R&D spending miss a considerable amount of informal or part-time innovative activity by small firms (Alfred Kleinknecht 1987). Third, within the subset of companies, or relatively homogeneous divisions of companies, that do perform R&D and/or receive patents, R&D spending and patenting tend to rise approximately linearly with size (measured by sales). There are deviations from this pattern across industries, with innovative activity increasing more than proportionately in some fields and less than proportionately in others, but proportionality provides the best global characterization (Scherer 1984, pp. 227–36). Fourth, company R&D/sales ratios vary widely from one industry to another—much more so than they vary within industries as a function of firm size. When industry differences are controlled through either disaggregation of data to the line of business level or the use of fixed industry effect variables, the proportionality result is strengthened. Fifth, increased size attained through diversification over a larger number of product lines is not generally conducive to higher R&D/sales ratios, although selective or "purposive" diversification into synergistic lines may support above-average R&D efforts. (John Scott and George Pascoe 1987).

R&D expenditures are inputs into the innovative process. Whether patents better characterize inputs or outputs is less clear. The conditions under which inferences about statistical relationships between R&D *inputs* and firm size can be extended to relationships between innovative *outputs* and firm size were questioned by Franklin Fisher and Peter Temin (1973) and clarified by Meir Kohn and John Scott (1982). Since then, richer data on companies' innovative outputs have become available. The most comprehensive data collection approach has been to identify and trace to their originators technological innovations important enough to warrant annotation in the vast array of trade journals covering particular industries. Such data for the United States have been mined particularly assiduously by Zoltan Acs and David Audretsch (1990). The results reveal that small firms generate significantly more innovations per thousand employees than their larger brethren. Reconciling this finding with other studies showing R&D spending to vary roughly linearly with firm size, Acs and Audretsch (1991) found that on average, small companies achieve more innovations per million dollars of R&D spending than large enterprises.

Thus, the weight of the existing statistical evidence goes against Schumpeter's 1942 argument that large corporations are particularly powerful engines of technological innovation. Doubts nevertheless remain. William Nordhaus (1969) showed that the larger is the base of sales to which a given innovation will be applied, the farther into the stage of diminishing marginal returns a profit-maximizing firm will carry its R&D, and hence, ceteris paribus, the higher will be the resulting innovation's "quality." Extending this reasoning, Wesley Cohen and Steven Klepper (1992) demonstrated that under plausible stochastic conditions, large and small firms will devote the same fraction of their sales revenues to R&D, but the larger entities will pursue each project more intensively so that there will be fewer innovations per million R&D dollars (as the Acs-Audretsch results suggest), but the average innovation will be of higher quality. Opposing their conjecture is fragmentary case study evidence by Arnold Cooper (1964) that for given technological challenges, small firms utilize their R&D resources more efficiently. If Cohen and Klepper are right, it is not sufficient to count innovations; their economic impact must also be measured. Research to test the Cohen-Klepper hypotheses is ongoing. We return to the issue from another perspective in Section IV.

Equally difficult conceptual and data problems arise in assessing statistically the relationship between market structure, ranging from atomistic to monopolized, and the propensity to innovate. Early interindustry and cross-sectional analyses revealed positive and often statistically significant correlations between industry seller concentration ratios and indices of R&D intensity, e.g., R&D employment as a percentage of total company employment. Several early tests for nonlinearity supported the "inverted U" theoretical conjecture, with maximum levels of innovative effort per unit of firm size occurring at four-seller concentration ratios in the range of 50 to 65—i.e., at middling levels of oligopoly. However, both the positive correlations and the inverted U have tended to vanish when adequate controls were included for interindustry differences in "technological opportunity." The mysterious concept of "technological opportunity" was originally construed to reflect the richness of the scientific knowledge base tapped by firms. It has been proxied through fixed industry effects and, more recently, using responses collected through a survey of 650 industrial R&D managers (Levin et al. 1987). Studies using the survey data suggest that effects originally attributed to technological opportunity may also include differences among industries in firms' ability to appropriate the economic benefits from their innovations (Cohen, Levin, and David Mowery 1987).

The question remained, is there some causal connection between "opportunity," more or less broadly defined, and market structure that might impart either omitted variable or simultaneity biases? Industries with a high potential for technological innovation might contain few sellers because successful innovators have in the past erected patent or other barriers to new entry. Or high required R&D expenditures may create a form of scale economy, limiting the number of viable market participants. Or rapid innovation may increase the turbulence of intraindustry structural changes, acting under some variant of Gibrat's law to propel concentration levels upward over time. See Scherer and David Ross (1990, pp. 141–46), and Nelson and Winter (1982, ch. 14). Alternatively, if rich innovation possibilities make it easier for new firms to enter, opportunity may be concentration-decreasing, not concentration-increasing. Failure to control for differences in opportunity in the former set of cases will impart positive biases to coefficients relating innovative activity to seller concentration. In the latter set of cases, negative biases are likely to emerge.

The most careful exploration of these issues has been undertaken by Paul Geroski (1990) using British data on significant technological innovation counts by industry for the years 1970–79. One important finding is that on average, high rates of innovation were concentration-reducing, not concentration-increasing.[7] Using five-year lagged innovation counts rather than fixed effects to control for interindustry differences in technological opportunity, Geroski observed the association between seller concentration and industry innovativeness to move from positive and statistically insignificant without such controls to *negative* and significant with controls. In the same model, Geroski included a further indirect effect of monopoly power: higher concentration led to wider future price-cost margins, whose expectation *increased* industry members' innovative output. However, when the two industry structure effects, one direct and one indirect, were evaluated simultaneously, the direct effect predominated, and so innovation was found to be less vigorous in more highly concentrated industries. Thus, the results did not support the 1942 Schumpeterian conjectures.

Geroski's "indirect" monopoly power test extends still another Schumpeterian tradition that has been subjected to extensive, more narrowly focused, empirical testing. The argument is that enterprises with monopoly power command supranormal profits usable in discretionary ways, among other things, to support innovation more liberally than it would be if risky R&D activities had to be financed by tapping financial markets (posing inter alia asymmetric information problems). It is important here to distinguish two quite different chains of causation: excess cash flows lead to innovation, while successful innovation leads to excess cash flow. If innovation occurs more or less continuously, powerful time series methods are needed to disentangle the

[7] U.S. counterpart studies with partly similar and partly conflicting or more complex results are Arun Mukhopadhyay (1985), John Lunn (1986), and Richard Levin and Peter Reiss (1988).

chains. It should not be surprising, therefore, that tests of the hypothesis have yielded equivocal results.

New light has been shed by Charles Himmelberg and Bruce Petersen (1990). Proceeding from the observation that large, well-established firms enjoy an abundance of financing sources and are therefore unlikely to find R&D outlays (modest compared to total investment programs) significantly constrained by current cash flow conditions, they sample a subset of firms lacking such advantages—notably, 179 U.S. companies that were both small and specialized in high-technology industries. For that group—to be sure, more like the new firms of Schumpeter's 1912 vision than the corporate giants emphasized in *CSD*—R&D expenditures were found to rise significantly with higher ratios of internal cash flow to beginning-of-period asset values.

C. *The State of the Art*

The theoretical and empirical work that has accumulated in recent decades provides at best meager support for Schumpeter's 1942 theses concerning the most favorable structural climate for industrial innovation. As Cohen and Levin (1989, p. 1078) conclude in their survey of the empirical evidence, "[T]he effects of firm size and concentration on innovation, if they exist at all, do not appear to be important." Other surveys encompassing a wider array of nations sustain similar views. See Kamien and Schwartz (1982, ch. 3), and Baldwin and Scott (1987, pp. 63–113). To be sure, threshold effects exist. Even though idea-rich small firms originate a disproportionate share of innovations, most small enterprises are not particularly innovative. Large companies may carry their new technologies to a higher degree of perfection than small firms, and, as we shall see in the next section, they may excel at certain *kinds* of innovative activities. But neither giant company size nor a high degree of seller concentration appears necessary to maintain a vigorous pace of technological advance. Keeping markets open to new entrants with novel ideas—a notion closer to the Schumpeterian vision of 1912 than to his 1942 view—seems a more important condition for progress.

IV. *The Renaissance of Schumpeterian Policy*

These findings from economic research on Schumpeter's 1942 conjectures seem strangely at odds with recent developments in national policy. Theory and empirical evidence suggest that *CSD* provided faulty guidance concerning the industrial structures most conducive to technological innovation. Yet especially in the United States, it has been argued with increasing frequency that domestic enterprises are too small to maintain technological leadership in an increasingly global marketplace, and that antitrust policies aimed at maintaining competitive *domestic* market structures discourage innovation. Merger law changes proposed by the Reagan Administration but rejected by Congress were premised on an argument that U.S. corporations needed to "combine and restructure . . . to meet new competition from abroad" (U.S. House of Representatives 1986, pp. 5–42). In 1984, the antitrust laws were amended to facilitate cooperative R&D joint ventures among competing companies. More recently, bills have been introduced to permit joint ventures in the production and marketing of new products as well (U.S. Senate 1991).

A. *European and Japanese Parallels*

Since the 1960s, the economic policies of France, England, and West Germany have reflected similar views. At first, giant U.S. firms such as General

Motors, United States Steel Corporation, and IBM were viewed there as having a decisive advantage at technological innovation over the European companies that led their national markets—much smaller, of course, than the U.S. market. Many policy actions of both national governments and the European Community since then have followed the logic articulated by Jean-Jacques Servan-Schreiber (1968, p. 159):

> The first problem of an industrial policy for Europe consists in choosing 50 to 100 firms which, once they are large enough, would be the most likely to become world leaders of modern technology in their fields. At the moment we are simply letting European industry be gradually destroyed by the superior power of American corporations.

More recently, Japan has replaced the United States as the most dynamic challenger to established technological positions. In some respects, Japan's strategy has been distinctly Schumpeterian. Schumpeter's writings evoked great enthusiasm among Japanese scholars, and his personal library resides at Hitotsubashi University in Tokyo (Allen 1991, vol. 1, pp. 270–74). In the late 1940s an intense debate over Japanese development strategy was resolved in favor of MITI, which advocated a Schumpeterian emphasis on building high-technology industries, over the contention of Bank of Japan staff that Japan should cultivate low-technology industries and pursue comparative advantage through its low labor costs (Christopher Freeman 1987, pp. 34–35). The terms of this debate were essentially the terms set out in Schumpeter's 1912 *Theory of Economic Development*, although the Japanese industrial "motorcars" were also provided with brakes—crisis cartels, initial protection against imports, and similar measures if their investments were threatened by structural or macroeconomic setbacks. Whether Japan's technological success is attributable also to *industry* structures like those espoused in the 1942 *CSD* is less clear. Two dimensions are relevant: firm size per se, and cooperation as an alternative or supplement to competition.

During the 1950s and 1960s, when Japan moved rapidly toward the technological frontier, its companies were not large in an absolute sense. Owing to the wartime devastation of Japanese industry, Ken-ichi Imai (1990, p. 192) argues, "Even managers of relatively large firms . . . had to virtually become operators of small firms again." Thus, entrepreneurship in the early stages of Japan's postwar development conformed more to Schumpeter's 1912 paradigm than to the 1942 *CSD* model. However, successful growth has made Japan's leading enterprises large in an absolute sense, whether size is measured in terms of individual operating units or the more encompassing multiindustry *keiretsu* groups.

For our present purposes, there are two remarkable facts about the more recent Japanese experience. First, Japanese enterprises not only accomplished their initial task of coming from far behind to reach the technological frontier. They have taken the lead in such fields as consumer electronics, DRAM semiconductor chips, automobile quality, automated manufacturing, ceramics, and much else. Second, frontier-extending innovations have come for the most part from large enterprises, not from startup firms. In sharp contrast to the U.S. situation, Japan's high-technology venture startup sector is tiny, and for top students at the best engineering schools, the most sought-after goal continues to be a lifetime career in a large enterprise. Thus, despite their humble origins, Japan's large enterprises have successfully made the transition to becoming powerful innovators of the sort idealized by Schumpeter in *Capitalism, Socialism, and De-*

mocracy. This success has precipitated doubts among Americans as to whether Schumpeter might have been right in *CSD* after all, whatever the statistical analyses of past and present structures might reveal.

Japanese companies have also led the way in cooperative research and development ventures. The best known of many examples was the MITI-sponsored VLSI (very large scale integration) project during the late 1970s, in which the top five Japanese semiconductor manufacturers participated, and which helped them surge from a position of inferiority to being the technological leaders in producing 64 kilobit and larger dynamic random access memory chips (DRAMs). Concern that Japanese firms were gaining an advantage through interfirm cooperation to achieve larger scale, reduce competitive duplication, and facilitate information transfer in R&D projects stimulated passage of the U.S. National Cooperative Research Act of 1984. (See Albert Link and Laura Bauer 1989; Michael Katz and Janusz Ordover 1990; and the symposium on cooperative research in the *Journal of Economic Perspectives*, Summer 1990.) In Europe too, numerous cooperative R&D programs have been mounted with explicit European Community backing, beginning in atomic energy and aerospace and extending more recently to semiconductors, information technology, and much else. See Todd Watkins (forthcoming).

Although the admonition is often breached, it is important not to exaggerate the significance of Japanese and European cooperative R&D. The Japanese VLSI project appears to have been atypical. More often, Japanese R&D joint ventures "tend to pursue nonthreatening and often low priority projects . . . remote from the commercialization process" (George Heaton 1988, p. 37; see also Daniel Okimoto 1989, pp. 57 and 71). The technological outcomes from cooperative R&D arrangements initiated in Europe during the 1980s have thus far proved disappointing. (See "European Research: Still Looking for the Answer," *The Economist*, June 22, 1991, pp. 73–74; and "Europe's Electronics Rescue Plan," *New York Times*, Sept. 5, 1991, p. D1.)

Sorting out whether the renewed attractiveness of the *CSD* paradigm is solidly grounded or a mere fad requires answers to three rather fundamental questions: (1) Have the structural conditions, on both the demand and supply sides, under which industrial innovation occurs changed in ways that might make what was untrue in earlier decades true now? (2) Is there a clear "best" (e.g., Schumpeterian) way to organize industry for research and development, or are the approaches to innovation culture-bound, and hence apt to differ among national regimes? (3) Does the best organizational structure for innovation vary with the particular set of technical problems to be solved?

B. *Demand and Supply Changes*

From pioneering contributions by Jacob Schmookler (1966) and Richard Nelson (1959), economists have recognized that demand and supply conditions determine not only prices, but also the pace of technological change.

On the demand side, there have definitely been major structural shifts. Markets, and especially markets for high-technology goods, have become increasingly global in scope. Seller concentration ratios, meaningfully defined, have declined as the relevant playing field widened. If there was something approximating a seller concentration level most conducive to innovation when markets were preponderantly national, new worldwide concentration ratios could have fallen below the optimal range, ceteris paribus. However, other things have not remained equal. With the broadening of market horizons and acceleration of the international diffusion of new product sales, the quasi-rent potential that an individual innovator can tap has increased apace. It is not obvious that a smaller share of a larger pie is innovation-retarding.

On the supply side, the costs of innovation might conceivably have risen relative to market potential, causing optimal market structures to become more concentrated and/or raising the cost-saving benefits of interfirm cooperation. It is easy to find supporting examples. R&D costs to the two-prototype flight stage were roughly $20 million for 1950s-vintage supersonic fighter aircraft, while the teams contending in the U.S. Air Force's Advanced Tactical Fighter competition concluded in April 1991 spent $1.7 billion each to reach a comparable benchmark.[8] In constant dollars, R&D costs appear to have increased by a factor of 20. The costs of developing and clinically testing approved new drug chemical entities in the late 1950s averaged about $534,000 per NCE (Mansfield et al. 1971, p. 67). By the 1980s, the comparable average was $48 million (Joseph DiMasi et al. 1991). Here, constant-dollar costs increased by about 24 times. However, these examples may be atypical—one drawn from a domain in which government contract support is ubiquitous, the other from a field in which safety and efficacy regulation have had a marked cost-increasing impact.

Attempting to obtain more general insights on cost trends, the author tapped data on the average R&D costs of innovations winning *Research and Development* magazine's competition to honor the hundred most significant technological innovations in each year from 1969 through 1986 (Scherer 1991, pp. 30–31). Any such effort reveals a formidable problem: the distribution of costs is highly skewed, so that outlying values have a disproportionate influence on central tendency measures. To minimize the problem, the largest R&D cost observation from each year's cohort was eliminated, leaving an overall average R&D cost per prize-winning innovation (in 1982 dollars) of $1.35 million. Within the truncated annual cohorts, average constant-dollar costs rose at an annual rate of 1.0 percent, but the time trend was not statistically significant. Thus, although there may have been dramatic changes on the supply side in some extreme instances, it remains unproven that R&D costs in the more typical case have risen appreciably over time, requiring across-the-board structural adaptations.

C. *Cultural Differences*

Very large Japanese firms appear to have overcome the barriers to becoming innovative powerhouses according to the 1942 Schumpeterian paradigm. In addition to the attraction they hold for university graduates, their R&D success may stem from unique facets of business culture not easily imitated in other nations. Masahiko Aoki (1990, p. 3) reports that

> Japanese firms have cultivated an ability for rapid response by developing an internal scheme in which emergent information is utilized effectively on-site and in which operating activities are coordinated among related operating units on the basis of information sharing.

He attributes Japanese *keiretsu* members' innovative successes to their rich in-house information networks and to the growth orientation shared by owner and employee interests because of corporate stock cross-holdings and employees' lifetime tenure. Although exceptions exist,

[8] On 1950s aircraft developments, see Thomas Marschak (1967).

large U.S. firms appear to have greater difficulty achieving high levels of intra-organizational communication, in part because managers' and engineers' job tenures are shorter and less secure, warranting less investment in developing rich interpersonal relationships and a broad organizational view through protracted job rotation. To the extent that organizational cultures do differ, the Schumpeterian large firm–innovator paradigm may apply with greater force in Japan than in America. Merely increasing U.S. company size without changing the underlying culture might fail to spur innovativeness, or even degrade it.

Cultural differences may also affect the extent to which cooperative interfirm R&D efforts enhance innovativeness. In all national environments, cooperation has proved difficult because the desire to gain a market advantage over one's rivals remains strong, and because companies tend to free-ride by keeping their top technical talent at home and assigning the second team to joint ventures. Even in Japan's VLSI project, intense rivalry and tendencies toward secrecy persisted. Success was achieved in part by limiting the focus to process infrastructure developments that were precursors to commercial chip design—a tactic that can be emulated in any nation—and also by MITI's active involvement to ensure that top individuals were seconded from companies to the joint research facility (Donna Doane 1984, ch. VI; Okimoto 1989, ch. 2; and Fumio Kodama 1991, pp. 84–91). Building a culture of cooperation for its own sake appears to be as important a goal in the European Community authorities' backing of R&D joint ventures as the belief that the projects will solve significant innovation blocking structural problems (Watkins forthcoming). Experience with the most closely comparable U.S. R&D joint ventures suggests that cooperation is indeed not easily cultivated.[9]

D. *Meshing Structure with the Technical Challenge*

Innovation entails a complex array of economic activities. It is conceivable that concentrated market structures are more conducive to some kinds of innovative activity than to others. This insight is emphasized among others in a controversial article by Charles Ferguson (1988). He begins from the widely accepted premise that the U.S. high-technology venture capital industry, characterized by easy access to capital and high mobility of creative scientists and engineers, has been enormously effective at generating innovative new *products*. But the typical venture firm, he claims, has failed to follow through on its first mover advantage by sustaining patient, continuing process development work aimed at reducing costs to minimum feasible levels while retaining high quality standards. As a result, U.S. innovators are said to have lost their initial lead to larger, more far-sighted Japanese firms willing to do the process work that permits low-cost production for mass markets.

Ferguson's thesis is consistent with Edwin Mansfield's remarkable finding (1988), from a survey of 100 matched Japanese and American corporations, that on average Japanese firms devoted 64 percent of their R&D budgets to internal process development and improvement, while their U.S. counterparts spent only 32 percent. Ferguson carries the argument a step farther, insisting that the U.S. high-technology venture industry

[9] See Katz and Ordover (1990, pp. 181–90); "Sematech Policies Assailed," *New York Times*, July 24, 1991, p. D24; "New Drive by Sematech for Funding," *New York Times*, Dec. 20, 1991, p. D1; and "A Founding Member Leaves Sematech Chip Consortium," *New York Times*, Jan. 7, 1992, p. D1.

has on the whole *impaired* the nation's technological strength by draining technical talent from large, well-established firms with stronger incentives to support extensive cost reduction R&D. It is readily shown that the larger the market share a firm can expect to tap, the greater will be the absolute savings from a process innovation reducing unit costs by some given percentage. And with diminishing marginal returns in process R&D, firms with sizable market shares will carry their efforts farther, achieving larger cost reductions (Cohen and Klepper 1992).

Data dividing U.S. companies' R&D expenditures into "product" and "internal process" categories for 1,819 product lines covered by the Federal Trade Commission's 1974 Line of Business survey were used to conduct a stylized test of the Ferguson hypothesis (Scherer 1991, pp. 32–33). Across all lines, process R&D averaged 26 percent of total R&D outlays. With additional controls for field of technology and industry concentration, the share of R&D outlays devoted to process improvement was found to increase by ten percentage points on average with each tenfold increase in a product line's sales. The size effect was highly significant statistically, suggesting that larger business units do, as Ferguson claims, devote relatively more attention to cost-reducing R&D (and, it follows, less to work on new products).

Whether it would be desirable to reallocate U.S. innovative activity away from venture firms like those of the 1912 Schumpeterian paradigm to the well-established giants lauded in *CSD* remains questionable. The answer could depend upon the specific technical challenge. Thus, it may be no accident that the United States retains a strong lead in microprocessor semiconductor chips, where bold product design advances can capture the market, while it lags in memory chips, where patient attention to yield-increasing, cost-reducing detail is more important.[10]

Culture may matter here too. Even if there were no high-technology venture capital industry, R&D staff turnover in sizable corporations is likely to be higher under accepted U.S. career path norms than it is in Japan. Japan differs also in the extent to which R&D engineers go through training programs with a strong on-the-shop-floor production orientation. U.S. R&D staff are more likely to move directly from their universities to ivory tower R&D labs, remaining aloof from factory operations. Thus, as the M.I.T. Commission on Industrial Productivity concluded, innovation in the United States can be strengthened by altering attitudes within companies and in the universities that train their staffs (Michael Dertouzos et al. 1989, chs. 11 and 12).

V. *Conclusion*

Half a century after the publication of *Capitalism, Socialism, and Democracy*, Schumpeter's vision of the industrial structure most conducive to technological progress and hence to economic growth remains both relevant and controversial. The book's publication stimulated a growing stream of theoretical and empirical research. Most of that research supports a conclusion that Schumpeter overstated the advantages of large, monopolistic corporations as engines of technological change. But doubts have intruded, and important policy relevant questions remain unanswered.

Several issues rank high on the priority list for future research on the Schum-

[10] See the quotation from the head of a Japanese semiconductor development group: " 'I have many friends in the U.S.,' he said, 'but every time I meet them their job is changed. My job has not changed for 18 years. My staff stays with me; they do not jump to other jobs or companies. I think that lifetime employment in Japan is very suitable for memory-chip making'." "Hitachi's Quest for Super Chips," *New York Times*, Sept. 10, 1990, p. C4.

peterian agenda. First, at what kinds of innovative activities do large organizations have comparative advantage? Comparative disadvantage? Second, how does the innovative capacity of large organizations vary with intrafirm organization, and especially with the nature of communication networks within firms? Third, do differences in national market size, market openness, and "culture" modify the relationships between market structure and innovation? (Current efforts to build a data base using a common methodology to tally innovation output indicators across ten or more nations are likely to facilitate analyses on this point.)[11] Fourth, the size distribution of R&D project costs must be delineated more precisely. Only then can we estimate the incidence of projects too costly to be undertaken by any but companies exceeding substantial size thresholds. Last but not least is a question salient to current policy debates: what has been achieved by the hundreds of cooperative R&D ventures formed in the United States during the 1980s? Under what industrial and technological conditions have they excelled, and where have they failed? Since the proper counter-factual is what would have been accomplished through independent R&D activity, case study and statistical research at least as rich and well structured as the best prior work testing Schumpeterian hypotheses will be needed.[12] As always, much remains to be learned.

[11] The work is directed by Alfred Kleinknecht under the auspices of the European Community's Institute for Prospective Technological Studies at Ispra, Italy.

[12] The most careful statistical work to date is by Link and Bauer (1989, pp. 86–89), who found that 62 U.S. companies engaging in cooperative R&D experienced significantly higher productivity growth than 30 nonparticipants. It is unclear, however, whether the difference stemmed from company characteristics that made cooperators more open to outside ideas, or whether cooperation per se was the cause. For a review of extant case studies, see Katz and Ordover (1990).

References

ACS, ZOLTAN J. AND AUDRETSCH, DAVID B. *Innovation and small firms*. Cambridge: MIT Press, 1990.

———. "R&D, Firm Size and Innovative Activity," in *Innovation and technological change*. Eds.: ZOLTAN J. ACS AND DAVID B. AUDRETSCH. N.Y.: Harvester Wheatsheaf, 1991, pp. 39–59.

ALLEN, ROBERT LORING. *Opening doors: The life & work of Joseph Schumpeter*. 2 volumes. New Brunswick: Transaction, 1991.

AOKI, MASAHIKO. "Toward an Economic Model of the Japanese Firm," *J. Econ. Lit.*, Mar. 1990, *28*(1), pp. 1–27.

ARROW, KENNETH J. "Economic Welfare and the Allocation of Resources for Invention," in *The rate and direction of inventive activity*. National Bureau of Economic Research conference report. Princeton: Princeton U. Press, 1962, pp. 609–25.

BALDWIN, WILLIAM L. AND CHILDS, GERALD L. "The Fast Second and Rivalry in Research and Development," *Southern Econ. J.*, July 1969, *36*(1), pp. 18–24.

BALDWIN, WILLIAM L. AND SCOTT, JOHN T. *Market structure and technological change*. Chur, Switzerland: Harwood, 1987.

BARZEL, YORAM. "Optimal Timing of Innovations," *Rev. Econ. Statist.*, Aug. 1968, *50*(3), pp. 348–55.

BOUND, JOHN ET AL. "Who Does R&D and Who Patents?" in *R&D, patents, and productivity*. Ed.: ZVI GRILICHES. Chicago: U. of Chicago Press, 1984, pp. 21–54.

COHEN, WESLEY M. AND KLEPPER, STEVEN. "The Anatomy of Industry R&D Intensity Distributions," *Amer. Econ. Rev.*, Sept. 1992, 82(4).

COHEN, WESLEY M. AND LEVIN, RICHARD C. "Empirical Studies of Innovation and Market Structure," in *Handbook of industrial organization*. Vol. 2. Eds.: RICHARD SCHMALENSEE AND ROBERT WILLIG. Amsterdam: North-Holland, 1989, pp. 1059–1107.

COHEN, WESLEY M.; LEVIN, RICHARD C. AND MOWERY, DAVID. "Firm Size and R&D Intensity: A Reexamination," *J. Ind. Econ.*, June 1987, *35*(4), pp. 543–63.

COOPER, ARNOLD C. "R&D Is More Efficient in Small Companies," *Harvard Bus. Rev.*, May/June 1964, *42*(3), pp. 75–83.

DASGUPTA, PARTHA AND STIGLITZ, JOSEPH. "Uncertainty, Industrial Structure, and the Speed of R&D," *Bell J. Econ.*, Spring 1980, *11*(1), pp. 1–28.

DERTOUZOS, MICHAEL L.; LESTER, RICHARD K. AND SOLOW, ROBERT M. *Made in America: Regaining the productive edge*. Cambridge: MIT Press, 1989.

DIMASI, JOSEPH A. ET AL. "Cost of Innovation in the Pharmaceutical Industry," *J. Health Econ.*, July 1991, *10*(2), pp. 107–42.

DOANE, DONNA L. "Two Essays on Technological Innovation: Innovation and Economic Stagnation, and Interfirm Cooperation for Innovation in Japan." Ph.D. dissertation, Yale U., 1984.

FELLNER, WILLIAM J. "The Influence of Market Structure on Technological Progress," *Quarterly J. Econ.*, Nov. 1951, *65*(4), pp. 556–77.

FERGUSON, CHARLES H. "From the People Who

Brought You Voodoo Economics," *Harvard Bus. Rev.*, May–June 1988, *88*(3), pp. 55–62.

FISHER, FRANKLIN M. AND TEMIN, PETER. "Returns to Scale in Research and Development: What Does the Schumpeterian Hypothesis Imply?" *J. Polit. Econ.*, Jan.–Feb. 1973, *81*(1), pp. 56–70.

FREEMAN, CHRISTOPHER. *Technology, policy, and economic performance: Lessons from Japan.* London: Pinter, 1987.

GALBRAITH, JOHN KENNETH. *American capitalism: The concept of countervailing power.* Boston: Houghton-Mifflin, 1952.

GEROSKI, PAUL A. "Innovation, Technological Opportunity, and Market Structure," *Oxford Econ. Pap.*, July 1990, *42*(3), pp. 586–602.

GILBERT, RICHARD J. AND NEWBERY, DAVID M.G. "Preemptive Patenting and the Persistence of Monopoly," *Amer. Econ. Rev.*, June 1982, *72*(3), pp. 514–26.

HARHOFF, DIETMAR. "Strategic Spillover Production, Vertical Organization, and Incentives for Research and Development." Ph.D. Dissertation, M.I.T., 1991.

HEATON, GEORGE R., JR. "The Truth About Japan's Cooperative R&D," *Issues in Science and Technology*, Fall 1988, *5*(1), pp. 32–40.

HEERTJE, ARNOLD, ed. *Schumpeter's vision: Capitalism, socialism and democracy after 40 years.* NY: Praeger, 1981.

HIMMELBERG, CHARLES P. AND PETERSEN, BRUCE C. "R&D and Internal Finance: A Panel Study of Small Firms in High-Tech Industries." Working paper, Fed. Res. Bank of Chicago, Mar. 1990.

VON HIPPEL, ERIC. *The sources of innovation.* NY: Oxford U. Press, 1988.

IMAI, KEN-ICHI. "Patterns of Innovation and Entrepreneurship in Japan," in *Evolving technology and market structure.* Eds.: ARNOLD HEERTJE AND MARK PERLMAN. Ann Arbor: U. of Michigan Press, 1990, pp. 187–201.

JEWKES, JOHN; SAWERS, DAVID AND STILLERMAN, RICHARD. *The sources of invention.* NY: St. Martin's, 1959.

KAMIEN, MORTON I. AND SCHWARTZ, NANCY L. "Timing of Innovations under Rivalry," *Econometrica*, Jan. 1972, *40*(1), pp. 43–60.

______. "On the Degree of Rivalry for Maximum Innovative Activity," *Quart. J. Econ.*, May 1976, *90*(2), pp. 245–60.

______. *Market Structure and Innovation.* Cambridge: Cambridge U. Press, 1982.

KATZ, MICHAEL L. AND ORDOVER, JANUSZ. "R&D Cooperation and Competition," *Brookings Pap. Econ. Act.: Microeconomics*, 1990, pp. 137–203.

KLEINKNECHT, ALFRED. "Measuring R&D in Small Firms: How Much Are We Missing?" *J. Ind. Econ.*, Dec. 1987, *36*(2), pp. 253–56.

KODAMA, FUMIO. *Analyzing Japanese high technologies.* London: Pinter, 1991.

KOHN, MEIR AND SCOTT, JOHN T. "Scale Economies in Research and Development: The Schumpeterian Hypothesis," *J. Ind. Econ.*, Mar. 1982, *30*(3), pp. 239–49.

LEVIN, RICHARD C. AND REISS, PETER C. "Cost-Reducing and Demand-Creating R&D with Spillovers," *Rand J. Econ.*, Winter 1988, *19*(4), pp. 538–56.

LEVIN, RICHARD C. ET AL. "Appropriating the Returns from Industrial Research and Development," *Brookings Pap. Econ. Act.*, 1987, (3), pp. 783–820.

LINK, ALBERT N. AND BAUER, LAURA L. *Cooperative research in U.S. manufacturing: Assessing policy initiatives and corporate strategies.* Lexington: Heath-Lexington, 1989.

LOURY, GLENN C. "Market Structure and Innovation," *Quart. J. Econ.*, Aug. 1979, *93*(3), pp. 395–410.

LUNN, JOHN. "Research and Development and the Schumpeterian Hypothesis: Alternate Approach," *Southern Econ. J.*, July 1982, *49*(1), pp. 209–17.

______. "An Empirical Analysis of Process and Product Patenting: A Simultaneous Equation Framework," *J. Ind. Econ.*, Mar. 1986, *34*(3), pp. 319–30.

MACLAURIN, W. RUPERT. *Invention and innovation in the radio industry.* NY: Macmillan, 1949.

MANSFIELD, EDWIN J. "Size of Firm, Market Structure, and Innovation," *J. Polit. Econ.*, Dec. 1963, *71*(6), pp. 556–76.

______. "Industrial R&D in Japan and the United States: A Comparative Study," *Amer. Econ. Rev.*, May 1988, *78*(2), pp. 223–28.

MANSFIELD, EDWIN J. ET AL. *Research and innovation in the modern corporation.* NY: Norton, 1971.

MARKHAM, JESSE W. *Competition in the rayon industry.* Cambridge: Harvard U. Press, 1952.

MARSCHAK, THOMAS. "The Role of Project Histories in the Study of R&D," in *Strategy for R&D.* Eds.: THOMAS MARSCHAK, THOMAS K. GLENNAN, AND ROBERT SUMMERS. NY: Springer, 1967, pp. 49–139.

MUKHOPADHYAY, ARUN. "Technological Progress and Change in Market Concentration in the U.S., 1963–77," *Southern Econ. J.*, July 1985, *52*(1), pp. 141–49.

NELSON, RICHARD R. "The Economics of Invention: A Survey of the Literature," *J. Bus.*, Apr. 1959, 32(2), pp. 101–27.

NELSON, RICHARD R. AND WINTER, SIDNEY G. *An evolutionary theory of economic change.* Cambridge: Harvard U. Press, 1982.

New York Times, various issues.

NORDHAUS, WILLIAM D. *Invention, growth, and welfare: A theoretical treatment of technological change.* Cambridge: MIT Press, 1969.

OKIMOTO, DANIEL I. *Between MITI and the market: Japanese industrial policy for high technology.* Stanford: Stanford U. Press, 1989.

PAKES, ARIEL AND SCHANKERMAN, MARK. "An Exploration into the Determinants of Research Intensity," in *R&D, patents, and productivity.* Ed.: ZVI GRILICHES. Chicago: U. of Chicago Press, 1984, pp. 209–32.

REINGANUM, JENNIFER F. "Uncertain Innovation and the Persistence of Monopoly," *Amer. Econ. Rev.*, Sept. 1983, *73*(4), pp. 741–48.

______. "The Timing of Innovation: Research, Development, and Diffusion," in *Handbook of industrial organization.* Vol. 1. Eds.: RICHARD SCHMALENSEE

AND ROBERT WILLIG. Amsterdam: North-Holland, 1989, pp. 849–908.

SCHERER, F. M. "Research and Development Resource Allocation under Rivalry," *Quart. J. Econ.*, Aug. 1967, *81*(3), pp. 359–94.

______. "The Propensity To Patent," *Int. J. Ind. Organ.*, Mar. 1983, *1*(1), pp. 107–28.

______. *Innovation and growth: Schumpeterian perspectives*. Cambridge: MIT Press, 1984.

______. "Changing Perspectives on the Firm Size Problem," in *Innovation and technological change*. Eds.: ZOLTAN ACS AND DAVID AUDRETSCH. NY: Harvester Wheatsheaf, 1991, pp. 24–38.

SCHERER, F. M. AND ROSS, DAVID. *Industrial market structure and economic performance*. 3rd ed. Boston: Houghton-Mifflin, 1990.

SCHMOOKLER, JACOB. *Invention and economic growth*. Cambridge: Harvard U. Press, 1966.

SCHUMPETER, JOSEPH A. *Theorie der wirtschaftlichen Entwicklung*. Leipzig: Duncker & Humblot, 1912. Revised English translation by REDVERS OPIE, *The theory of economic development*. Cambridge: Harvard U. Press, 1934.

______. *Business cycles*. NY: McGraw-Hill, 1939.

______. *Capitalism, socialism, and democracy*. NY: Harper, 1942.

______. "The March into Socialism," *Amer. Econ. Rev.*, May 1950, *40*(2), pp. 446–56.

SCOTT, JOHN T. AND PASCOE, GEORGE. "Purposive Diversification of R&D in Manufacturing," *J. Ind. Econ.*, Dec. 1987, *36*(2), pp. 193–205.

SERVAN-SCHREIBER, JEAN-JACQUES. *The American challenge*. NY: Atheneum, 1968.

U.S. HOUSE OF REPRESENTATIVES, Committee on Banking, Finance and Urban Affairs, Subcommittee on Economic Stabilization, Hearings, *Structuring American industry for global competition*. Washington: U.S. GPO, Mar. 1986.

U.S. SENATE, COMMITTEE ON THE JUDICIARY. *The National cooperative research act extension of 1991*. Report 102–146, Washington: U.S. GPO, Sept. 1991.

WATKINS, TODD A. "The Evolution of European Technological Cooperation: The Institutional Economics of Cooperative R&D." Ph.D. Dissertation, Harvard U. forthcoming.

Part IV
Intellectual Property Rights

[12]

Journal of Economic Perspectives—Volume 5, Number 1—Winter 1991—Pages 29–41

Standing on the Shoulders of Giants: Cumulative Research and the Patent Law

Suzanne Scotchmer

Sir Isaac Newton himself acknowledged, "If I have seen far, it is by standing on the shoulders of giants." Most innovators stand on the shoulders of giants, and never more so than in the current evolution of high technologies, where almost all technical progress builds on a foundation provided by earlier innovators. For example, most molecular biologists use the basic technique for inserting genes into bacteria that was pioneered by Herbert Boyer and Stanley Cohen in the early 1970s, and many use a technique for causing bacteria to express human proteins that was pioneered at Genentech. In pharmaceuticals, many drugs like insulin, antibiotics, and anti-clotting drugs have been progressively improved as later innovators bettered previous technologies. Computer text editors are similar to one another, as are computer spreadsheets, in large part because innovators have inspired each other. An early example of cumulative research was Eli Whitney's cotton gin, which was quickly modified and improved by other innovators who seriously curtailed his profit.[1]

[1]Eli Whitney was very generous in disclosing details of his gin to other innovators, even beyond what was required by patent law. Other innovators patented improvements, but after much litigation the new patents were held to infringe Whitney's underlying patent. Whitney and his partner did not recover sufficient damage awards to compensate them for their litigation and the time that it took to enforce the patent. For an extensive discussion, see Jeanette Mirsky and Allan Nevins, *The World of Eli Whitney* (Macmillan Co, 1952). Klemperer (1990) also discusses the cotton gin.

■ *Suzanne Scotchmer is Professor of Public Policy, University of California, Berkeley, California.*

Most economics literature on patenting and patent races has looked at innovations in isolation, without focusing on the externalities or spillovers that early innovators confer on later innovators. But the cumulative nature of research poses problems for the optimal design of patent law that are not addressed by that perspective. The challenge is to reward early innovators fully for the technological foundation they provide to later innovators, but to reward later innovators adequately for their improvements and new products as well. This paper investigates the use of patent protection and cooperative agreements among firms to protect incentives for cumulative research.

The Available Tools

The breadth of patent protection is a key consideration in the incentives to innovate. Patent applicants protect themselves against competition from derivative products by claiming broad protection. Patent law would provide no protection at all if it did not protect against trivial changes, like color or size. The allowable breadth of claims is determined by patent examiners and the judiciary. If broad protection is granted, then a derivative or second generation product will likely infringe the prior patent, so a license on the original patent is required to market it. If patent protection is narrow, then many derivative products and applications can be patented and marketed without infringement.[2] We might be tempted to conclude that broad protection encourages firms to find fundamental technologies but discourages them from seeking out second-generation applications and derivative products. However, these two conclusions may be inconsistent, since proper incentives to find fundamental technologies may require that the first patent holder earn profit from the second generation products that follow. There will be no such profit if no second-generation products follow.

Patent protection would be an unnecessary policy tool if the government had the same information about the costs and benefits of individual research projects as firms have. In that case, the government could simply select the research projects that would be socially efficient and commission research from the lowest cost firms. However, the government will generally have less information than firms,[3] and I will therefore assume that the length and breadth of patent protection and other aspects of the government's policy toward R&D cannot depend on firms' private information about their expected costs.[4]

[2] U.S. patent protection is broader than that in most other countries, particularly Japan, partly due to the "doctrine of equivalents," which can broaden protection beyond the claims in the patent according to similarity of function.

[3] Wright (1983) discusses the private information of firms as the main justification for patent protection, rather than using prizes or contracts as incentive instruments.

[4] A policy of reimbursing costs to the successful innovator would not be adequate, since a project that was a "good bet" at the beginning might nevertheless fail.

Given that the length and breadth of patent protection cannot depend on the expected costs of an R&D project, the only way to ensure that firms undertake every research project that is efficient is to let the firms collect as revenue all the social value they create. Otherwise, some projects that are socially desirable will not be undertaken. If an innovation is a reduction in the cost of producing a good, then the social value is the saved costs. If the innovation is an improvement to a product, the social value is the difference in consumers' willingnesses to pay for the improved and unimproved products. When research firms collect all the social value as profit, households still benefit, but in their capacity as shareholders rather than as consumers.

But there are at least two problems with allowing research firms to collect all the social surplus as profit (or as much as possible). First, strong patent protection leads to socially inefficient monopoly pricing. Second, firms in a patent race may overinvest in research if the patent is worth more than the (minimum) cost of achieving it (Loury, 1979). This problem is related to the problem of the commons: An increase in one firm's rate of investment transfers some probability of becoming the patentholder from other firms to itself. Because of this transfer, all firms might overinvest.[5] These points are well-recognized in the R&D literature.

When an initial innovation facilitates later ones, as is the case with basic research, another issue arises. Part of the first innovation's social value is the boost it gives to later innovators, which can take at least three forms. If the second generation could not be developed without the first, then the social value of the first innovation includes the incremental social surplus provided by second generation products. If the first innovation merely reduces the cost of achieving the second innovation, then the cost reduction is part of the social surplus provided by the first innovation. And if the first innovation accelerates development of the second, but at the same cost, then its social value includes the value of getting the second innovation sooner.

Because of these externalities provided to later innovators, developing the first innovation may be efficient even if its expected cost exceeds its value as a stand-alone product. First innovators will have correct incentives to invest only if they receive some of the social surplus provided by second generation products. But at the same time, enough profit must be left for the second innovators so that they will invest if investing is efficient. This essay asks how close patent incentives can come to accomplishing that goal.

A premise in much of what follows is that firms other than the first innovator should participate in development of second generation products. Since the first innovator might not have expertise in all applications, more second generation products are likely to arise if more researchers have incen-

[5]But Gandal and Scotchmer (1989) show that prior agreements among firms that would otherwise race can overcome the incentive for overinvestment, even if their research costs are private information. When such prior agreements are allowed, the firms will invest at the efficient rates if and only if the private value of the patent is equal to the social value.

tive to consider them. In this view, contrary to the premise of much of the patent race literature, creativity is largely serendipitous. Not every R&D firm sees the same opportunities for new products.

However, outside research firms can integrate with initial patentholders in at least two ways: the firms can form cooperative ventures to research and develop new products, and they can form licensing agreements after products have been developed and patents have been awarded. I will call these two types of contracts prior agreements and licenses, respectively. Prior agreements permit firms to share the costs, as well as the proceeds, of research.[6] Licenses are negotiated after research costs are sunk and patents have been awarded. Both types of agreements can increase profit by improving efficiency and possibly by reducing product market competition. Although many authors have discussed cooperation in research, they have not focussed, as I will, on how the breadth of patent protection and cooperation among research firms work together in protecting incentives to innovate. In this view of how incentives to innovate are protected, a key role of patent protection is that it sets bargaining positions for the prior agreements and licenses that will form, and therefore determines the division of profit in these contracts.

Patent Protection and Licensing

A system of property rights that might seem natural would be to protect the first innovator so broadly that licensing is required from all second generation innovators who use the initial technology, whether in research or in production. But such broad protection can lead to deficient incentives to develop second generation products. When the licensing agreement is negotiated after a patent has been granted, research costs have already been sunk. The bargaining surplus to be split between the first and second innovators at that time is the incremental market value of the second product, *not* net of research costs. A second innovator who cannot market the next generation product without a license has a very weak bargaining position. If the second innovator does not get all the surplus being bargained over, he will earn only a fraction of the new product's market value and presumably only a fraction of its social value, and this fraction may be less than the cost of developing it. Hence the incentive for an outside firm to develop second generation products can be too weak.[7] Under such broad patent protection, the incentive for the first innovator to develop a second generation product will be stronger than for an

[6]Prior cooperation in research has been treated leniently by antitrust law and the authorities. For example, the National Cooperative Research Act (1984) established that joint ventures are not per se illegal, but will be treated according to a "rule of reason." The Act also reduced damages in civil suits from treble to single, provided the firms follow the proper notification procedure. Perspectives on this act are discussed in the symposium on "Collaboration, Innovation and Antitrust" in the Summer 1990 issue of this journal.

[7]See Green and Scotchmer (1990) for an elaboration of this idea.

outside firm (provided the first innovator has expertise to develop the new product, and thinks of it), since the first innovator will earn the entire incremental profit.

As well as offering deficient incentives for second innovators, broad patent protection might inefficiently inflate incentives for the first innovation. In licensing agreements, the first innovator will earn a share of the market value of each infringing later product. If the first innovation reduces the cost of achieving later innovations, but is not the only possible vehicle to achieve them, the first innovator's share should not exceed the cost reduction. If it does, the first innovator will be overrewarded.

In what follows I explore two solutions to these defective incentives. The remainder of this section investigates what happens if the first innovator's patent protection is narrowed so that a different enough second generation product does not infringe and thus can be marketed without a license from the first innovator. In the following section, I investigate prior agreements in which second innovators can "sell" their ideas to the first innovator or integrate with the first innovator. Neither solution is perfect, as we shall see.

The inadequacies of narrowing patent protection are most easily exposed if we first suppose that first and second generation products do not compete in the market, although second generation products build on the first generation technology; for example, many new pharmaceuticals that are therapies for different illnesses all build on a few basic techniques of bio-engineering. Second innovators cannot have excessive incentive to invest, since they cannot earn more than consumers' willingness to pay in the markets they serve. Licensing from the first innovator would transfer away some of the second innovators' revenue and hence reduce their incentive to invest. To provide efficient incentives to the second innovator, society should protect the first innovation so narrowly that a new product never infringes and therefore second innovators never have to license. But such a scheme does not sufficiently reward the first innovator, since the first innovator does not profit from the cost reduction conferred on the second innovators.

The first innovator's incentive to invest becomes still weaker under narrow patent protection if the second generation product is a substitute for the first. Competition between the two patentholders would erode their joint profit, transferring some of the social surplus of the combined innovations to consumers. As an example, suppose that the second generation product is a superior version of a drug, and that the two patent-holders compete on price. Then the second generation product will survive in the market and its price will equal the difference in consumers' willingness to pay for the two drugs plus the marginal cost of producing the drug. In this outcome, the second innovator earns as profit exactly the incremental social value of the newer drug, while the first innovator's profit falls to zero.

Such profit erosion could be mitigated if the antitrust authorities permitted collusive licensing among patentholders who would otherwise compete. For example, licensing with per-unit royalties can lead to collusive outcomes, since the royalty raises the licensees' private production cost and therefore keeps the equilibrium price high.[8] In ordinary antitrust law, collusion through licensing would violate the spirit of the Sherman Act and subsequent legislation. But where incentives to innovate are at stake and where later technology builds on an earlier technology, such collusion allows the first innovator to profit from the externality conferred on later innovators. Of course, firms would be tempted to exploit any leniency by the antitrust authorities in contexts where incentives to innovate are not at stake. This problem should not be minimized.

There is something quite general economists can say about the combined effects of patent law with licensing: No such policy can achieve fully efficient incentives, even if society permits collusive licensing between patent holders who would otherwise compete and the firms jointly collect all the social surplus as profit. This is essentially because of "double marginalization." To give the second innovator an incentive to invest whenever social benefits exceed R&D costs, the second innovator must earn the entire social surplus of his innovation. But to compensate the first innovator for the externality or spillover she provides, she too must earn part of this surplus. It is impossible to give the surplus to both parties.[9]

When both first and second generation products are developed, the division of profit between the two innovators depends on the breadth of patent protection. To see this, assume that there is a random component to the outcome of a research project, so that when a research firm invests in a second generation product, it does not know whether its product will infringe the prior patent. The breadth of the prior patent determines the probability that the second generation product will infringe. If the second product turns out to infringe, the second innovator must license and this will force him to share the profit of the improvement with the first innovator. The second innovator is in a better position if its product turns out not to infringe, since the second innovator can profitably compete with the prior patentholder in the market. Thus, if breadth of the first patent could be interpreted to depend on the expected costs and benefits of a second generation product, we could ensure that the second innovator's expected profit would be zero. If not, some second generation products will be stymied even though they would contribute positively to joint profit and to social welfare, and the second innovators who invest will typically make positive profit.

To summarize, the "natural" system of property rights—requiring every later innovator to license any underlying technology—will on average give deficient incentives for outside firms to develop second generation products.

[8]If royalties are permitted, then licensing is similar to permitting the initial patent-holder to buy up the patents on later products that use the initial technology.

[9]Green and Scotchmer (1990) argue this in a model where the second innovation would be impossible without the first innovation. It is also true in the less drastic case when the first innovation merely reduces the cost of achieving the second innovation.

This is because the second product infringes and therefore the second innovator must transfer some of the innovation's revenue to the first innovator by licensing. If the first innovator can be relied upon to develop all second generation products, this would not matter. Second, no system of narrower patent protection and licensing can give the right incentives to both the first innovator and other firms that develop improvements, even if collusive licensing among noninfringing products were allowed. The latter result depends on my premise that the breadth of an underlying patent cannot be separately tailored to the costs and benefits of each second generation product.

In the next section, I ask to what extent these inadequacies of patent protection and licensing can be overcome with prior agreements reached before some or all the patents have been obtained. Incentives with licensing are defective mainly because firms negotiate after all costs have been sunk and patents have been issued. A prior agreement integrates the potential second innovator into the firm of the first innovator before investing in the second innovation. Such prior agreements can indeed guarantee efficient investment in second generation products, but cannot perfectly solve the incentive problem unless the negotiation is before *all* costs are sunk, including the costs of the first innovator, or unless the first innovator has all the bargaining power.

Prior Agreements

Prior agreements among research firms are often called research joint ventures. Joint ventures presumably form to increase the joint profit of the members, but they do not necessarily increase social welfare, since the cooperation is among firms only and does not include consumers.[10] Joint ventures increase profit both by providing incentives to the members to invest more efficiently, and by finding ways to transfer social surplus from consumers to firms. The greater efficiency might result from exploiting economies of scale (Katz, 1986), from sharing technological know-how (Bhattacharya, Glazer and Sappington, 1988), or from undoing the inefficiencies of a patent race (Gandal and Scotchmer, 1989).

One solution to the incentive problem would be to integrate all possible innovators into one firm before even the first innovator has invested. Then, provided the integrated firm gets most of the social surplus from the joint innovations, it should invest (close to) efficiently. Although research firms do not know with certainty what projects they will think of after the first generation technology has been developed, they have expectations about the possible benefits and costs of such projects. Provided all researchers have similar expectations, an agreement negotiated before the first investment could ensure

[10]The Coase theorem would conclude that bargains increase the joint welfare of all the parties. Thus, if consumers and firms could jointly cooperate, prior agreements would inevitably benefit both groups. But since some of the parties are excluded—namely consumers—there is no guarantee that prior agreements increase total welfare.

that the first innovation is undertaken if and only if efficient, where efficiency is defined relative to the prior judgments about costs and benefits. But the more serendipitous is the discovery of second generation products, and the more difficult it is to include all potential second innovators, the less feasible such an agreement seems. I therefore consider the more limited prospects when integration occurs after the first innovation. The difficulties in transferring profit to the first innovator are clearest if we assume that the innovators can jointly collect all the social surplus as profit, provided they do not compete in the market.

After the first patent has issued, a potential second innovator could approach the first patent-holder with an idea for an improvement or new product, and suggest that they share both the costs and proceeds of research. Such an agreement can increase joint profit by increasing investment in profitable second generation products and by preventing market competition among firms that would otherwise own competing patents. If patent protection is broad, without this prior agreement the second innovator could have deficient incentive to invest, as explained above. With a prior agreement, the initial patent-holder can agree to share both the costs and the proceeds of the second innovation, and will do so whenever benefits exceed costs.

Prior agreements are a social improvement over licensing because they can improve incentives to invest in second generation products, whatever the breadth of patent protection. With licensing, the breadth of patent protection serves two purposes: It determines investment in second generation products and determines how the firms' joint expected profits will be divided. With prior agreements, the breadth of patent protection serves one purpose instead of two: The two innovators have an incentive to invest efficiently in second generation products whatever the breadth of patent protection. The breadth of protection determines only the bargaining positions, hence the division of profit.[11]

Whether a prior agreement can provide efficient incentives for the first investment as well as the second depends on two factors: how much social surplus must be transferred from the second innovator to the first (how big the externality is), and the second innovator's bargaining power. A second innovator who has a strong bargaining position will earn positive profit in a prior agreement, thus limiting how much social surplus the first innovator can collect. The second innovator's bargaining position is strongest if there is a high probability the second innovation would not infringe and if the second generation product is itself patentable. The second innovator will also have a strong bargaining position if no other firm is capable of developing the second generation product.

[11]On the other hand, approaching the first innovator with the idea for the second innovation might give away the idea of the potential second innovator, and thereby undermine its bargaining position. The law has remedies for this problem, but presumably they do not work perfectly.

To bid the second innovators' profits to zero, it is not necessarily enough that each potential second innovator has many competitors. Suppose the second generation product is itself patentable and the first innovator agrees with one potential second innovator that that firm will develop the improved product and have an exclusive license on the initial patent. Despite this agreement, the parties will be forced to negotiate with any other firm that gets the improvement first since they could not duplicate the second invention without infringement. Anticipating this, potential second innovators might not bid away their profit in trying to make an agreement with the first innovator (Scotchmer, 1990).

Some authors who have written on research joint ventures have pointed out that firms' incentives to cooperate at the research stage depend partly on whether the members can also collude in using the resulting patents (Ordover and Baumol, 1985; Katz, 1986; d'Aspremont and Jacquemin, 1988; Choi, 1989). My discussion has assumed they can, and that such collusion is an important way to protect incentives to innovate. It would be difficult to implement a rule that permitted the consolidation of property rights through a prior agreement only if the eventual patents would otherwise be infringing. When research outcomes are random, patent authorities cannot know whether an R&D project will result in new technology that infringes an earlier patent.

Conclusion

A main conclusion of this essay is that it is misleading to ask how broad patent protection should be without simultaneously asking whether research firms can integrate or otherwise cooperate. Similarly, it is misleading to ask how leniently society should treat cooperation among research firms without simultaneously asking how broad patent protection should be. Prior agreements have the advantage of leading to more efficient investment in second-generation products. But if the patent authorities disallow collusive licensing of patents that are noninfringing, they might also want to disallow prior agreements. A prior agreement merges the interests of the two firms whether or not the patents would be noninfringing, and could circumvent the antitrust authority's desire to force competition. Of course, competition has a disadvantage where incentives to innovate are at stake: It erodes the joint profit of the firms, which undermines at least one firm's incentive to invest in R&D when such investment is efficient.

There are no simple conclusions to draw about the optimal breadth of patents. It is not necessarily optimal to protect the first innovation so broadly that every derivative or second generation product infringes. If prior agreements are disallowed or ineffective for some reason, then broad patent protection could discourage the development of second generation products, as explained above. And if the first innovator does not expect to profit by licensing to second generation innovators, broad protection could inhibit the first innovation as well, thus undermining the entire research line.

Broad protection might also be undesirable when prior agreements are allowed. To encourage researchers to invest in second generation products, the first innovator might have to make prior agreements with firms that have bargaining power. Their bargaining power derives from the fact that, without an agreement, they might have a credible threat not to invest, and from patentability of the second generation product. Because of their bargaining power, they may get a share of the bargaining surplus. Suppose patent protection is narrowed enough so that second innovators will invest without a prior agreement, but their profit is kept low. This would increase the first patentholder's profit, as may be necessary to compensate the first innovator for the externality or spillover conferred on second innovators.[12]

Patent law is limited in its instruments: the main ones are the patent life and the breadth of protection.[13] The private value of patent protection is linked to the social value of the technology through market demand, but is not linked to firms' research costs. The optimal rule for the breadth of a patent can only use information that is available to patent examiners and courts.[14] Thus, the patenting rule can depend on observable aspects of discovered technologies, but not on prior expectations regarding technological outcomes and costs of research. This restriction greatly reduces the effectiveness of patent law in protecting incentives.

Before investing in a second generation technology, the researcher must evaluate the probability that the new technology will not infringe the prior patent. This probability depends on the breadth of the prior patent and on the distribution of possible outcomes of the second investment. The probability of infringing the first patent is lower if the distribution of outcomes places greater weight on outcomes that lie outside the allowed claims of the first patent. A project with low probability of infringing will look more profitable to the second generation researcher than a project that places greater prior weight on outcomes that infringe the first patent. But both projects could have the same expected social value if the expected costs of one project were sufficiently higher than the other. The patent policy should equally encourage two projects

[12]See Green and Scotchmer (1990) for an elaboration of this argument.

[13]Of course, details like priority rules also matter. Everywhere except the United States, a disputed patent issues to the first applicant. In the United States, a disputed patent issues to the first inventor, regardless of when application occurs. See Scotchmer and Green (1990) for a discussion of the incentive effects of these two rules. The anomalous American rule is now being reconsidered.

[14]Except for Wright (1983), authors have not focussed explicitly on what information is available to patent authorities. Gandal and Scotchmer (1989), Green and Scotchmer (1990), Klemperer (1990) and Scotchmer and Green (1990) assume that patent protection cannot depend on costs. Klemperer (1990) and Gilbert and Shapiro (1990) assume that the breadth of patent protection can depend on aspects of market demand. Green and Scotchmer (1990) assume that the breadth of patent protection can depend on the level of previous technical advance.

with the same social value, but that cannot be accomplished with a patent rule that depends only on technological outcomes.

A disadvantage of narrow patent protection that I have not yet discussed is that it might discourage first innovators from patenting and disclosing their technologies. Patent law requires disclosure for the same reason that innovators dislike it: it is the vehicle by which technical knowledge is passed from the patenting firm to its competitors.[15] The first innovator would rather develop second generation products than let other firms develop them, since that would be more profitable. As a consequence, the first innovator has a strategic incentive not to patent the innovation. Instead, the first innovator could hold the product off the market until it develops the more valuable second generation products, or it could market the first product and rely for protection on the law of trade secrets, which does not require disclosure, but also does not protect against independent invention.

The first innovator's incentive to patent the initial technology depends on: (i) the profitability of marketing the first technology prior to the development of second generation products; (ii) the extent of disclosure that patenting entails;[16] (iii) the ease with which the technology could be reverse-engineered if marketed but not patented; and (iv) the breadth of patent protection. The incentive not to patent is especially strong when patent protection is narrow, since a second generation product is then more likely to damage the first innovator's profit.

The problem of cumulative research is especially acute when the first technology has very little value on its own, but is a foundation for valuable second generation technologies. Even with licensing, the first innovator might not capture the full social value that it facilitates and may have deficient incentive to invest. This is presumably why governments fund basic research. The branches of government that fund research are not those that set patent policy, and the decision to support basic research might be interpreted as a recognition that patents and licensing are inadequate.

Governments have taken different views of whether publicly sponsored research should also be patentable. The U.S. government permits and even encourages patenting of results from government sponsored research; for example, the Boyer-Cohen patent. In contrast, the British government forbade the Cambridge Molecular Biology Lab from patenting monoclonal antibodies

[15] The disclosure requirement in section 112 of the patent law states that "the specification shall contain a written description ... in such full, clear, concise and exact terms as to enable any person skilled in the art ... to make and use the same ..." Scotchmer and Green (1990) show that a leading firm might not want to patent a patentable technology even if this means holding it off the market until the next innovation in order to avoid reverse engineering.

[16] Disclosure of some technologies, such as chemical compositions, teaches competitors much that is valuable. For example, after the basic material of superconductors was disclosed, many other researchers developed aspects of it (*The New York Times*, January 2, 1989, p. 34). Disclosure of other technologies, like bio-engineered proteins, gives away much less that is useful to competitors, and therefore innovators will not fear disclosure as much.

in the mid-1970s. Permitting patents on government sponsored research rewards successful innovators twice, once through government funding and again through the patents.

I have focussed in this essay on how to divide joint profit among innovators when one innovator's technology builds on another's. I have not focussed on the length of patent protection as a policy tool. My simplifying assumption that the innovators can jointly get all the social surplus as profit if they do not compete in the market essentially means that the patent continues forever. When owners of noninfringing patents cannot collude through licensing, a noninfringing second generation product will undermine the profitability of the first patent, and this may happen before the end of the patent life. Mansfield (1984) surveyed research firms and found that the effective lives of most patents are much shorter than the stipulated 17 years. In the extreme case where a second product is an improved version of the first, the effective length of the first patent may simply end when the second product is introduced. The effective patent life is determined directly by its breadth, since the breadth determines how long it takes until this happens. On the other hand, when second generation products serve different markets, as when consumers have different tastes and a second generation product is a variant that serves a different set of consumers, length and breadth can be chosen independently, although the monopoly power conferred by a patent may be eroded over the patent's life as similar but noninfringing products are invented.

Gilbert and Shapiro (1990) and Klemperer (1990) have discussed breadth of patent protection in the context of single innovations without focussing on the problem of dividing profit among sequential innovators. In their conceptualization, breadth and length are substitute ways to provide a fixed profit to a single innovator, and they ask what combination of length and breadth minimizes the cost of monopoly distortions. They do not ask how large the fixed profit should be and therefore do not focus on how to preserve incentives to innovate. The prescription for patent breadth that comes out of that perspective may well conflict with the prescription for patent breadth that comes from considering incentives to innovate when research is cumulative.

It appears that patent policy is a very blunt instrument trying to solve a very delicate problem. Its bluntness derives largely from the narrowness of what patent breadth can depend on, namely the realized values of the technologies. As a consequence, the prospects for fine-tuning the patent system seem limited, which may be an argument for more public sponsorship of basic research.

■ *Prepared for the Symposium on Intellectual Property Law, funded by the RAND Corporation and the John P. Olin Foundation, which took place in Washington D.C. on October 24, 1989. I thank Stan Besen, Joe Farrell, Jerry Green, Leo Raskind, Steve Salop, Carl Shapiro, Eugene Smolensky, Joe Stiglitz and Timothy Taylor for useful comments. I thank Lucette Decorde for her able research assistance and the NSF, Grant SES 89 09503, for financial support.*

References

d'Aspremont, C., and A. Jacquemin, "Cooperative and Noncooperative R & D in Duopoly with Spillovers," *American Economic Review*, 1988, *78*, 1133–1137.

Bhattacharya, S., J. Glazer, and D. Sappington, "Motivating Exchange of Knowledge in R & D Ventures: First-Best Implementation," Bell Communications Research, Technical Memorandum, 1988.

Choi, Jay P., "An Analysis of Cooperative R & D," mimeo, Dept. of Economics, Harvard University, 1989.

Gandal, N., and S. Scotchmer, "Coordinating Research Through Research Joint Ventures," GSPP Working Paper #171, University of California, Berkeley, 1989.

Gilbert, Richard and Carl Shapiro, "Optimal Patent Length and Breadth," *The RAND Journal of Economics*, 1990, *21*, 106–112.

Green, J., and S. Scotchmer, "Antitrust Policy, the Breadth of Patent Protection and the Incentive to Develop New Products," GSPP Working Paper #171, (revised), University of California, Berkeley, 1990.

Jacquemin, A., "Cooperative Agreements in R & D and European Antitrust Policy," *European Economic Review*, 1988, *32*, 551–560.

Katz, M., "An Analysis of Cooperative Research and Development," *The RAND Journal of Economics*, 1986, *17*, 527–543.

Klemperer, Paul, "How Broad should the Scope of Patent Protection Be?" *The RAND Journal of Economics*, Spring 1990, *21*, 113–130.

Loury, Glenn C., "Market Structure and Innovation," *Quarterly Journal of Economics*, 1979, *XCIII*, 395–410.

Mansfield, Edwin, "R & D and Innovation: Some Empirical Findings." In Griliches, Zvi, ed., *R & D, Patents and Productivity*. Chicago: University of Chicago Press for the National Bureau of Economic Research, 1984.

Ordover, Janusz, and William J. Baumol, "Antitrust for High-Technology Industries: Assessing Research Joint ventures and Mergers," *Journal of Law and Economics* 1985, *XXVIII*, 331–331.

Scotchmer, S., "Protecting Early Innovators: Should Accessory Products, Bundled Improvements and Applications be Patentable?," GSPP Working Paper #183, University of California, Berkeley, 1990.

Scotchmer, S., and J. Green, "Novelty and Disclosure in Patent Law," *The RAND Journal of Economics*, 1990, *21*, 131–146.

Wright, Brian, "The Economics of Invention Incentives: Patents, Prizes and Research Contracts," *American Economic Review*, 1983, *73*, 691–707.

[13]

Journal of Economic Perspectives—Volume 5, Number 1—Winter 1991—Pages 43–60

A Patent System for Both Diffusion and Exclusion

Janusz A. Ordover

Public policy analysis of optimal patent regimes is often framed as a tradeoff between static and dynamic efficiency. In this context, static efficiency considerations mandate that the knowledge-asset, resulting from research and development investments undertaken by firms and individuals, be made widely available to all those who are willing to pay the low marginal cost of dissemination. This perspective implies that public policy should encourage the widespread use of knowledge-assets, and hence suggests that property rights in such assets should be minimal. But if the owners of knowledge-assets have only minimal property rights, their returns may not be high enough to recoup their initial investment, much less attain a sufficient return to compensate them for undertaking a risky project. As a result, the initial investment may not be undertaken, thereby harming dynamic efficiency. This perspective argues for public policies that provide strong and exclusive property rights for knowledge-assets, even if this is to the detriment of diffusion.

In this analytical framework, weak patent protection and strict antitrust policy are taken to be directed toward static concerns, while protection of intellectual property through strong patent laws is taken as a reflection of broader social concerns for long-run growth and technological progress. This characterization has some truth, but the magnitude of the conflict between static and dynamic efficiency, or between diffusion and exclusion, can easily be exaggerated.

■ *Janusz A. Ordover is Professor of Economics, New York University, New York, New York. This article was written while he was Visiting Professor at Yale University School of Organization and Management, New Haven, Connecticut.*

In this article, I want to argue that weak patent protection need not be inimical to economic growth and, conversely, that strong patent protection need not be an enemy of diffusion.[1] Appropriately structured patent law and antitrust rules can together ensure incentives for R&D and also induce cooperation among firms in diffusing R&D results through licensing and other means. At the same time, cooperation among firms at the R&D stage can counterbalance weak patent protection by internalizing spillovers from ongoing R&D programs, and such cooperation may also produce additional spillovers from the existing knowledge.

This overview of the issues leads to several important observations. First, the short term vs. long term, and exclusion vs. diffusion dichotomies are much too crude for designing overall public policy towards intellectual property. Second, as has been increasingly recognized in the technical economic literature, patents can be "adjusted" among many different dimensions, with differing implications for social welfare and inventive activity. Third, a "cross-sectional" review of property rights and antitrust regimes in the United States, Japan, and the European Community points to significant differences in the relevant institutional frameworks and in the choices on exclusion/diffusion tradeoffs. This cross-sectional evidence suggests that the current U.S. policy thrust of strengthening patent protection while reducing antitrust restrictions on the exercise of intellectual property rights is not a precondition for high rates of growth and for the maintenance of R&D incentives. The importance of this observation is especially pertinent now when new GATT disciplines on intellectual property rights are being negotiated. The United States is quite correct, however, on insisting that the worst excesses of intellectual piracy should be stopped and that patent systems should be nondiscriminatory against foreign applicants.

Designing Patent Law for Cooperation: The Case of Japan

To an American economist, the notion that exclusion is inherent in patent law may seem obvious. After all, the right that a patent bestows on its holder is clearly the right to exclude those who do not own the patent. This exclusion occurs despite the fact that others may have invested in an R&D project that

[1] A related issue is that strong patent protection may not necessarily be conducive to growth. I do not discuss this issue in the article, but the economics of the situation is reasonably clear. Strong intellectual property rights regimes (extensive patent protection, for example) can generate "competitive spillovers" that affect the payoffs to R&D-market competition and product market competition among rival firms. For example, in a strong intellectual property regime, R&D investment by one firm can significantly raise the costs of doing R&D for other firms thereby potentially discouraging them from making their own investments. In another context, tightening of copyright protection for computer software has been seen by some as potentially harmful to the evolution of the software industry. For more on competitive spillovers, see Katz and Ordover (1990), and Scotchmer's article in this issue.

could have resulted in the same patentable process or product, or even invented the same process or product.

But this characterization of the patent law purely as an exclusionary mechanism is simplistic; after all, a patent-holder has to reveal the content of the patent to the public, which includes revealing it to actual and potential rivals. As has been noted by Horstmann, McDonald, and Sliwinski (1985), and subsequently developed by Scotchmer, disclosure could undermine the competitive advantage that an innovation confers. To avoid disclosure, an innovator may decide to protect an invention as a trade secret, rather than with a patent. Such a strategy may be especially reasonable for process innovations, which are not easily reverse-engineered.

The Japanese patent system, combined with weak trade secret law, is designed to induce innovators to disclose strategic information sooner than does the American system. Moreover, its institutional features encourage diffusion by creating strong incentives for licensing and cross-licensing of patents. The Japanese patent system has several especially salient differences from the American system.[2]

First to File vs. First to Invent

In Japan, patents are granted to those who filed first, and not to those who have invented first, as in the United States.[3] Indeed, the United States and Canada are the only countries which use the first-to-invent priority rule, although it is quite possible that the United States will adopt the standard first-to-file priority rule as a part of the ongoing international negotiations on intellectual property rights. The first-to-file rule clearly pushes innovators into the Patent Office sooner than they otherwise might elect. Thus, information is disseminated earlier than it would be under the alternative priority rule. Moreover, the rush to patent means that many applications are incomplete and vague, which increases the possibility of conflicting claims once patents are granted. During the first 15 months after the filing, the applicant can modify and fine-tune the application. As a result, firms often resort to licensing to settle the differences between patent claims. Scotchmer and Green (1990) explore some economic effects of the two priority rules.

Pre-Grant Disclosure

A second aspect of the Japanese patent system which contributes to diffusion is the requirement to "lay open" the application for 18 months after filing. This step entails publishing the (photo)copy of the application in the Official Gazette for all to examine; in addition, the Japan Invention Society maintains a computerized data base of laid-open applications, which also increases their

[2]Readers interest in more detail on the Japanese patent system might consult Doi (1980), Amemiya (1989), and Amemiya and Nishimoto (1989).

[3]I should note, however, that article 123 of the Japanese Patent Law allows for invalidation of patents granted to other than the innovator in some situations.

accessibility. As is well-known, the U.S. patent systems also requires disclosure. However, under the U.S. system, disclosure occurs only after the patent is granted and not before. Thus, plausibly, in Japan inventions become public knowledge sooner than in the United States.

Pre-Grant Opposition

The Japanese patent system allows third parties to oppose the granting of patent. (Pre-grant opposition also used to be possible in Europe. Germany, at least, has now dropped this institutional feature.) In the United States, the patent examiner makes the initial determination of whether the application meets the statutory requirements for granting a patent. Other interested parties can intervene only after the patent has been granted. In Japan, on the other hand, after the patent application has been filed and laid open, opposing parties can attempt to demonstrate that the application does not meet the requirements of novelty or nonobviousness and industrial applicability.

There are several explanations for pre-grant opposition. Perhaps the most important is that the applicant need not demonstrate "prior art" that informs the invention. Prior art consists of all pertinent past patents and other relevant information used in the inventive process. In Japan, the opposing parties strive to show during the pre-grant opposition phase that the novelty requirement is not met. This institutional feature of the Japanese patent system does reduce the critical role of the patent examiner, which helps to explain why the number of applications per examiner is so much higher in Japan. In fact, many of those applications are never actually examined.

Pre-grant opposition enhances incentives to license the innovation, for at least two reasons. First, a quick grant of a license can discourage the interested parties from opposing the application. Inasmuch as there is a 20-year cap on the life of the patent, as measured from the date of application, drawn out opposition can reduce the economic value of the patent. The opposition phase creates strategic incentives for early bargaining between the innovator and those potential rivals who would be disadvantaged, if the patent were granted.

Similar bargains occur in the United States, but only after patent has been granted and in the shadow of possible or pending patent litigation. It should be kept in mind that in the United States, courts now uphold over 80 percent of patents tested through litigation. I think that bargaining prior to receiving the patent increases the likelihood of settlement through licensing. See Maurer (1990) for a technical discussion of patent licensing and litigation.

A second reason why pre-grant opposition creates strong licensing incentives is that during the opposition phase, the applicant can be opposed by a large number of filings which have been prepared by expert staffs over many months (or even years). The applicant has only a few months to respond and if the examiner concludes that even a single opposition has not been answered adequately, the application can be denied. In the U.S. courts, the patent-holder is confronted with only one opponent at the time, and in the atmosphere that strongly favors the owner of the patent.

Licensing and Royalties

A Japanese inventor is entitled to royalties only from those who utilized the potentially patentable invention *knowingly* after the application has been laid open. Therefore, an innovator has an incentive, albeit not a duty, to notify all likely users of its contents. One might surmise from this that innovators' rivals would prefer not to "know" that the innovation has been laid open for examination. The incentive to obtain the information is created by the rivals' right to oppose the application before the patent is granted, as discussed above. Moreover, someone who finds a use for an invention during the laying-open phase is generally entitled to use the invention at (reasonable) royalties. This requirement to license is a policy response, implemented in 1970, to an extremely long lag between the filing and granting of the patent. This lag, which may be as long as six to seven years on average, is a result of legal and staffing decisions. Allegedly, the Japan Patent Office (JPO) is severely understaffed. See Hearings (1988) for more on the delay issue.[4] The presence of the delay also underscores the importance of laying open the application to ensure diffusion before the innovation loses some of its social value.

Ministry of International Trade and Industry (MITI) can also compel a patent-holder to cross-license the innovation if its subject matter pertains to technology of national importance (article 93). Such compulsory cross-licensing is not frequent, but it does nonetheless create problems for firms on the technological frontier. Finally, a difficulty in obtaining injunctive relief in Japan against an alleged infringer creates additional incentives for licensing and reasonable royalties.

I now turn to another difference between the Japanese patent system and the U.S. system, which pertains to the *scope of patent claim*. This is a rather complex and, to a nonspecialist, a highly confusing issue, which only recently has been given theoretical treatment by economists (Gilbert and Shapiro, 1990; Klemperer, 1990). There are at least two aspects to the issue, which can be easily confounded. Firstly, there is a question of the number of claims that can be made in a single patent application. Until the beginning of 1988, the Japanese patent law limited each application to a single *independent claim*, commonly referred to as a "head claim." (The same restriction existed in Europe, as well.) Thus, to use Klemperer's (1990, p. 114) example, a bicycle inventor, besides patenting a bicycle, would also have to patent separate bicycle parts that comprise a bicycle—a basic invention. Or an inventor of an oversized tennis racket would have to file separate patents for a frame and, possibly, for the method of stringing such a racket. In contrast, in the United States, a single

[4]A recent U.S.-Japan dispute illustrates the problem well. Allied Signal alleged in its complaint to the U.S. Trade Representative that it waited eleven years for a patent on amorphous metal transformers. Allegedly, while Allied was waiting for the patent, the Japanese government subsidized domestic firms to copy Allied's technology. Moreover, Allied alleges that the industry formed a licensing cartel designed to deny Allied its profits from a new technology. The Japanese deny all the allegations.

application can include a number of independent claims that together comprise the invention.

It is not clear how the restriction to a single independent (head) claim affects R&D incentives or dissemination. It clearly leads to a multitude of patents, with all the negative costs that this imposes on the innovator and the society.

The second aspect of the scope of the patent claim is more relevant to our concerns here. In particular, it consists in the notion that the Japanese patent law, and the Japanese patent examiners, sharply delimit the *breadth* of patent coverage.[5] Unsurprisingly, this has the effect of reducing patent-holders' market power. It can create a situation whereby an owner of a basic Japanese patent can be severely restricted in the ability to exercise the property right by the presence of possibly conflicting and closely related patents (Hearings, 1988). It is also important to recall that in Japan, potential competitors who have "reverse engineered" the invention during the laying-open and opposition phases can file their own patent applications for possibly quite small modifications of the original invention and thereby box in the early innovator.[6] The narrow scope of patent coverage and very relaxed novelty requirement (Scotchmer and Green, 1990) severely reduce the exclusionary value of the patent right. The upshot of this is that patent-holders have an incentive to cross-license, an effect which does not materialize, however, in the existing studies of welfare implications of adjustments in patent scope (in the sense of breadth).

The Japanese patent system is a complex web of policy choices more or less consciously structured to affect R&D diffusion while maintaining overall incentives for R&D investment. The Japanese patent system subordinates the short-term interests of the innovator in the creation of exclusionary rights to the broader policy goals of diffusion of technology. Because the scope of the patent is (apparently) narrow under the Japanese patent law and the novelty requirement is quite weak, it rewards those who reverse engineer and modify, often in minor ways, the existing inventions and penalizes those who wish to protect their major technological breakthroughs.

It would be foolish to argue that the Japanese have struck an ideal balance between appropriability, exclusion and diffusion. The Japanese patent system evolved to meet the needs of an economy that imported R&D. Japanese firms

[5]Scope (breadth) of the patent can be interpreted as the portion of an abstract product/technology space covered by the innovation (Klemperer, 1990).

[6]I have been told that these subsequent filers have been known to threaten the original innovator with infringement suits. A subsequent innovator incorporates the basic innovation in the improvement (however marginal) and then patents the improvement. As a result, the owner of the basic patent may be foreclosed from incorporating its invention in subsequent products without violating the improvement patent. This creates incentives for cross-licensing. For example, Mitsubishi (Japan) apparently incorporated a basic patent by Fusion Systems (U.S.) for a special type of an ultraviolet lamp into over 200 patents of its own filed with the Japanese Patent Office. Mitsubishi then suggested a cross-licensing agreement to obviate whatever intellectual property rights conflicts may arise. Fusion Systems so far has refused. See Fallows (1990).

have become extremely adept at adapting, modifying, and commercializing this imported knowledge (for example, the VCR technology). The Japanese system is also well-suited to an industrial structure in which a preponderance of R&D is performed by large corporations, which have large staffs that monitor the activities of the Patent Office, engage in the pre-grant opposition process, and conduct frequent licensing and cross-licensing negotiations.

In my view, these elements of the Japanese patent system are less well-designed to meet the intellectual property protection requirements of an economy that has become a technological leader, as opposed to a technological borrower (Doe, 1988; Dryden, 1988; Vogel, 1989). A technological leader is likely to place greater weight on the ability to exclude and to garner market power from its innovative efforts. This is especially true if innovator's market power is dissipated by competition from foreign rivals (Ordover and Willig, 1990). Consequently, both internal pressure from domestic technology leaders and, for different reasons, from the United States has already led to "strengthening" the Japanese system of intellectual property rights.

I do not dispute the role of the patent system and other intellectual property protection to provide incentives for technological progress. However, the Japanese experience, and the experience of the United States prior to the 1980s, suggests that incentives for innovation can flourish under the regimes of relatively "weak" patent protection. It is easy to overemphasize the importance of intellectual property rights in the innovative process.

Diffusion In A Regime of Strong Patent Protection: Licensing

The Japanese experience illustrates how a patent system that offers a narrow patent scope and weak novelty requirement can induce licensing, especially if it also creates institutional obstacles for obtaining a patent by means of an expensive and possibly protracted pre-grant opposition process. In Japan, diffusion through licensing and cross-licensing is a result of a "weak" patent system and occurs often even before the patent is granted. A weak patent system is one that maximizes the *unappropriated* portion of the social benefit from the innovation during the statutory life of the patent, subject to a target discounted present value of profit flow to the innovator. Under this definition, the Japanese patent system is weaker than the American system, despite the fact that the Japanese system is also predicated on the notion that the innovator should have the right to exclude others from using the fruits of R&D.

In this section, I want to consider how strong patent protection can affect diffusion through licensing after the intellectual property right has been established. The usual story would argue that strong patent protection reduces diffusion because it reduces spillovers. However, a full story must reflect two additional considerations.

The first consideration is that when the patent regime is weak, innovators must rely on other means to achieve appropriability and excludability, and those other means may be even worse for diffusion of ideas than are patents. For example, an important study by Levin, Klevorick, Nelson, and Winter (1984, 1987) showed that the effectiveness of product and process patents is quite low in many industries. As a result, firms rely on non-patent mechanisms, like secrecy and first-mover advantages, to protect their intellectual property. Secrecy is especially inimical to diffusion, even as compared to patent protection. When secrecy is the main means of protection, the sale of information through licensing may become quite difficult, if not impossible. In Japan, however, reliance on trade secrets is risky because the country lacks the equivalent of the Uniform Trade Secrets Act that controls unauthorized use of proprietary information in the United States.

The second consideration, somewhat related to the first, is that since markets for information are notoriously prone to failure, social policies that make it easier for the potential licensor to threaten other firms will tend to encourage licensing.[7] Here the idea is that one important obstacle to licensing is the licensor's concern about the inability to control the licensee's use of the information transferred, both during the license negotiation process and after the license has been signed. For example, take the worst-case scenario where the potential licensee demands to see the information before taking the license and then simply appropriates it without any compensation to the innovator. In a case like this, a strong patent regime improves the licensor's ability to preclude the potential licensee from utilizing the leaked information. Since strong patents reduce the risks associated with licensing, they will increase incentives to license and hence diffusion.

A licensor may also be concerned that the licensee could use the licensed innovation as an input into its own R&D and subsequently obtain an improvement patent that could afford it a competitive advantage over the original licensor. (This concern is especially pronounced if the novelty and nonobviousness requirements are weak.) To minimize this risk, the licensor may demand a grant-back provision in the license contract, or it may insist on compulsory cross-licensing of derivative patents. Essentially, these types of provisions provide the licensor with some property right in the subsequent innovations that build on the original innovator's R&D.

Because of high transaction costs and the potential for market failure in the licensing process, licensors and licensees may have to resort to complex licensing schemes to realize intrinsic gains from trade. However, antitrust considerations may limit the ability of the parties to strike an efficient bargain. Antitrust treatment of licensing agreements is an important element of the overall strength (or weakness) of patent rights (Ordover, 1984; Gilbert and Shapiro, 1990). A regime of strong (or broad) patent protection allows the patent-holder significant latitude in structuring the terms of the license. A regime of weak

[7]For more on why licensing may be difficult, see Caves, Crookell, and Killig (1983) and Shapiro (1985). Some additional points are reviewed in Ordover and Baumol (1988) and Katz and Ordover (1990).

(narrow) patent protection, on the other hand, places severe restrictions on permissible licensing provisions.

How much latitude should licensors have in structuring licenses?[8] This issue has had a long history in the antitrust/patent literature.[9] The underlying tradeoffs are these: if licensing contracts are severely restricted, an innovator may choose not to license at all so that patent disclosure and spillovers provide the only sources of information; if licensing is less restricted, it could provide rivals with access to technology, albeit at prices that could be too high from the standpoint of social welfare. Moreover, having gotten an access to the technology, the licensees could opt not to do their own R&D, which may be inimical to technological progress. For antitrust purposes, rather than full-blown social welfare calculations, a minimal benchmark for testing a licensing provision should be whether the provision weakens inter- and intra-technology competition relative to what it would have been if the license had not been issued. Such a test would catch "sham" licensing schemes for implementing downstream collusion, since such agreements would lead to relatively weaker competition. This test would also catch provisions that elevate entry barriers into R&D competitions over and above those barriers that arise from prior existence of pertinent intellectual property rights. For example, cross-licensing could be used to discriminate against new entrants who do not have attractive portfolios to cross-license; or a license could be used to bribe an efficient rival to avoid entering with an alternative technology.

At present, U.S. antitrust policy has taken a very pro-licensing attitude.[10] Licensing provisions are scrutinized under the rule of reason standard, except for those that restrict the pricing freedom of the licensee. The latter restrictions may be found to be illegal per se, despite the fact that reasonable arguments can be made for their pro-competitive rationale. The pro-patent antitrust policy mirrors and reinforces the general pro-intellectual property rights policy stance that has emerged in the United States over the last decade.

European Community (EC) policies on licensing are more restrictive than the current U.S. standard, for two distinct reasons. First, the EC has never accepted the current U.S. antitrust policy view that vertical restrictions in a contrast between a seller (licensor) and a buyer (licensee) are, in general, a response to some sort of market failure.[11] Thus, the EC holds that such

[8]Five years ago, I was inclined to the view that the latitude should be large, especially if the patent life is foreshortened (Ordover, 1984). I have now seen the error of my ways (Gilbert and Shapiro, 1990).

[9]Early discussions of alternative tests are in Bowman (1973) and Baxter (1966). Kaplow (1984) critiques their approaches and offers an interesting but informationally demanding alternative, which is discussed in Ordover (1984) and Ordover and Baumol (1988).

[10]For the latest statement of enforcement guidelines in this area, see "Antitrust Enforcement Policy for International Operations," Antitrust Division, Dept. of Justice, November 10, 1989. It appears, however, that the Division has taken a more sanguine view of licensing restrictions than found in the existing case law.

[11]See Comm'n Reg. No. 2349/84 (1984) for patent licensing agreements and Comm'n Doc. No. 88/EM-4 for know-how licensing agreements. OECD (1989) contains an excellent comparative assessment.

restrictions are more likely to be efficiency-reducing restraints on competition than efficiency-enhancing incentive schemes designed to overcome a market failure. In Europe, the licensee cannot be obliged to assign the licensor exclusive rights in or patents for improvements or for new applications of the licensed patents. See Scotchmer (in this issue) for more on this point. Interestingly, a similar hostility to such exclusive grantbacks is evidenced in the new licensing guidelines issued by the Japan Fair Trade Commission. The Japanese attitude could perhaps be attributed to the policy goal of maximizing diffusion. Obviously, exclusive licensing agreements have the effect of constricting diffusion.

Second, EC policy on licensing reflects the overarching political goal of promoting intra-community trade and removing barriers to such trade. As a result, licensing provisions that unduly restrict such trade flows—for example, by limiting the territory in which the licensee may sell the goods produced under the license—may not be accorded "block exemptions." A block exemption removes the firm's conduct (like the license and its terms) from under the scrutiny of Articles 85 and 86 of the Treaty of Rome, which regulate business conduct in the Community.

Cooperative Licensing and Conflicts within Patent Systems

Within any patent system, however carefully designed, conflicts may arise among patent holders. These conflicts can be resolved through litigation or through cooperation. Cooperative solutions will entail cross-licensing or patent pooling, both of which enable patent-holders to benefit from their exclusionary rights.

Let me illustrate the issue with a simple example (see Fox and Sullivan, 1989, p. 353, for antitrust analysis). Assume that four leading chocolate manufacturers, controlling 80 percent of U.S. output, have received separate patents for a new type of chocolate (chip). Each patentee claims: (1) that its own invention is nonobvious and that it was not anticipated by any prior art and that its patent is thus valid; (2) that the patent held by each of the other three firms is invalid for lack of invention; and (3) that the technology embodied in the manufacture of chocolate (chips) made by each of the other three firms infringes its own.

As the example shows, an owner of an exclusionary right may end up with an asset whose value may significantly depend on the state of other conflicting, or complementary, exclusionary rights. It's conceivable that an owner of an exclusionary right may be blocked from exercising its exclusionary right. In the example above, if two patents with overlapping claims are granted, infringement is inevitable. It can only be resolved through litigation that adjusts the scope of the conflicting patents or through cross-licensing. Assume that chocolate produced with the new technology can come in three versions—sweet, bittersweet, and bitter—and that only bittersweet is subject to overlapping claims. Then, litigation would resolve the conflicting claims by granting the monopoly to bittersweet to one of the two patentees. A restricted cross-license would enable the two firms to manufacture bittersweet, leading to a duopoly, to the benefit of the consumer.[12] Would the two firms cross-license prior to the outcome of the litigation? Possibly yes, if there is an uncertainty as to the outcome of litigation.

If the two patents do not necessarily conflict, their owners could end up competing with each other to license their technologies to the remaining firms. Such competition is more likely to break out if, for example, by diffusing its technology through licensing, the licensor can hope to establish it as the industry standard. Control of the industry standard may, in turn, advantage its "sponsor" in a current round of product competition and in the future rounds of technological competition. For example, Sun Microsystems licenses its vital computer technologies very widely, in the hope of stimulating demand for its workstations and ensuring a steady flow of technological upgrades from its licenses that would further enhance the appeal of Sun's products.

In the example, when the patented technologies are complementary and block each other, licensing will be inevitable. Here two alternatives arise. First, the firms can cross-license to each other for a mutually agreed fee or for no fee at all. Second, the firms can form a separate entity and assign patents to it. The new entity then licenses back the patents to the innovators and it may also license the package of patents to third parties.

A competitive problem can arise if firms with complementary or alternative non-conflicting patents make a strategic claim of conflict, with the aim of colluding in downstream markets. This collusion might be done by putting restraints in the patent pool's licensing package or, perhaps less visibly, through royalty terms that restrict output (Shapiro, 1985). Still, potential welfare losses from the misuse of cross-licensing could be partly offset by welfare gains from cooperative diffusion of patented R&D. Moreover, cooperative cross-licensing reduces the uncertainties and costs of patent enforcement and saves on transaction costs of multiple bilateral licensing negotiations, thereby generating additional welfare gains.

Cooperative exchanges of patents designed to overcome potentially conflicting intellectual property rights (such as patents) strengthen R&D incentives by increasing expected profits from doing R&D on the average, relative to a benchmark where cross-licensing is tightly circumscribed. But this is not the end of the story: the hitch comes from the fact that private and social incentives for licensing do not invariably line up. This creates potential competitive concerns, one of which has already been identified: if the cooperative exchange of patents implements collusion in downstream markets it may actually harm social welfare. (But remember that if constricting downstream competition is a

[12]It is well-known that only extra bittersweet or bittersweet chocolate is worth eating. So the example may be unrealistic.

way of increasing appropriability, and if the returns to R&D are below the social rates of return, then less competition might be in the social interest.)

The second concern is that cross-licensing and patent-pooling can create entry barriers into R&D markets and thus may stifle future rounds of technological competition. For example, the pool may issue fewer licenses that would become available if firms were making licensing decisions noncooperatively (Katz and Shapiro, 1986); or the exchange of patents could give a significant head start to a handful of firms, thus discouraging others from participating in R&D competition.

When cross-licensing and patent-pooling agreements cover not only existing patents but also future patents, additional welfare considerations arise. Consider this scenario: firms A and B race to develop technologies for fabrication of a new generation of semiconductor chips. Assume that (Bertrand-type) product market competition between the two firms would dissipate the profits from the new generation of chips. If A and B were to agree to cross-license future technologies (patents) at no fee, which is not uncommon, neither would have an incentive to engage in the race, especially if profits can still be made from the current generation of chips.[13] But would A and B be willing to enter into a prospective cross-licensing agreement designed to slow down the R&D race? The answer depends on the additional details that are missing from my stripped-down example. One important consideration might be the cross-licensors' ability to monitor each other's R&D expenditures. In absence of verifiability, each firm would fear that its rival might get a head start into the next round of technological race. The global web of technological agreements that has emerged in the information technology sector, and which has been attributed to the growing costs of independent R&D, has also improved verifiability and may ultimately weaken technological and product competition in the sector. These sorts of considerations become central in the analysis of research joint ventures, which are the subject of the next section.

Creating Spillovers to Cure Them? Research Joint Ventures

Research joint ventures and research consortia have acquired a mystique as a cure for, if not all then at least many, market failures that befall private R&D investment decisions. This type of R&D cooperation is alleged to mitigate the ambient spillover problem by internalizing leakage from R&D; improve coordi-

[13]For a more detailed discussion, see Ordover and Willig (1985), Grossman and Shapiro (1986), and Katz and Ordover (1990). See Baumol (1990) for a more sanguine view. Katz and Ordover also point out that prospective cross-licensing agreements can be a response to problems of commitment posed by the existence of bargaining power that can arise when some firms have innovated successfully and others have not. In particular, a winner of a patent race cannot precommit to leaving some surplus for the loser in the patent licensing agreement. As a result, both firms may enter a race, as long as there are expected positive profits to be realized. From this perspective, each firm's incentive to enter into a *prospective* cross-licensing agreement would most likely be driven by its assessment of its R&D capability relative to that of its rivals.

nation among independent research projects and thereby reduce wasteful duplication of R&D efforts and thus economize on scarce R&D resources; spread the risks associated with large-scale R&D programs that cannot be spread through capital markets; assure access to necessary complementary know-how and R&D assets more readily or more cheaply than what could be obtained through contracts; and take advantage of scale economies in R&D. This view has been most fully developed by David Teece and Tom Jorde, and they explain it in the symposium in the Summer 1990 issue of this journal on whether production joint ventures should be encouraged as a way of stimulating innovation. Along with the dissident views presented in that symposium, Porter (1990) questions usefulness of joint ventures for competitive survival of firms.

Theoretical economic research has shown that when spillovers are moderately high and overall protection of property rights is weak, then research joint ventures can be an effective means for internalizing spillover externalities. Obviously, when spillovers are too large, every firm may be reluctant to join a venture. By staying out, the firm avoids making its pro rata contribution towards the venture's R&D costs and can free ride on the R&D efforts of the putative joint venture. And when spillovers are quite low, their internalization is usually not a key reason underlying a formation of a venture. Ordover and Willig (1985), for example, construct a model in which there are no spillovers and a venture is formed to economize on R&D costs.

A common assumption in the theoretical models of R&D joint ventures is that new knowledge diffuses completely among venture participants. This assumption hides the fact that intra-venture spillovers are a result of complex strategic decisions by venture participants. Consequently, the next vintage of joint venture models should pay closer attention to information exchanges among co-venturers. Bhattacharya et al. (1987) and Gandal and Scotchmer (1989) offer preliminary discussions of this issue, while Chesnais (1988) shows how the institutional form of cooperation depends on the desire to protect core technology.

Two sets of problems pertaining to diffusion and spillovers must be addressed by any venture. The first concerns protecting unauthorized access to pre-existing, or "background" knowledge held by co-venturers. Background information includes any intellectual property that partners possess and that may be brought to the joint venture or that may somehow be accessed as a side effect of cooperation. Participants in an R&D joint venture must contain and control the spillovers from background information, which may frequently include the most important strategic knowledge possessed by the participants. The fears of such enhanced spillovers prompted Reich and Mankin (1986) to warn that joint ventures with foreign (read Japanese) firms would give away America's future.

In general, background information or knowledge remains a property of its owners in any joint venture. However, this background knowledge may have to be licensed to co-participants in the venture, if access to it is necessary for the participants' R&D activities undertaken within the joint venture. For some R&D joint ventures, the operating guidelines specify that the necessary "background" has to be licensed-in without a royalty. Thus, the guidelines for the Alvey Programme, a major research program in England in various areas of information technology, specify that the "background" necessary for undertaking an Alvey project must be licensed without a fee to the members of the project. Similar requirements are stated in the guidelines for the European Strategic Programme for R&D in Information Technology (ESPRIT), organized under the aegis of the European Community (OECD, 1989).

IBM and AT&T both contributed proprietary technology to Sematech, a large-scale U.S. consortium whose mission is to improve domestic semiconductor fabrication technology and strengthen the supplier base to the semiconductor industry (CBO, 1990). However, both IBM's and AT&T's contributions were voluntary and designed to get the venture under way. Admittedly, the technologies contributed from these two companies were not state-of-the-art. They were nevertheless valuable to the chip fabrication projects at Sematech and to the rival semiconductor firms.

Companies have developed ways to protect themselves from unauthorized access to their core background knowledge in ventures. For example, in the Institute for New Generation Computer Project (ICOT), which is a Japanese consortium focused on the development of the "Fifth Generation" computer, "the companies deposit sealed envelopes with relevant background to be opened in case of subsequent dispute" (OECD, 1989, p. 125). If background knowledge leaks out, or is contributed to the project, the participants will be able to determine the source of the "input" and either prohibit its further unauthorized use or give preferential treatment to the source of the input. In an earlier Japanese consortium, the Very Large Scale Integration (VLSI) consortium, which lasted from 1976 to 1979, the participants reduced unauthorized spillovers by organizing R&D projects in separate laboratories, each of which was dominated by managers and scientists from a single participating company. However, unlike earlier Japanese consortia, there was ongoing exchange of technical information among corporate participants and, most importantly, the consortium's research was located in a special facility dedicated to semiconductor technology development.

Besides dealing with property rights in "background" knowledge, ventures must also resolve legal and other problems relating to "foreground" knowledge, which encompasses new knowledge generated by the venture. These problems are less straightforward than they may appear. First, often there are technical difficulties in transferring R&D results from the researchers in the venture to the participants. These difficulties tend to increase with the number of participants in the venture. This is because participants may differ regarding the venture's research objectives, or in their capabilities in absorbing the results of the venture's R&D; or in their abilities to commercialize the results, and for other reasons.

For example, it is well-known that the Microelectronics and Computer Technology Corp. (MCC) initially experienced serious difficulties in transferring ("handing-off") technologies developed by the research staff in a form that could be readily commercialized by the co-venturers. Partly this happened because MCC's research staff pursued its own research agendas, despite the complex (and cumbersome) oversight from the Board of Directors and numerous technical advisory boards, and despite the fact that some personnel on loan from the parents participated in the venture's R&D activities (Lineback, 1987, 1988; Dunn, 1989). To improve the "hand-off," MCC modified some of its research plans to serve the commercial needs of members and improved informational links between research staffs in the parent companies and its own. In the process, the consortium made itself into an open-market supplier of contract R&D in computer-aided design and engineering and in computing technology that competes with other stand-alone firms (Katz and Ordover, 1990).

The already mentioned VLSI consortium avoided the problems of unfocused and run-away research staff pursuing its own research agenda by extensive pre-planning and extremely well-focused detailed R&D objectives, which were substantially adapted from the plans of IBM and other U.S. computer and semiconductor firms. On the other hand, the technology transfer problem persisted during the earlier phases of the project. Separate VLSI laboratories were initially quite reluctant to share the flow of research results among themselves, and their directors took to locking the doors to the laboratories at night. (So much for the vaunted cooperative spirit among Japanese firms.) Later on, under pressure from MITI, intra-venture transfer and technology "hand-off" from the venture to the corporate participants improved perceptibly. Experience from small private ventures also indicates that the participants frequently manipulate strategically the flow of "foreground," as well as of the "background" knowledge and information (for examples, see Harrigan, 1985, 1986). These examples show that the intensity of spillovers within a venture is a result of complex decisions regarding organizational design and private incentives of venture participants.

With respect to "foreground" knowledge, the venture must also determine how intellectual property rights will be allocated. A number of alternatives are possible. A venture may hold the patents and then license them to the participants. MCC, for example, holds the intellectual property rights internally and licenses their use to participants on various terms. Participants in a specific project receive a non-exclusive royalty-free license; however, not all MCC shareholders participate in all MCC projects, and there are also "associate" members who have no equity in MCC but participate on a project-by-project basis. After three years, the technology can be licensed to other MCC shareholders and to non-members. The interesting feature of MCC's licensing scheme is that "shadow royalty rates" are attached to the licenses. Each participant is entitled to a royalty-free license for the amount up to three times the contribution to the project. Similar staggered licensing schemes have been adopted by the Alvey Programme and by ESPRIT.

In Japan, the locus of ownership of intellectual property rights flowing from a consortium is determined by the nature and degree of governmental subsidy. Under the *hojokin* formula, the government provides 40–60 percent financing, using conditional loans whose repayments are tied to profits. This formula was used for VLSI, for example, and under it patents belong to the research association or to the participants, but can be assigned to MITI. Under the *itakuhi* formula, the government provides full contract financing of research. This formula was used in the case of ICOT, and under it patents belong to MITI and can be licensed to the members of the consortium and foreign firms.

Clearly, joint ventures are not without their own special problems when it comes to creating and protecting patents and other types of intellectual property rights. A 1989 OECD report (p. 126) points out that large-scale joint ventures and multinational patent licensing networks present difficult legal questions respecting the ownership of intellectual property. Consequently, joint ventures with a high R&D component cannot be relied upon to ameliorate the spillover problem significantly. Such ventures can, perhaps, be more effective in aggregating complementary (background) knowledge than would be possible by means of an arms-length contract.

Concluding Comment

I have focused on mechanisms for ameliorating difficulties in the production and dissemination of knowledge. The main conclusion that emerges from this essay is that a patent regime can be constructed from a wide menu of policies affecting intellectual property rights. Each policy mix has different consequences for diffusion through both unrecompensed spillovers and licensing, and it permits a different share of the total social return from the innovation to be appropriated by the innovator. This suggests that possibilities exist for selecting different points on the trade off between dissemination, R&D incentives, and share of social benefit appropriated by the innovator. Even though the focus in this discussion was on R&D, patents, and their results, my basic conclusions apply also to other types of intellectual property rights.

Whether the underlying models and the available data are good enough to provide sufficient guidance for the welfare-maximizing choice is, at this stage, more doubtful.[14] My own hunch is that the current U.S. policy stance that

[14] For example, Klemperer (1990) shows that depending on the structure of preferences, patents should be either very narrow or very broad and their length should be very long or very short. Gilbert and Shapiro (1990) argue that patents should be infinitely long-lived but that their scope should be very narrow. They also argue that their model suggests severe restrictions on the patent-licensing agreements. I am not convinced. I have noted that in many instances restrictions are necessary if licensing is to occur. Insofar as licensing is preferable on welfare grounds to no licensing, restrictions increase both the licensor's profits and social welfare.

advocates very strong intellectual property rights may have gone too far in protecting the interests of the inventor.

■ *Some of the material in this article is treated more extensively in Ordover and Baumol (1988) and Katz and Ordover (1990). I should like to thank Stan Besen, Eleanor Fox, Carl Shapiro, Joseph Stiglitz, Timothy Taylor, and the participants at the Symposium on Intellectual Property held in Washington, D.C. on October 24, 1989, sponsored by the John M. Olin Foundation and the RAND Corporation. Alannah Orrison and Ken Rogoza provided excellent research assistance. Finally, the C.V. Starr Center for Applied Economics provided financial support.*

References

Amemiya, S., "Intellectual Property Protection." In Kitagawa, Z., ed., *Doing Business in Japan*. New York: Mathew Bender, 1989.

Amemiya, S., and K. Nishimoto, "Patents and Utility Models. In Kitagawa, Z., ed., *Doing Business in Japan*. New York: Mathew Bender, 1989.

Baumol, W. J., "Technology-Sharing Cartels," working paper, C. V. Starr Center, New York University, 1990.

Baxter, William F., "Legal Restrictions on Exploitation of Patent Monopoly: An Economic Analysis," *Yale Law Journal*, 1966, *76*, 267–370.

Bhattacharya, S., J. Glazer, and D. Sappington, "Motivating Exchange of Knowledge in R&D Ventures: First-Best Implementation," technical memorandum, Bell Communications Research, 1987.

Besen, S., and G. Saloner, "The Economics of Telecommunications Standards." In Crandall, R., and K. Flamm, eds., *Changing the Rules: Technological Change, International Competition, and Regulation in Communications*. Washington, D.C.: Brookings Institution, 1989.

Bowman, W., *Patent and Antitrust Law*. Chicago: University of Chicago Press, 1973.

Chesnais, F., "Technical Cooperation Agreements Between Firms," *STI Review*, 1988, *4*, 52–117.

Caves, R., H. Crookell, and J. P. Killig, "The Imperfect Market for Technology Licenses," *Oxford Bulletin of Economics and Statistics*, 1983, *45*, 249–267.

Congressional Budget Office (CBO), *Using R&D Consortia for Commercial Innovation: SEMATECH, X-Ray Lithography, and High-Resolution Systems*. Washington, D.C.: Congress of the United States, 1990.

Doe, P., "Japan Starts Fighting for Its Own Patent Rights," *Electronic Business*, September 1, 1988, 36–39.

Doi, T., *The Intellectual Property Law of Japan*. Alpen aan de Rijn, The Netherlands: Sijthoff & Noordhoff, 1980.

Dryden, S., "The U.S. and Japan Look for a Patent Medicine: the Trade Talks Turn to 'Intellectual Property,'" *Business Week*, September 5, 1988, 28–29.

Dunn, D., "MCC, DEC Sign Joint Research Accord," *Electronic News*, January 16, 1989, p. 35.

Fallows, James, "The Great Japanese Misunderstanding," *The New York Review of Books*, November 8, 1990, 33–39.

Fox, E., and L. Sullivan, *Antitrust*. St. Paul: West Publishing, 1989.

Gandal, N., and S. Scotchmer, "Coordinating Research Through Research Joint Ventures," Working Paper E-89-27, Hoover Institution, Stanford University, 1989.

Gilbert, R., and C. Shapiro, "Optimal Patent Protection and Breadth," *RAND Journal of Economics*, 1990, *21*, 106–12.

Green, J., and S. Scotchmer, "Antitrust Policy, the Breadth of Patent Protection and the Incentive to Develop New Products," Graduate School of Public Policy Working Paper, University of California at Berkeley, 1989.

Grossman, G., and C. Shapiro, "Research Joint Ventures: An Antitrust Analysis," *Journal of Law, Economics, and Organization*, 1986, 2, 315–37.

Harrigan, K. R., *Strategies for Joint Ventures*. Lexington, Mass.: Lexington Books, 1985.

Harrigan, K. R., *Managing for Joint Venture Success.* Lexington, Mass.: Lexington Books, 1986.

Hearings before the Subcommittee on Foreign Commerce and Tourism, Committee on Commerce, Science, and Transportation, U.S. Senate, 100th Congress, "Effect of the Japanese Patent System on the American Business," 1988.

Horstman, I., G. M. MacDonald, and A. Sliwinski, "Patents as Information Transfer Mechanisms: To Patent or (Maybe) Not to Patent," *Journal of Political Economy,* 1985, *93,* 837–56.

Jorde, T., and D. Teece, "Innovation and Cooperation: Implications for Competition and Antitrust," *Journal of Economic Perspectives,* Summer, 1990, *4,* 75–96.

Kaplow, L., "The Patent-Antitrust Intersection: A Reappraisal," *Harvard Law Review,* 1984, *97,* 1813–1892.

Katz, M., and J. A. Ordover, "R&D Cooperation and Competition," *Brookings Papers on Economic Activity: Microeconomics,* 1990, 137–91.

Katz, M., and C. Shapiro, "How to License Intangible Property," *Quarterly Journal of Economics,* 1986, *101,* 567–89.

Klemperer, P., "How Broad Should the Scope of Patent Protection Be?" *RAND Journal of Economics,* 1990, *21,* 113–30.

Levin, R. C., A. K. Klevorick, R. R. Nelson, and S. G. Winter, "Survey Research on R&D Appropriability and Technological Opportunity: Part I," working paper, Department of Economics, Yale University, 1984.

Levin, R. C., A. K. Klevorick, R. R. Nelson, and S. G. Winter, "Appropriating the Returns from Industrial Research and Development," *Brookings Papers on Economic Activity: Microeconomics,* 1987, *3,* 783–820.

Lineback, J. R., "It's Time for MCC to Fish or Cut Bait," *Electronics,* June 1987, 32–33.

Lineback, J. R., "MCC, After Five Years of R&D, Refocuses to Earn Its Keep," *Electronics,* December 1988, 89–91.

Maurer, M. J., "Designing an Optimal System of Patent Licensing and Litigation," working paper, Duke University, 1990.

Organization for Economic Cooperation and Development (OECD), *Major Research Programmes for Information Technology.* Paris: OECD, 1989.

Ordover, J. A., "Economic Foundation and Considerations in Protecting Industrial and Intellectual Property," *Antitrust Law Journal,* 1984, *53,* 503–518.

Ordover, J. A., and W. J. Baumol, "Antitrust Policy for High-Technology Industries," *Oxford Review of Economic Policy,* 1988, *4,* 13–35.

Ordover, J. A., and R. D. Willig, "Antitrust for High-Technology Industries: Assessing Research Joint Ventures and Mergers," *Journal of Law and Economics,* 1985, *28,* 311–333.

Ordover, J. A., and R. D. Willig, "Economics of Section 337 of the Tariff Act: The 1988 Amendment," mimeo, February 1990.

Porter, M. E., *The Competitive Advantage of Nations.* New York: Free Press, 1990.

Reich, R. B., and E. D. Mankin, "Joint Ventures Give Away Our Future," *Harvard Business Review,* 1986, *86,* 78–86.

Scotchmer, S., and J. Green, "Novelty and Disclosure in Patent Law," *RAND Journal of Economics,* 1990, *21,* 131–46.

Shapiro, C., "Patent Licensing and R&D Rivalry, *American Economic Association Papers and Proceedings,* 1985, *75,* 25–30.

Spence, M., "Cost Reduction, Competition and Industry Performance," *Econometrica,* 1984, *52,* 101–121.

Vogel, S. K., "Japanese High Technology, Politics, and Power," Research Paper No. 2, Berkeley Roundtable on the International Economy, Berkeley, CA, 1989.

[14]

The Economic Journal, 91 (*December* 1981), 907–918
Printed in Great Britain

IMITATION COSTS AND PATENTS: AN EMPIRICAL STUDY*

Edwin Mansfield, Mark Schwartz and Samuel Wagner

It has long been recognised that the costs of imitating new products have an important effect on the incentives for innovation in a market economy.[1] As Arrow (1962) and others have pointed out, if firms can imitate an innovation at a cost that is substantially below the cost to the innovator of developing the innovation, there may be little or no incentive for the innovator to carry out the innovation. In their discussions of the innovation process, economists frequently have called attention to the major role played by the costs of imitation, but there has been little or no attempt to measure these costs, to test various hypotheses concerning the factors influencing them, or to estimate their effects. In this paper, we report some findings of what seems to be the first study of this topic. In addition, we present data regarding the effects of patents on imitation costs and on the rate of innovation.

I. IMITATION COSTS AND TIMES

Many economists, including Scherer (1977) and Mansfield *et al.* (1977*b*), have cited the need for empirical studies of imitation costs and imitation times. To carry out such a study, we obtained data from firms in the chemical, drug, electronics, and machinery industries concerning the cost and time of imitating (legally) 48 product innovations. By imitation cost we mean all costs of developing and introducing the imitative product, including applied research, product specification, pilot plant or prototype construction, investment in plant and equipment, and manufacturing and marketing startup.[2] (If there was a patent on the innovation, the cost of inventing around it is included.) By imitation time we mean the length of time elapsing from the beginning of the imitator's applied research (if there was any) on the imitative product to the date of its commercial introduction.[3]

The firms were chosen more or less at random from among the major firms in these four industries in the Northeast of the United States, and the new pro-

* The research on which this paper is based was supported by grants to Mansfield by the National Science Foundation which, of course, is not responsible for the views expressed here. We appreciate this support, as well as the cooperation of the firms that provided data, without which this study could not have been done. A preliminary version of this paper was presented by Mansfield at the 1980 meetings of the American Economic Association. Parts of this paper were contained in the 1981 Andersen Lectures that he presented at the University of Brussels, and in invited lectures he gave at Middlebury College and at the International Institute of Management in Berlin.

[1] For example, see Arrow (1962) and Freeman (1974).

[2] For definitions of these terms, see Mansfield *et al.* (1971) and Mansfield (1980).

[3] For many products, the estimate of imitation cost (and of imitation time) was an average of two estimates provided by the firms, each reflecting a somewhat different set of assumptions.

Table 1

Imitation Cost (Divided by Innovation Cost) of 48 New Products, by Industry and Cost of Innovation

Imitation cost (divided by innovation cost)	Innovations costing more than $1 million			Innovations costing less than $1 million		
	Chemicals	Drugs	Electronics and machinery	Chemicals	Drugs	Electronics and machinery
(A) Number of new products						
Less than 0·20	1	1	1	1	0	0
0·20 and under 0·40	0	3	0	0	0	2
0·40 and under 0·60	1	1	2	5	0	0
0·60 and under 0·80	2	5	0	2	0	4
0·80 and under 1·00	2	3	1	2	1	1
1·00 and over	2	2	1	1	0	1
Total	8	15	5	11	1	8
(B) New products weighted by innovation cost* (%)						
Less than 0·20	3	3	17	15	0	0
0·20 and under 0·40	0	11	0	0	0	34
0·40 and under 0·60	‡	1	53	46	0	0
0·60 and under 0·80	44	54	0	13	0	36
0·80 and under 1·00	15	21	9	22	100	18
1·00 and over	38	9	22	4	0	11
Total†	100	100	100	100	100	100

* The weighted number of new products is expressed as a percentage of the column total.
† Because of rounding errors, items may not sum to column total.
‡ Less than 0·5.

ducts were chosen more or less at random from among those introduced recently by these firms.[1] In 34 of the cases, the new product had already been imitated, so the data are based on actual experience. In the remaining 14 cases, no imitator had appeared as yet, but the innovating firm provided us with detailed estimates that were regarded as being reliable.[2] Also, in all 48 cases,

[1] We say that the firms and innovations were chosen 'more or less' at random because a table of random numbers really was not used. In some cases, the sample was a systematic sample from a list; in other cases, there were so few innovations in the time frame we selected that the choice was automatic. Both successful and unsuccessful new products were included in the sample. About half of the innovations turned out to be successful (in the sense that they were relatively profitable to the innovator). A high percentage of the firms we approached agreed to co-operate so there was little problem of nonresponse. A considerable number of in-depth interviews were held with major officials of each firm to insure that the data were as accurate as possible. Nineteen of the innovations occurred in the chemical industry (including petroleum refining); sixteen occurred in the drug industry; ten occurred in the electronics industry; and three occurred in the machinery industry. Five of the new products were first introduced before 1960; fifteen were first introduced during the 1960s; and 28 were first introduced during 1970–6.

[2] In each such case, the innovating firm provided an estimate of how much money and time it would have cost the most likely (and most efficient) imitator to have imitated the product. In some cases, it was assumed that the time-cost combination that would have been chosen by the imitator was midway between the least-cost combination and the least-time combination. According to the firms, these estimates are unlikely to be very wide of the mark. If we had dropped these cases from the sample, we would have risked the introduction of a serious bias. Thus, this seemed to be the best feasible procedure. This procedure also was used in other cases, and seemed accurate enough for present purposes.

Table 2

Imitation Time (Divided by Innovation Time) of 48 New Products, by Industry and Cost of Innovation

Imitation time (divided by innovation time)	Innovations costing more than $1 million: Chemicals	Drugs	Electronics and machinery	Innovations costing less than $1 million: Chemicals	Drugs	Electronics and machinery
(A) Number of new products						
Less than 0·30	1	2	1	2	0	0
0·30 and under 0·50	1	5	1	2	0	2
0·50 and under 0·70	1	3	1	4	1	2
0·70 and under 0·90	3	0	0	1	0	1
0·90 and under 1·10	1	3	1	1	0	1
1·10 and over	1	2	1	1	0	2
Total	8	15	5	11	1	8
(B) New products weighted by innovation cost* (%)						
Less than 0·30	3	10	35	16	0	0
0·30 and under 0·50	2	28	17	36	0	16
0·50 and under 0·70	35	24	19	27	100	23
0·70 and under 0·90	22	0	0	16	0	14
0·90 and under 1·10	18	16	22	4	0	18
1·10 and over	19	23	9	2	0	30
Total†	100	100	100	100	100	100

* The weighted number of new products is expressed as a percentage of the column total.
† Because of rounding errors, items may not sum to column total.

data were obtained from the innovating firm concerning the costs of the innovation, as well as the time it took to bring the innovation to market (from the beginning of applied research to the date of its commercial introduction).

The innovations included in the sample vary widely in importance, but practically all are major new products that are central to the innovators' activities, not peripheral to their main business. For 30 of the 48 products, the innovation cost exceeded $1 million; for 12 products, it exceeded $5 million. About 70% of the innovations were patented. In only one case did the innovator license the imitating firm. In all other cases, the imitator received no help from the innovator. In general, these firms did not appear to be very interested in licensing these innovations, at least in the relevant time period, often because they did not feel it was in their interest to encourage potential rivals.

On the average, the ratio of the imitation cost to the innovation cost was about 0·65, and the ratio of the imitation time to the innovation time was about 0·70. As shown in Tables 1 and 2, there is considerable variation about these averages. In about half of the cases, the ratio of imitation cost to innovation cost was either less than 0·40 or more than 0·90. In about half of the cases, the ratio of imitation time to innovation time was either less than 0·40 or more than 1·00. Products with a relatively high (low) ratio of imitation cost to innovation cost tended to have a relatively high (low) ratio of imitation time to innovation time.[1]

It may come as a surprise that imitation cost was no smaller than innovation cost in about one seventh of the cases. This was not due to any superiority of the imitative product over the innovation. Instead, in a substantial percentage of these cases, it was due to the innovator's having a technological edge over its rivals in the relevant field. Often this edge was due to superior 'know-how' – that is, better and more extensive technical information based on highly specialised experience with the development and production of related products and processes. Such know-how is not divulged in patents and is relatively inaccessible (at least for a period of time) to potential imitators.

Based on these results, it appears that innovators routinely introduce new products despite the fact that other firms can imitate these products at about two thirds (often less) of the cost and time expended by the innovator.[2] In some cases, this is because, although other firms could imitate these products in this way, there are other barriers to entry (for example, lack of a well-known brand name) that discourage potential imitators. But to a greater extent (at least in this sample), it seems to be due to a feeling on the part of the innovators that, even if imitators do begin to appear in a relatively few years, the innovation still will be profitable.

II. THE IMITATOR'S TIME-COST TRADEOFF FUNCTION

Clearly, the time it takes a firm to imitate a new product can generally be reduced by spending more money. Each product's imitator was confronted by a time-cost tradeoff function, which is the relationship between the amount spent by the imitator and the length of time it would take to imitate this new product.[3] The time-cost combination chosen by the imitator is one point on this tradeoff function. In most cases, we obtained an estimate of the minimum time that the imitator could have taken to carry out the project, and the corresponding cost, as well as an estimate of the minimum cost and the corresponding time. Together with other information, this enabled us to determine the approximate position of the imitator's time-cost tradeoff function for most products. If imitation cost is measured as a percentage of innovation cost and if imitation time is measured as a percentage of innovation time, all of these time-cost tradeoff functions can be plotted on the same graph. When this is done, we find that a product's ratio of imitation cost to innovation cost is a good indicator of how high and how far to the right the product's time-cost tradeoff function is located.

[1] The correlation coefficient between a product's ratio of imitation cost to innovation cost and its ratio of imitation time to innovation time is about 0·8.

[2] The available data pertain only to innovations that were introduced, not to those where the incentives were insufficient for their introduction. Although they cannot measure the effects of imitation cost on how many innovations were introduced, they provide valuable evidence concerning the characteristics of those that were introduced.

[3] For further discussion of time-cost tradeoff functions, see Scherer (1967, 1980), Mansfield *et al.* (1971, forthcoming), and Baldwin and Childs (1969).

Using the time-cost tradeoff function for each product, we computed the arc elasticity of cost with respect to time. The median value of this elasticity is about 0·7, which means that a 1 % reduction in time results in about a 0·7 increase in cost, on the average.[1] These elasticities pertain to the imitation time that is midway between the minimum time and the time corresponding to minimum cost. Since the time-cost tradeoff function is convex, this elasticity increases as imitation time falls (and approaches its minimum value), and decreases as imitation time rises (and approaches the time corresponding to minimum cost). For most products, the minimum-time imitation cost is less than 50 % greater than the minimum imitation cost, but in some cases the difference is 100 % or more. Ordinarily, the minimum-cost imitation time is less than 75 % greater than the minimum imitation time, but in over one-fourth of the cases the difference is 100 % or more (Table 3).

Table 3

Ratio of Minimum-Time Imitation Cost to Minimum Imitation Cost, and Ratio of Minimum-Cost Imitation Time to Minimum Imitation Time, Percentage Distribution of New Products in the Sample

Value of ratio	Minimum-time imitation cost ÷ minimum imitation cost	Minimum-cost imitation time ÷ minimum imitation time
	(% of new products)	
1·00 and under 1·25	36	18
1·25 and under 1·50	33	24
1·50 and under 1·75	6	24
1·75 and under 2·00	6	3
2·00 and under 2·25	12	15
2·25 and under 2·50	6	6
2·50 and over	0	9
Total	100*	100*

* Because of rounding errors, the percentages do not sum to the column total.

III. DETERMINANTS OF IMITATION COSTS

The following hypotheses help to explain the substantial variation among products in the ratio of imitation cost to innovation cost.[2] First, we would

[1] Estimates of the elasticity of cost with respect to time could be obtained for 39 products. In 5 cases, it was less than 0·25; in 10 cases, it was 0·25 and under 0·50; in 12 cases, it was 0·50 and under 1·00; in 10 cases, it was 1·00 and under 2·00; and in 2 cases, it was 2·00 and over.

[2] These hypotheses also help to explain the variation among products in the ratio of imitation time to innovation time, since (as pointed out in note 1 on p. 910) a product's ratio of imitation time to innovation time is positively correlated with its ratio of imitation to innovation cost.

For some purposes, one might want to explain the variation among products in the ratio of imitation cost to innovation cost when the ratio of imitation time to innovation time is held constant (at its mean value). Based on our estimates of the time-cost tradeoff functions, we were able to make rough estimates of most products' ratio of imitation cost to innovation cost under these circumstances. When these estimates are used (in place of C_i) in equation (2), the results are relatively unchanged. Thus, the hypotheses discussed in this section can explain the variation among products in both the unadjusted ratio of imitation cost to innovation cost and the ratio of imitation cost to innovation cost when the ratio of imitation time to innovation time is held constant.

expect the ith product's ratio of imitation cost to innovation cost (denoted by C_i) to be inversely related to the proportion of the ith product's innovation cost that goes for research (rather than product specification, pilot plant or prototype, plant and equipment, or manufacturing or marketing startup). An imitator frequently can spend much less time and money on research than the innovator because the product's existence and characteristics provide the imitator with a great deal of information that the innovator had to obtain through its own research. On the other hand, an imitator often has to go through many of the same steps as the innovator with respect to pilot plant or prototype construction, investment in plant and equipment, and manufacturing and marketing startup.

Second, we would expect C_i to be relatively large if the ith product is a new drug where the Food and Drug Administration requires that an imitative product must be tested in much the same way as the innovative product with which it competes. As Kitch (1973) and others have pointed out, this requirement increases the money and time that an imitator must spend. (How much it increases them is a matter of some debate; results on this score are given below.) Third, we would expect C_i to be relatively small if the ith product consists of a new use for an existing material and if some firm other than the innovator has patents on this material. In such cases, the patent holder often can imitate the innovation relatively cheaply and quickly.[1]

To test these hypotheses, we assume that

$$C_i = \alpha_0 + \alpha_1 R_i + \alpha_2 D_i + \alpha_3 G_i + z_i, \tag{1}$$

where R_i is the percentage of the innovation cost that went for applied research in the case of the ith product, D_i is a dummy variable that equals 1 if the ith innovation was an ethical drug (subject to regulation of the sort described above) and 0 otherwise, G_i is a dummy variable that equals 1 if the ith product was a new use for an existing material on which a firm other than the innovator holds patents and 0 otherwise, and z_i is a random error term. Data were collected regarding R_i, D_i, and G_i for 28 of the products in our sample,[2] and we computed least-squares estimates of the α's. The results are

$$\underset{(11{\cdot}8)}{C_i = 0{\cdot}838} - \underset{(3{\cdot}80)}{0{\cdot}00684 R_i} + \underset{(2{\cdot}31)}{0{\cdot}310 D_i} - \underset{(2{\cdot}41)}{0{\cdot}536 G_i}. \quad (\bar{R}^2 = 0{\cdot}49) \tag{2}$$

The t-ratios are given in parentheses.

[1] In one case in our sample, an agricultural chemical firm found a new use for a particular chemical. This firm's supplier held a patent on the broad group of chemicals of which this chemical was part, but was unaware of this new use. When the innovator marketed the chemical for this new use, the supplier could (and did) imitate the innovation quickly and at a low cost.

Another factor we would like to include in equation (1) is the extent of the imitator's technological capabilities and know-how, relative to those of the innovator, but we could find no way to measure this factor adequately.

[2] The sample was composed of two subsamples, each of which was collected quite independently from the other. In the subsample for which Schwartz was particularly responsible, we obtained the data needed to test these hypotheses. In the other sub-sample, for which Wagner was particularly responsible, we did not obtain these data, but focused attention on other questions instead. Equation (4) is also based only on the former subsample. One product was not included in equation (2) because the imitator received a license from the innovator, but if this product is included, the results do not change in any significant way. Interaction terms were also included in equation (2), but were not significant.

Each of the regression coefficients in equation (2) has the expected sign, is statistically significant, and is large enough so that the effect of each independent variable is substantial. If R_i increases by 20 percentage points, C_i decreases by about 0·14. If the *i*th product is a new drug, C_i is about 0·31 larger than would otherwise be the case. If the *i*th product is a new use for an existing material on which a firm other than the innovator holds the patents, C_i is about 0·54 smaller than would otherwise be the case. These three independent variables can explain about half of the variation in C_i.

IV. PATENTS AND IMITATION COSTS

Still another factor that affects the ratio of imitation cost to innovation cost is whether or not the innovator has patents on the new product. Contrary to popular opinion, patent protection does not make entry impossible, or even unlikely. Within 4 years of their introduction, 60% of the patented successful innovations in our sample were imitated. Nonetheless, patent protection generally increases imitation costs. To obtain information concerning the size of this increase, the firms in our sample were asked to estimate how much the value of C_i for each patented product increased because it was patented. The median estimated increase in C_i was 11%. They also were asked to estimate how much the value of C_i for each unpatented product would have increased if it had been patented. The median estimated increase in C_i was only about 6%. (Indeed, for 2 of these products, patent protection would have reduced the money and time required for imitation because in these cases the innovator could keep secret the essential information underlying the product, whereas if it patented it, some of the information would have been disclosed.) The fact that a patent resulted in a larger increase in the imitation costs of the patented products than of the unpatented products was, of course, a major reason why some products were patented and others were not.[1]

In the ethical drug industry, patents had a bigger impact on imitation costs than in the other industries, which helps to account for survey results[2] indicating that patents are regarded as more important in ethical drugs than elsewhere. The median estimated increase in C_i due to patent protection was about 30% in ethical drugs, in contrast to about 10% in chemicals and about 7% in electronics and machinery. Without patent protection, it frequently would have been relatively cheap (and quick) for an imitator to determine the composition of a new drug and to begin producing it. However, for many of these electronics and machinery innovations, it would have been quite difficult for imitators to determine from the new product how it is produced, and patents would not add a great deal to imitation cost (or time).[3] These results are in accord with Taylor

[1] Since some of the products in the sample were inherently unpatentable, they had to be omitted, so the results reported in this paragraph are based on 43 products. Using the Wilcoxon test, the difference between the median increase in C_i for patented and unpatented products is significant at the 0·05 probability level. In the entire sample of 48 products, the percentage that was patented was 81 (drugs), 63 (chemicals), and 69 (electronics and machinery).

[2] For a description of these surveys and their results, see Taylor and Silbertson (1973) and Scherer (1977).

[3] Taylor and Silberston (1973) point out that a high proportion of patents in electronics are thought to be of doubtful validity, and that, even if an electronics patent is valid, it may afford little

and Silbertson's (1973) conclusion that the lack of patent protection would reduce the rate of expenditure on innovative activity to a greater extent in drugs than in other industries.

V. IMITATION COSTS, ENTRY, AND CONCENTRATION

We turn now from the determinants of imitation costs to their effects on entry and concentration. Holding constant the discounted profit (gross of the imitation cost) that the imitator expects to earn by imitating a new product, the new product is more likely to be imitated if the imitation cost is small. To discourage entry, the innovator may adopt pricing (and other) policies to reduce the imitator's expected discounted gross profit if the imitation cost is low. Taking this into account, is it still true that the probability of entry is inversely related to the size of the imitation cost? To find out, we determined whether each product in the sample was imitated within 4 years[1] after it was first introduced. (Innovations that had been on the market less than 4 years and unsuccessful innovations clearly had to be omitted.) Then we carried out a logit analysis to determine whether C – the ratio of imitation cost to innovation cost – influences the probability that entry of this sort occurred within 4 years. Letting P be the probability that such entry did *not* occur, we found that

$$\ln\left(\frac{P}{1-P}\right) = \underset{(2\cdot 04)}{-3\cdot 10} + \underset{(1\cdot 97)}{3\cdot 92C}. \qquad (3)$$

Thus, imitation cost seems to be related in the expected way to whether or not entry occurs.[2]

Imitation cost may also affect an industry's level of concentration. We would expect an industry's concentration level to be relatively low if its members' products and processes can be imitated easily and cheaply.[3] For each of the 16 detailed industries included in the sample, we calculated the mean value of C_i, the mean for the jth industry being denoted by $\bar{C}_j$. Then to estimate the relationship between the mean imitation cost and the concentration level, we

---

protection. See Taylor and Silberston (1973), chapter 12. For some ways (other than patents) in which electronics firms protect their products from imitation, see *Science*, 202, November 24, 1978, pp. 848–9. Also see Scherer (1977).

Although the median increase in imitation cost due to patents (outside the drug industry) is rather small, the increase in some cases is very substantial. Outside drugs, patents increased imitation cost by 100 % or more in about one-quarter of the cases. This helps to explain our findings in Section VI below.

[1] In the bulk of the cases, the new product could have been imitated in 2 years or less even if the imitator carried out the project at the most leisurely pace. In practically all cases it could be imitated in 3 years or less. Thus, 4 years was plenty of time for an imitator to enter.

[2] Note that a one-tailed test is appropriate here. The methods used to estimate this equation are described in Berkson (1953). For an analysis of entry rates in the chemical industry that estimates the effects of certain kinds of imitation costs, see Mansfield *et al.* (1977*b*), pp. 119–22. For some evidence that imitation costs influence the gap between social and private rates of return from innovations, see Mansfield *et al.* (1977*a*).

[3] In their simulation model, Nelson and Winter (1978) use a hypothetical 'ease of imitation factor' which is defined differently from our C_i, but it seems clear that this factor is closely related to what our C_i measures.

regressed each industry's 4-firm concentration ratio, K_j, on $\bar{C}_j$, the result being[1]

$$K_j = 6{\cdot}22 + 61{\cdot}5\bar{C}_j. \quad (4)$$
$$(8{\cdot}80)\ (12{\cdot}8)$$

This finding, which seems to be the first empirical evidence regarding the relationship between the ease of imitation and the level of concentration, is entirely consistent with this hypothesis. Given the large number of factors influencing an industry's concentration level, it is interesting that this relationship is relatively close ($r^2 = 0{\cdot}60$). Differences among industries in the technology transfer process (including transfers that are both voluntary and involuntary from the point of view of the innovator) may be able to explain much more of the interindustry variation in concentration levels than is generally recognised.

VI. PATENTS AND THE RATE OF INNOVATION

Finally, we turn to one of the most important and controversial questions concerning the patent system: what proportion of innovations would be delayed or not introduced at all if they could not be patented? To shed light on this question, we asked each innovating firm whether it would have introduced each of its patented innovations in our sample if patent protection had not been available. Although answers to such questions have obvious limitations and must be treated with caution, they should shed some light on this topic, about which so little is known.[2] According to the firms, about one-half of the patented innovations in our sample would not have been introduced without patent protection. The bulk of these innovations occurred in the drug industry. Excluding drug innovations, the lack of patent protection would have affected less than one-fourth of the patented innovations in our sample.

Taylor and Silberston (1973) have provided estimates of the proportion of R and D expenditures in various U.K. industries that would not have been carried out without patent protection. Applying the industry weights in our sample to their data, one would expect that about 36% of the R and D expenditures of the firms in our sample would not have been carried out under these circumstances, if U.S. and British firms are alike in this regard. This is somewhat less[3] than the proportion of innovations that, according to the firms, would not have been introduced without patent protection.[4]

[1] To prevent confusion, note that each concentration ratio refers to all the detailed industry's products, not just the new products in our sample. In effect, we treat the products in our sample in each detailed industry as a sample of all the detailed industry's products, and use $\bar{C}_j$ as a measure of the ease with which all the latter products can be imitated.

[2] These data could be obtained for 31 innovations. Some innovations could not be included because they were unpatentable or because no reliable information could be obtained from the firm. For a good discussion of what currently is known about this topic, see Scherer (1977 and 1980) and Taylor and Silberston (1973).

[3] This difference is statistically significant at the 0·05 probability level.

[4] There is no reason to expect the proportion of R and D expenditures that would not be carried out without patent protection to equal the proportion of innovations that would not be carried out under these circumstances. If the innovations that would not be carried out tend to be less R and D-intensive than those that would be carried out under these circumstances, the former proportion would be less

One important reason why patents frequently are not regarded as crucial is that they often have only a limited effect on the rate of entry. For about half of the innovations, the firms felt that patents had delayed the entry of imitators by less than a few months.[1] Although patents generally increased the imitation costs, they did not increase the costs enough in these cases to have an appreciable effect on the rate of entry. But although patent protection seems to have only a limited effect on entry in about half of the cases, it seems to have a very important effect in a minority of them. For about 15% of the innovations, patent protection was estimated to have delayed the time when the first imitator entered the market by 4 years or more.

According to many economists, patent protection tends to be more important to smaller firms than to larger ones. However, although this proposition seems reasonable, the existing evidence on this score is weak and sometimes contradictory.[2] To test this proposition, we carried out a logit analysis to see whether π, the probability that an innovation would have been introduced without patent protection, is related to S, the 1976 sales of the firm that carried out the innovation. Since practically none of the drug innovations would have been introduced without patent protection, such innovations were excluded.[3] The results are:

$$\ln\left(\frac{\pi}{1-\pi}\right) = \underset{(1{\cdot}03)}{0{\cdot}878} + \underset{(0{\cdot}030)}{0{\cdot}00012}S. \tag{5}$$

Since the coefficient of S is far from statistically significant, there is no evidence that patent protection was more likely to be deemed essential for innovations carried out by smaller firms than for those carried out by larger ones.[4]

VII. CONCLUSION

At a purely theoretical level, imitation costs have long played a major role in economic models of innovation and technological change, but systematic empirical investigations have been lacking. While the present study is only a relatively small-scale undertaking with obvious limitations, we believe that it

than the latter. Also, there is no reason to expect that American and British firms would behave in exactly the same way. Thus, if patents are regarded as more important in the United States than in the United Kingdom, this would result in the sort of difference we observe. In addition, there are sampling errors in both our estimates and those of Taylor and Silberston.

[1] Each innovating firm was asked to estimate how much sooner the first imitating firm would have entered the market if the new product had not had patent protection. The resulting estimates are rough, but, according to the firms, they are reasonably accurate. They seem consistent with the findings in the first paragraph of this section. Outside drugs, patents were estimated to delay the entry of imitators by 2 years or more in 25% of the cases.

[2] See Scherer (1977).

[3] Had the drug innovations been included, a bias might have resulted, since, as noted above, patents are regarded as particularly important in the drug industry. Thus, since the drug firms were larger than the average in the sample, this might have resulted in a spurious inverse relationship between π and S. S is measured in billions of dollars.

[4] These results may be due in part to the fact that practically all of the innovating firms in the sample are quite large. If a larger number of very small firms were included, the results might be more in line with the proposition in the text.

demonstrates both that empirical investigations of this topic can be carried out and that, unless such investigations are carried out, we are unlikely to understand many fundamental aspects of the economics of technological change. In our view, there is a considerable need for more studies of the size, determinants, and effects of imitation costs.

Our findings should help to promote the search for more realistic and useful models of the innovative process. In recent years, there has been a tendency for such models to assume that the innovator receives all of the benefits from an innovation and that imitation can be ignored. (For example, see Dasgupta and Stiglitz (1980) and Loury (1979).) And in studies of optimal patent life, it is often assumed that the patent holder is free from imitation for the life of the patent. Although we understand how convenient such assumptions may be, our results suggest how considerably they depart from reality. (Recall, for example, that about 60% of the patented innovations in our sample were imitated within 4 years.) We hope that the excellent theorists who are working in this area will soon be able to relax these assumptions.[1]

This study also sheds new light on the effects of the patent system. Contrary to the assumption of many economic models, a patent frequently does not result in a 17-year monoply over the relevant innovation. Patents do tend to increase imitation costs, particularly in the drug industry, but excluding drugs, patent protection did not seem essential for the development and introduction of at least three-fourths of the patented innovations studied here. From the point of view of public policy, this obviously is an interesting finding.

University of Pennsylvania

Date of receipt of final typescript: May 1981

References

Arrow, K. (1962). 'Economic welfare and the allocation of resources for invention.' In *The Rate and Direction of Inventive Activity*. New York: National Bureau of Economic Research.

Baldwin, W. and Childs, G. (1969). 'The fast second and rivalry in research and development.' *Southern Economic Journal*, July, pp. 18–24.

Berkson, J. (1953). 'A statistically precise and relatively simple method of estimating the bioassay with quantal response, based on the logistic function.' *Journal of the American Statistical Association*, pp. 565–99.

Dasgupta, P. and Stiglitz, J. (1980). 'Industrial structure and the nature of innovative activity.' ECONOMIC JOURNAL, June, pp. 266–93.

Freeman, C. (1974). *The Economics of Industrial Innovation*. Middlesex: Penguin.

Kitch, E. (1973). 'The patent system and new drug application.' In *Regulating New Drugs* (ed. R. Landau), pp. 81–108. Chicago: University of Chicago.

Loury, G. (1979). 'Market structure and innovation.' *Quarterly Journal of Economics*, (August), pp. 395–410.

[1] The innovator may receive far less than other firms that subsequently introduce the innovation in a somewhat different way or at a more propitious time. Even if the innovator is very successful, imitators may gain a substantial share of the benefits. Models that ignore considerations of this sort should be viewed with caution, since (among other reasons) they omit factors that are important impediments to firms' appropriation of the benefits from their innovations. By doing so, they may exaggerate the incentive for firms to invest in R and D and other innovative activities.

Mansfield, E. (1980). 'Basic research and productivity increase in manufacturing.' *American Economic Review*, (December).

——, Rapoport, J., Schnee, J., Wagner. S. and Hamburger, M. (1971). *Research and Innovation in the Modern Corporation*. New York: W. W. Norton.

——, Rapoport, J., Romeo, A., Wagner, S. and Beardsley, G. (1977*a*). 'Social and private rates of return from industrial innovations.' *Quarterly Journal of Economics*, vol. 16, May, pp. 221–40.

——, Rapoport, J., Romeo, A., Villani, E., Wagner, S. and Husic, F. (1977*b*). *The Production and Application of New Industrial Technology*. New York: W. W. Norton.

——, Romeo, A., Schwartz, M., Teece, D., Wagner, S. and Brach, P. (forthcoming). *Technology Transfer, Productivity and Economic Policy*. New York: W. W. Norton.

Nelson, R. and Winter, S. (1978). 'Forces generating and limiting concentration under Schumpeterian competition.' *Bell Journal of Economics*, vol. 9, (Autumn), pp. 524–48.

Scherer, F. M. (1967). 'Research and development resource allocation under rivalry.' *Quarterly Journal of Economics*, vol. 81, August, pp. 359–94.

——, (1977). *The Economic Effects of Compulsory Patent Licensing*. New York: New York University Monograph 1977–2 in Finance and Economics.

——, (1980). *Industrial Market Structure and Economic Performance*, 2nd ed. Chicago: Rand McNally.

Taylor, C. and Silberston, Z. (1973). *The Economic Impact of the Patent System*. Cambridge University Press.

[15]

RICHARD C. LEVIN
Yale University

ALVIN K. KLEVORICK
Yale University

RICHARD R. NELSON
Columbia University

SIDNEY G. WINTER
Yale University

Appropriating the Returns from Industrial Research and Development

TO HAVE the incentive to undertake research and development, a firm must be able to appropriate returns sufficient to make the investment worthwhile. The benefits consumers derive from an innovation, however, are increased if competitors can imitate and improve on the innovation to ensure its availability on favorable terms. Patent law seeks to resolve this tension between incentives for innovation and widespread diffusion of benefits. A patent confers, in theory, perfect appropriability (monopoly of the invention) for a limited time in return for a public

We are grateful for the support of the National Science Foundation and especially to Rolf Piekarz of the NSF's Division of Policy Research and Analysis.

We also wish to thank the 650 respondents to our survey and the R&D executives who helped us pretest it—especially Ralph Gomory, Bruce Hannay, and Lowell Steele. Donald DeLuca, Wendy Horowitz, and other members of the Roper Center for Survey Research helped manage the survey. Robert W. Wilson, and Margaret Blair, Marc Chupka, Emily Lawrance, Constance Helfat, Andrew Joskow, Kathleen Rodenrys, Somi Seong, Andrea Shepard, and Hal Van Gieson also provided valuable assistance.

disclosure that ensures, again in theory, widespread diffusion of benefits when the patent expires.

Previous investigations of the system suggest that patents do not always work in practice as they do in theory.[1] On the one hand, appropriability is not perfect. Many patents can be circumvented; others provide little protection because of stringent legal requirements for proof that they are valid or that they are being infringed. On the other hand, public disclosure does not always ensure ultimate diffusion of an invention on competitive terms. For example, investments to establish the brand name of a patented product may outlive the patent itself.[2] And patents may not always be necessary. Studies of the aircraft and semiconductor industries have shown that gaining lead time and exploiting learning curve advantages are the primary methods of appropriating returns. Other studies have emphasized the importance of complementary investments in marketing and customer service.[3]

Evidence on the nature and strength of conditions for appropriability and on the working of the patent system is, however, scattered and unsystematic. Because imperfect appropriability may lead to underinvestment in new technology, and because technological progress is a primary source of economic growth, it would be useful to have a more comprehensive empirical understanding of appropriability, in particular, to identify those industries and technologies in which patents are effective in preventing competitive imitation of a new process or product. It would also be desirable to know where patents can be profitably licensed. Where patents are not effective, it would be useful to understand why they are not and whether other mechanisms are.

1. F. M. Scherer and others, *Patents and the Corporation: A Report on Industrial Technology under Changing Public Policy*, 2d ed. (privately published, 1959); and C. T. Taylor and Z. A. Silberston, *The Economic Impact of the Patent System: A Study of the British Experience* (Cambridge University Press, 1973).

2. See, for example, Meir Statman, "The Effect of Patent Expiration on the Market Position of Drugs," in Robert B. Helms, ed., *Drugs and Health: Economic Issues and Policy Objectives* (Washington, D.C.: American Enterprise Institute, 1981), pp. 140–51.

3. The importance of lead time and learning curve advantages is documented in Almarin Phillips, *Technology and Market Structure: A Study of the Aircraft Industry* (Lexington Books, 1971); and John E. Tilton, *International Diffusion of Technology: The Case of Semiconductors* (Brookings, 1971). For the importance of marketing and customer service, see Marie-Therese Flaherty, "Field Research on the Link between Technological Innovation and Growth: Evidence from the International Semiconductor Industry," working paper 84-83 (Harvard University, Graduate School of Business Administration, no date).

This paper describes the results of an inquiry into appropriability conditions in more than one hundred manufacturing industries. We discuss how this information has been and might be used to cast light on important issues in the economics of innovation and public policy. Our data, derived from a survey of high-level R&D executives, are informed opinions about an industry's technological and economic environment rather than quantitative measures of inputs and outputs.

Although our use of semantic scales to assess, for example, the effectiveness of alternative means of appropriation introduces considerable measurement error, more readily quantifiable proxies would probably not serve as well. Remarkable progress has been made toward developing a methodology to estimate the economic value of patents.[4] But suitable data are as yet unavailable in the United States, and European data lack sufficiently reliable detail to support inferences about interindustry differences in the value of patents. Our judgment was that asking knowledgeable respondents about the effectiveness of patents and alternative means of appropriation was at least as likely to produce useful answers as asking for quantitative estimates of the economic value of a typical patent.

We have taken considerable care to establish the robustness of our findings in the presence of possibly substantial measurement error, but ultimately the value of the data will depend on their contribution to better empirical understanding of technological change and more discriminating discussion of public policy. To view the empirical contribution of the data from the simplest perspective, consider their potential for improving the quality of research that uses patent counts to measure innovative activity.[5] This line of inquiry has shown, among other results, that industries vary significantly in the average number of patents generated by each dollar of R&D investment.[6] Our findings on industry

4. See, especially, Ariel Pakes, "Patents as Options: Some Estimates of the Value of Holding European Patent Stocks," *Econometrica*, vol. 54 (July 1986), pp. 755–84.

5. For a summary of the best of this work, see Zvi Griliches, Ariel Pakes, and Bronwyn H. Hall, "The Value of Patents as Indicators of Inventive Activity," working paper 2083 (Cambridge, Mass.: National Bureau of Economic Research, November 1986). For other perspectives on the usefulness of patent data, see the special issue of *Research Policy*, vol. 16 (August 1987).

6. F. M. Scherer, "The Propensity to Patent," *International Journal of Industrial Organization*, vol. 1 (March 1983), pp. 107–28; and John Bound and others, "Who Does R&D and Who Patents?" in Zvi Griliches, ed., *R&D, Patents and Productivity* (University of Chicago Press for National Bureau of Economic Research, 1984), pp. 21–54.

differences in patent effectiveness may help explain this variation in the apparent productivity of R&D.

More fundamentally, large and persistent interindustry differences in R&D investment and innovative performance have resisted satisfactory explanation, in part for lack of data that adequately represent the theoretically important concepts of appropriability and technological opportunity. Promising but ultimately unsatisfactory results have been obtained in exploratory work that used crude proxy variables and econometric ingenuity to capture the influence of appropriability and opportunity conditions.[7] Our desire to provide a stronger basis for this line of inquiry was a prominent motive for our survey research and helped to shape its design.

Finally, gathering better information on the nature and strength of appropriability is particularly timely in view of the prominence of current debates on the adequacy of laws and institutions to protect intellectual property. One impetus for change has been the need to clarify and perhaps strengthen the system of property rights at various new frontiers of technology. Thus, for example, recent legislation has adapted copyright law to protect the rights of the creator of new computer software, a new legal framework has been constructed to protect intellectual property embodied in semiconductor chip designs, and important court decisions and administrative actions have shaped the development of a property rights system in biotechnology.[8]

Another spur to change has been the need to resolve conflicts between the aims of social regulation and the exercise of intellectual property rights. For example, the Drug Price Competition and Patent Term Restoration Act of 1984 extended patent lives of pharmaceuticals to compensate for regulatory requirements that delay the introduction of new drugs.

7. Richard C. Levin, "Toward an Empirical Model of Schumpeterian Competition," working paper 43 (Yale University, School of Organization and Management, 1981); Richard C. Levin and Peter C. Reiss, "Tests of a Schumpeterian Model of R&D and Market Structure," in Griliches, ed., *R&D, Patents and Productivity*, pp. 175–204; and Ariel Pakes and Mark Schankerman, "An Exploration into the Determinants of Research Intensity," in Griliches, ed., *R&D, Patents and Productivity*, pp. 209–32.

8. See Computer Software Act of 1980; Semiconductor Chip Protection Act of 1984; *Diamond* v. *Chakrabarty*, 447 U.S. 305 (1980), holding that plant and animal life is patentable under U.S. patent law; and D. J. Quigg, memorandum of April 7, 1987, explaining the policies of the U.S. Patent and Trademark Office concerning applications to patent life forms.

Intellectual property rights also figure prominently among policy issues milling under the banner of competitiveness. Recent annual reports of the U.S. trade representative have focused on the difficulties U.S. manufacturers encounter in protecting intellectual property rights in foreign markets. The trade bill passed in 1987 by the House of Representatives contains several provisions that increase the scope of protection and the opportunities for relief available to U.S. manufacturers confronted with imports that infringe these rights.[9] Proposed antitrust legislation, motivated by a concern that courts have kept inventors from reaping rewards that patent laws are intended to provide, stipulates that patent license agreements and similar contracts relating to use of intellectual property "shall not be deemed illegal per se under any of the antitrust laws."[10]

To the extent that all this activity attempts to rectify obvious inadequacies in existing institutions, the case for reform appears strong and straightforward. It is easy to deplore the blatant copying of innovative integrated circuit designs, the importation of "knock off" copies of trademarked or patented U.S. products, and the piracy of copyrighted written matter and audio and video cassettes. But reforms may yield unintended consequences. In its simplest form, this concern translates into wariness about Trojan horses: provisions brought into the law by the rhetorical tug of "competitiveness" and "intellectual property" may harbor instruments of protectionism and price fixing. Other potential consequences are subtler but no less important. For example, seemingly uniform adjustments of intellectual property, antitrust, or trade law may affect some industries quite differently than others.

And it should not be taken for granted that more appropriability is better, that better protection necessarily leads to more innovation, which yields better economic performance—higher standards of living, better competitiveness, and so on. Better protection may yield more innovation at the cost of incrementally increasing resources devoted to producing the innovation: the larger prize may merely encourage duplicative private effort to capture it.[11] Alternatively, better protection may induce inno-

9. See H.R. 3, the Omnibus Trade and Competitiveness Reform Act of 1987, which is currently under consideration by a House-Senate conference committee.

10. H.R. 557 and S. 438, 100 Cong., 1 sess.

11. This is the "free access" externality, first emphasized in the context of innovation in Yoram Barzel, "Optimal Timing of Innovations," *Review of Economics and Statistics*, vol. 50 (1968), pp. 348–55. For a survey of the literature on patent races, see Jennifer

vation of the wrong kind, or it may buy the innovation by further delaying access to it on competitive terms.[12]

The premise that stronger protection will always improve the incentives to innovate is also open to challenge. Unimpeded diffusion of existing technology is immediately beneficial not only for consumers but also for those who would improve that technology. Because technological advance is often an interactive, cumulative process, strong protection of individual achievements may slow the general advance. This would not occur in a hypothetical world without transaction costs, in which efficient contracts to share information would be made. In reality, however, markets for rights to information are subject to major transactional hazards, and strong protection of a key innovation may preclude competitors from making socially beneficial innovations. The semiconductor industry of the 1950s and 1960s provides an excellent example of rapid progress in a cumulative technology that might have been impossible under a regime that strongly protected intellectual property.[13]

The remainder of this paper discusses our survey instrument, the construction of the sample, and the interpretation of the data, then turns to our findings concerning the effectiveness of patents and other means of appropriating the returns from R&D. The results of related work that employs the survey data to reexamine central questions in the empirical literature on R&D are summarized, and we discuss how our findings might contribute to a more discriminating discussion of patent law, antitrust law, and trade policy.

Questionnaire Design and Survey Methods

The content of our questionnaire was shaped with guidance from the conceptual literature on technological change, empirical literature on

Reinganum, "The Timing of Innovation: Research, Development and Diffusion," in Richard Schmalensee and Robert Willig, ed., *Handbook of Industrial Organization* (North-Holland, 1988).

12. Richard R. Nelson, "Assessing Private Enterprise: An Exegesis of Tangled Doctrine," *Bell Journal of Economics*, vol. 12 (Spring 1981), pp. 93–111; and William D. Nordhaus, *Invention, Growth, and Welfare: A Theoretical Treatment of Technological Change* (MIT Press, 1969).

13. Richard C. Levin, "The Semiconductor Industry," in Richard R. Nelson, ed., *Government and Technical Progress: A Cross-Industry Analysis* (Pergamon Press, 1982), pp. 9–100.

the economic impact of the patent system, the work of Mansfield and his associates on imitation costs, and numerous case studies.[14] The questionnaire was aimed at high-level R&D managers with knowledge of both the relevant technology and market conditions. To check the interpretability of the questions and the likely validity and reliability of the responses, we pretested the questionnaire with twelve managers representing diverse businesses.[15]

To understand how appropriability differs across industries, we asked each respondent to report typical experiences or central tendencies within a particular industry. Respondents were thus treated as informed observers of a line of business rather than as representatives of a single firm, an approach that encouraged cooperation (they were not placed in the position of possibly divulging practices or policies of their own firms), but led inevitably to heterogeneity in the responses within a given industry.

The questionnaire contained four parts. Parts 1 and 2 concerned appropriability; parts 3 and 4 concerned technological opportunity and technological advance. Questions in part 1 asked about the effectiveness of alternative means of protecting the competitive advantages of R&D, limits on the effectiveness of patents, and ways of acquiring knowledge of a competitors' technology. Part 2 asked about the cost and time required to imitate innovations of rivals; we distinguished process from

14. Among the sources of ideas for the questions are Paul Allan David, *Technical Choice, Innovation and Economic Growth: Essays on American and British Experience in the Nineteenth Century* (Cambridge University Press, 1975); Richard R. Nelson and Sidney G. Winter, "In Search of Useful Theory of Innovation," *Research Policy*, vol. 6 (Winter 1977), pp. 36–76; Nathan Rosenberg, "Science, Invention and Economic Growth," *Economic Journal*, vol. 84 (March 1974), pp. 90–108; and Devandra Sahal, *Patterns of Technological Innovation* (Addison-Wesley, 1981). For empirical literature on the economic effects of the patent system, see Scherer and others, *Patents and the Corporation;* and Taylor and Silberston, *The Economic Impact of the Patent System*. For imitation costs, see Edwin Mansfield, Mark Schwartz, and Samuel Wagner, "Imitation Costs and Patents: An Empirical Study," *Economic Journal*, vol. 91 (December 1981), pp. 907–18.

15. These managers had experience in communications equipment, industrial inorganic chemicals, metal cutting machine tools, shoe machinery, household electrical appliances, processed foods, computing equipment, semiconductors, copper smelting and refining, radio and TV sets, and industrial organic chemicals. They were asked to complete the questionnaire for a specific line of business, but to keep in mind the suitability of the questions for other lines of business with which they were familiar. After completing the questionnaire, they were interviewed face-to-face or by telephone. Interviews typically lasted one-half hour or more, and each question was discussed to eliminate sources of ambiguity.

product innovations, major from typical, and patented from unpatented.[16] Part 3 explored the links between an industry's technology and other sources of technological contribution. We asked about the importance of scientific research in general and university-based research in particular. We also asked about the extent to which interindustry spillovers are an important source of technological opportunity. Part 4 asked some broad questions about the pace and character of technological advance.[17] This paper analyzes responses to the questions in parts 1 and 2.[18]

SAMPLE CONSTRUCTION

As a sampling frame, we used the lines of business defined by the Federal Trade Commission. In the manufacturing sector, these chiefly correspond to four-digit SIC industries, although some are defined as groups of four-digit or even three-digit industries. The FTC lines provide the most disaggregated level at which data on R&D expenditures are available. An additional consideration was that F. M. Scherer's technology flow matrix, which classifies patents by industry of origin and industry of use, was also constructed at this level of aggregation.[19]

Ultimately, we received responses from 650 individuals representing 130 lines of business, with ten or more responses from eighteen industries and five to nine from twenty-seven industries. The sample was reason-

16. The questions were similar to those in Mansfield, Schwartz, and Wagner, "Imitation Costs and Patents," but covered typical rather than specific innovations. Our industry sample was also broader.

17. One objective was to examine "natural trajectories" of the sort described in Nelson and Winter, "In Search of Useful Theory," p. 56.

18. Data from responses to questions in parts 3 and 4 have been used in Richard C. Levin, Wesley M. Cohen, and David C. Mowery, "R&D Appropriability, Opportunity, and Market Structure: New Evidence on Some Schumpeterian Hypotheses," *American Economic Review*, vol. 75 (May 1985, *Papers and Proceedings, 1984*), pp. 20–24; Cohen, Levin, and Mowery, "Firm Size and R&D Intensity: A Re-examination," *Journal of Industrial Economics*, vol. 35 (June 1987), pp. 543–65; and Richard C. Levin and Peter C. Reiss, "Cost-Reducing and Demand-Creating R&D with Spillovers" (Stanford University, Graduate School of Business, 1986). See Richard R. Nelson, "Institutions Supporting Technical Advance in Industry," *American Economic Review*, vol. 76 (May 1986, *Papers and Proceedings, 1985*), pp. 186–89, for findings on the importance of external sources of technological knowledge.

19. F. M. Scherer, "Inter-industry Technology Flows in the United States," *Research Policy*, vol. 11 (August 1982), pp. 227–45.

ably representative of firms performing R&D, though the exclusion of those without publicly traded securities undoubtedly means that small start-up ventures, important sources of innovation, were underrepresented. The number of respondents in a line of business was positively correlated with the line's R&D spending, sales volume, and R&D intensity. The number of respondents did not increase in strict proportion to the level of industry R&D or sales, but the rate of response within a line of business was not significantly correlated with industry R&D spending, sales, or R&D intensity. The Appendix presents further details of sample construction.

METHODOLOGICAL ISSUES

Given our interest in identifying differences in the appropriability of R&D, it is reassuring that analysis of variance confirmed the presence of significant interindustry variation in the responses to most questionnaire items.[20] There was, however, also substantial intraindustry variation in the responses.

There are several potential sources of intraindustry heterogeneity in the responses to any given question. First, the lines of business as defined by the FTC may be objectively heterogeneous in their products and technologies. For example, if two firms classified as manufacturers of industrial inorganic chemicals produce different products using different technologies, they might differ markedly in their perception of the effectiveness of patents or the time required for imitation in their "industry." To eliminate this source of heterogeneity, we asked respondents to identify two major innovations—a process and a product—within their industries during the past ten to fifteen years. For most industries with ten or more respondents, more than half the respondents agreed on at least one such innovation. We thus believe it unlikely that overly aggregated industry definition was a major source of intraindustry heterogeneity.[21]

20. Interindustry differences are significant at the 0.05 level for approximately 60 percent of the questions in parts 1 and 2 of the questionnaire. If a higher level of aggregation is used to measure industry effects, such as the level at which the National Science Foundation reports R&D spending (a hybrid of two- and three-digit level industries), interindustry differences are significant at the 0.05 level for 70 percent of the questions.

21. Heterogeneity, as anthropologists have long insisted, is, however, in the eye of the

A respondent's perception of the central tendencies within an industry may also be affected by his firm's policies or strategies. Respondents in the same line of business may thus have different perceptions of the common technological environment that they were asked to characterize. A two-way analysis of variance of the responses on the effectiveness of patents, for example, revealed that both firm and industry effects are statistically significant. A representative multi-industry firm, however, tends to be involved in technologically related industries, and thus what appear to be effects attributable to the firm in the data may simply reflect the correlation in responses from related industries.

The third, and probably most important, source of intraindustry heterogeneity is the inherently subjective nature of the semantic scales used in the survey. Most answers were reported on a seven-point Likert scale. The effectiveness of patents in preventing duplication was, for instance, evaluated on a scale ranging from "not at all effective" to "very effective." There is no natural or objective anchor for such evaluative ratings. Individuals may perceive the same environment but simply use the scale differently. Some might systematically favor high scores; others might concentrate responses in the center of the scale; still others might frequently use extreme values.

The numerous techniques available to control for differences among respondents in means and variances generally require abandoning one or more dimensions along which the data might be informative. For example, we were interested in interindustry comparisons of answers to a single question; controlling for fixed effects among respondents would vitiate such comparisons, since we expected a respondent's mean score over all questions to depend on his industry. Standardizing the variance of each respondent's answers raised similar problems: the distribution of "correct" responses was unknown and it almost certainly differed systematically among industries. Rather than impose an arbitrary standardization, therefore, we examined the results for each group of questions using a variety of techniques and perspectives to assess the

beholder. One R&D manager, asked to inform us about the air and gas compressor industry, inquired whether we were interested in large, medium, or small compressors. In his view the technologies were fundamentally different. We asked him to note on his questionnaire where the answers to our questions differed across these size categories. The booklet he returned contained no such notation.

robustness of our principal conclusions. There was undeniably much noise in the data, but several important signals were robust to alternative weightings of the observations, alternative partitions of the sample, and the use of alternative summary statistics.[22]

We sidestepped one methodological difficulty by treating ratings along a seven-point semantic continuum as if they were interval data. The data were, of course, more properly to be regarded as ordinal. It would have been straightforward to treat them as ordinal if we had been interested only in interindustry comparisons of responses to a single question. We also sought, however, to make comparisons among questions (for example, are patents more or less effective than secrecy in protecting process innovations from duplication?), and we therefore treated the data as if they were interval.[23]

One additional methodological concern was whether our level of industry aggregation was appropriate for the problems being studied. The FTC line-of-business level was chosen to facilitate merging the data with disaggregated R&D data and Scherer's classification of patents by industries of origin and use. Our analysis indicated, however, that most of the interesting interindustry distinctions among the 130 lines defined at the FTC level were robust to an aggregation of the data into the 25 industry groups used by the National Science Foundation in its annual survey of R&D spending and employment patterns.

Patents and Other Means of Appropriation

Table 1 shows the pattern of responses, based on a seven-point scale, to questions on the effectiveness of alternative means of capturing and protecting the competitive advantages of new or improved processes and products. The first two columns report the mean response for the

22. One notable consequence arising from the measurement error in the data was that industry mean responses from lines of business with only one or two respondents tended to be disproportionately located near the extremes of the distribution of mean responses to any given question. Most conclusions based on the full sample of 130 lines, and virtually all those emphasized in this paper, were replicated in the smaller sample of 75 lines that had more than two respondents.

23. We designed the questionnaire to ensure that cross-question comparisons would arise naturally in the minds of the respondents. The items were arranged in blocks, with each item in a block rated on the same semantic scale.

Table 1. Effectiveness of Alternative Means of Protecting the Competitive Advantages of New or Improved Processes and Products[a]

Method of appropriation	Overall sample means: Processes	Overall sample means: Products	Distribution of industry means[b]: Processes	Distribution of industry means[b]: Products
Patents to prevent duplication	3.52 (0.06)	4.33 (0.07)	2.6–4.0[c]	3.0–5.0[c]
Patents to secure royalty income	3.31 (0.06)	3.75 (0.07)	2.3–4.0[c]	2.7–4.8[c]
Secrecy	4.31 (0.07)	3.57 (0.06)	3.3–5.0	2.7–4.1
Lead time	5.11 (0.05)	5.41 (0.05)	4.3–5.9[c]	4.8–6.0[c]
Moving quickly down the learning curve	5.02 (0.05)	5.09 (0.05)	4.5–5.7	4.4–5.8
Sales or service efforts	4.55 (0.07)	5.59 (0.05)	3.7–5.5	5.0–6.1

Source: Authors' calculations.
a. Range: 1 = not at all effective; 7 = very effective. Standard errors in parentheses.
b. From the upper bound of the lowest quintile of industries to the lower bound of the highest quintile.
c. Differences in means significant at the .01 level.

entire sample of 650 respondents to each question, as well as the standard error of each estimated mean. These statistics, of course, give equal weight to each respondent and consequently weight each industry in proportion to its number of respondents. The overall pattern across questions, however, is robust to the use of alternative summary statistics, such as the mean of industry means or the median of industry means. This is apparent in columns 3 and 4, which summarize the distribution of industry mean responses to each question. Each pair of numbers represents the range of industry means from the upper bound of the lowest quintile to the lower bound of the highest quintile of industries: 20 percent of the 130 industries had mean responses at or below the bottom of the range indicated for each question, and 20 percent had mean responses at or above the top of the range. Mean responses for the remaining 60 percent (or 78 industries) fell within the reported range.

The picture is striking. For new processes (columns 1 and 3), patents were generally rated the least effective of the mechanisms of appropriation: only 20 percent of the lines of business surveyed rated process patent effectiveness in excess of 4.0. Eighty percent scored the effectiveness of lead time and learning curve advantages on new processes in excess of 4.3. Secrecy, though not considered as effective as lead time and learning advantages, was still considered more effective than patents in protecting processes.

Patents for products were typically considered more effective than those for processes, and secrecy was considered less effective in protecting products than processes. Generally, lead time, learning curves, and sales or service efforts were regarded as substantially more effective than patents in protecting products. Eighty percent of the sample businesses rated the effectiveness of sales and service efforts above 5.0, but only 20 percent considered product patents this effective.[24]

The tendency to regard secrecy as more effective than process patents but less effective than product patents probably reflects the greater ease and desirability of maintaining secrecy about process technology. Firms may sometimes refrain from patenting processes to avoid disclosing either the fact or the details of an innovation.[25] But firms have every incentive to advertise the advantages of new or improved products and to get them into the hands of customers, thereby facilitating direct observation of the product and the technology it embodies. Maintaining secrecy about product innovations is thus likely to be both difficult and undesirable.

Respondents also tended to regard patents to prevent duplication as more effective than patents to secure royalty income. This finding was consistent with the view that licensing arrangements are beset with transactional difficulties.

Only 3 of 130 lines of business rated process patents higher than five on a seven-point scale of effectiveness in preventing duplication. Two of these were concrete and primary copper; the other had only a single respondent.[26] Only 5 of 130 industries rated product patents to prevent

24. This view of the efficacy of sales and service efforts is consistent with the emphasis given to investment in "cospecialized assets" as a means of appropriation in David J. Teece, "Profiting from Technological Innovation: Implications for Integration, Collaboration, Licensing and Public Policy," *Research Policy*, vol. 15 (December 1986), pp. 285–305.

25. See Ignatius Horstmann, Glenn M. MacDonald, and Alan Slivinski, "Patents as Information Transfer Mechanisms: To Patent or (Maybe) Not to Patent," *Journal of Political Economy*, vol. 93 (October 1985), pp. 837–58, for a theoretical treatment of the issue.

26. To preserve confidentiality, we do not identify any industry in which there was only one response. Hereafter, we refer to such cases as singletons.

duplication higher than six points. Two of these were singletons; the other three were drugs, pesticides, and industrial organic chemicals. Twenty other lines rated product patents between five and six. Of those with more than two responses, almost all fell neatly into chemical products (including inorganic chemicals, plastic materials, synthetic fibers, synthetic rubber, and glass) or relatively uncomplicated mechanical equipment (air and gas compressors, power-driven hand tools, and oilfield machinery). The only anomalies were roasted coffee and products of steel rolling and finishing mills.

Table 2 shows additional industry-level detail—the mean rating given for the effectiveness of patents in preventing duplication in eighteen industries with ten or more respondents. These industries tend to be much more research-intensive than the sample average, yet the pattern of interindustry variation was similar to that in the full sample. Except for petroleum refining, product patents were considered more effective than process patents. Only four chemical industries (drugs, plastic materials, inorganic chemicals, and organic chemicals) and petroleum refining rated process patent effectiveness higher than four on a seven-point scale, and only these four chemical industries and steel mills rated product patents higher than five.[27]

The data on these eighteen most heavily sampled industries help to establish the robustness of our conclusion about the limited effectiveness of patents as a means of appropriation. In none did a majority of respondents rate patents—either to prevent duplication or to secure royalty income—as more effective than the most highly rated of the other four means of appropriating returns from new processes, although in drugs and petroleum refining a majority regarded process patents as at least the equal of the most effective alternative mechanism of appropriation. In only one industry, drugs, were product patents regarded by a majority of respondents as strictly more effective than other means of appropriation.[28] In three others—organic chemicals, plastic materials,

27. The same pattern appears when the survey data are aggregated up to the level (roughly two and one-half digit) at which the National Science Foundation reports detailed data on the extent and composition of research and development expenditures. Of the twenty-five industries into which the manufacturing sector is divided, only industrial chemicals, drugs, and petroleum refining rated process patents higher than four points, and only industrial chemicals and drugs rated product patents higher than five.

28. Our results were reinforced by Edwin Mansfield's finding that among the twelve broadly defined industries he studied only in the drug industry were patents considered

Table 2. Effectiveness of Process and Product Patents in Industries with Ten or More Survey Responses

Industry	*Process patents*		*Product patents*	
	Mean	*Standard error*	*Mean*	*Standard error*
Pulp, paper, and paperboard	2.6	0.3	3.3	0.4
Cosmetics	2.9	0.3	4.1	0.4
Inorganic chemicals	4.6	0.4	5.2	0.3
Organic chemicals	4.1	0.3	6.1	0.2
Drugs	4.9	0.3	6.5	0.1
Plastic materials	4.6	0.3	5.4	0.3
Plastic products	3.2	0.3	4.9	0.3
Petroleum refining	4.9	0.4	4.3	0.4
Steel mill products	3.5	0.7	5.1	0.6
Pumps and pumping equipment	3.2	0.4	4.4	0.5
Motors, generators, and controls	2.7	0.3	3.5	0.5
Computers	3.3	0.4	3.4	0.4
Communications equipment	3.1	0.3	3.6	0.3
Semiconductors	3.2	0.4	4.5	0.4
Motor vehicle parts	3.7	0.4	4.5	0.4
Aircraft and parts	3.1	0.5	3.8	0.4
Measuring devices	3.6	0.3	3.9	0.3
Medical instruments	3.2	0.4	4.7	0.4
Full sample	3.5	0.06	4.3	0.07

Source: Authors' calculations. Mean score on a scale of 1 to 7.

and steel mill products—most respondents rated patents as no less effective than the best alternative.

The exclusion from our sample of firms that offered no publicly traded securities may have biased our findings. For small, start-up ventures, patents may be a relatively effective means of appropriating R&D returns, in part because some other means, such as investment in complementary sales and service efforts, may not be feasible. The patents held by a small, technologically oriented firm may be its most marketable asset. Although our respondents were asked to describe the typical experience of firms in their industries, they may well have overlooked aspects of appropriability that are particularly relevant for new firms.

essential to developing and marketing most inventions. Chemicals was the only other industry that considered patents essential for as many as 30 percent of inventions. See "Patents and Innovation: An Empirical Study," *Management Science*, vol. 32 (February 1986), pp. 173–81.

The most probable explanation for the robust finding that patents are particularly effective in chemical industries is that comparatively clear standards can be applied to assess a chemical patent's validity and to defend against infringement. The uniqueness of a specific molecule is more easily demonstrated than the novelty of, for example, a new component of a complex electrical or mechanical system. Similarly, it is easy to determine whether an allegedly infringing molecule is physically identical to a patented molecule; it is more difficult to determine whether comparable components of two complex systems "do the same work in substantially the same way." To the extent that very simple mechanical inventions approximate molecules in their discreteness and easy differentiability, it is understandable that industries producing such machinery rank just after chemical industries in the perceived effectiveness of patent protection.

The perceived ineffectiveness of patents in most industries raises the question of why firms use them. Further work is needed here, but we offer some speculations informed by the comments of our pretest subjects and by several survey respondents at a conference we held to report preliminary findings. These executives identified two motives for patenting that have little connection with appropriating returns from investment. One is to measure the performance of R&D employees, which is a significant problem because these workers are typically engaged in team production. Legal standards for identifying inventors on a patent application are, however, reasonably rigorous. The second motive is to gain access to certain foreign markets. Some developing countries require, as a condition of entry, that U.S. firms license technology to a host-country firm, and some patents are filed primarily to permit such licensing.[29]

29. Yet another motive discussed in the literature is to gain strategic advantage in negotiation. In the semiconductor industry, for instance, the cumulative nature of the technology makes it difficult to participate legally without access to the patents of numerous firms. In consequence, there is widespread cross-licensing. Established firms, however, rarely license a new entrant until it has established a significant position in the market. As a defense against infringement suits, a prudent new entrant will establish a patent portfolio of its own, thus compelling established firms to negotiate cross-license agreements. See Eric von Hippel, "Appropriability of Innovation Benefit as a Predictor of the Source of Innovation," *Research Policy*, vol. 11 (January 1982), pp. 95–115; and Levin, "Semiconductor Industry," pp. 80–81.

Conditions Affecting Appropriability

Thus far we have focused on the overall strength of various mechanisms of appropriation and on interindustry variations in the effectiveness of patents. The patterns of covariation in the responses, however, suggested that interindustry differences in conditions affecting appropriability might be summarized by a limited number of factors. Moreover, the clear indications that patents are effective in only a few industries suggested that it might be fruitful to classify industries into clusters distinguishable by a primary means of appropriation and perhaps by the overall ease of appropriating returns. Such clusters could prove useful in examining links between appropriability conditions and measures of R&D, innovation, and productivity growth.

Correlations among responses to questions on the effectiveness of alternative means of appropriation revealed some interesting patterns.[30] When patents effectively prevent competitors from duplicating processes and products, they tend also to be effective in securing royalty income. But neither form of effectiveness was strongly correlated with the effectiveness of other means of appropriation. For processes, there was a strong connection among three other mechanisms: lead time, learning curve advantages, and secrecy. For products, superior sales and service efforts were strongly linked to lead time and learning advantages, though not to secrecy.

The correlations suggested that the mechanisms of appropriation may be reduced to two dimensions: one associated with the use of patents, the other related to secrecy, lead time, and learning curve advantages. For product innovations, sales and service efforts may be involved in the second of these dimensions. We investigated this possibility by reducing the data to principal components and employing a variety of factor-analytic techniques. Principal factor analysis and several methods

30. Simple correlation coefficients were calculated using the individual respondent and industry mean responses as the units of observation. Correlations among industry means for the entire sample of 130 lines of business were qualitatively similar to those obtained when the sample was restricted to those with more than two responses. These and other correlation matrices discussed in this paper are available from the authors upon request.

Table 3. Principal Components Analysis of Methods of Appropriation

	Processes and products separately		*Processes and products together*	
Method of appropriation	*Coefficients of 1st principal component*	*Coefficients of 2d principal component*	*Coefficients of 1st principal component*	*Coefficients of 2d principal component*
New Processes				
Patents to prevent duplication	.04	.86	.01	.73
Patents to secure royalties	.12	.86	.08	.78
Secrecy	.59	−.12	.54	.04
Lead time	.84	−.09	.79	−.04
Moving down the learning curve	.84	−.05	.80	−.04
Sales and service efforts	.51	.11	.45	−.06
Cumulative variance explained	.34	.59	n.a.	n.a.
New Products				
Patents to prevent duplication	.06	.87	.06	.73
Patents to secure royalties	.06	.87	.07	.80
Secrecy	.51	.01	.51	.06
Lead time	.84	.00	.79	−.03
Moving down the learning curve	.84	−.07	.82	−.04
Sales and service efforts	.69	−.09	.62	−.11
Cumulative variance explained	.36	.61	.31	.50

Source: Authors' calculations.

of rotation did little to alter the picture presented by the principal components, which are shown in table 3.[31]

The first two columns of the table show the weights associated with the first two principal components when the six questions relating to process appropriability are analyzed separately from the six questions relating to product appropriability. The next two columns report the results of a principal components analysis on the entire set of twelve questions. With both approaches, the first principal component gives near-zero weight to the two patent-related methods of appropriation and heavy weight to the other mechanisms. The weighting is reversed for the second principal component. Thus the first two principal components (and, in the factor analysis, the first two factors) are readily interpreted, respectively, as nonpatent- and patent-related dimensions of appropriability. Despite this clear interpretation, the data do not reduce very

31. The results reported in table 3 are based on a principal components analysis undertaken at the level of individual responses. An analysis at the level of industry mean responses produced similar results.

Table 4. Cluster Analysis of Mechanisms of Appropriation

Method of appropriation	*Cluster*		
	1	*2*	*3*
New Processes			
Number of industries	38	67	25
Mean score			
Patents to prevent duplication	3.1	3.0	4.7
Patents to secure royalties	2.9	2.9	4.8
Secrecy	2.8	4.6	4.7
Lead time	4.2	5.4	5.6
Learning curves	4.3	5.3	5.1
Superior sales or service	4.7	4.5	4.9
New Products			
Number of industries	20	68	42
Mean score			
Patents to prevent duplication	3.1	3.8	5.3
Patents to secure royalties	3.2	3.1	5.0
Secrecy	2.6	3.5	4.0
Lead time	4.0	5.6	5.7
Learning curves	4.2	5.3	5.2
Superior sales or service	5.2	5.7	5.6

Source: Authors' calculations.

satisfactorily to just two dimensions. As table 3 indicates, when the process and product questions are analyzed separately, the first two components explain only 60 percent of the variance in the responses to six questions, and when the two sets of questions are combined, two components explain only 50 percent of the variance.

Our interpretation that the means of appropriation can be grouped into patent and nonpatent mechanisms was nonetheless reinforced by a cluster analysis that classified industries according to mean responses to the relevant questions. The best clustering results were achieved by dividing the industries into three groups, as shown in table 4. Industries assigned to cluster 1 tended to have relatively low scores for all mechanisms of appropriation. Sales and service effort was the most highly rated mechanism and was, in fact, regarded as reasonably effective in capturing returns from new products. Industries in cluster 2 rated lead time and learning curves as relatively effective, but not patents. Secrecy was important in appropriating process returns, and sales and service efforts complemented lead time and learning advantages for products. Only for cluster 3 were product and process patents deemed effective, but still the effectiveness of lead time and learning was no lower than for the industries in cluster 2. Those few industries in which patents were rated as more effective than other mechanisms were all in the third cluster.

The cluster analysis suggested that there was a group of industries in which no appropriation mechanism was particularly effective. As an alternative approach to identifying settings with low appropriability, we considered the maximum score an industry assigned to any of the six mechanisms on the questionnaire. Only 11 of the 130 failed to rate at least one means of appropriating returns from product innovation higher than five on the seven-point scale. The industries in this group with more than two responses were all drawn from the food products and metalworking sectors: milk, meat products, iron and steel foundries, boiler shops, and screw machine products (nuts, bolts, and screws). Many more industries (34 of 130) rated no means of appropriating process returns higher than five. This group contained all the industries (except milk) that ranked low on product appropriability but was otherwise a diverse lot. The heaviest concentration was in fabricated metals and machinery. But several chemical industries were also represented, including the three industries in which product patents were viewed as most effective—organic chemicals, pesticides, and drugs.

The urge to find patterns in the data should not be carried too far. The associations among mechanisms of appropriation revealed by the correlation, principal components, and cluster analyses are suggestive, but there is substantial heterogeneity in the underlying data. As noted, the first two principal components, though readily interpretable, explained an unsatisfactory fraction of the overall variance. A similar lack of good fit characterized the cluster analyses of process and product appropriability. Despite the fairly clear interpretation that could be given to each cluster, the variance within the clusters was almost twice that between clusters.

Limitations on Effectiveness of Patents

To understand why patent protection might be weak in some industries, we asked respondents to rate the importance of possible limitations on patent effectiveness. Table 5 summarizes the responses. The ability

Table 5. Limitations on Effectiveness of Patents for New or Improved Processes and Products[a]

	Overall sample means		*Distribution of industry means[b]*	
Limitation	*Processes*	*Products*	*Processes*	*Products*
New processes or products patentable	4.32 (0.07)	3.75 (0.07)	3.6–5.4[c]	2.8–4.8
Patents unlikely to be valid if challenged	4.18 (0.06)	3.92 (0.07)	3.5–5.0[c]	3.0–5.0[c]
Firms do not enforce patents	4.29 (0.06)	3.84 (0.07)	3.5–5.0[c]	3.0–4.8[c]
Competitors legally "invent around" patents	5.49 (0.05)	5.09 (0.06)	4.9–6.0	4.4–5.9[d]
Technology moving so fast that patents are irrelevant	3.40 (0.07)	3.34 (0.07)	2.0–4.3[d]	2.0–4.0[d]
Patent documents disclose too much information	4.19 (0.07)	3.65 (0.07)	3.2–5.0	2.8–4.5[c]
Licensing required by court decisions	2.96 (0.06)	2.79 (0.06)	2.0–3.8	2.0–3.3
Firms participate in cross-licensing agreements with competitors	3.08 (0.06)	2.93 (0.06)	2.2–3.9[d]	2.1–3.9[d]

Source: Authors' calculations.
a. Range: 1 = not an important limitation; 7 = very important limitation. Standard errors in parentheses.
b. From the upper bound of the lowest quintile of industries to the lower bound of the highest quintile.
c. Interindustry differences significant at the .10 level.
d. Interindustry differences significant at the .01 level.

of competitors to "invent around" both process and product patents was rated higher than five on a seven-point scale of importance by 60 percent of the responding industries. Only one other constraint—the lack of ready patentability for new processes—was rated this important by more than 20 percent. Limitations on patents were generally considered more severe for processes than for products, which was consistent with our finding that product patents tend to be more effective than process patents. In particular, the lack of patentability was more serious for processes than for products, and so was the disclosure of information through patent documents.[32]

32. Additional evidence of the internal consistency of the survey results was provided by the pattern of negative correlation between responses concerning limitations on patent effectiveness and responses concerning the effectiveness of patents. Using either individual respondents or industry means as the unit of observation, all such correlation coefficients were negative except in the case of compulsory licensing. Most correlations were significant at the .01 level.

The responses concerning limits on patent effectiveness may illuminate and focus policy discussion. In recent years there has been considerable interest in making patent protection more effective. One initiative has been to make the legal requirements for a valid patent claim less stringent.[33] Another has been to vacate court decrees that compel licensing. Our data identified industries in which stringent requirements for patent validity or compulsory licensing were perceived as important limitations on the usefulness of patents in appropriating returns.

Respondents from twenty-two lines of business, mostly in the food processing and fabricated metals sectors, considered the likely inability to withstand challenges to validity as significantly limiting the effectiveness of process patents (scoring the importance higher than five on a seven-point scale); for fourteen of these industries the mean response was six or higher on the scale. This group and the nineteen industries citing invalidity as a constraint on the effectiveness of product patents (again assigning a score higher than five) overlapped considerably. Further investigation would be required to determine just why these two sectors appear to have difficulty establishing valid claims. Perhaps because they are mature industries, opportunities may be limited or novelty may be difficult to achieve or simply difficult to prove.

Compulsory licensing was rarely judged a significant limit on the effectiveness of patents. Only one industry with one respondent rated this constraint higher than five on the scale for products, and only six cited compulsory licensing of process patents as of comparable importance. Two of these industries were not singletons—metal containers and electron tubes. Compulsory licensing decrees were thus perceived as important in only a small subset of the industries that F. M. Scherer indicated were subject to such decrees.[34] The overall lack of impact from compulsory licensing requirements was consistent with Scherer's finding that they did not discourage R&D spending.

33. For example, P.L. 98–622, passed in 1984, modified the previous requirement that each coinventor listed in a patent application also had to be a coinventor on every claim of the patent. The new law allows inventors to apply jointly, even though they may not have physically worked together, made the same level of contribution, or contributed individually to the subject matter of each claim. For a thorough discussion, see Patrick Kelley, "Recent Changes in the Patent Law Which Affect Inventorship and the Ownership of Patents," unpublished manuscript (1985).

34. F. M. Scherer, *The Economic Effects of Compulsory Patent Licensing* (New York University, Graduate School of Business Administration, 1977).

The choice between obtaining a patent and maintaining secrecy may be influenced by the extent to which the disclosures made in the patent document facilitate inventing around the patent. Our data provided some support for this theory. The effectiveness of secrecy was positively correlated with the extent to which disclosures limited the effectiveness of patents. The link was stronger for product patents than for process patents. But patent disclosures represented a substantial limitation on the effectiveness of product patents for only 4 of the 130 industries (scoring as high as six on the scale), and only 16 regarded process disclosures as comparably important. In only one line of business of those with five or more respondents—metal cutting machine tools—did disclosures constrain so substantially the effectiveness of both process and product patents.

Channels of Information Spillover

To the extent that a rival can learn easily about an innovator's technology, the incentive to invest in R&D is attenuated. But to the extent that learning is easy, wasteful duplication or near duplication of R&D effort by rival firms may be avoided. Also, knowledge of an innovator's new technology may complement rival R&D effort by enhancing its productivity. Richard Nelson and Sidney Winter, Michael Spence, and Richard Levin and Peter C. Reiss have developed models that begin to disentangle these offsetting effects, called by Spence the incentive and efficiency effects of interfirm spillovers.[35] A sharper characterization of interindustry differences in the nature and strength of the mechanisms by which firms learn about their competitors' technology should advance these modeling efforts.

Table 6 summarizes the responses to questions about the effectiveness of alternative ways of learning. There is little difference between the pattern of responses for processes and for products, except that, as one would expect, reverse engineering is markedly more effective in yielding

35. Richard R. Nelson and Sidney G. Winter, "The Schumpeterian Tradeoff Revisited," *American Economic Review*, vol. 72 (March 1982), pp. 114–32; Michael Spence, "Cost Reduction, Competition, and Industry Performance," *Econometrica*, vol. 52 (January 1984), pp. 101–21; and Levin and Reiss, "Tests of a Schumpeterian Model," and "Demand-Creating and Cost-Reducing R&D."

Table 6. Effectiveness of Alternative Methods of Learning about New Processes and Products[a]

Method of learning	Overall sample means: Processes	Overall sample means: Products	Distribution of industry means[b]: Processes	Distribution of industry means[b]: Products
Licensing technology	4.58 (0.07)	4.62 (0.07)	3.4–5.6[c]	3.5–5.5[c]
Patent disclosures	3.88 (0.05)	4.01 (0.06)	3.0–4.6[c]	3.0–4.8[c]
Publications or technical meetings	4.07 (0.05)	4.07 (0.05)	3.4–4.7	3.3–4.6[d]
Conversations with employees of innovating firm	3.64 (0.06)	3.64 (0.06)	2.9–4.7[d]	2.9–4.5[d]
Hiring R&D employees from innovating firm	4.02 (0.07)	4.08 (0.07)	2.7–5.0[c]	2.8–5.0[c]
Reverse engineering of product	4.07 (0.07)	4.83 (0.06)	3.0–5.0[c]	4.0–5.7[d]
Independent R&D	4.76 (0.06)	5.00 (0.05)	4.0–5.5	4.4–5.6[c]

Source: Authors' calculations.
a. Range: 1 = not at all effective; 7 = very effective. Standard errors in parentheses.
b. From the upper bound of the lowest quintile of industries to the lower bound of the highest quintile.
c. Interindustry differences in means significant at the .01 level.
d. Interindustry differences in means significant at the .05 level.

information about product technology. On average, independent R&D was rated as the most effective means of learning about rival technology.[36] This may appear to be wasteful duplication, but it need not be. One pretest subject said that R&D effort devoted to determining what a competitor has done may have strong complementarities with a firm's own research program in areas not directly imitative of the innovating competitor. Licensing was also rated, on average, an important way of gaining access to a rival's new technology.

The correlations among individual and industry mean responses show that mechanisms relying on interpersonal communication (publications and technical meetings, informal conversations, and hiring away employees) are strongly intercorrelated. Learning through licensing technology is uncorrelated with nearly all other learning mechanisms except disclosure through patent documents. There are two possible interpre-

36. Wesley Cohen and Daniel Levinthal have studied the incentives to engage in R&D that is directed toward developing absorptive capacity, the ability to make use of technology developed by others. See "Innovation and Learning: The Two Faces of R&D" (Carnegie-Mellon University, Department of Social and Decision Sciences, March 1987).

tations of this last connection. Potential licensees may learn about the opportunity to license through patent documents, or the documents may prove useful in employing new technology once it is licensed. We cannot tell whether the "announcement" effect or the "complementary information" effect of disclosures predominates.[37]

The pattern of correlation suggested that there might be three or four clusters of industries, distinguished in turn by an emphasis on learning through licensing, interpersonal channels, and reverse engineering or independent R&D, or both. The results obtained from cluster analysis were not entirely satisfactory.[38] Nonetheless, table 7 presents the results of grouping the lines of business into three clusters on the basis of responses to the questions on channels of spillover.

For both new processes and products, the largest group of industries typically relied on licensing and independent R&D to learn about competitive technology. Interpersonal channels were relatively unimportant, and reverse engineering was important for products. For both processes and products, there was a second cluster of industries in which interpersonal channels of spillover were most important. In the case of learning about new products, only ten industries were classified in this cluster, and in the case of learning about new processes, other channels—independent R&D and reverse engineering—were nearly as valuable. For processes, a third cluster appeared to find all mechanisms of learning relatively unproductive. For products this group found all mechanisms moderately effective.

Cost and Time Required for Imitation

As part of our investigation we asked respondents to estimate typical costs and time required to duplicate several categories of innovations if

37. The correlations between the effectiveness of particular learning mechanisms and the effectiveness of alternative methods of appropriation are interesting and internally consistent. In particular, when patent protection is effective, learning tends to take place primarily through licensing and patent disclosures. The effectiveness of patents is essentially uncorrelated with the effectiveness of interpersonal channels of learning and of independent R&D, and it is negatively correlated with the effectiveness of reverse engineering.

38. With three clusters the ratio of variance among clusters to variance within clusters was low, but attempts to find more than three clusters were thwarted by the persistent appearance of clusters containing only one or two lines of business.

Table 7. Cluster Analysis of Channels of Learning

Learning mechanism	*Cluster* 1	2	3
New Processes			
Number of industries	68	43	19
Mean score			
Licensing technology	5.0	4.3	2.5
Patent disclosures	4.0	4.0	3.2
Publications or technical meetings	3.8	4.6	3.9
Conversations with employees of innovating firm	3.2	4.8	3.0
Hiring R&D employees from innovating firm	3.7	5.1	2.4
Reverse engineering of product	3.8	4.6	4.0
Independent R&D	5.0	4.6	4.3
New products			
Number of industries	68	10	52
Mean score			
Licensing technology	4.7	2.5	4.5
Patent disclosures	3.9	2.9	4.3
Publications or technical meetings	3.7	5.1	4.3
Conversations with employees of innovating firm	3.0	4.6	4.5
Hiring R&D employees from innovating firm	3.2	4.4	4.9
Reverse engineering of product	4.7	3.0	5.2
Independent R&D	5.1	3.7	5.0

Source: Authors' calculations.

they were developed by a competitor. For each category, respondents were asked to identify (within a range) the cost of duplication as a percentage of the innovator's R&D cost. Intervals measured in months or years were used to classify the time required. In light of evidence that there is a time-cost trade-off in certain industries, we asked respondents to estimate the cost and time required "to have a significant impact on the market."[39]

Tables 8 and 9 show frequency distributions of industry median responses.[40] The dispersion of industry medians suggests substantial variations among industries in both the cost and time required to duplicate

39. See Edwin Mansfield, *Industrial Research and Technological Innovation: An Econometric Analysis* (Norton, 1968), for evidence on the time-cost trade-off.

40. Qualitatively identical results and interpretations are obtained from frequency distributions of individual responses and from the distribution of industry means.

Table 8. Cost of Duplicating an Innovation as a Percentage of Innovator's R&D Cost, Frequency Distribution of Median Responses

Type of innovation	*Less than 25 percent*	*26 to 50 percent*	*51 to 75 percent*	*76 to 100 percent*	*More than 100 percent*	*Timely duplication not possible*
New process						
Major patented new process	1	5	19	66	26	10
Major unpatented new process	5	10	55	49	6	2
Typical patented new process	2	15	61	41	6	2
Typical unpatented new process	8	43	58	14	4	0
New product						
Major patented new product	1	4	17	63	30	12
Major unpatented new product	5	13	58	40	7	4
Typical patented new product	2	18	64	32	9	2
Typical unpatented new product	9	58	40	15	5	0

Source: Survey of 127 lines of business.

all categories of innovation. If, however, individual responses to the questions on cost are coded on a six-point interval scale, there is sufficient intraindustry variation to render interindustry differences insignificant at the 0.01 level. Interindustry differences in the time required for duplication are, by contrast, significant at the 0.01 level in every instance except the time required to duplicate a typical patented new process.

Several conclusions are apparent. First, duplicating major innovations tends to cost more and take longer than duplicating typical innovations. (In a sense, this confirms that respondents correctly interpreted the distinction between typical and major innovations.) Second, for a given category of innovation, the cost and time required to duplicate are distributed very similarly for products and processes. Products tend to be slightly cheaper and quicker to duplicate than processes, though this generalization does not hold for major patented innovations. Finally, patents tend to raise imitation costs and time for each category of innovation. These increases can be regarded as alternative indicators of the relative effectiveness of patents in different industries.

To explore this point further, we coded the individual responses to the imitation costs and time questions on a six-point interval scale,

Table 9. Time Required to Duplicate an Innovation, Frequency Distribution of Median Responses

Type of innovation	*Less than 6 months*	*6 months to 1 year*	*1 to 3 years*	*3 to 5 years*	*More than 5 years*	*Timely duplication not possible*
New process						
Major patented new process	0	4	72	37	9	7
Major unpatented new process	2	20	84	17	2	4
Typical patented new process	0	40	73	13	0	3
Typical unpatented new process	8	66	47	6	1	1
New product						
Major patented new product	2	6	64	40	8	9
Major unpatented new product	3	22	89	12	1	2
Typical patented new product	5	39	72	6	4	3
Typical unpatented new product	18	67	39	4	1	0

Source: Survey of 129 lines of business.

calculated the individual and industry mean increases in costs and time associated with the presence of patents, and correlated these, respectively, with individual and industry mean responses to our questions on the effectiveness of patents in preventing duplication. For each category of innovation, the reported effectiveness of patents was positively correlated with the increase in duplication costs and time associated with patents, although the correlations tended to be stronger for products than for processes. We also found some evidence, at the level of the individual respondent, that patent effectiveness was associated with the absolute level of duplication costs for patented processes and products. We found a much stronger association, however, between reported patent effectiveness and the amount of time required to duplicate both patented process and product innovations.

These broad-brush patterns of association conceal some striking anomalies. For particular categories of innovation, at least two and as many as fourteen industries reported that patents actually reduced the costs or time required for duplication. A partial explanation is that a disproportionate number of these industries also reported that disclosure of information through patent documents was a significant limitation on patent effectiveness.

A second anomaly is that, despite the positive correlation between patent effectiveness and the costs of imitating patented products, in several industries patents were relatively ineffective and duplication costs were nonetheless very high, whether or not the innovation was patented. Among these were guided missiles and several types of industrial machinery (food products machinery, electric welding apparatus, and speed changers, drives, and gears). In these instances the relative complexity of the products presumably makes reverse engineering inherently costly despite relatively weak patent protection.

It is interesting to compare our findings with those of Edwin Mansfield, Mark Schwartz, and Samuel Wagner, who studied the effects of patents on imitation costs in three industries.[41] They concluded that patents generally raised imitation costs by 30 percentage points in drugs, 20 points in chemicals, and 7 points in electronics. To render our data comparable, we evaluated each respondent's answer at the mean of the relevant range and computed crude industry average imitation costs for each type of innovation.[42] Our results were consistent with those of Mansfield, Schwartz, and Wagner. We found that patents raise imitation costs by 40 percentage points for both major and typical new drugs, by 30 points for major new chemical products, and by 25 points for typical chemical products. In electronics, our results differed somewhat for semiconductors, computers, and communications equipment, but the range was 7 to 15 percentage points for major products and 7 to 10 for typical products.[43]

41. Mansfield, Schwartz, and Wagner, "Imitation Costs and Patents."

42. The ranges are shown in the headings of table 8. The fifth and sixth column headings are not readily quantified. To permit the comparison discussed in the text, we assigned these categories the values of 112.5 percent and 137.5 percent, respectively, thereby maintaining a constant spacing of 25 percentage points between each pair of categories.

43. Our results on the time required to duplicate a rival's new products or processes were also roughly consistent with recent findings of Edwin Mansfield. In all but one of the ten industries he surveyed, the median respondent indicated that six to twelve months usually elapsed before the nature and operation of a new product were known to a firm's rivals. Effective duplication, as we have defined it, should take as long or longer, and table 9 shows that it typically does. The median and modal industries require one to three years to duplicate a major innovation or a typical patented innovation. A typical unpatented innovation, however, is more often duplicated within six to twelve months. See "How Rapidly Does New Industrial Technology Leak Out?" *Journal of Industrial Economics*, vol. 34 (December 1985), pp. 217–24.

Although the costs and time required for duplication are related to the effectiveness of patents, they do not seem to be linked strongly to any other mechanism of appropriability. In particular, most imitation time and cost measures are uncorrelated with lead time and learning curve advantages, and where such correlations are statistically significant (at the level of the individual respondent), the correlation coefficient is invariably below .15. These results make sense. Lead time and learning advantages may permit appropriation of returns even when duplication is relatively quick and inexpensive. Effective patents, however, presumably require considerable time and expense to be invented around.

Finally, most of our respondents believed only a few firms were capable of duplicating new processes and products. As table 10 shows, the median and modal number of firms judged capable of duplicating a major process or product innovation was three to five. The median and modal number of firms regarded as capable of duplicating a typical process or product innovation was six to ten. The data revealed only the slightest tendency toward a smaller number of capable duplicators for processes than for products.

R&D and Innovation

In this section, we summarize how data derived from our survey have been employed to understand better the sources of interindustry differences in R&D spending and the rate of technological advance. In the first such effort Richard Levin, Wesley Cohen, and David Mowery used several survey-based measures to explain variations in the published Federal Trade Commission data on industry-level R&D spending as a percentage of sales.[44] They also sought to explain interindustry differences in the rate at which new processes and new products were introduced during the 1970s, as reported by our survey respondents.[45]

44. The ratio of company-financed R&D to sales (R&D intensity) varies considerably among industries defined at the FTC line-of-business level of aggregation. In the 1976 data used by Levin, Cohen, and Mowery, R&D intensity ranged from 0.08 percent to 8.5 percent; both the mean and standard deviation were 1.7 percent. See "R&D Appropriability, Opportunity, and Market Structure."

45. Respondents were asked to identify, on a seven-point Likert scale ranging from "very slowly" to "very rapidly," the rate at which new processes and products had been

Table 10. Number of Firms Capable of Duplicating an Innovation, Frequency Distribution of Median Responses

Type of innovation	*None*	*1 or 2*	*3 to 5*	*6 to 10*	*More than 10*
Major new or improved process	2	32	75	18	2
Typical new or improved process	1	7	41	58	22
Major new or improved product	2	25	73	25	4
Typical new or improved product	1	5	33	63	26

Source: Survey of 129 lines of business.

In a subsequent paper, Cohen, Levin, and Mowery studied the extent to which the same survey-based measures explained the powerful industry effects in the confidential FTC data on R&D intensity at the level of the business unit.[46]

The first paper focused on the Schumpeterian hypothesis that R&D intensity and innovation rates are significantly influenced by the level of industry concentration. One common rationale for this hypothesis is that industry concentration enhances the potential for appropriation of R&D returns. A different view is that, in the long run, concentration tends to be a consequence of industry evolution in a regime of abundant technological opportunity and a high degree of uncertainty associated with investment in R&D. Both perspectives suggest that there is no simple, causal relationship between concentration per se and R&D. Concentration may be statistically significant in simple regression specifications because it reflects the influence of the unobserved appropriability and opportunity conditions that directly affect R&D spending and the rate of innovation.

In ordinary least squares and two-stage least squares specifications that included only the four-firm concentration ratio and its square as

introduced in their industries since 1970. Industry mean responses were highly correlated with total factor productivity growth, and the plausibility of the responses was reinforced by the identity of the highest and lowest industries in the sample. Excluding singletons, the lines of business reporting the slowest rates of product introduction were concrete, cement, boiler shops, milk, gypsum, primary copper, grain mill products, and sawmills. Those reporting the most rapid rates of product introduction were electrical equipment for internal combustion engines, radio and TV sets, computers, semiconductors, communications equipment, photographic equipment and supplies, engineering and scientific instruments, and guided missiles. Levin, Cohen, and Mowery used as a dependent variable the average of each industry's reported rates of process and product introduction.

46. Cohen, Levin, and Mowery, "Firm Size and R&D Intensity."

regressors, Levin, Cohen, and Mowery replicated with the industry-level FTC data the familiar inverted-U relationship between concentration and R&D intensity, and they found a strong relationship of the same form between concentration and the rate of innovation.[47] Adding two-digit industry fixed effects weakened slightly the effect of concentration on R&D, but the innovation-rate equation was unaffected.

The results changed dramatically with the addition of measures of appropriability and technological opportunity derived from the survey.[48] Whether or not two-digit industry fixed effects were included, the coefficients on concentration and its square fell by an order of magnitude in the R&D equation, and the effect of concentration was no longer statistically significant at the .05 level in either the R&D intensity or the innovation-rate equation. The vector of survey-based opportunity variables was significant at the .05 level in all specifications, and the opportunity and appropriability variables were jointly significant. The appropriability variables, however, were not individually significant in the R&D equation, although the rate of innovation was positively related to the effectiveness of an industry's most effective means of appropriation.[49]

The paper by Cohen, Levin, and Mowery used the disaggregated FTC data at the level of the business unit to investigate the Schumpeterian hypothesis linking size and R&D intensity. The authors found that when either fixed industry effects (at the level of the line of business) or survey-based industry characteristics were taken into account, firm size had a very small and statistically insignificant effect on R&D intensity. The size of the business unit did have a significant effect on the probability

47. All coefficients in the R&D and innovation-rate equations were statistically significant at the .01 level.

48. To represent appropriability conditions, Levin, Cohen, and Mowery used two survey-based measures: the maximum of the mean scores an industry's respondents assigned to the effectiveness of the six methods of appropriation and the time required to duplicate effectively a patented major product innovation. To represent opportunity conditions, they used a measure of an industry's closeness to science as well as variables summarizing the importance of four other external sources of knowledge for an industry's technological advance: material suppliers, equipment suppliers, users of the industry's products, and government agencies and research labs.

49. It may seem anomalous that the effectiveness of appropriation was positively related to innovation but not to R&D, but the relationship was observed at the level of the industry. Better appropriability may discourage R&D directed toward imitation to an extent that more than compensates for its stimulus to innovative R&D. Such a reallocation of effort would be entirely consistent with the observed positive relationship between appropriability and the rate of innovation.

of engaging in R&D, but there was no perceptible tendency for R&D intensity to increase with size within the group of R&D performers. Size effects, however, explained only two-tenths of 1 percent of the variance in R&D intensity, while industry effects at the line-of-business level explained half this variance.

Cohen, Levin, and Mowery found that industry-level measures of appropriability, opportunity, and demand conditions were consistently significant in ordinary least squares, GLS, and Tobit regressions explaining business unit R&D intensity. Moreover, these industry characteristics explained approximately half the variance in R&D intensity explained by fixed industry effects. When attention was focused on those lines for which there were at least three survey respondents, measured industry characteristics explained 56 percent of the variation in R&D intensity among industries. Within particular two-digit industries (chemicals, machinery, and electrical equipment), measured characteristics explained 78 to 86 percent of the variance explained by fixed effects.

The results obtained in the two papers indicated that survey-based measures can contribute substantially to an explanation of interindustry differences in R&D intensity and innovative performance. Measures derived from the survey, despite their imperfections, have also been found useful for various other purposes.[50]

Remarks on Policy

Our findings suggested some general principles relevant to policies that affect the incentives to engage in innovative activity.

A first principle is that the patent system and related institutions to

50. Levin and Reiss have used survey-based measures of appropriability and opportunity in a simultaneous equation model of R&D spending and market structure that builds on their work in "Tests of a Schumpeterian Model." Cohen and Levinthal use survey-based variables in their work on R&D as investment in absorptive capacity; see "Innovation and Learning." Iain Cockburn and Zvi Griliches are studying the usefulness of our survey measures in estimating the value of patents from stock market data; see "Industry Effects and Appropriability Measures in the Stock Market's Valuation of R&D and Patents," *American Economic Review* (forthcoming, May 1988). Meryl Finkel, "Overseas Research and Development by U.S. Multinationals: Ownership Structure Decisions" (Ph.D. dissertation, Harvard University, 1986), explored the effect of our appropriability variables on the investment decisions of multinational corporations. Franco Malerba is using the survey data to explain interindustry differences in the extent and effectiveness of learning mechanisms.

protect intellectual property should be understood as social structures that *improve* the appropriability of returns from innovation. They are not the only nor necessarily the primary barriers that prevent general access to what would otherwise be pure public goods. Lead time accrues naturally to the innovator, even in the absence of any deliberate effort to enhance its protective effect. Secrecy, learning advantages, and sales and service efforts can provide additional protection, though they require the innovator's deliberate effort. The survey confirmed that these other means of appropriation are typically more important than the patent system. Hence in examining a proposed adjustment of the patent system or related institutions, it is important to recognize that the incremental effect of the policy change depends on the protection other mechanisms provide.

The survey results also confirmed substantial interindustry variation in the level of appropriability and in the mechanisms that provide it. From this follows our second major principle, which is that the incremental effects of policy changes should be assessed at the industry level. For example, in the aircraft industry, where other mechanisms provide considerable appropriability, lengthening the life of patents would tend to have little effect on incentives for innovation. In the drug industry the effect of a longer lifetime would matter more.[51]

Finally, improving the protection of intellectual property is not necessarily socially beneficial. Empirical work has so far indicated a positive cross-sectional relationship between strong appropriability, as measured by variables constructed from our survey, and innovative performance. But the social cost-benefit calculation is not straightforward. Stronger appropriability will not yield more innovation in all contexts and, where it does, innovation may come at excessive cost.

To illustrate how our survey results and general perspective might inform policy discussion, consider the 1987 proposal (S. 438, H.R. 557) that patent license agreements and other contracts relating to the use of intellectual property "shall not be deemed illegal per se under any of the antitrust laws." One consequence would be to eliminate the per se illegality of tie-in arrangements (those in which purchase of one product,

51. For a calculation of the impact of the Drug Price Competition and Patent Term Restoration Act of 1984, see Henry Grabowski and John Vernon, "Longer Patents for Lower Imitation Barriers: The 1984 Drug Act," *American Economic Review*, vol. 76 (May 1986, *Papers and Proceedings, 1985*), pp. 195–98.

the "tying product," is dependent on purchase of other products) where the tying product is covered by a patent or otherwise protected as intellectual property.[52] Our findings have suggested some issues a court should consider in evaluating such a tying arrangement under the rule of reason.

When the rule of reason is applied to tying cases, a relevant consideration is the firm's power in the market for the tying good. Courts have often presumed that intellectual property protection is itself evidence for such power. To the other good reasons for rejecting such a presumption,[53] we add that the mere existence of a patent or other legal protection says nothing about its efficacy in a competitive context. As the survey results showed, the effectiveness of protection varies widely among industries. Thus in deciding a case, a court should inquire into the actual competitive significance of intellectual property protection in the particular market.

Suppose, for example, that a pharmaceutical company were to tie hospital sales of supplies or equipment to its sale of a patented drug. Since patent protection of drugs is generally strong and effective, and a drug is often uniquely suited for particular purposes, skepticism about the reasonableness of the tie-in would be in order. The arrangement could not plausibly be regarded as a straightforward means of appropriating returns to which the firm was entitled as owner of the patent. Given the typical effectiveness of drug patents, the price of the drug should suffice for that purpose. There might, of course, be benign explanations for the tie; for example, if the supplies or equipment were complementary to the use of the drug, the arrangement might be explicable as an attempt to control the quality of treatment. But if no such explanation were

52. We focus on this particular consequence of the proposed legislation and set aside two major considerations regarding its merits in its present form. First, without amendment the legislation is likely to undercut severely the per se treatment of price fixing. Second, it might be more appropriate to consider eliminating per se treatment of all tying arrangements rather than just those involving intellectual property. On this point, see the concurring opinion in *Jefferson Parish* v. *Hyde*, 466 U.S. 2 (1984).

53. See E. W. Kitch, "Patents: Monopolies or Property Rights," in John Palmer, ed., *Research in Law and Economics: The Economics of Patents and Copyrights*, vol. 8 (JAI Press, 1986), pp. 31–47; and the associated commentary of F. M. Scherer, p. 51. *Digidyne Corp.* v. *Data General Corp.*, 743 F. 2d 1336 (9th Cir. 1984) is an example of an application of per se doctrine in a context where the intellectual property (software) does not convincingly convey market power.

supported by the evidence, the tie would seem an unreasonable restraint of trade.

By contrast, consider a producer of a patented product in an industry where no mechanism of appropriability functions particularly well—plywood, for example, where patents, secrecy, lead time, and learning advantages are all rated no higher than four on a seven-point scale of effectiveness. In this instance the low level of appropriability in general and the ineffectiveness of patents in particular should weigh against any presumption that a patent confers market power. The patentee in such an industry should be entitled to some scope for ingenuity in constructing arrangements that maximize the return to the patent, provided that these arrangements are not open to antitrust objections on grounds independent of the role played by the patent.

The intellectual property provisions of the Omnibus Trade and Competitiveness Reform Act also serve to illustrate the relevance of the survey results. One provision requires the U.S. trade representative to identify countries that have been particularly insensitive, as a matter of law or de facto policy, to the need for protection of intellectual property and to initiate unfair trade practice (section 301) investigations against them.[54] This provision of the trade bill would complement the administration's diplomatic efforts to strengthen intellectual property rights throughout the world and particularly in countries that permit firms to copy patented or copyrighted products from the United States.

Since the impact of legal protection of intellectual property depends on the strength of other appropriability mechanisms and varies widely among industries, focused efforts to solve problems in specific markets would be more prudent than a broad attempt to upgrade protection. There is little point in expending diplomatic capital to compel foreign countries to pass or enforce laws that, in most industries, would have minimal impact on the competitive process. By contrast, in those specific industries such as pharmaceuticals—in which patent protection is effective, other means of appropriation are poor substitutes, and foreign governments often restrict, officially or tacitly, the ability of U.S. firms to exploit patents—a more persuasive case could be made for the United States to pressure its trading partners to change their behavior.

54. The trade representative may at his discretion escape this requirement by finding that such an investigation would not be in the national interest.

Appendix: Details of Sample Construction

Our review of the FTC data indicated that several lines of business did not report any R&D activity, and several others were aggregated to prevent violating confidentiality rules. Anticipating difficulty in finding knowledgeable respondents in industries without formal R&D activity, and wishing to avoid industry categories that included technologically disparate products, we eliminated those lines of business from our sampling frame.

The industries eliminated on grounds of heterogeneity were either the FTC's aggregations of technologically disparate industries or those corresponding to SIC industries with four-digit codes ending with 9. Such industries are residual categories within the relevant three-digit groups; their titles usually contain the words "miscellaneous, not elsewhere classified."

Confidentiality requirements prohibited us from using the FTC data as a means of identifying the firms that conduct R&D in each line of business. Instead, we used the *Business Week* annual R&D survey to identify all publicly traded firms that reported R&D expenses in excess of either 1 percent of sales or $35 million. This constitutes a nearly comprehensive list of private firms performing significant R&D. There were 746 such firms in 1981, when our survey design efforts commenced.

We used the information in Dun and Bradstreet's *Million Dollar Directory* to assign each of the *Business Week* firms to its major lines of business. Dun and Bradstreet's does not provide a complete list of each firm's lines of business, but it indicates as many as six four-digit SICs for each firm, in rough order of sales. Since some firms operate in nonmanufacturing industries, in manufacturing industries absent from our sample, or in two or more industries that fall into only one FTC line of business, we had substantially fewer than 746 × 6 observations. Within our sample lines of business, we found a total of 1,928 units operated by 688 firms.

A major design problem was how to obtain responses for business units within the same firm. Of our 688 firms, 470 participated in more than one of our sample lines of business. We initially attempted to identify relevant respondents using *Industrial Research Laboratories of the United States*. But our pretest subjects told us that more than half the people in such a sample were inappropriate. Some had been assigned to the wrong line of business; others had been promoted or had left the relevant division or the firm.

We therefore adopted a two-stage approach in which each firm's senior R&D vice president or chief executive officer was asked to furnish the names of employees with the knowledge to complete the questionnaire for specific lines of business. We sent first-round requests to 470 firms representing 1,710 business units. There was attrition of 332 business units from this sample for three reasons: the firm did not do R&D in the specified line of business, the industry definition did not fit any of its activities, or a respondent could not be located. From this adjusted sample frame of 1,378 business units in firms with multiple units, we received names of respondents for 716. We sent questionnaires to each of these potential respondents as well as to representatives of the 218 firms operating in only one line of business. At this stage, there was some further attrition in the sample. Ultimately, we received 650 completed questionnaires from an overall adjusted sample frame of 1,562—an overall response rate of 41.6 percent.

Comments and Discussion

Richard Gilbert: The authors' research program will have lasting value for people interested in R&D markets and markets for intellectual property. They are correct in focusing on appropriability as a key factor in the incentive to undertake R&D. And their findings are generally consistent with those of other studies, for example, those by Christopher T. Taylor and Z. A. Silberston and Edwin Mansfield, Mark Schwartz, and Samuel Wagner.[1] While this consistency may take a bit of the drama out of what Levin and his colleagues have done, the convergence of knowledge on this subject gives us some reason to believe we might be getting to the truth.

One of the authors' main conclusions is that there are very large differences, both among industries and within them, in the effectiveness of various means of appropriating intellectual property and also in the cost of imitation. It is an important result, but one that may cause some consternation. A main function of microeconomic theory is to form testable generalizations about the way the world works. Some of the work implies—at least, in the market for R&D and intellectual property—that such generalizations are extremely risky. We might be inventing a new field of microeconomic analysis, or "picoeconomics." Picoeconomics would keep us busy for a long, long time. But if we go down that path, our models will soon become as complicated as the world we are trying to explain.

These particular authors do not seem to want to lead us toward

1. Christopher T. Taylor and Z. A. Silberston, *The Economic Impact of the Patent System: A Study of the British Experience* (Cambridge University Press, 1973); and Edwin Mansfield, Mark Schwartz, and Samuel Wagner, "Imitation Costs and Patents: An Empirical Study," *Economic Journal*, vol. 91 (December 1981), pp. 907–18.

picoeconomics. It is apparent from their follow-up work that they intend to draw some general conclusions about how appropriability varies across industries. Clearly, we would like to know how market structure and capital intensity in different industries influence the degree of appropriation and affect incentives to innovate. But merely adding appropriation as another explanatory variable in these regressions does little good. We really want to know whether there are systematic relations between the degree of appropriation and other observable economic variables.

With regard to methodology, I suggest using a weighting scheme based on the amount of R&D a firm does and, perhaps, the number of patents it has produced. The purpose would be to weight responses according to the quality of the information. Some industries seem not to have performed any R&D for twenty years. Although it is important to know why these firms have not been active, their responses should be adjusted to reflect the information they possess. Also, I suggest that in their survey work the authors include a definition of R&D. That is not a trivial task because there is much variation in what is called research and development.

I would have liked the survey to address more directly some of the theoretical issues in the economics of R&D. Various models in the R&D literature have different implications for the simultaneous determination of R&D intensity and market structure. For example, models such as that of Partha Dasgupta and Joseph Stiglitz imply that the current rate of R&D spending should be independent of cumulative R&D expenditures by a firm or the rivals of the firm.[2] This is a consequence of the constant-hazard rate model. Other researchers such as Drew Fudenberg and his colleagues imply that past R&D is crucial to current and future R&D expenditures.[3] The dynamic implications of these models are very different. The preemption-type models also suggest that a history of successful R&D gives a firm a technological advantage that provides some protection from future R&D competition and tends to increase concentration in a market.

2. Partha Dasgupta and Joseph E. Stiglitz, "Uncertainty, Industrial Structure, and the Speed of R&D," *Bell Journal of Economics*, vol. 11 (Spring 1980), pp. 1–28.

3. Drew Fudenberg and others, "Preemption, Leapfrogging and Competition in Patent Races," *European Economic Review*, vol. 22 (June 1983), pp. 3–31.

I would be interested in knowing if the survey could have elicited some kind of response about the way R&D success alters the competitive environment of the firms and, conversely, how the competitive environment influences R&D spending. The questions in which the investigators ask how many firms are viable competitors with a given firm and how many could replicate its R&D bear on this question. It is interesting that the number of serious rivals for each firm was small, somewhere between three and six.

There is an empirical problem with surveys of the relationship between competition and R&D. If R&D really does have an effect on entry and competition, then the sample is necessarily biased. There are potential competitors who were not represented in the sample because the firms failed. How one accounts for the failures and puts them back in the sample is a difficult empirical problem.

In terms of patent policy, the diversity and effectiveness of patent protection across industries raises an obvious and interesting question. How should patent policy deal with the large differences in the values of patents among various industries? The authors point to the example of the semiconductor industry and suggest that the industry need not be any worse off as a consequence of the limited patent protection it has received. But one might also question whether the performance of the pharmaceuticals industry might have been improved if patent protection had been circumscribed to some extent.

I have experimented with a very simple model of optimal patent life with limited appropriability. It is basically a Nordhaus-type model with entry and spillovers.[4] It shows that the optimal patent life is not a function of the size of the innovation, so one does not have to worry that there are big innovations in some industries and small innovations in others. The optimal patent life does, however, depend on elasticities of R&D and its benefits and costs. It also depends on the degree of appropriability, and there is the intuitive answer that the optimal life is inversely related to the extent of private appropriation of the social value of the invention. This result suggests that we need either more protection in semiconductors or less protection in the patent drug industry.

The authors' survey provides a basis for contrasting patents with

4. See William D. Nordhaus, *Invention, Growth, and Welfare: A Theoretical Treatment of Technological Change* (MIT Press, 1969).

other approaches to protect intellectual property. A patent is a peculiar policy instrument. It represents an unnatural barrier to market entry that is erected to facilitate private appropriation. The survey suggests that other factors may be more important as a means of appropriation, and that other unnatural barriers might be more effective in stimulating R&D. As an extreme example (which I am not proposing), a tax on capital could arguably make entry more difficult and therefore stimulate R&D. It would be useful to do a survey of the effectiveness of different laws governing rights for intellectual property in different countries in an industry such as pharmaceuticals, which is one of the few in which patents really do seem to matter.

The survey results suggest that patents are important as a barrier to entry in the semiconductor industry not because they protect an individual innovation or invention but because they provide a hurdle for potential entrants, who have to acquire a package of marketable processes and products that they can cross-license to other firms. While this seems an inappropriate or at least unintended outcome of the patent grant, it might well be that by increasing ordinary barriers to entry in the semiconductor industry, the returns to research and development would be enhanced. This is another illustration of the Schumpeterian hypothesis and the tensions between strong enforcement of the antitrust laws and the desire to provide a stable platform for encouraging investment in R&D. If these observations with regard to the semiconductor industry generalize to other markets, they provide a starting point for further reexamination of the antitrust laws in the context of industrial R&D policy.

Thus this survey has raised some very interesting questions. Now we have to get on with their resolution. Thanks to this project, we have some of the data we need for the job.

Zvi Griliches: We should be grateful to Richard Levin and his associates for providing us with a new and detailed glimpse into a subject that is both very important and also lacking in good data. Far too little fresh economics data is collected, and we all have much to learn from the effort of this endeavor. That I am going to quarrel with some of the authors' assessments does not diminish in my eyes the basic value of this enterprise.

The authors have collected a large set of responses from many individuals located in different industries. This multipurpose survey will have many uses as we learn more about the responses and how to interpret them. I will focus on how these answers can help us learn which industries find patents an effective mechanism for appropriating returns from innovative effort, which ones do not, and whether mechanisms are available instead of or in addition to patents.

"Conditions of appropriability" determine the returns from a given innovative effort and hence the incentive to engage in it. One would expect that in industries in which appropriability is easy, there would be more innovative effort, higher returns, and a faster rate of technological progress. Such conditions may not be a fixed, unchanging characteristic of an industry, however. As more inventive effort is pursued, projects may become less easily appropriable, information may be leaked, and conditions may actually equalize among industries. The problem here is the same as in most empirical research programs: What is exogeneous and what is endogenous?

There are two generic problems with using the responses from such a survey: Are the responses comparable among individuals and do they reflect real differences among industries? Given the use of a scale of one to seven, I remain unsure about whether one person's response of five is equivalent to another's of four or six. Most questions do not have an objective anchor and could, therefore, differ greatly in the meanings attached to them by different respondents. This may account for the large dispersion in responses to most questions even within the same industry. It also leads to the difficulty of deciding whether the responses reflect real differences across industries or just random fluctuations among individuals.

There is a surprising amount of variability within industries in responses to the same question. Some questions, such as those on the effectiveness of patents, are reasonably objective and seem to have a variance among industries. Other questions, such as whether secrecy is effective, do not seem to be particularly industry-specific and do not discriminate well among industries. Questions about lead time, secrecy, sales effort, and service quality are really questions about different ways of succeeding, not about properties of an industry. It is well to have a long lead time or to achieve secrecy, but how is that to be accomplished? These are not characteristics over which either the firm or the policy-makers have clear control. Patents are at least a somewhat better defined

Table 1. Analysis of Variance of Differences among Industries in Responses to Questions on the Effectiveness of Different Appropriability Mechanisms

Question	Individual response[a]		Aggregated industry response[b]	
	Processes	Products	Processes	Products
Patents to prevent duplication	1.9	2.8	3.4	5.5
Patents to secure royalties	1.6	1.4	2.4	2.3
Secrecy	1.2	1.2	1.3	1.3
Lead time	1.7	1.6	2.2	2.5
Quickly down the learning curve	1.0	1.0	.9	1.2
Superior sales or service	1.2	.9	.7	.9
Average of questions 1 and 2	2.1		4.8	
Average of questions 3 through 6	1.1		1.6	

Source: Authors' calculations.
a. Covers 541 responses for 130 industries; approximate .05 significance level is 1.3.
b. Covers 620 responses for 24 industries; approximate .05 significance level is 1.6.

instrument, and we have some ideas about how the patent system could be tinkered with.

Table 1 presents my analysis of variance results for some of the responses derived from the Yale survey. (I am grateful to Levin for providing me with the original survey data.) It shows that there is more variance among industries in the responses to questions on the effectiveness of patents, especially for products, and very little variance in the other questions on conditions for appropriability, especially for process innovations. Process innovations are clearly less industry-specific and so is the importance of superior sales and service efforts.

Another way of seeing this problem is to look at the authors' table 6, which shows that for many of the nonpatent mechanisms the cross-correlation among methods of appropriation is lower at the aggregated industry level than at the level of the individual response. If industrial classification mattered, one would expect higher correlations for the aggregated variables.[1] This point is illustrated by a very simple model. Assume that two questions effectively measure the same thing. Then a variance components model for responses to these questions would be

$$y_{qij} = m_i + a_j + e_{qij},$$

where m_i is the "true" industry effect perceived by all individuals, a_j is

1. Y. Grunfeld and Zvi Griliches, "Is Aggregation Necessarily Bad?" *Review of Economics and Statistics*, vol. 42 (February 1960), pp. 1–13.

the individual deviation from the average respondent independent of the industry he is in, and e_{qij} is the random response error associated with the particular question q and individual ij. Taking these effects as independent from each other, and assuming that the first component does not average out as one aggregates within industries but that the other ones do in proportion to the average number of respondents per industry, gives us a little model that can be fit to the observed variances and covariances at the micro and macro industry levels. The following material shows the relevant numbers for two pairs of questions: whether product patents are effective against duplication and in securing royalty income, and whether moving quickly down the learning curve and superior sales and service efforts are effective in protecting the competitive advantage of new products. The two-question expected variance-covariance matrix is

Individual level	*Industry level*
$\begin{bmatrix} \sigma_m^2 + \sigma_a^2 + \sigma_1^2 & \sigma_m^2 + \sigma_a^2 \\ \dots & \sigma_m^2 + \sigma_a^2 + \sigma_2^2 \end{bmatrix}$	$\begin{bmatrix} \sigma_m^2 + \dfrac{(\sigma_A^2 + \sigma_1^2)}{N_i} & \sigma_m^2 + \sigma_A^2/N_i \\ \dots & \sigma_m^2 + \dfrac{(\sigma_A^2 + \sigma_2^2)}{N_i} \end{bmatrix}$

where m_i is the "true" industry effect perceived by all individuals, a_j is the individual deviation from the average respondent independent of the

	Product patents IB1	*Effective IB2*	*Learning curve IB5*	*Secrecy IB6*
Individual	2.860	1.435	1.810	.803
(N = 643)	. . .	2.748	. . .	1.662
Industry	.618	.384	.148	.013
(N = 24)	. . .	.376	. . .	.075
Implied estimates				
($\bar{N}$ = 27)				
σ_A^2		1.09		.820
σ_m^2		.34		<0

The numbers imply that the common variance between industries accounts for about one-eighth of the variance at the level of the individual response and more than half at the aggregated industry level. For the two other questions the implied "true" variance between industries is negative. (For the patents question the correlation rises from .51 at the level of the individual respondent to .80 for averages at the NSF industry level, while for the two appropriability questions the numbers go from .46 to .12, implying that such averaging attenuates rather than strengthens the relationship between the responses to such questions.) In short, while these questions might be interesting, they do not seem to be able to pick out significant differences among industries. Coming quickly down the learning curve and providing superior services are about equally effective.

Do the results of this survey help us explain other phenomena besides the relationship between answers to different sets of related questions within the survey? Looking at the responses to the questions evaluating patents and other appropriability mechanisms, the evidence appears to be mixed. Levin, Cohen, and Mowery did not find the appropriability variables significant in explaining differences among industries in R&D intensity, even in the absence of industry dummies.[2] In a forthcoming paper Iain Cockburn and Zvi Griliches use the Yale survey responses, aggregated to fifty-five industries at approximately a 3-digit SIC level, to see whether the stock market values the accumulated patents and the current R&D policy of a firm more or less in industries where the appropriability conditions are better in some sense.[3] Table 2 reproduces typical results from this study. Patent effectiveness measures help in some sense. The equations seem to imply that both accumulated past patents and current R&D moves are valued more by the market when patent protection is effective. Other appropriability measures do not help. But neither set of measures does better than just an interaction with ten higher-level (2-digit) industries dummies. The greater detail available in the Yale survey appears to be counterbalanced by the (inevitably?) greater imprecision of these measures at the detailed industry level. So there is something there but not as much as might be wished. But we should be thankful for there is hope that a more detailed study of these and other responses in this survey will help us understand our world better. In particular, the information on the differential

2. Richard C. Levin, Wesley M. Cohen, and David C. Mowery, "R&D Appropriability, Opportunity, and Market Structure: New Evidence on Some Schumpeterian Hypotheses," *American Economic Review*, vol. 75 (May 1985, *Papers and Proceedings, 1984*), p. 23.

3. Iain Cockburn and Zvi Griliches, "Industry Effects and Appropriability Measures in the Stock Market's Valuation of R&D and Patents," working paper 2465 (National Bureau of Economic Research, December 1987).

Table 2. Stock Market's Valuation of R&D and Patents, 722 U.S. Manufacturing Corporations, 1980[a]

Variables	*Coefficients*						
	(1)	*(2)*	*(3)*	*(4)*	*(5)*	*(6)*	*(7)*
SP/A[b]	.165	.380	.107	.249	.360	.077	.199
	(.100)	(.171)	(.167)	(.155)	(.170)	(.183)	(.161)
K/A[c]	...	...	.932	.335	...	.898	.342
			(.201)	(.178)		(.224)	(.175)
NR[d]	...	...	...	11.96	...	...	12.24
				(1.37)			(1.38)
PPP[e]	...	.034	.019	.019	.035	.023	.019
		(.024)	(.024)	(.023)	(.024)	(.025)	(.024)
PPP · (*SP/A*)	...	.236	.075	.098	.267	.115	.164
		(.116)	(.110)	(.101)	(.133)	(.142)	(.128)
PPP · (*K/A*)	...	...	.365	...	...	.432	...
			(.130)			(.172)	
PPP · *NR*	...	...	...	2.788	...	...	2.60
				(1.231)			(1.39)
NPP[f]	...	...	...	...	.039	.100	.054
					(.079)	(.089)	(.075)
NPP · (*SP/A*)	...	...	...	...	.127	.174	.290
					(.388)	(.432)	(.293)
NPP · (*K/A*)	...	...	...	...	...	.263	...
						(.636)	
NPP • *NR*	...	...	...	...	...	...	−1.89
							(4.77)
$\bar{R}^2$	.166	.172	.200	.310	.170	.198	.309

Source: Adapted from Iain Cockburn and Zvi Griliches, "Industry Effects and Appropriability Measures in the Stock Market's Valuation of R&D and Patents," working paper 2465 (National Bureau of Economic Research, December 1987), tables 3b and 5.

a. Dependent variable is log *Q* (market value divided by replacement value). All equations contain also ten 2-digit SIC industry dummy variables and a logarithm of total assets variable whose coefficient is small but consistently significant, on the order of −.03 (.01).

b. Stock of patents (30 percent depreciation rate) divided by total net assets.

c. Cumulated R&D "capital" stock (15 percent depreciation rate) divided by total net assets.

d. Net R&D investment divided by net assets $(R - .15K)/A$.

e. Sum of responses to "patents provide protection against duplication" questions for both process and product innovations. Averages at a fifty-five industries (approximately 3.5 digit level) aggregation.

f. Average of responses to all other "effectiveness of nonpatent appropriability mechanisms" questions.

connectedness of science in different industries is very intriguing and may be of help in future analyses of the contribution of science to technological advance.

General Discussion

Richard Levin agreed with Zvi Griliches that the appropriability variables could not discriminate effectively among more than about ten industry groupings, but he suggested that this may be a good thing, especially in light of Richard Gilbert's concern that studies such as this amount to "picoeconomics," from which no generalizations can be drawn. Sidney Winter noted that the results suggest there may be a relatively short list of variables to consider in an analysis of appropriability and incentives for R&D, and that ten industry groupings may provide all the information needed. In other dimensions, a finer division may be important. For example, Levin pointed out that, relative to questions on lead time, learning curves, and duplication costs, the questions on patent effectiveness discriminate better, as do the questions on learning and information spillover and those on duplication time.

Moreover, he added, a principal conclusion, that patents do not matter very much except in the chemical industries and in semiconductors, comes through regardless of problems with questions about other mechanisms of appropriation. In these two industries, the meaning and role of patent protection is different. Chemical products are easy to patent because the structure of the molecule of each product is unique, but patents are easy to invent around because it is often possible to create a discrete but structurally similar product with similar properties. With semiconductors, however, the innovation process is cumulative, with each invention built very distinctly on the previous one. The innovation provided by one firm makes the product invented by another firm more valuable. So the role patents play is to define the property rights (usually through the licensing process) so that the proceeds of this cumulative process can be shared and innovation can be encouraged.

Griliches also raised questions about whether the variables measured in this study are appropriately regarded as exogenous. Sales and service effort, for example, is a choice variable for the firms, and hence effectiveness should be endogenous. Likewise, lead time should be thought of as an outcome of the technology race, rather than an exogenous condition of it. Winter pointed out that even the distinction of product and process may be somewhat endogenous in the sense that firms take into consideration the importance of secrecy and the possibility of reverse engineering in designing their products. Firms often work to make their high-technology products inaccessible to reverse engineering, he noted, which tends to make these products more like processes from the standpoint of appropriability.

Several participants seemed concerned about problems of measurement and scaling biases in the data. Levin responded that various techniques, such as weighting the responses by the inverse of the variances or standard deviations of the individual responses, were tried to correct for these biases. The principal findings were robust to efforts to stretch or squeeze the distribution of responses, he noted, but it was unclear what such "corrections" mean since no one knows what the true distribution should be.

Joseph Farrell took issue with Gilbert's argument that it is not particularly interesting or helpful to worry about the determinants of R&D in industries that do not perform R&D. In fact, he suggested, it would be very useful to know why some industries seem to do so little research while others do so much. Gilbert agreed, but argued that it was still important to assign some sort of weights to individual responses to particular questions, based on the respondent's experience with those issues. Martin Baily argued that weighting the responses by how much R&D the responding firms do would be inappropriate, however, since the amount of R&D is what the authors are trying to explain.

Richard Schmalensee suggested that some of the within-industry variance in the responses about sales and service and other mechanisms of appropriation may be due to the fact that the R&D executives who responded to the questionnaire are less knowledgeable about what happens to the product after it leaves their jurisdiction in the organization.

Responding to a question from Paul Joskow, Levin said if he were doing the survey over again, he would want to do more pretesting. For example, the authors might have picked up on the issue of intraindustry variance earlier if they had pretested multiple respondents from the same industry. Or they might have learned ways to restructure questions to discriminate more carefully between exogenous and endogenous factors.

Robert Litan noted that one implication of the study for trade policy is that issues of intellectual property rights should be dealt with industry by industry. This is, in effect, how section 301 of the Trade Act already works, he added. The section provides a procedure for pursuing complaints about unfair trade practices abroad, but these complaints must be brought product by product. Litan also suggested that mechanisms of appropriating returns from R&D might vary between large and small firms. Levin agreed, but noted that the sampling procedure surveyed only publicly held firms, so that start-up ventures were completely excluded. Patents may be much more important for a start-up company because they provide something tangible to sell if the firm tries to sell out later.

Part V
The Diffusion of Innovations

[16]

Econometrica, Vol. 29, No. 4 (October 1961)

TECHNICAL CHANGE AND THE RATE OF IMITATION*

BY EDWIN MANSFIELD

This paper investigates the factors determining how rapidly the use of a new technique spreads from one firm to another. A simple model is presented to help explain differences among innovations in the rate of imitation. Deterministic and stochastic versions of this model are tested against data showing how rapidly firms in four industries came to use twelve important innovations. The empirical results seem quite consistent with both versions of the model.

1. INTRODUCTION

ONCE AN innovation is introduced by one firm, how soon do others in the industry come to use it? What factors determine how rapidly they follow? The importance of these questions has long been recognized. For example, Mason, Clark, Dunlop, and others pointed out in 1941 the need for studies to investigate how rapidly "... an innovation spreads from enterprise to enterprise."[1] Eighteen years later, the need is perhaps even more obvious than it was then.

This paper summarizes some theoretical and empirical findings regarding the rate at which firms follow an innovator. A simple model is presented to help explain differences among innovations in the rate of imitation. The model is then tested against data showing how rapidly firms in four quite different industries came to use twelve innovations. The results—though by no means free of difficulties—seem quite encouraging, and should help to fill a significant gap in the literature concerning technical change.

The plan of the paper is as follows. Section 2 lists the twelve innovations and shows how rapidly firms followed the innovator in each case. Sections 3–5 present and test a deterministic model constructed to explain the observed differences among these rates of imitation. A stochastic version is discussed

* The work on which this report is based was supported initially by research funds of the Graduate School of Industrial Administration and then by a contract with the National Science Foundation. It is part of a broader study of research and technological change that I am conducting under this contract. The paper has benefitted from discussions with a number of my colleagues at Carnegie Institute of Technology, particularly A. Meltzer, F. Modigliani, J. Muth, and H. Wein (now at Michigan State University). A version of it was presented at the December, 1959 meeting of the Econometric Society. My thanks also go to the many people in industry who provided data and granted me interviews. Without their cooperation, the work could not have been carried out.

[1] Committee on Price Determination [7, p. 169].

in Section 6. Some of the study's limitations are pointed out in Section 7 and its conclusions are summarized in Section 8.

2. RATES OF IMITATION

This section describes how rapidly the use of twelve innovations spread from enterprise to enterprise in four industries—bituminous coal, iron and steel, brewing, and railroads. The innovations are the shuttle car, trackless mobile loader, and continuous mining machine (in bituminous coal); the by-product coke oven, continuous wide strip mill, and continuous annealing line for tin plate (in iron and steel); the pallet-loading machine, tin container, and high-speed bottle filler (in brewing); and the diesel locomotive, centralized traffic control, and car retarders (in railroads).

These innovations were chosen because of their outstanding importance and because it seemed likely that adequate data could be obtained for them. Excluding the tin container, all were types of heavy equipment permitting a substantial reduction in costs. The most recent of these innovations occurred after World War II; the earliest was introduced before 1900. In practically all cases, the bulk of the development work was carried out by equipment manufacturers and patents did not impede the imitation process.[2]

Figure 1 shows the percentage of major firms that had introduced each of these innovations at various points in time. To avoid misunderstanding, note three things regarding the data in Figure 1. (1) Because of difficulties in obtaining information concerning smaller firms and because in some cases they could not use the innovation in any event, only firms exceeding a certain size (given in the Appendix) are included.[3] (2) The percentage of firms

[2] For descriptions of these innovations (and in some cases historical material), see Association of Iron and Steel Engineers [2] for the strip mill, Camp and Francis [5] for continuous annealing and the by-product coke oven, annual issues of *Coal Age* on mechanization and American Mining Congress [1] for the innovations in coal, *Fortune* (January, 1936) for the tin container, *American Brewer* (September, 1954) for pallet loaders and other issues for high speed fillers, Jewkes, *et al.* [12] for the diesel locomotive, Mansfield and Wein [17, pp. 292–3] for car retarders, and Union Switch and Signal Co. [24] for centralized traffic control. Actually, continuous annealing was often no cheaper than previous techniques, but it was required by customer demands.

[3] For the innovations in the steel industry, we imposed the particular size limits cited in the Appendix because, according to interviews, it seemed very unlikely that firms smaller than this would have been able to use them. For the innovations in the coal and brewing industries, there were no adequate published data concerning the dates when particular firms first introduced them, and we had to get the information directly from the firms. It seemed likely that "non-response" would be a considerable problem among the smaller firms. This consideration, as well as the fact that the smallest firms often could not use them, led us to impose the rather arbitrary lower limits on size shown in the Appendix. For the railroad innovations, we took firms

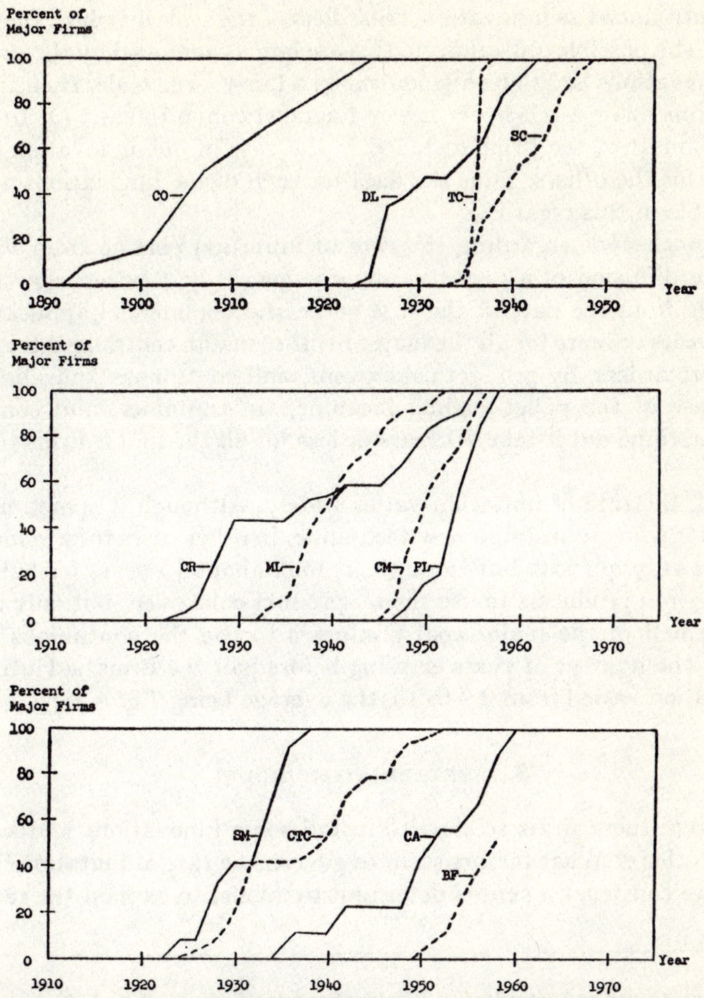

FIGURE 1.—Growth in the Percentage of Major Firms that Introduced Twelve Innovations, Bituminous Coal, Iron and Steel, Brewing, and Railroad Industries, 1890–1958.

a. By-product coke oven (CO), diesel locomotive (DL), tin container (TC), and shuttle car (SC).

b. Car retarder (CR), trackless mobile loader (ML), continuous mining machine (CM), and pallet-loading machine (PL).

c. Continuous wide-strip mill (SM), centralized traffic control (CTC), continuous annealing (CA), and highspeed bottle filler (BF).

Source: See the *Appendix*.

Note: For all but the by-product coke oven and tin container, the percentages given are for every two years from the year of initial introduction. Zero is arbitrarily set at two years prior to the initial introduction in these charts (but not in the analysis). The length of the interval for the by-product coke oven is about six years and for the tin container, it is six months. The innovations are grouped into the three sets shown above to make it easier to distinguish between the various growth curves.

having introduced an innovation, regardless of the scale on which they did so, is given. The possible objections to this are largely removed by the fact that these innovations had to be introduced on a fairly large scale. By using them at all, firms made a relatively heavy financial commitment.[4] (3) In a given industry, most of the firms included in the case of one innovation are also included for the others. Thus the data for each of the innovations are quite comparable in this regard.

Two conclusions regarding the rate of imitation emerge from Figure 1. First, the diffusion of a new technique is generally a rather slow process. Measuring from the date of the first successful commercial application,[5] it took 20 years or more for all the major firms to install centralized traffic control, car retarders, by-product coke ovens, and continuous annealing. Only in the case of the pallet-loading machine, tin container, and continuous mining machine did it take 10 years or less for all the major firms to install them.

Second, the rate of imitation varies widely. Although it sometimes took decades for firms to install a new technique, in other cases they followed the innovator very quickly. For example, it took about 15 years for half of the major pig-iron producers to use the by-product coke oven, but only about 3 years for half of the major coal producers to use the continuous mining machine. The number of years elapsing before half the firms had introduced an innovation varied from 0.9 to 15, the average being 7.8.

3. A DETERMINISTIC MODEL

Why were these firms so slow to install some innovations and so quick to install others? What factors seem to govern the rate of imitation? In this section, we construct a simple deterministic model to explain the results in

---

large enough to use car retarders and centralized traffic control and, to insure comparability, the same size limits were used for the diesel locomotive.

[4] The only alternative would be to take the date when a firm first used the innovation to produce some specified percentage of its output. In almost every case, such data were not published and it would have been extremely difficult, if not impossible, to obtain them from the firms. To install a strip mill, by-product coke ovens, continuous annealing, or car retarders, a firm had to invest many millions of dollars. Even for shuttle cars, trackless mobile loaders, canning equipment, and pallet loaders, the investment (although less than $100,000 usually) was by no means trivial in these industries.

[5] Note that we measure how quickly other firms followed the one that first *successfully* applied the technique. Others may have tried roughly similar things before but failed. By a successful application, we mean one where the equipment was used commercially for years, not installed and quickly withdrawn.

Figure 1. In Sections 4 and 5, we test this model and see the effects of introducing some additional variables into it.

The following notation is used. Let n_{ij} be the total number of firms on which the results in Figure 1 for the jth innovation in the ith industry are based[6] ($j = 1,2,3$; $i = 1,2,3,4$). Let $m_{ij}(t)$ be the number of these firms having introduced this innovation at time t, π_{ij} be the profitability of installing this innovation relative to that of alternative investments, and S_{ij} be the investment required to install this innovation as a per cent of the average total assets of these firms. More precise definitions of π_{ij} and S_{ij} are provided in Section 4. Let $\lambda_{ij}(t)$ be the proportion of "hold-outs" (firms not using this innovation) at time t that introduce it by time $t+1$, i.e.,

$$(1) \qquad \lambda_{ij}(t) = \frac{m_{ij}(t+1) - m_{ij}(t)}{n_{ij} - m_{ij}(t)}.$$

Our basic hypothesis can be stated quite simply. We assume that the proportion of "hold-outs" at time t that introduce the innovation by time $t+1$ is a function of (1) the proportion of firms that already introduced it by time t, (2) the profitability of installing it, (3) the size of the investment required to install it, and (4) other unspecified variables. Allowing the function to vary among industries, we have

$$(2) \qquad \lambda_{ij}(t) = f_i\left(\frac{m_{ij}(t)}{n_{ij}}, \pi_{ij}, S_{ij}, \ldots\right).$$

In the following few paragraphs, we take up the presumed effects of variation in $m_{ij}(t)/n_{ij}$, π_{ij}, and S_{ij} on $\lambda_{ij}(t)$ and the reasons for interindustry differences in the function.

First, one would expect that increases in the proportion of firms already using an innovation would increase $\lambda_{ij}(t)$. As more information and experience accumulate, it becomes less risky to begin using it.[7] Competitive pressures

[6] That is, the total number of firms for which we have data. See Table I for the n_{ij}.

[7] The profitability of installing each of these innovations was viewed at first with considerable uncertainty. For example, there was great uncertainty about maintenance costs for diesel locomotives, "down-time" for continuous mining machines, the safety of centralized traffic control, and the useful life of by-product coke ovens. The perceived risks seldom disappeared after only a few firms had introduced them; in some cases according to interviews, it took many years. This helps to account for the fact (noted above) that the imitation process generally went on rather slowly.

This also suggests the potential importance of two variables not recognized explicitly in (2): (1) the extent of the initial uncertainty concerning the profitability of an innovation (depending in part on the extent of prior field testing by manufacturers), and (2) the rate at which this uncertainty declined. For some types of innovations, a few installations and a relatively short period of use can cut the risks to little or nothing. For other types, installations under many sorts of conditions and a long period of use

mount and "bandwagon" effects occur. Where the profitability of using the innovation is very difficult to estimate, the mere fact that a large proportion of its competitors have introduced it may prompt a firm to consider it more favorably. Both interviews with executives in the four industries and the data in Figure 1 indicate that this is the case.[8]

Second, the profitability of installing the innovation would also be expected to have an important influence on $\lambda_{ij}(t)$. The more profitable this investment is relative to others that are available, the greater is the chance that a firm's estimate of the profitability will be high enough to compensate for whatever risks are involved and that it will seem worthwhile to install the new technique rather than to wait. As the difference between the profitability of this investment and that of others widens, firms tend to note and respond to the difference more quickly. Both the interviews and the few other studies regarding the rate of imitation suggest that this is so.[9]

Third, for equally profitable innovations, $\lambda_{ij}(t)$ should tend to be smaller for those requiring relatively large investments. One would expect this on

---

(to be sure of useful life, maintenance costs, etc.) are required. See Section 7 for further discussion.

Note that all these innovations are new processes or (in the case of the tin container) a packaging innovation. This model would not be applicable to some innovations, like an entirely new product, where, as more firms produce it, it becomes less profitable for others to do so. There is no evidence of significant decreases of this sort in these cases. (E.g., as more firms introduced diesel locomotives, this did not make it less profitable for others to do so.) Moreover, the model presumes that π_{ij} is appreciably greater than unity (which is the case here). Finally, the data support the hypothesis in the text (see note 8).

[8] Beginning with the date when $m_{ij}(t) = 1$, we computed $\lambda_{ij}(t)$ and $m_{ij}(t)/n_{ij}$, using the intervals in the footnote to Figure 1 as time units and stopping when $m_{ij}(t) = n_{ij}$. Then we calculated the correlation between $\lambda_{ij}(t)$ and $m_{ij}(t)/n_{ij}$. The correlation coefficients were .77 (continuous mining machine), .85 (by-product coke oven), .49 (tin container)., 65 (centralized traffic control), .85 (continuous wide strip mill), .66 (diesel locomotive), .52 (car retarders), .55 (bottle fillers), .96 (pallet loaders), .81 (continuous annealing), .46 (trackless mobile loader), and .94 (shuttle car). Using a one-tailed test (which is appropriate here), all coefficients but those for the continuous mining machine, bottle filler, and trackless mobile loader are significant (0.05 level). All are positive.

Of course, others have stated similar hypotheses before. E.g., Schumpeter [20] noted that "accumulating experience and vanishing obstacles" smooth the way for imitators, and Coleman, *et al.* [6] noted a "snow-ball" effect. For what it is worth, almost all the executives we interviewed considered this effect to be present. The number of installations of the innovation or the number of firms using it might have been used rather than the proportion of firms, but the latter seems to work quite well.

[9] See Griliches [9]. Another recent study of the rate of imitation is found in Yance [25]. Both papers focus on only one innovation (hybrid corn and the diesel locomotive). Of course, the interfirm variation in profitability, as well as the average, could influence $\lambda_{ij}(t)$.

the grounds that firms tend to be more cautious before committing themselves to such projects and that they often have more difficulty in financing them. According to the interviews, this factor is often important.

Finally for equally profitable innovations requiring the same investment, $\lambda_{ij}(t)$ is likely to vary among industries. It might be higher in one industry than in another because firms in the former industry have less aversion to risk, because markets are more keenly competitive, because the attitude of the labor force toward innovation is more favorable, or because the industry is healthier financially. Casual observation suggests that such interindustry differences may have a significant effect on $\lambda_{ij}(t)$.

Returning to equation (2), we act as if the number of firms having introduced an innovation can vary continuously rather than assume only integer values, and we assume that $\lambda_{ij}(t)$ can be approximated adequately within the relevant range by a Taylor's expansion that drops third and higher order terms. Assuming that the coefficient of $(m_{ij}(t)/n_{ij})^2$ in this expansion is zero (and the data in Figure 1 generally support this),[10] we have

$$(3)\quad \lambda_{ij}(t) = a_{i1} + a_{i2}\frac{m_{ij}(t)}{n_{ij}} + a_{i3}\pi_{ij} + a_{i4}S_{ij} + a_{i5}\pi_{ij}\frac{m_{ij}(t)}{n_{ij}} + a_{i6}S_{ij}\frac{m_{ij}(t)}{n_{ij}} + a_{i7}\pi_{ij}S_{ij} + a_{i8}\pi_{ij}^2 + a_{i9}S_{ij}^2 + \ldots,$$

where additional terms contain the unspecified variables in (2). Thus,

$$(4)\quad m_{ij}(t+1) - m_{ij}(t) = [n_{ij} - m_{ij}(t)]\left[a_{i1} + a_{i2}\frac{m_{ij}(t)}{n_{ij}} + \ldots + a_{i9}S_{ij}^2 + \ldots\right].$$

Assuming that time is measured in fairly small units, we can use as an approximation the corresponding differential equation[11]

$$(5)\quad \frac{dm_{ij}(t)}{dt} = [n_{ij} - m_{ij}(t)]\left[Q_{ij} + \phi_{ij}\frac{m_{ij}(t)}{n_{ij}}\right],$$

the solution of which is

$$(6)\quad m_{ij}(t) = \frac{n_{ij}[e^{l_{ij} + (Q_{ij} + \phi_{ij})t} - (Q_{ij}/\phi_{ij})]}{1 + e^{[l_{ij} + (Q_{ij} + \phi_{ij})t]}},$$

[10] To test this assumption for these innovations, we used $(m_{ij}(t)/n_{ij})^2$ as an additional independent variable in the regression described in note 8 and used the customary analysis of variance to determine whether this resulted in a significant increase in the explained variation. For all but continuous annealing, car retarders, and the diesel locomotive, it does not (and in these cases, the increase is often barely significant). Hence, in most cases, there is no evidence that this coefficient is nonzero.

[11] This is like some approximations commonly used in capital theory. E.g., one often replaces equations like $x(t+1) - x(t) = rx(t)$ with $\dot{x}(t) = rx(t)$.

where l_{ij} is a constant of integration, Q_{ij} is the sum of all terms in (3) not containing $m_{ij}(t)/n_{ij}$, and

$$\phi_{ij} = a_{i2} + a_{i5}\pi_{ij} + a_{i6}S_{ij} + \ldots . \tag{7}$$

Of course, ϕ_{ij} is the coefficient of $m_{ij}(t)/n_{ij}$ in (3).

To get any further, we must impose additional constraints on the way $m_{ij}(t)$ can vary over time. One simple condition we can impose is that, as we go backward in time, the number of firms having introduced the innovation must tend to zero,[12] i.e.,

$$\lim_{t \to -\infty} m_{ij}(t) = 0 . \tag{8}$$

Using this condition, it follows that

$$m_{ij}(t) = n_{ij}[1 + e^{-(l_{ij} + \phi_{ij}t)}]^{-1} . \tag{9}$$

Thus, the growth over time in the number of firms having introduced an innovation should conform to a logistic function, an *S*-shaped growth curve frequently encountered in biology and the social sciences.[13]

If (9) is correct, it can be shown that the rate of imitation is governed by only one parameter—ϕ_{ij}.[14] Assuming that the sum of the unspecified terms in (7) is uncorrelated with π_{ij} and S_{ij} and that it can be treated as a random error term,

[12] Of course, other conditions could be imposed in addition or instead. For example in Section 6, we take as given the date when a particular number of firms had installed an innovation and we force $m_{ij}(t)$ to equal that number at that date. But (8) is all we need for present purposes. Note that it implies that Q_{ij} is zero. The data described in note 8 are consistent with this, but even if Q_{ij} were nonzero but small (and it certainly could not be large), (9) should be a reasonably good approximation. Note too that, if the model holds, $\phi_{ij} > 0$.

[13] Note that things are simplified here by the fact that all firms we consider eventually introduced these innovations. (A few went out of business first, but not because of the appearance of the innovation. See the Appendix.) Had this not been the case, it would have been necessary either to provide a mechanism explaining the proportion that did not do so or to take it as given. Of course, by taking only the larger firms, we made sure that all could use these innovations. The smaller firms (that were not potential users) are omitted. For the high-speed bottle filler, we make the reasonable assumption that all the major firms will ultimately introduce it. Note that the argument here is different from that generally used in biology to arrive at the logistic function and that it explicitly includes variables affecting its shape.

[14] It seems reasonable to take, as a measure of the rate of imitation, the time span between (1) the date when 20 per cent (say) of the firms had introduced an innovation, and (2) the date when 80 per cent (say) had done so. According to the model, this time span equals 2.77 ϕ_{ij}^{-1} and is therefore independent of l_{ij}. If, rather than 20 and 80, we take P_1 and P_2, it can be shown that the time span equals $\phi_{ij}^{-1} \ln [(1-P_1)P_2/P_1(1-P_2)]$.

TABLE I

PARAMETERS, ESTIMATES, AND ROOT-MEAN-SQUARE ERRORS: DETERMINISTIC AND STOCHASTIC MODELS

Innovation	Parameters[a]						
	n_{ij}	π_{ij}	S_{ij}	d_{ij}	g_{ij}	t^*_{ij}	δ_{ij}
Diesel locomotive	25	1.59	.015	35	1.00	1925	1
Centralized traffic control	24	1.48	.024	0	1.50	1926	1
Car retarders	25	1.25	.785	0	1.50	1924	0
Continuous wide strip mill	12	1.87	4.908	30	4.50	1924	0
By-product coke oven	12	1.47	2.083	10	4.00	1894	1
Continuous annealing	9	1.25	.554	9	4.25	1936	1
Shuttle car	15	1.74	.013	9	1.25	1937	0
Trackless mobile loader	15	1.65	.019	6	2.50	1934	1
Continuous mining machine	17	2.00	.301	8	1.00	1947	1
Tin container	22	5.07	.267	0	6.50	1935	1
High-speed bottle filler	16	1.20	.575	10	2.25	1951	1
Pallet-loading machine	19	1.67	.115	0	2.25	1948	1

Innovation	Estimates and Root Mean Square Errors[b]					
	$\hat{l}_{ij}$	$\hat{\phi}_{ij}$	r_{ij}	$\hat{\theta}_{ij}$	Error (Det.)	Error (St.)
Diesel locomotive	− 6.64	.20	.89	.30	2.13	5.63
Centralized traffic control	− 7.13	.19	.94	.24	1.52	3.44
Car retarders	− 3.95	.11	.90	.17	2.08	5.02
Continuous wide strip mill	−10.47	.34	.95	.42	.83	.90
By-product coke oven	− 1.47	.17	.98	.18	.16	.84
Continuous annealing	− 8.51	.17	.93	.22	.74	1.42
Shuttle car	−13.48	.32	.95	.45	.86	2.03
Trackless mobile loader	−13.03	.32	.97	.39	.71	1.66
Continuous mining machine	−24.96	.49	.98	.59	.81	2.22
Tin container	−84.35	2.40	.96	2.64	1.34	3.00
High-speed bottle filler	−20.58	.36	.97	.40	.56	.95
Pallet-loading machine	−29.07	.55	.97	.63	1.10	1.58

[a] For definitions of these parameters, see Sections 3 (n_{ij}), 4 (π_{ij} and S_{ij}), and 5 (d_{ij}, g_{ij}, t^*_{ij}, and δ_{ij}).

[b] For definitions of these measures, see Sections 4 ($\hat{l}_{ij}$, $\hat{\phi}_{ij}$, r_{ij}, and error (det.)) and 6 ($\hat{\theta}_{ij}$ and error (st.)).

Source: See the *Appendix* and notes 6, 15, 16, 17, 20, 22, 25, 29, 32, and 34.

$$\phi_{ij} = b_i + a_{i5}\pi_{ij} + a_{i6}S_{ij} + z_{ij}\,, \tag{10}$$

where b_i equals a_{i2} plus the expected value of this sum and z_{ij} is a random variable with zero expected value. Hence, the expected value of ϕ_{ij} in a particular industry is a linear function of π_{ij} and S_{ij}.

To sum up, the model leads to the following two predictions. First, the number of firms having introduced an innovation, if plotted against time, should approximate a logistic function. Second, the rate of imitation in a particular industry should be higher for more profitable innovations and innovations requiring relatively small investments. More precisely, ϕ_{ij}, a measure of the rate of imitation, should be linearly related to π_{ij} and S_{ij}.

4. TESTS OF THE MODEL

We test this model in two steps: (1) by estimating ϕ_{ij} and l_{ij} and determining how well (9) fits the data, and (2) by seeing whether the expected value of ϕ_{ij} seems to be a linear function of π_{ij} and S_{ij}. The results of these tests suggest that the model can explain the results in Figure 1 quite well.

To carry out the first step, note that, if the model is correct, it follows from (9) that

$$\ln\left[\frac{m_{ij}(t)}{n_{ij} - m_{ij}(t)}\right] = l_{ij} + \phi_{ij}t\,. \tag{11}$$

Measuring time in years (from 1900), treating this as a regression equation, and using least squares (after properly weighting the observations [**4**]), we derive estimates of l_{ij} and ϕ_{ij} (Table I).[15] To see how well (9) can represent the data, we insert these estimates into (9) and compare the calculated increase over time in the number of firms having introduced each innovation with the actual increase.

Judging merely by a visual comparison, the calculated growth curves generally provide reasonably good approximations to the actual ones. However, the "fit" is not uniformly good. For the railroad innovations, there is evidence of serial correlation among the residuals, and the approximations are less satisfactory than for the other innovations. Table I contains two rough measures of "goodness-of-fit": the root-mean-square deviation of

[15] When ln $[m_{ij}(t)/n_{ij} - m_{ij}(t)]$ was infinite, the observation was omitted. The observations were one year apart for the continuous wide strip mill, continuous mining machine, and centralized traffic control, six years apart for the by-product coke oven, one month apart for the tin containers, and two years apart for the others. For the high-speed bottle filler, we used all the data available thus far, but the imitation process is not yet complete (cf. Figure 1).

actual from the computed number of firms having introduced the innovation[16] and the coefficient of correlation between ln $[m_{ij}(t)/n_{ij}-m_{ij}(t)]$ and t.[17] They seem to bear out the general impression that (9) represents the data for most of the innovations quite well.[18]

To carry out the second step, we assume that a_{i5} and a_{i6} do not vary among industries.[19] Thus (10) becomes

$$\phi_{ij} = b_i + a_5\pi_{ij} + a_6S_{ij} + z_{ij}\,, \tag{12}$$

and, if we assume that errors in the estimates of ϕ_{ij} are uncorrelated with π_{ij} and S_{ij}, we have

$$\hat{\phi}_{ij} = b_i + a_5\pi_{ij} + a_6S_{ij} + z'_{ij}\,, \tag{13}$$

where $\hat{\phi}_{ij}$ is the estimate of ϕ_{ij} derived above. Assuming that z'_{ij} is distributed normally with constant variance, standard tests used for linear hypotheses can be applied to determine whether a_5 and a_6 are nonzero.

Before discussing the results of these tests, we must describe how π_{ij} and S_{ij} are measured. We obtained from as many of the firms as possible estimates of the pay-out period for the initial installation of the innovation and the pay-out period required from investments. They were derived primarily from correspondence, but published materials were used where possible.[20]

[16] For each innovation, the difference between the computed and actual values of $m_{ij}(t)$ was obtained for $t = t^*_{ij}, t^*_{ij}+1, \ldots, t^{**}_{ij}$, where time is measured in the units described in note 15, t^*_{ij} is the first date when $m_{ij}(t) = 1$, and t^{**}_{ij} is the first date when $m_{ij}(t) = n_{ij}$. Then the root-mean-square of these differences was obtained. Note that this is a measure of absolute, not relative, error, and when comparing the results for different innovations, take account of differences in n_{ij}.

[17] This coefficient is labeled r_{ij} in Table I. Note that it is based on weighted observations (the weights being those suggested by Berkson [4]).

[18] Of course, the logistic function is not the only one that might represent the data fairly adequately. (E.g., the normal cumulative distribution function would probably do as well.) And, since the actual curve has to be roughly S-shaped, it is not surprising that it fits reasonably well. There seemed to be little point in attempting any formal "goodness-of-fit" tests here. The number of firms considered in each case is quite small, and the tests would not be very powerful. All that we conclude from Table I is that (9) provides a reasonably adequate description of the data and hence that ϕ_{ij} is generally reliable as a measure of the rate of imitation.

[19] Because of the small number of observations, we are forced to make this assumption. If data were available for more innovations, it would not be necessary. Of course, if interindustry differences in these coefficients are statistically independent of π_{ij} and S_{ij}, a_5 and a_6 can be regarded as averages and there is no real trouble.

[20] A letter was sent to each firm asking (1) how long it took for the initial investment in the innovation to pay for itself, and (2) what the required pay-out period was during the relevant period. (For a definition of the pay-out period, see Swalm [22]. Both the realized and required pay-out periods are before taxes.) Replies were received from 50 per cent of the rail roads and coal producers and 20 per cent of the steel companies and breweries. The estimates for the railroad innovations are probably most accurate since the firms referred us to fairly reliable published studies (see the Appendix). The data

Then the average pay-out period required by the firms (during the relevant period) to justify investments divided by the average pay-out period for the innovation was used as a measure of π_{ij}.[21] To measure S_{ij}, we used the average initial investment in the innovation as a percentage of the average total assets of the firms (during the relevant period).[22] Table I contains the results.

Using these rather crude data, we obtained least squares estimates of the parameters in (13) and tested whether they were zero. The resulting equation is

$$\hat{\phi}_{ij} = \begin{Bmatrix} -.29 \\ -.57 \\ -.52 \\ -.59 \end{Bmatrix} + \underset{(.015)}{.530\pi_{ij}} - \underset{(.014)}{.027S_{ij}} \qquad (r = .997)\,, \tag{14}$$

for the coal innovations are probably quite good too. The estimates for the steel innovations and the tin container are probably less accurate because they occurred so long ago and because fewer firms provided data. Despite the fact that the averages are probably more accurate than the individual figures and that the results were checked with other sources in interviews (see the Appendix), the estimates of the π_{ij} are fairly rough.

[21] For relatively long-lived investments, the reciprocal of the pay-out period is a fairly adequate approximation to the rate of return. See Gordon [8] and Swalm [22]. Hence this measure of π_{ij} is approximately equal to the average rate of return derived (ex post) from the innovation divided by the average rate of return firms required (ex ante) to justify investments. Of course, it would be more appropriate for firms to recognize that introducing it next year, the year after, etc., are the alternatives to introducing it now and to make an analysis like Terborgh's [23]. But as he points out, most firms seem to make these decisions on the basis of the "rate of return" or "pay-out period"; and thus it seems reasonable to use such measures in a study (like this) where we try to explain behavior, not prescribe it.

Even so, this measure is only an approximation. (1) The rate of return from the investment in an innovation is measured ex post, not ex ante. (2) The average rate of return that could have been realized by the "hold-outs" at a particular point in time probably differed from the average rate of return actually realized by all firms. Our estimates are based implicitly on the factor prices and age of old equipment (where replacement occurred) that prevailed when the innovation was installed, and these (and other) factors, as well as the innovation itself, do not remain unchanged. According to the interviews, the average return that could have been realized probably varied over time about an average that was highly correlated with our estimate, and hence the latter provides a fair indication of the level. But the parameters describing the temporal variation about this level are among the "other" variables in (2). See Section 7 and note 40.

[22] For the steel, coal, and brewing industries, we obtained the total assets of as many of these firms as possible (during the relevant period) from *Moody's*. For the railroads, the 1936 reproduction costs in Klein [13] were used for the firms. Estimates of the approximate investment required to install each innovation were obtained primarily from the interviews (described in the Appendix), and each was divided by the average total assets of the relevant firms. Since the investment could vary considerably, the results are only approximate.

where the top figure in the brackets pertains to the brewing industry, the next to coal, the following to steel, and the bottom figure pertains to the railroads. The coefficients of π_{ij} and S_{ij} have the expected signs (indicating that increases in π_{ij} and decreases in S_{ij} increase the rate of imitation), and both differ significantly from zero. As expected, there are significant interindustry differences, the rate of imitation (for given π_{ij} and S_{ij}) being particularly high in brewing. These differences seem to be broadly consistent with the hypothesis often advanced that the rate of imitation is higher in more competitive industries, but there are too few data to warrant any real conclusion on this score.[23]

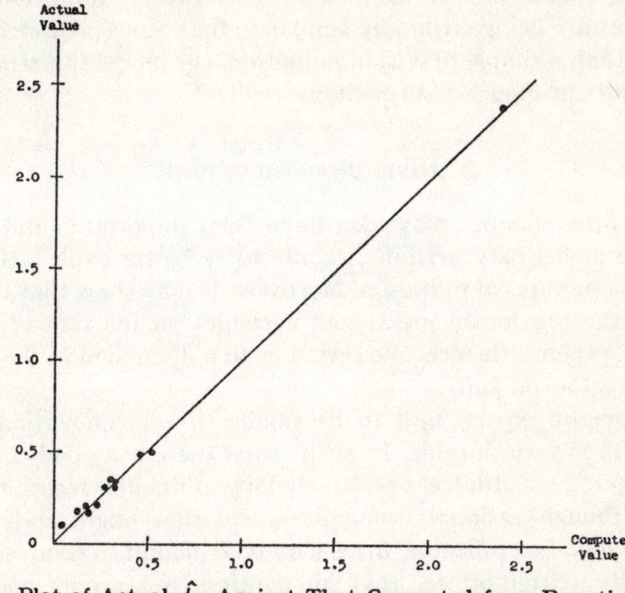

FIGURE 2.—Plot of Actual $\hat{\phi}_{ij}$ Against That Computed from Equation (14), Twelve Innovations.

Source: Table I and equation (14).
Note: The difference between an actual and computed value of $\hat{\phi}_{ij}$ is equal to the vertical difference between the point and the 45° line.

[23] For a more extended statement of this hypothesis, see Robinson [19]. We encounter here the usual difficulty in measuring the "extent of competition." But most economists would undoubtedly agree that brewing and coal are more competitive than steel and railroads (and the average value of b_i is larger in the former industries). When six of my colleagues were asked to rank them by "competitiveness," they put brewing first, coal second, iron and steel third, and railroads fourth. If these ranks are correlated with the estimates of b_i in (14), the rank correlation coefficient is positive (.80), but not statistically significant. (The .05 probability level is used throughout this paper.) There are too few industries to allow a reasonably powerful test of this hypothesis even if our procedures were refined somewhat, but for most rankings that seem sensible, the data are consistent with the hypothesis (the correlation is positive).

The scatter diagram in Figure 2 shows that (14) represents the data surprisingly well. When corrected for the relatively few degrees of freedom, the correlation coefficient is .997. Of course, one point (the tin container) strongly affects the results. But if that point is omitted, the interindustry differences remain much the same, the coefficients of π_{ij} and S_{ij} keep the same signs (the latter becoming non-significant), and the correlation coefficient, corrected for degrees of freedom, is still .97.[24]

Hence, the model seems to fit the data quite well. In general, the growth in the number of users of an innovation can be approximated by a logistic curve. And there is definite evidence that more profitable innovations and ones requiring smaller investments had higher rates of imitation, the particular relationship being strikingly similar to the one we predicted. Though it is no more than a simple first approximation, the model can represent the empirical results in Figure 1 surprisingly well.

5. ADDITIONAL FACTORS

Of course, other factors may also have been important, and their inclusion in the model may permit a significantly better explanation of the differences among rates of imitation. Moreover, it may show that the apparent effect of the previously mentioned variables on the rate of imitation is partly due to their influence. We turn now to a discussion of the influence of four other factors on $\lambda_{ij}(t)$.

First, one might expect $\lambda_{ij}(t)$ to be smaller if the innovation replaces equipment that is very durable. In such cases, there is a good chance that a firm's old equipment still has a relatively long useful life, according to past estimates. Although rational economic calculation might indicate that replacement would be profitable, firms may be reluctant to scrap equipment that is not fully written off and that will continue to serve for many years.

[24] Five points should be noted. (1) The tin container is a somewhat different type of innovation from the others and the profitability data for it are probably least reliable (see note 20). Thus, besides its being an extreme point in Figure 2, there are other possible grounds for excluding it, and it is reassuring to note that its exclusion has so little effect on the results. (2) Because of the way it is constructed, the model can only be expected to work if π_{ij} and S_{ij} remain within certain bounds. If π_{ij} is close to one or S_{ij} is very large or both, it is likely to perform poorly. (3) One-tailed tests are used to determine the significance of the coefficients of π_{ij} and S_{ij}. (4) It is noteworthy that $\hat{\phi}_{ij}$, not a very obvious measure of the rate of imitation, should be linearly related to π_{ij} and S_{ij}, as the model predicts. Using the result in note 14, the number of years elapsing between when 20 per cent and when 80 per cent introduced the innovation can easily be used instead as the dependent variable in (14). The resulting relationship is nonlinear. (5) This model may also help to explain the empirical results Griliches [9] obtained in his excellent study of the regional acceptance of hybrid corn, but for which he provided no formal theoretical underpinning.

If so, d_{ij}—the number of years that typically elapsed before the old equipment was replaced (before the innovation appeared)—may be one of the excluded variables in (2), and hence $\hat{\phi}_{ij}$ may be a linear function of π_{ij}, S_{ij}, and d_{ij}.[25] If d_{ij} is included in (14), we find that

$$(15)\quad \hat{\phi}_{ij} = \begin{Bmatrix} -.28 \\ -.56 \\ -.51 \\ -.57 \end{Bmatrix} + \underset{(.015)}{.528\pi_{ij}} - \underset{(.015)}{.020S_{ij}} - \underset{(.0014)}{.0017d_{ij}} \qquad (r = .997)\,.$$

Though there is some apparent tendency for the rate of imitation to be lower in cases where very durable equipment had to be replaced, it is not statistically significant. (For the values of d_{ij} and the other factors discussed below, see Table I.)

Second, one might expect $\lambda_{ij}(t)$ to be higher if firms are expanding at a rapid rate. If they are convinced of its superiority, the innovation will be introduced in the new plants built to accommodate the growth in the market. If there is little or no expansion, its introduction must often wait until the firms decide to replace existing equipment. If g_{ij}—the annual rate of growth of industry sales during the period—is included in (14),

$$(16)\quad \hat{\phi}_{ij} = \begin{Bmatrix} -.32 \\ -.56 \\ -.53 \\ -.58 \end{Bmatrix} + \underset{(.032)}{.484\pi_{ij}} - \underset{(.013)}{.025S_{ij}} + \underset{(.026)}{.042g_{ij}} \qquad (r = .998)\,.$$

Thus, there is some apparent tendency for the rate of imitation to be higher where output was expanding at a very rapid rate, but it is not statistically significant.[26]

[25] If we carry d_{ij} from (2) on and make the same assumptions we did with regard to π_{ij} and S_{ij}, it follows that ϕ_{ij} should be a linear function of π_{ij}, S_{ij}, and d_{ij}. This, of course, is also true for the other variables discussed in this section. Estimates of d_{ij} were obtained primarily from the interviews. If an innovation was largely a supplement or addition to old plant (like centralized traffic control and car retarders), if it served a different purpose than the old equipment (like canning equipment vs. bottling equipment), and if it displaced only labor (like pallet loaders), we let d_{ij} equal zero. In such cases, there appeared to be no equipment whose durability could influence the decision significantly. Note that the age distribution and number of units of old equipment are also important in determining how long it takes before the first one "wears out." It would also have been preferable to have included the profitability of replacing old equipment of various ages rather than just the average figure used. But the available data would not permit this. For further discussion, see Mansfield [**15**].

[26] Of course, the effect of g_{ij} may depend on whether old equipment must be replaced, how durable it is, the difference between the profitability of replacement and of installing the innovation in new plant, the extent of excess capacity at the beginning of the period, the size of a plant relative to the size of the market, etc. (For some relevant

Third, $\lambda_{ij}(t)$ may have increased over time. This hypothesis has been advanced by economists on numerous occasions.[27] Presumably, the reasons for such a trend would be the evolution of better communication channels, more sophisticated techniques to evaluate machine replacement, and more favorable attitudes toward technological change. If t_{ij}—the year (less 1900) when the innovation was first introduced[28]—is included in (14),

$$(17) \quad \hat{\phi}_{ij} = \begin{Bmatrix} -.37 \\ -.64 \\ -.55 \\ -.63 \end{Bmatrix} + \underset{(.015)}{.535\pi_{ij}} - \underset{(.016)}{.027S_{ij}} + \underset{(.0014)}{.0014t_{ij}} \qquad (r = .997)\,.$$

There is some apparent tendency for the rate of imitation to increase over time, but it is not statistically significant.

Finally, one might suppose that $\lambda_{ij}(t)$ would be influenced by the phase of the business cycle during which the innovation was first introduced. Let δ_{ij} equal one if the innovation is introduced in the expansion phase and zero if it is introduced in the contraction phase.[29] Including δ_{ij} in (14), we have

$$(18) \quad \hat{\phi}_{ij} = \begin{Bmatrix} -.26 \\ -.55 \\ -.48 \\ -.57 \end{Bmatrix} + \underset{(.017)}{.530\pi_{ij}} - \underset{(.020)}{.033S_{ij}} - \underset{(.045)}{.022\,\delta_{ij}} \qquad (r = .997)\,,$$

and the effect of δ_{ij} turns out to be non-significant.

Thus, our results regarding these additional factors are largely inconclusive. Though each might be expected to have some effect, their inclusion in the analysis does not lead to a significantly better explanation of the observed differences in $\hat{\phi}_{ij}$. Their apparent effects are in the expected direction, but data for more innovations will be required before one can be reasonably sure of the persistence and magnitude of their influence. There is no evidence

---

discussion, see Scitovsky [21].) The effect of this factor, like d_{ij}, reflects the possible unwillingness of firms to scrap existing equipment. It would have been preferable to have used figures on the profitability of using the innovation in new plant as well to replace equipment of various ages, but data were not available. Note too that g_{ij} may be affected by the appearance of the innovation. For a description of the data on g_{ij}, see the Appendix. If S_{ij} is omitted from (15)–(18), the coefficients of d_{ij}, g_{ij}, etc. remain non-significant.

[27] E.g., Mack [14, p. 295] and Jerome [11, p. XXV].

[28] I.e., $t_{ij} = t^*_{ij} - 1900$. See note 32.

[29] We used the National Bureau's reference dates (given in Moore [18]), to determine whether t^*_{ij} was a year of contraction or recovery. The residuals in Figure 2 do not seem to be affected by whether or not an innovation was being accepted sometime during the depression of the nineteen-thirties.

that the effects of the previously considered variables on the rate of imitation are due to the operation of these factors. When these other factors are included, the coefficients of π_{ij} and S_{ij} and the interindustry differences remain relatively unchanged (though the coefficient of S_{ij} becomes non-significant.)

6. A STOCHASTIC VERSION OF THE MODEL

In this section, we present and test a somewhat more sophisticated, stochastic version of the model. For the jth innovation in the ith industry, let $P_{ij}(t,k)$ be the probability that any one of the "hold-outs" at time t will introduce it by time $t+k$, and assume (for small k) that

$$P_{ij}(t,k) = \theta_{ij}\frac{m_{ij}(t)}{n_{ij}}k\,, \tag{19}$$

$$\theta_{ij} = b_i' + a_5'\pi_{ij} + a_6'S_{ij} + z_{ij}''\,, \tag{20}$$

where the coefficients in (20) are analogous to those in (12). To see how closely this resembles the deterministic model, note that the *expected* increase in the number of users between time t and time $t+1$,

$$[n_{ij} - m_{ij}(t)]\,\theta_{ij}\frac{m_{ij}(t)}{n_{ij}}\,,$$

is almost identical with the expression given before for the *actual* increase in the number of users (see (4) and (5)). The only (apparent) difference is that the terms corresponding to Q_{ij} are assumed to be zero from the start here, whereas in Section 3 this followed from (8).[30]

Whereas our task in the deterministic model was to determine how the actual number of users grows over time, our problem here is to see how the expected number grows. To do so, we first obtain an expression for $P_{ij}^r(t)$ —the probability that at time t there are exactly r firms in the ith industry that have not yet introduced the jth innovation. From (19),

$$P_{ij}^r(t+k) = P_{ij}^r(t)\left[1 - r\frac{\theta_{ij}}{n_{ij}}(n_{ij}-r)k\right] + P_{ij}^{r+1}(t)(r+1)\frac{\theta_{ij}}{n_{ij}}(n_{ij}-r-1)k + o(k)\,,$$

where $o(k)$ represents terms that, if divided by k, tend to zero as k vanishes.[31]

[30] To simplify things, we also assume from the beginning that the coefficients of π_{ij} and S_{ij} do not vary among industries (Cf. (20)). In the deterministic model we introduced this assumption later in the analysis.

[31] To derive this expression, note that $P_{ij}^r(t+k)) = \sum_{x=r}^{n_{ij}} P_{ij}^x(t)\, D_{ij}^{x-r}(t)$, where $D_{ij}^{x-r}(t)$ is the probability that $(x-r)$ firms begin using the innovation between time t and time $t+k$, given that $(n_{ij}-x)$ firms are using it at time t. From (19),

$$D_{ij}^0(t) = [1-(\theta_{ij}/n_{ij})\,(n_{ij}-r)k]^r;$$

And if we subtract $P^r_{ij}(t)$ from both sides, divide by k, and let k tend to zero, the following differential-difference equation results:

$$(21) \quad \dot{P}^r_{ij}(t) \begin{cases} = -\dfrac{\theta_{ij}}{n_{ij}} r(n_{ij}-r)P^r_{ij}(t) + \dfrac{\theta_{ij}}{n_{ij}}(r+1)(n_{ij}-r-1)P^{r+1}_{ij}(t), \text{ if } r < n_{ij}-1 , \\ = -\dfrac{\theta_{ij}}{n_{ij}} r\, P^r_{ij}(t) , \text{ if } r = n_{ij}-1 , \end{cases}$$

with the initial conditions that $P^r_{ij}(t^*_{ij})$ equals unity for $r = n_{ij}-1$ and zero otherwise, and where $\dot{P}^r_{ij}(t)$ is the time derivative of $P^r_{ij}(t)$ and t^*_{ij} is the date (taken as given)[32] when the innovation was first introduced. From (21), it follows [3] that $M_{ij}(t^*_{ij}+V)$—the expected number of firms using the innovation at time $t^*_{ij}+V$—equals

$$(22) \quad n_{ij} - \sum_{r=1} \frac{(n_{ij}-1)!}{(n_{ij}-r-1)!\,(r-1)!}\Big[(n_{ij}-2r)^2 - \frac{\theta_{ij}}{n_{ij}} V + 2 - (n_{ij}-2r) \sum_{u=r}^{n_{ij}-r-1} u^{-1}\Big] e^{-r(n_{ij}-r)(\theta_{ij}/n_{ij})V} ,$$

where r runs up to $(n_{ij}-1)/2$ for n_{ij} odd and up to $n_{ij}/2$ for n_{ij} even.[33]

Thus, the stochastic version of the model leads to the following two propositions. First, the expected number of firms having introduced an innovation at any date subsequent to t^*_{ij} should be given by (22). Second, θ_{ij}—the parameter that, for given n_{ij}, determines the expected rate of imita-

---

$$D^1_{ij}(t) = (r+1)\frac{\theta_{ij}}{n_{ij}}(n_{ij}-r-1)k\Big[1 - \frac{\theta_{ij}}{n_{ij}}(n_{ij}-r-1)k\Big]^{r-1} ;$$

etc. Thus,

$$P^r_{ij}(t+k) = P^r_{ij}(t)\Big[1 - \frac{\theta_{ij}}{n_{ij}}(n_{ij}-r)k\Big]^r$$
$$+ P^{r+1}_{ij}(t)\,(n+1)\frac{\theta_{ij}}{n_{ij}}(n_{ij}-r-1)k\Big[1 - \frac{\theta_{ij}}{n_{ij}}(n_{ij}-r-1)k\Big]^{r-1} + o(k)$$
$$= P^r_{ij}(t)\Big[1 - r\frac{\theta_{ij}}{n_{ij}}(n_{ij}-r)k\Big] + P^{r+1}_{ij}(t)\Big[(r+1)\frac{\theta_{ij}}{n_{ij}}(n_{ij}-r-1)k\Big] + o(k) .$$

Note two things, (1) The assumption that the probability in (19) is the same for all "hold-outs" is obviously only a convenient simplification. Without it, the analysis becomes extremely difficult. (2) We assume that the decisions of the "hold-outs" between time t and time $t+k$ are independent. Hence, the probability that one "hold-out" will introduce it between time t and time $t+k$ is $[n_{ij}-m_{ij}(t)]\theta_{ij}m_{ij}(t)k/n_{ij}+o(k)$, and the probability that none will introduce it is $1-[n_{ij}-m_{ij}(t)]\theta_{ij}m_{ij}(t)k/n_{ij}+o(k)$.

[32] For all but the tin container, we use December 31 of the year during which the innovation was first introduced by one of these firms (or if we know the month when it was first introduced, we use that December 31 which was closest in time to it). For the tin container, we use February 1, 1935. See Table I for the year in which each t^*_{ij} falls.

[33] When n_{ij} is even, some adjustment has to be made to this expression. See Bailey [3] for the adjustment and for a much more extensive discussion of the argument.

tion—should be a linear function of π_{ij} and S_{ij}, the intercept of the function differing among industries.

To see how well the first proposition seems to hold, we estimate θ_{ij}, insert it, n_{ij}, and t_{ij}^{*} into (22), and compare the growth over time in the computed number of users with the actual growth curve. Table I contains a rough measure of "goodness-of-fit": the root-mean-square deviation of the actual from the computed number of users.[34] Although the agreement between the actual and computed growth curves is sometimes quite good, Table I shows that (22) is never so accurate as (9) in representing the data. Perhaps this is due in part to differences in the way parameters are estimated as well as differences in the model.[35]

To test the second proposition, we assume that errors in the estimates of θ_{ij} are uncorrelated with π_{ij} and S_{ij} and hence that

$$\hat{\theta}_{ij} = b_i' + a_5'\pi_{ij} + a_6'S_{ij} + z_{ij}''' , \tag{23}$$

where $\hat{\theta}_{ij}$ is our estimate of θ_{ij} and z_{ij}''' is a random error term. Then, proceeding as we did in Sections 4 and 5, we get much the same sort of results as those obtained there. (1) We estimate the parameters in (23), find that these estimates are almost identical with those given in (14) for the analogous parameters,[36] and conclude that the resulting equation represents the data very well ($r = .997$). (2) We introduce the additional variables discussed in Section 5 into (23) and find that their coefficients have the expected signs, but that (except for the coefficient of t_{ij}) all are non-significant.

Thus, the model, in either its deterministic or stochastic form, seems to represent the empirical results in Figure 1 quite well. Although (22) does not fit the data as well as (9) in the deterministic version, the purpose and interpretation of these equations are quite different and hence this may not be very surprising.[37] With regard to the explanation of differences in the

[34] To estimate θ_{ij}, we used the statistic suggested by Bailey [**3**, p. 41] to estimate θ_{ij}/n_{ij} and multiplied the result by n_{ij}. This estimate of θ_{ij} is biased, but we can show that the bias is approximately equal to $\theta_{ij}/(n_{ij}-1)$, which is usually quite small. These estimates, together with the tables in Mansfield and Hensley [**16**], were used to calculate the growth over time in the expected number of users. The root-mean-square errors are computed in the way described in note 16.

[35] Two parameters (l_{ij} and ϕ_{ij}) are fitted to the data in the deterministic model whereas only (θ_{ij}) is fitted here. However, this is only part of the explanation (see note 37). Note too that the differences between the computed and actual curves are by no means random for a number of these innovations. The computed curves sometimes lie below the actual ones.

[36] The only appreciable difference is that the coefficient of π_{ij} is .580 rather than .530.

[37] In the deterministic model, we fit (9) to the data in such a way as to minimize differences between the actual and computed curves. In this model, we use the data to estimate the curve of *averages* that would result if the process were repeated indefinite-

rate of imitation, (23), like (13) in the deterministic model, provides an excellent fit. The stochastic version of the model seems somewhat more reasonable to me, but it is difficult to tell at this point whether it is a significant improvement over the simpler, deterministic version.

Before concluding, one final point should be noted regarding the dispersion about the expected value in (22). This dispersion is often relatively large and hence a prediction of the rate of imitation based on (22) would be fairly crude, even if the model were correct and θ_{ij} were known in advance. To illustrate this, consider an innovation where $n_{ij} = 12$, $t_{ij}^* = 1924$, and $\theta_{ij} = .42$. Table II shows the expected number of firms having introduced it at various points in time and the relatively large standard deviation of the distribution about this expectation. Of course, despite this variation, the model could set useful upper and lower bounds on how long the entire process would take.[38]

TABLE II

EXPECTED VALUE AND STANDARD DEVIATION OF THE NUMBER OF FIRMS HAVING INTRODUCED AN INNOVATION, ASSUMING THAT $n_{ij} = 12$, $t_{ij}^* = 1924$, AND $\theta_{ij} = .42$, 1924–39

Date (Dec. 31)	Expected Number	Standard Deviation
1924	1.0	—
1925	1.5	.9
1926	2.1	1.4
1927	2.7	1.6
1928	3.6	2.3
1929	4.5	2.7
1930	5.5	2.9
1931	6.4	3.1
1932	7.2	3.2
1933	8.0	3.1
1934	8.8	2.9
1935	9.4	2.6
1936	10.0	2.4
1937	10.5	2.1
1938	10.8	1.8
1939	11.1	1.5

Source: Tables in Mansfield and Hensley [16]. Small errors due to interpolation are present in both the expected values and standard deviatious.

ly. Whereas the point in the former case was to minimize differences between the computed and actual curves, this was not the point here and hence it is not surprising that the deterministic model seems superior in this respect.

[38] For further discussion of the distribution about these expected values, see Mansfield and Hensley [16] and the literature cited there.

7. LIMITATIONS

In evaluating these results, the limitations of the data, methods, and scope of the investigation must be taken into account. For example, although its scope far exceeds that of the few studies previously conducted, it nonetheless is limited to only four industries and to only a few innovations in each one. Before concluding, it seems worthwhile to point out some of these limitations in more detail.

First, there is the matter of scope. Although the four industries included here vary with respect to market structure, size of firm, type of customer, etc., they can hardly be viewed as a cross-section of American industry. For example, none is a relatively young industry with a rapidly changing technology. In addition, the innovations included here are all important, they generally required fairly large investments, and the imitation process was not impeded by patents. To check and extend these results, similar studies should be carried out for other industries and other types of innovations. Moreover, international comparisons might be attempted.

Second, the data are not always as precise as one would like. For example, much of the data regarding the profitability of installing an innovation had to be obtained from questionnaires and interviews with firms. Although these estimates were checked against others obtained from suppliers of the equipment, trade journals, etc., they are probably fairly rough. Similarly, the estimates of the pay-out period required for investments and the durability of old equipment are only rough. Because of these errors of measurement, the regression coefficients in (14)–(18) are probably biased somewhat toward zero.[39]

Third, the data measure the rate of imitation among large firms only. As noted above, firms not exceeding a certain size (specified in the Appendix) were excluded. Moreover, the rate at which firms imitated an innovator, not the rate at which a new technique displaced an old one or the rate at which investment in a new technique mounted, is considered here. Although these are related topics also being studied, they were not taken up in this paper.

Fourth, various factors other than those considered here undoubtedly exerted some influence on the rate of imitation. Variation over time in the profitability of introducing an innovation (due to improvements, the business

[39] It can easily be shown that measurement errors, if random, have this effect. Another limitation of the analysis is that the model can only be expected to hold for relatively profitable innovations. Certainly, it will not hold in cases where $\pi_{ij} < 1$ and it may do poorly if π_{ij} is not appreciable greater than unity. But this limitation is not very serious because if π_{ij} does not exceed unity, the innovation almost certainly will not become generally accepted, and if π_{ij} is not much greater than unity, the innovation is probably not very important.

cycle, etc.) is one potentially important factor that is omitted.[40] Others are the sales and promotional efforts made by the producers of the new equipment and the extent of the risks firms believed they assumed at first by introducing an innovation.[41] I could find no satisfactory way to measure them, but perhaps further research will obviate this difficulty and disclose the effects of these and other such factors on the rate of imitation. I suspect that the results underestimate their influence and that data for more innovations would show that the unexplained variation (due to their effects) is greater than is indicated in Figure 2.

8. CONCLUSION

In their discussions of technological change, economists have often cited the need for more research regarding the rate of imitation. Because an innovation will not have its full economic impact until the imitation process is well under way, the rate of imitation is of considerable importance. It is unfortunate that so little attention has been devoted to it and that consequently we know so little about the mechanism governing how rapidly firms come to use a new technique.

My purpose in this paper has been to present and test a simple model designed to help explain differences among process innovations in the rate of imitation. This model is built largely around one hypothesis—that the probability that a firm will introduce a new technique is an increasing func-

[40] One cannot tell with any accuracy how, and to what extent, the profitability varied in each of these cases. But, to illustrate the factors at work, note a few broad changes that occurred in the case of three of the innovations. (1) The appearance of improved diesel locomotives in the Thirties and Forties increased the profitability of introducing them and hastened their acceptance. In addition, first costs declined, the relative prices of coal and oil changed, and the composition of the "hold-outs" varied. (2) The demand for toluene during World War I undoubtedly accelerated the imitation process in the case of the by-product coke oven. (3) The Depression reduced the profitability of introducing centralized traffic control, whereas the wartime traffic boom increased its profitability. For further discussion, see note 21. Of course, the relative profitability may have varied less than the absolute profitability.

[41] In each of these cases, there was considerable promotional effort by the producers of the new equipment and considerable help given to the firms that used it. E.g., Semet-Solvay and Koppers helped to finance by-product coke plants and provided personnel, General Motors helped to train diesel operators, kept maintenance men at hand, etc. For some discussion of the perceived risks, see note 7.

Another factor that could be very important for some innovations is the presence of patents. Still another is the development of a marked improvement in the innovation. E.g., an innovation may be applicable at first to only a few firms and a significant improvement is required before others use it. In a case like this, the model is clearly not applicable. See Mansfield [15]. In each of these cases, improvements occurred, but they were of a more gradual and less significant nature.

tion of the proportion of firms already using it and the profitability of doing so, but a decreasing function of the size of the investment required.[42] When confronted with data for twelve innovations, this model seems to stand up surprisingly well. As expected, the rate of imitation tended to be faster for innovations that were more profitable and that required relatively small investments. An equation of the form predicted by the model can explain practically all of the variation among the rates of imitation.

As expected, there were also interindustry differences in the rate of imitation. We have too few industries really to test the hypothesis often advanced that the rate of imitation is faster in more competitive industries, but the differences seem to be generally in that direction. When several other factors are included in the model, the empirical results are largely inconclusive. There was some apparent tendency for the rate of imitation to be higher when the innovation did not replace very durable equipment, when the firms' output was growing rapidly, and when the innovation was introduced in the more recent past. But it was almost always statistically non-significant.

In view of the relatively small number of innovations and the uneven quality of the data, these results are quite tentative. But even so, they should be useful to economists concerned with the dynamics of firm behavior and the process of economic growth. To understand how economic growth is generated, we must know more about the way innovations occur and how they become generally accepted. Further attempts should be made to check and extend the results and to pursue other approaches to the particular area considered here. We need to know much more about this and other aspects of technological change.

Carnegie Institute of Technology and Yale University

APPENDIX

Basic Data

Iron and steel industry: For the continuous wide strip mill, all firms having more than 140,000 tons of sheet capacity in 1926 were included; for the by-product coke oven, all firms with over 200,000 tons of pig iron capacity in 1901 were included; and for continuous annealing of tin plate, the nine major producers of tin plate in 1935 were included. In the case of the strip mill and coke oven, a few of these firms merged or went out of business before installing them, and there was no choice but to exclude them.

The date when each firm first installed a continuous wide strip mill was taken from the Association of Iron and Steel Engineers [2]. Similar data for the coke oven were obtained from various editions of the *Directory of Iron and Steel Works* of the American

[42] This is the stochastic version of the model. In the deterministic version, we use the proportion of "hold-outs" that introduce it rather than this probability.

Iron and Steel Institute and issues of the *Iron Trade Review* and *Iron Age*. The date when each firm installed continuous annealing lines was obtained from correspondence with the firms.

The size of the investment required and the durability of replaced equipment came from interviews. (The estimates of the pay-out periods were also checked there. Cf. note 20.) The interviews (each about 2 hours long) were with major officials of three steel firms, the president and research manager of a firm that builds strip mills and continuous annealing lines, officials of a firm that builds coke ovens, and representatives of a relevant engineering association and of a trade journal. The data on growth of output were annual industry growth rates and were for sheets during 1926–37 (strip mill), pig iron during 1900–25 (coke oven), and tin plate during 1939–56 (continuous annealing). They were taken from the *Bituminous Coal Annual* of the Bituminous Coal Institute, Association of Iron and Steel Engineers [**2**], and *Annual Statistical Reports* of the American Iron and Steel Institute.

Railroad industry: For centralized traffic control, all Class I line-haul roads with over 5 billion freight ton-miles in 1925 were included. For the diesel locomotives and car retarders, essentially the same firms were included. (The Norfolk and Western, a rather special case, was replaced by the New Haven and Lehigh Valley in the case of the diesel locomotive. Some important switching roads were substituted in the case of car retarders, an innovation in switching techniques.) An entire system is treated here as one firm.

The date when a firm first installed centralized traffic control was usually derived from a questionnaire filled out by the firm. For those that did not reply, estimates by K. Healy [**10**] were used. The date when each firm first installed diesel locomotives was determined from various editions of the Interstate Commerce Commission's *Statistics of Railways*. The date when each firm first installed car retarders was taken from various issues of *Railway Age*.

For centralized traffic control and car retarders, some of the pay-out periods were estimates published in the *Signal Section, Proceedings of Association of American Railroads*, and the rest were obtained from questionnaires filled out by the firms. All estimates of the pay-out period for the diesel locomotive and the pay-out period required for investment were obtained from questionnaires. Information regarding the size of the investment required and the durability of old equipment was obtained primarily from interviews with eight officials of six railroads (ranging from president to chief engineer) and three officials of a signal manufacturing firm and a locomotive manufacturing firm. The data on growth of output were annual growth rates for total freight ton-miles during 1925–41 (diesel locomotive) and 1925–54 (centralized traffic control and car retarders.) They were taken from the *Statistics of Railways*.

Bituminous coal industry: Practically all firms producing over 4 million tons of coal in 1956 (according to McGraw-Hill's *Keystone Coal Buyers Manual*) were included. A few firms that did strip mining predominantly were excluded in the case of the continuous mining machine, and a few had to be excluded in the case of the shuttle car and trackless mobile loader because they would not provide the necessary data. The date when each firm first introduced these types of equipment was usually obtained from questionnaires filled out by the firms, but in the case of the continuous mining machine, data for two firms that did not reply were derived from the *Keystone Coal Buyers Manual*.

Data regarding the size of the investment required and the durability of old equipment were obtained from interviews with two vice-presidents of coal firms, several executives of firms manufacturing the equipment, employees of the Bureau of Mines, and representatives of an independent coal research organization. The data on growth of output were annual growth rates for bituminous coal production during 1934–51 (trackless mobile loader), 1937–51 (shuttle car), and 1947–56 (continuous mining machine). They were taken primarily from the *Bituminous Coal Annual*.

Brewing industry: We tried to include all breweries with more than $1 million in assets in 1934 (according to the *Thomas Register*), but several would not provide the necessary data and they could not be obtained elsewhere. The date when a firm first installed each type of equipment was usually taken from a questionnaire that it filled out, but in a few cases it was provided by manufacturers of the equipment or articles in the *Brewers' Journal*. The size of the necessary investment and the durability of equipment that was replaced were determined from interviews with a number of officials in two breweries and sales executives of two can companies. The data on growth of output were annual growth rates for beer production during 1935–37 (tin containers) and 1950–58 (pallet loading machine and high-speed bottle filler). They were taken from the *Brewers' Almanac* and *Business Week* (June 20, 1959). The 1950–58 figures refer only to half of these larger firms.

REFERENCES

[1] American Mining Congress: *Coal Mine Modernization Yearbook* (annual).

[2] Association of Iron and Steel Engineers: *The Modern Strip Mill* (Pittsburgh, 1941).

[3] Bailey, N.: "Some Problems in the Statistical Analysis of Epidemic Data," *Journal of the Royal Statistical Society*, B, 17 (1955).

[4] Berksen, J.: "A Statistically Precise and Relatively Simple Method of Estimating the Bio-Assay with Quantal Responses, Based on the Logistic Function," *Journal of the American Statistical Association* (September, 1953).

[5] Camp, J. and C. Francis: *The Making, Shaping, and Treating of Steel* (U. S. Steel Co., 1940).

[6] Coleman, J., E. Katz, and H. Menzel: "The Diffusion of an Innovation Among Physicians," *Sociometry* (December, 1957).

[7] Committee on Price Determination: *Cost Behavior and Price Policy* (National Bureau of Economic Research, 1943).

[8] Gordon, M.: "The Payoff Period and the Rate of Profit," *Journal of Business* (October, 1955).

[9] Griliches, Z.: "Hybrid Corn: An Exploration in the Economics of Technological Change," *Econometrica* (October, 1957).

[10] Healy, K.: "Regularization of Capital Investment in Railroads," *Regularization of Business Investment* (National Bureau of Economic Research, 1954).

[11] Jerome, H.: *Mechanization in Industry* (National Bureau of Economic Research, 1934).

[12] Jewkes, J., D. Sawers, and R. Stillerman: *The Sources of Invention* (St. Martin's Press, 1958).

[13] Klein, L.: "Studies in Investment Behavior," *Conference on Business Cycles* (National Bureau of Economic Research, 1951).

[14] MACK, R.: *The Flow of Business Funds and Consumer Purchasing Power* (Columbia University, 1941).

[15] MANSFIELD, E.: "Acceptance of Technical Change: The Speed of Response of Individual Firms," (Multilithed, Carnegie Institute of Technology, 1959).

[16] ——— AND C. HENSLEY: "The Logistic Process: Tables of the Stochastic Epidemic Curve and Applications," *Journal of the Royal Statistical Society*, Series B (forthcoming).

[17] ——— AND H. WEIN: "A Model for the Location of a Railroad Classification Yard," *Management Science* (April, 1958).

[18] MOORE, G.: "Measuring Recessions," *Journal of the American Statistical Association* (June, 1958).

[19] ROBINSON, J.: *The Accumulation of Capital* (Irwin, 1956).

[20] SCHUMPETER, J.: *Business Cycles* (McGraw-Hill, 1939).

[21] SCITOVSKY, T.: "Economies of Scale and European Integration," *American Economic Review* (March, 1956).

[22] SWALM, R.: "On Calculating the Rate of Return of an Investment," *Journal of Industrial Engineering* (March, 1958).

[23] TERBORGH, G.: *Dynamic Equipment Policy* (McGraw-Hill, 1949).

[24] UNION SWITCH AND SIGNAL CO.: *Centralized Traffic Control* (Swissvale, 1931).

[25] YANCE, J.: "Technological Change as a Learning Process: The Dieselization of the Railroads" (Unpublished, 1957).

[17]

Clio and the Economics of QWERTY

By PAUL A. DAVID*

Cicero demands of historians, first, that we tell true stories. I intend fully to perform my duty on this occasion, by giving you a homely piece of narrative economic history in which "one damn thing follows another." The main point of the story will become plain enough: it is sometimes not possible to uncover the logic (or illogic) of the world around us except by understanding how it got that way. A *path-dependent* sequence of economic changes is one of which important influences upon the eventual outcome can be exerted by temporally remote events, including happenings dominated by chance elements rather than systematic forces. Stochastic processes like that do not converge automatically to a fixed-point distribution of outcomes, and are called *non-ergodic.* In such circumstances "historical accidents" can neither be ignored, nor neatly quarantined for the purpose of economic analysis; the dynamic process itself takes on an *essentially historical* character. Standing alone, my story will be simply illustrative and does not establish how much of the world works this way. That is an open empirical issue and I would be presumptuous to claim to have settled it, or to instruct you in what to do about it. Let us just hope the tale proves mildly diverting for those waiting to be told if and why the study of economic history is a necessity in the making of economists.

*Department of Economics, Encina Hall, Stanford University, Stanford, CA 94305. Support provided for this research, under a grant to the Technological Innovation Program of the Center for Economic Policy Research, Stanford University, is gratefully acknowledged. Douglas Puffert supplied able research assistance. Some, but not the whole, of my indebtedness to Brian Arthur's views on QWERTY and QWERTY-like subjects is recorded in the References. I bear full responsibility for errors of fact and interpretation, as well as for the peculiar opinions abbreviated herein. A fuller version with complete references, entitled "Understanding the Economics of QWERTY or Is History Necessary?," is available on request.

I. The Story of QWERTY

Why does the topmost row of letters on your personal computer keyboard spell out QWERTYUIOP, rather than something else? We know that nothing in the engineering of computer terminals requires the awkward keyboard layout known today as "QWERTY," and we all are old enough to remember that QWERTY somehow has been handed down to us from the Age of Typewriters. Clearly nobody has been persuaded by the exhortations to discard QWERTY, which apostles of DSK (the Dvorak Simplified Keyboard) were issuing in trade publications such as *Computers and Automation* during the early 1970's. Why not? Devotees of the keyboard arrangement patented in 1932 by August Dvorak and W. L. Dealey have long held most of the world's records for speed typing. Moreover, during the 1940's U.S. Navy experiments had shown that the increased efficiency obtained with DSK would amortize the cost of retraining a group of typists within the first ten days of their subsequent full-time employment. Dvorak's death in 1975 released him from forty years of frustration with the world's stubborn rejection of his contribution; it came too soon for him to be solaced by the Apple IIC computer's built-in switch, which instantly converts its keyboard from QWERTY to virtual DSK, or to be further aggravated by doubts that the switch would not often be flicked.

If as Apple advertising copy now says, DSK "lets you type 20–40% faster," why did this superior design meet essentially the same rejection as the previous seven improvements on the QWERTY typewriter keyboard that were patented in the United States and Britain during the years 1909–24? Was it the result of customary, nonrational behavior by countless individuals socialized to carry on an antiquated technological tradition? Or, as Dvorak himself once suggested, had there been a conspiracy among the members of the typewriter oligopoly to suppress an invention which they feared would so increase typewriter efficiency as ultimately to curtail the demand for their products? Or perhaps we should turn instead to the other popular "Devil Theory," and ask if political regulation and interference with the workings of a "free market" has been the cause of inefficient keyboard regimentation? Maybe it's all to be blamed on the public school system, like everything else that's awry?

You can already sense that these will not be the most promising lines along which to search for an economic understanding of QWERTY's present dominance. The agents engaged in production and purchase decisions in today's keyboard market are not the prisoners of custom, conspiracy, or state control. But while they are, as we now say, perfectly "free to choose," their behavior, nevertheless, is held fast in the grip of events long forgotten and shaped by circumstances in which neither they nor their interests figured. Like the great men of whom Tolstoy wrote in *War and Peace*, "(e) very action of theirs, that seems to them an act of their own free will, is in an historical sense not free at all, but in bondage to the whole course of previous history..." (Bk. IX, ch. 1).

This is a short story, however. So it begins only little more than a century ago, with the fifty-second man to invent the typewriter. Christopher Latham Sholes was a Milwaukee, Wisconsin printer by trade, and a mechanical tinkerer by inclination. Helped by his friends, Carlos Glidden and Samuel W. Soule, he had built a primitive writing machine for which a patent application was filed in October 1867. Many defects in the working of Sholes' "Type Writer" stood in the way of its immediate commercial introduction. Because the printing point was located underneath the paper carriage, it was quite invisible to the operator. "Non-visibility" remained an unfortunate feature of this and other up-stroke machines long after the flat paper carriage of the original design had been supplanted by arrangements closely resembling the modern continuous roller-platen. Consequently, the tendency of the typebars to clash and jam if struck in rapid succession was a particularly serious defect. When a typebar stuck at or near the printing point, every succeeding stroke merely hammered the same impression onto the paper, resulting in a string of repeated letters that would be discovered only when the typist bothered to raise the carriage to inspect what had been printed.

Urged onward by the bullying optimism of James Densmore, the promoter-venture capitalist whom he had taken into the partnership in 1867, Sholes struggled for the next six years to perfect "the machine." From the inventor's trial-and-error rearrangements of the original model's alphabetical key ordering, in an effort to reduce the frequency of typebar clashes, there emerged a four-row, upper case keyboard approaching the modern QWERTY standard. In March 1873, Densmore succeeded in placing the manufacturing rights for the substantially transformed Sholes-Glidden "Type Writer" with E. Remington and Sons, the famous arms makers. Within the next few months QWERTY's evolution was virtually completed by Remington's mechanics. Their many modifications included some fine-tuning of the keyboard design in the course of which the "*R*" wound up in the place previously allotted to the period mark "." Thus were assembled into one row all the letters which a salesman would need to impress customers, by rapidly pecking out the brand name: TYPE WRITER

Despite this sales gimmick, the early commercial fortunes of the machine, with which chance had linked QWERTY's destiny remained terrifyingly precarious. The economic downturn of the 1870's was not the best of times in which to launch a novel piece of office equipment costing $125, and by 1878, when Remington brought out its Improved Model Two (equipped with carriage shift key), the whole enterprise was teetering on the edge of bankruptcy. Consequently, even though sales began to pick up pace with the lifting of the depression and annual typewriter production climbed to 1200 units in 1881, the market position which QWERTY had acquired during the course of its early career was far from deeply entrenched; the entire stock of QWERTY-embodying machines in the United States could not have much exceeded 5000 when the decade of the 1880's opened.

Nor was its future much protected by any compelling technological necessities. For, there were ways to make a typewriter without the up-stroke typebar mechanism that had called forth the QWERTY adaptation, and rival designs were appearing on the American scene. Not only were there typebar machines with "down-stroke" and "front-stroke" actions that afforded a visible printing point; the problem of typebar clashes could be circumvented by dispensing with typebars entirely, as young Thomas Edison had done in his 1872 patent for an electric print-wheel device which later became the basis for teletype machines. Lucien Stephen Crandall, the inventor of the second typewriter to reach the American market (in 1879) arranged the type on a cylindrical sleeve: the sleeve was made to revolve to the required letter and come down onto the printing-point, locking in place for correct alignment. (So much for the "revolutionary" character of the IBM 72/82's "golf ball" design.) Freed from the legacy of typebars, commercially successful typewriters such as the Hammond and the Blickensderfer first sported a keyboard arrangement which was more sensible than QWERTY. Then so-called "Ideal" keyboard placed the sequence DHIATENSOR in the home row, these being ten letters with which one may compose over 70 percent of the words in the English language.

The typewriter boom beginning in the 1880's thus witnessed a rapid proliferation of competitive designs, manufacturing companies, and keyboard arrangements rivalling the Sholes-Remington QWERTY. Yet, by the middle of the next decade, just when it had become evident that any micro-technological rationale for QWERTY's dominance was being removed by the progress of typewriter engineering, the U.S. industry was rapidly moving towards the standard of an upright front-stroke machine with a four-row QWERTY keyboard that was referred to as "the Universal." During the period 1895–1905, the main producers of non-typebar machines fell into line by offering "the Universal" as an option in place of the Ideal keyboard.

II. Basic QWERTY-Nomics

To understand what had happened in the fateful interval of the 1890's, the economist must attend to the fact that typewriters were beginning to take their place as an element of a larger, rather complex system of production that was technically interrelated. In addition to the manufacturers and buyers of typewriting machines, this system involved typewriter operators and the variety of organizations (both private and public) that undertook to train people in such skills. Still more critical to the outcome was the fact that, in contrast to the hardware subsystems of which QWERTY or other keyboards were a part, the larger system of production was nobody's design. Rather like the proverbial Topsy, and much else in the history of economies besides, it "jes' growed."

The advent of "touch" typing, a distinct advance over the four-finger hunt-and-peck method, came late in the 1880's and was critical, because this innovation was from its inception adapted to the Remington's QWERTY keyboard. Touch typing gave rise to three features of the evolving production system which were crucially important in causing QWERTY to become "locked in" as the dominant keyboard arrangement. These features were *technical interrelatedness*, *economies of scale*, and *quasi-irreversibility* of investment. They constitute the basic ingredients of what might be called QWERTY-nomics.

Technical interrelatedness, or the need for system compatibility between keyboard "hardware" and the "software" represented by the touch typist's memory of a particular arrangement of the keys, meant that the expected present value of a typewriter as an instrument of production was dependent upon the availability of compatible software created by typists' decisions as to the kind of keyboard they should learn. Prior to the growth of the personal market for typewriters, the purchasers of the hardware typically were business firms and therefore distinct from the owners of typing skills. Few incentives existed at the time, or later, for any one business to invest in providing its employees with a form of general human capital which so readily could be taken elsewhere. (Notice that it was the wartime U.S. Navy, not your typical employer, that undertook the experiment of retraining typists on the Dvorak keyboard.) Nevertheless the purchase by a potential employer of a QWERTY keyboard conveyed a positive pecuniary externality to compatibly trained touch typists. To the degree to which this increased the likelihood that subsequent typists would choose to learn QWERTY, in preference to another method for which the stock of compatible hardware would not be so large, the overall user costs of a typewriting system based upon QWERTY (or any specific keyboard) would tend to *decrease* as it gained in acceptance relative to other systems. Essentially symmetrical conditions obtained in the market for instruction in touch typing.

These decreasing cost conditions—or *system scale economies*—had a number of consequences, among which undoubtedly the most important was the tendency for the process of intersystem competition to lead towards de facto standardization through the predominance of a single keyboard design. For analytical purposes, the matter can be simplified in the following way: suppose that buyers of typewriters uniformly were without inherent preferences concerning keyboards, and cared only about how the stock of touch typists was distributed among alternative specific keyboard styles. Suppose typists, on the other hand, were heterogeneous in their preferences for learning QWERTY-based "touch," as opposed to other methods, but attentive also to the way the stock of machines was distributed according to keyboard styles. Then imagine the members of this heterogenous population deciding in random order what kind of typing training to acquire. It may be seen that, with unbounded decreasing costs of selection, each stochastic decision in favor of QWERTY would raise the probability (but not guarantee) that the next selector would favor QWERTY. From the viewpoint of the formal theory of stochastic processes, what we are looking at now is equivalent to a generalized "Polya urn scheme." In a simple scheme of that kind, an urn containing balls of various colors is sampled with replacement, and every drawing of a ball of a specified color results in a second ball of the same color being returned to the urn; the probabilities that balls of specified colors will be added are therefore increasing (linear) functions of the proportions in which the respective colors are represented within the urn. A recent theorem due to W. Brian Arthur et al. (1983; 1985) allows us to say that when a generalized form of such a process (characterized by unbounded increasing returns) is extended indefinitely, the proportional share of one of the colors will, with probability one, converge to unity.

There may be many eligible candidates for supremacy, and from an *ex ante* vantage point we cannot say with corresponding certainty which among the contending colors —or rival keyboard arrangements—will be the one to gain eventual dominance. That part of the story is likely to be governed by "historical accidents," which is to say, by the particular sequencing of choices made close to the beginning of the process. It is there that essentially random, transient factors are most likely to exert great leverage, as has been shown neatly by Arthur's (1983) model of the dynamics of technological competition under increasing returns. Intuition suggests that if choices were made in a forward-looking way, rather than myopically on the basis of comparisons among the currently prevailing costs of different systems, the final outcome could be influenced strongly by expectations. A particular system could triumph over rivals merely because the purchasers of the software (and/or the hardware) expected that it would do so. This intuition seems to be supported by recent formal analyses by Michael Katz and Carl Shapiro (1983), and Ward Hanson (1984), of markets where purchasers of rival products benefit from externalities conditional upon the size of the compatible system or "network" with which they thereby become joined. Although the initial lead acquired by QWERTY through its association with the Remington was quantitatively very slender, when magnified by expectations it may well have been quite sufficient to guarantee that the industry eventually would lock in to a de facto QWERTY standard.

The occurrence of this "lock in" as early as the mid-1890's does appear to have owed something also to the high costs of software "conversion" and the resulting *quasi-irreversibility of investments* in specific touch-typing skills. Thus, as far as keyboard conversion costs were concerned, an important asymmetry had appeared between the software and the hardware components of the evolving system: the costs of typewriter software conversion were going up, whereas the costs of typewriter hardware conversion were coming down. While the novel, non-typebar technologies developed during the 1880's were freeing the keyboard from technical bondage to QWERTY, typewriter makers were by the same token freed from fixed-cost bondage to any particular keyboard arrangement. Non-QWERTY typewriter manufacturers seeking to expand market share could cheaply switch to achieve compatibility with the already existing stock of QWERTY-programmed typists, who could not. This, then, was a situation in which the precise details of timing in the developmental sequence had made it privately profitable in the short run to adapt machines to the habits of men (or to women, as was increasingly the case) rather than the other way around. And things have been that way ever since.

III. Message

In place of a moral, I want to leave you with a message of faith and qualified hope. The story of QWERTY is a rather intriguing one for economists. Despite the presence of the sort of externalities that standard static analysis tells us would interfere with the achievement of the socially optimal degree of system compatibility, competition in the absence of perfect futures markets drove the industry prematurely into standardization *on the wrong system*—where decentralized decision making subsequently has sufficed to hold it. Outcomes of this kind are not so exotic. For such things to happen seems only too possible in the presence of strong technical interrelatedness, scale economies, and irreversibilities due to learning and habituation. They come as no surprise to readers prepared by Thorstein Veblen's classic passages in *Germany and the Industrial Revolution* (1915), on the problem of Britain's undersized railway wagons and "the penalties of taking the lead" (see pp. 126–27); they may be painfully familiar to students who have been obliged to assimilate the details of deservedly less-renowned scribblings (see my 1971, 1975 studies) about the obstacles which ridge-and-furrow placed in the path of British farm mechanization, and the influence of remote events in nineteenth-century U.S. factor price history upon the subsequently emerging bias towards Hicks' labor-saving improvements in the production technology of certain branches of manufacturing.

I believe there are many more QWERTY worlds lying out there in the past, on the very edges of the modern economic analyst's tidy universe; worlds we do not yet fully perceive or understand, but whose influence, like that of dark stars, extends nonetheless to shape the visible orbits of our contemporary economic affairs. Most of the time I feel sure that the absorbing delights and quiet terrors of exploring QWERTY worlds will suffice to draw adventurous economists into the systematic study of essentially historical dynamic processes, and so will seduce them into the ways of economic history, and a better grasp of their subject.

REFERENCES

Arthur, W. Brian, "On Competing Technologies and Historical Small Events: The Dynamics of Choice Under Increasing Returns," Technological Innovation Program Workshop Paper, Department of Economics, Stanford University, November 1983.

Arthur, W. Brian, Ermoliev, Yuri M. and Kaniovski, Yuri M., "On Generalized Urn Schemes of the Polya Kind," *Kibernetika*, No. 1, 1983, *19*, 49–56 (translated from the Russian in *Cybernetics*, 1983, *19*, 61–71).

______, ______, **and** ______, "Strong Laws for a Class of Path-Dependent Urn Processes," in *Proceedings of the International Conference on Stochastic Optimization, Kiev*, Munich: Springer-Verlag, 1985.

David, Paul A., "The Landscape and the Machine: Technical Interrelatedness, Land Tenure and the Mechanization of the Corn Harvest in Victorian Britain," in D. N.

McCloskey, ed., *Essays on a Mature Economy: Britain after 1840*, London: Methuen, 1971, ch. 5.

______, *Technical Choice, Innovation and Economic Growth: Essays on American and British Experience in the Nineteenth Century*, New York: Cambridge University Press, 1975.

Hanson, Ward A., "Bandwagons and Orphans: Dynamic Pricing of Competing Technological Systems Subject to Decreasing Costs," Technological Innovation Program Workshop Paper, Department of Economics, Stanford University, January, 1984.

Katz, Michael L. and Shapiro, Carl, "Network Externalities, Competition, and Compatibility," Woodrow Wilson School Discussion Paper in Economics No. 54, Princeton University, September, 1983.

Veblen, Thorstein, *Imperial Germany and the Industrial Revolution*, New York: MacMillan, 1915.

Technology Adoption in the Presence of Network Externalities

Michael L. Katz and Carl Shapiro

Princeton University

We analyze technology adoption in industries where network externalities are significant. The pattern of adoption depends on whether technologies are sponsored. A sponsor is an entity that has property rights to the technology and hence is willing to make investments to promote it. Key findings include the following: (1) compatibility tends to be undersupplied by the market, but excessive standardization can occur; (2) in the absence of sponsors, the technology superior *today* has a strategic advantage and is likely to dominate the market; (3) when one of two rival technologies is sponsored, that technology has a strategic advantage and may be adopted even if it is inferior; (4) when two competing technologies both are sponsored, the technology that will be superior *tomorrow* has a strategic advantage.

I. Introduction

The benefit that a consumer derives from the use of a good often depends on the number of other consumers purchasing compatible items. Consider, for example, the owner of a videocassette recorder (VCR). At present, there are two competing technologies, beta and VHS. These technologies are incompatible with one another; cassettes produced for use on one type of VCR cannot be played on the other type. Because of the increasing returns to scale in the provision of the complementary good (rental services of prerecorded cassettes),

This material is based on work supported by the National Science Foundation under grant nos. SES-8209412 (Katz) and SES-8408622 (Shapiro). We thank Gene Grossman and an anonymous referee for valuable comments on an earlier draft.

[*Journal of Political Economy*, 1986, vol. 94, no. 4]

the owner of a VCR will find a greater variety of programming available for his machine if more VCRs are sold using his technology. He also benefits from the ability to exchange cassettes with other users of compatible machines and from superior service that may be available for the VCR technology with the larger in-place base of machines. We call these positive external consumption benefits "network externalities."

Network externalities are significant in many important industries. They have long been recognized in the context of physical networks such as the telephone and railroad industries.[1] Most examples of network externalities entail the provision of a durable good and a complementary good or service. In the computer industry, software and hardware must be used together. The question of compatibility is whether different brands or technologies of the durable good (e.g., hardware) can utilize identical units of the complementary good (e.g., software). In the communications area, all television sets are compatible (i.e., they receive and process the same signals), but incompatible AM stereo technologies are currently competing for survival. In 1950, the Federal Communications Commission (FCC) adopted a color television system that was incompatible with the existing black and white system, although production soon was stopped by an order by the Director of Defense Mobilization.[2] Currently, there is a policy debate over standards for the emerging technology of high-definition television. Any technology requiring specific training is subject to network externalities: the training is more valuable if the associated technology is more widely adopted.

When network externalities are significant, so too are the benefits of having compatible products. There are two ways in which compatibility may be achieved. First, even products utilizing different technologies may be designed to work with one another. For example, the use of standardized interfaces allows a variety of components in a personal computer system to be used together; several different brands of personal computer may utilize a given brand of printer. In a variety of manufacturing industries, industrywide standards are designed to encourage compatibility.[3] Of course, producing compatible designs requires interfirm cooperation and may raise production costs. In Katz and Shapiro (1985, 1986), we study the private and social incentives to achieve technical compatibility, or standardization.

[1] Carlton and Klamer (1983) discuss compatibility in these industries as well as in the electronic funds transfer industry. Rohlfs (1974) and Oren and Smith (1981) analyze network externalities in the context of a monopoly telecommunications network.

[2] We are grateful to Stanley Besen for bringing this example to our attention.

[3] The American National Standards Institute establishes domestic standards and represents the United States in the International Organization for Standardization.

For some products (e.g., VCRs), the competing technologies may be inherently incompatible. In such cases, there is another way to enjoy the full benefits of network externalities: achieve de facto standardization by having all consumers purchase the same technology. The adoption of the QWERTY keyboard arrangement in the typewriter industry followed this pattern of standardization-by-sheer-force-of-numbers.[4] The costs of this type of standardization may be twofold. First, there may be a loss of variety, reducing the flow of services generated by the product. Computer users, for example, point out that Apple computers, while incompatible with IBM machines, perform a number of tasks that IBM-compatible machines cannot. These services would be lost if only IBM-compatible technologies were manufactured and consumed. Second, the costs of each technology may vary across consumers, so that, if only one technology is offered, some consumers are forced to purchase what is for them the more expensive technology. Cost differences across consumers are commonplace when there is technological progress and consumers purchase at different dates.

In this paper, we study the dynamics of industry evolution in a market with technological change where there are two inherently incompatible technologies subject to network externalities. Network externalities have two fundamental effects on the dynamics of industry evolution. First, the relative attractiveness today of rival technologies is influenced by their sales histories. In effect, there are "demand-side economies of scale"; a given product is more attractive the larger is the in-place base of consumers using that product. Second, and perhaps more important, in the presence of network externalities, a consumer in the market today also cares about the *future* success of the competing products. As long as the good is durable (or there is technology-specific human capital investment), the total benefits derived from it will depend, in part, on the number of consumers who adopt compatible products in the future.

Network externalities share the first type of increasing return to scale with learning by doing.[5] The second source of demand-side scale economies is, however, peculiar to industries with network externalities. As a result of these two effects, the dynamics of industries subject to network externalities are fundamentally different from those of conventional industries.

We examine two basic questions regarding the process of industry evolution: (1) Will the market achieve de facto standardization? (2)

[4] David (1985) provides a very entertaining description of how the QWERTY standard was adopted.

[5] Arthur (1985) analyzes technology choice in the presence of embodied technical progress, i.e., learning by doing.

When the market does choose a standard, is the choice the socially optimal one? The answers to these questions depend on whether there are well-defined property rights to the technologies.

In the absence of well-defined property rights, free entry into the supply of a technology will lead to marginal cost pricing.[6] In the light of the external benefits, one would expect the competitive equilibrium to be inefficient, and our formal results confirm this intuition. Distortions arise because each consumer ignores the network effects on other consumers when making his consumption decision. Hence, the market is biased toward nonstandardization. When standardization does occur, the wrong standard may be chosen. Our analysis supports the conventional view that the technology that is superior today has a strategic, first-mover advantage: it can become locked in as the standard.

If a single firm controls the property rights to a given technology or if there are other entry barriers into the supply of that technology, then a supplier will be willing to make investments in the form of penetration pricing to establish the technology because such investments can later be recouped by pricing in excess of marginal costs. We call such a firm a "sponsor." In the VCR industry, each of the two competing technologies is sponsored because patent protection prevents free entry into the VCR market. In the typewriter industry, on the other hand, property rights to specific keyboard designs were not established, and there were no sponsors in our sense of the term.

Sponsorship can internalize some of the externalities through below-cost pricing at the beginning of a technology's life. But sponsorship can create problems of its own. When one technology is sponsored and the other is not, the sponsored technology tends to be adopted too much. The market outcome still may entail splitting in cases in which standardization is socially optimal. But the market outcome also may entail excessive standardization or standardization on the wrong technology. In fact, we find that a sponsored technology may dominate the market even when *all* consumers agree that a rival, nonsponsored technology is superior.

For the case of two sponsors, the most dramatic effect of sponsorship is that a second-mover advantage arises. The technology that will be superior tomorrow has a strategic advantage. If consumers are forward looking (i.e., form rational expectations about future sales), we obtain exactly the opposite of the view that industries become locked in to older, inferior technologies. In a sense, the market exhibits "excessive foresight."

[6] Farrell and Saloner (1985) study the evolution of an industry in which one technology initially has been adopted and neither technology is proprietary.

In the following section we outline our formal model. In Section III we study the pattern of adoption that would be induced if a welfare maximizer chose all prices. The bulk of our analysis is in Section IV, where we describe market equilibria for several different patterns of sponsorship and compare the market outcomes with the socially optimal ones. Section V discusses the time consistency constraint faced by firms in our model and is followed by a brief discussion of the directions in which to extend the analysis.

II. Technology Choice by Consumers

We first develop the demand side of a simple model of technology choice in the presence of network externalities. We show how consumers respond to a given time path of prices. The model is kept as simple as possible in order to focus most clearly on the effects of network externalities and sponsorship.

There are two time periods, or generations of consumers, $t = 1, 2$, with N_t consumers in period t. A period t consumer has a completely inelastic demand for one unit of the good in period t. There are two technologies, A and B, from which to choose. The prices of technologies A and B in period t are denoted by p_t and q_t, respectively. In this section we take all prices as known and exogenously given. Let δ_t denote the discount for technology B (relative to technology A) during period t: $\delta_t \equiv p_t - q_t$.

The benefits that a consumer derives from consumption of one unit of the good depend on how many other consumers ultimately purchase compatible units, that is, units utilizing the same technology. In other words, the extent of consumption externalities depends only on the final network sizes. Let x_t and y_t denote the quantities of technologies A and B, respectively, that are sold in period t. Given our assumption that all period t consumers make purchases, $y_t = N_t - x_t$. A consumer who purchases technology A in period t derives gross benefits of $v(x_1 + x_2)$ and net benefits (or surplus) of $v(x_1 + x_2) - p_t$. The corresponding values for a consumer who purchases technology B in period t are $v(y_1 + y_2)$ and $v(y_1 + y_2) - q_t$.[7] By having gross consumption benefits depend only on the number of consumers purchasing the *same* technology, we are assuming that the two technologies are incompatible. Consumers in our model are homogeneous in that all of them have the same benefit function $v(\cdot)$.

First-period consumers rationally forecast second-period sales in order to make their purchase decisions. First-period consumers rec-

[7] As we shall see, allowing first-period consumers to derive intermediate benefits, i.e., have benefit functions of the form $u(x_1) + v(x_1 + x_2)$, would not change our results.

ognize that the consumption decisions of second-period buyers must be optimal given the pattern of first-period purchases (i.e., we look for a perfect equilibrium). Hence, we begin our analysis by considering the purchasing decisions of the N_2 consumers who choose technologies in period 2. We look for a Nash equilibrium in second-period technology choices, where each second-period consumer makes his technology choice taking the purchasing decisions of all other (first- and second-period) consumers as given. For a network to have positive second-period sales in equilibrium it must be the case that no consumer purchasing that technology wants to switch to the other technology. For technology A, this condition is $v(x_1 + x_2) - p_2 \geq v(y_1 + y_2 + 1) - q_2$, or $v(x_1 + x_2) - v(y_1 + y_2 + 1) \geq \delta_2$. For technology B, the condition is $v(x_1 + x_2 + 1) - v(y_1 + y_2) \leq \delta_2$.

Given that $v(\cdot)$ is an increasing function, these two conditions cannot be satisfied simultaneously. Therefore, one technology or the other will dominate in the second period: either $x_2 = 0$ or $x_2 = N_2$.[8] A similar argument shows that all first-period consumers will purchase the same technology as one another.[9]

Suppose that all of the first-period consumers have chosen technology A. Then an equilibrium exists in which all second-period consumers choose A if and only if $v(N_1 + N_2) - v(1) \geq \delta_2$. An equilibrium exists with all second-period consumers purchasing technology B if and only if $v(N_1 + 1) - v(N_2) \leq \delta_2$. These last two inequalities may be satisfied simultaneously.

Given that all period 2 consumers are identical, they will agree on which outcome yields higher consumer surplus. We assume that, when there are multiple equilibria, consumers buying in the same period can coordinate their purchase decisions to choose the outcome that is Pareto preferred. When the two outcomes are compared, second-period consumers will match first-period consumers and adopt technology A if and only if

$$v_b - v_2 \geq \delta_2, \tag{1}$$

where we have adopted the notation $v_b \equiv v(N_1 + N_2)$, with b for both, and $v_t \equiv v(N_t)$, $t = 1, 2$. The second-period outcome depends on a comparison of technology A's installed-base advantage, $v_b - v_2$, and technology B's price advantage, δ_2.

[8] If consumers ignored their influence on network size (i.e., if there were a continuum of consumers), then an unstable interior equilibrium might exist. Of course, our strong within-period bandwagon results rely on consumer homogeneity.

[9] It is for this reason that intermediate benefits do not alter the analysis. They would not affect the comparison between technologies since the intermediate benefits would be $v(N_1)$ in all cases. Allowing for intermediate benefits only strengthens the tendency of first-period consumers to cluster on one technology.

It can be shown in a similar fashion that, if first-period consumers all choose technology B, the second-period consumers will match if and only if

$$v_b - v_2 \geq -\delta_2. \tag{2}$$

Now consider first-period consumers. Each first-period consumer makes his technology choice taking the purchase decisions of other first-period consumers as given and having calculated the resulting second-period equilibrium purchasing decisions. If inequalities (1) and (2) both hold, that is, $|\delta_2| \leq v_b - v_2$, then first-period consumers know that, no matter which technology they choose, second-period consumers will follow. In this case, first-period consumers choose the standard by comparing $v_b - p_1$ and $v_b - q_1$. Technology A will be chosen if and only if $p_1 \leq q_1$, that is, $\delta_1 \leq 0$.

When (1) holds but (2) does not, that is, $\delta_2 < v_2 - v_b$, second-period consumers will purchase technology A no matter what first-period consumers choose to do. Thus first-period consumers compare $v_b - v_1$ with δ_1. Finally, when (1) is violated but (2) holds, that is, $\delta_2 > v_b - v_2$, second-period consumers will choose technology B no matter what first-period consumers do. In this case, first-period consumers compare $v_1 - v_b$ with δ_1. Figure 1 graphs the consumer equilibrium as a function of the price differences, δ_1 and δ_2. The notation AB in a region of the figure, for example, indicates that first-period consumers adopt technology A and second-period consumers choose technology B if δ_1 and δ_2 lie in the relevant region.

III. Second-best Pricing

We turn now to an analysis of the socially optimal price paths for the two technologies. We take total surplus as our welfare measure.[10] If a perfectly informed central planner had the power to order any given consumer to choose a particular technology, then there would be cases in which it would be socially optimal to have consumers making purchases within a single period choose different technologies.[11] We consider a more limited planner who can set prices in each period (using taxes and subsidies) but cannot mandate adoption. The analy-

[10] We do not address the interesting question whether an actual planner would, in fact, maximize total surplus. One might expect that the government would instead serve only the current generation of voters.

[11] For example, suppose that technology B is prohibitively expensive in period 1 but is cheaper in period 2. It may be socially optimal for most period 2 consumers to use B but for some to use A in order to increase the period 1 consumers' benefits. This pattern of splitting is most likely to be optimal if $N_2 > N_1$ and if there is a threshold network size (larger than N_1) at which $v(\cdot)$ increases sharply.

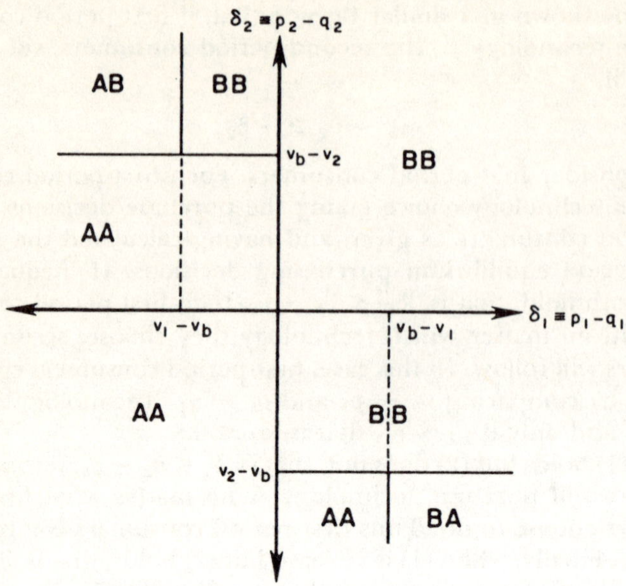

FIG. 1.—Technology choice by consumers

sis of the previous section shows that such a planner is restricted to choosing a single technology for all the consumers in any one period.

When consumers all agree on which technology is better (i.e., cheaper), then clearly the optimum entails all consumers adopting that technology. When the relative costs of the two technologies shift over time, one must compare the production cost savings of heterogeneous products with the loss of network externalities in deciding whether to standardize. Let c_t and d_t denote the constant marginal costs of technologies A and B, respectively, in period t. The costs are taken to be exogenously given and known to all agents. Let Δ_t denote technology B's cost advantage at time t: $\Delta_t \equiv c_t - d_t$.

Consider the case in which technology A is cheaper than B in the first period but more expensive in the second period: $\Delta_1 < 0$ and $\Delta_2 > 0$. It is useful to think of technology B as the newer technology; initially it is more expensive, but in time it overtakes its older rival. There are three outcomes to compare. Let W_{AB} denote the total surplus level that obtains when first-period consumers purchase technology A and second-period consumers purchase technology B. Define W_{AA} and W_{BB} analogously. We then have

$$W_{AA} = (N_1 + N_2)v_b - c_1N_1 - c_2N_2, \tag{3}$$

$$W_{BB} = (N_1 + N_2)v_b - d_1N_1 - d_2N_2, \tag{4}$$

$$W_{AB} = N_1v_1 + N_2v_2 - c_1N_1 - d_2N_2. \tag{5}$$

If all consumers purchase the same technology, the network benefits are identical whether the purchase pattern is AA or BB. The socially optimal decision is made on the basis of which technology is cheaper on average:

$$W_{BB} - W_{AA} = \Delta_1 N_1 + \Delta_2 N_2. \tag{6}$$

Suppose that $W_{AA} > W_{BB}$. There still is a question whether it is socially optimal to split consumers across the two technologies. Subtracting equation (5) from equation (3) yields

$$W_{AA} - W_{AB} = (N_1 + N_2)v_b - (N_1 v_1 + N_2 v_2) + d_2 N_2 - c_2 N_2.$$

Hence W_{AA} is greater than W_{AB} if and only if

$$\left(\frac{N_1}{N_2}\right)(v_b - v_1) + (v_b - v_2) > \Delta_2. \tag{7}$$

The right-hand side of (7) is the period 2 per capita advantage of splitting consumers across technologies—second-period consumers can purchase the technology that is cheaper at that time. The left-hand side of (7) measures the disadvantage of splitting—consumers are divided between incompatible technologies, reducing the extent to which the network externalities are enjoyed.

A similar line of argument shows that $W_{BB} > W_{AB}$ if and only if

$$\left(\frac{N_2}{N_1}\right)(v_b - v_2) + (v_b - v_1) > -\Delta_1. \tag{8}$$

Recall that Δ_1 is negative.

When the cost difference between the two technologies is near zero in at least one period, at least one of inequalities (7) and (8) is satisfied, and standardization is socially optimal. When the cost differences are large in both periods, splitting is optimal. Figure 2 illustrates the socially optimal outcome for all possible values of Δ_1 and Δ_2. In drawing figure 2, we have made use of the fact that the social problem is completely symmetric. The same type of analysis that we have applied to $\Delta_1 < 0$ and $\Delta_2 > 0$ also holds for the case in which $\Delta_1 > 0$ and $\Delta_2 < 0$.

IV. Industry Evolution and Technology Sponsorship

A. *Unsponsored Technologies*

We turn now to examination of the market outcome. Initially, we consider the case of competition between unsponsored technologies. In other words, there is free entry into the supply of products em-

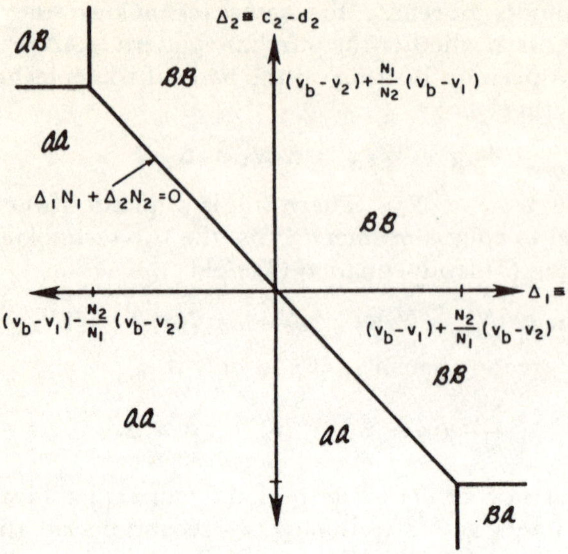

FIG. 2.—Socially optimal technology choice

bodying each technology. Thus each technology is offered at its marginal cost during each period: $p_t = c_t$ and $q_t = d_t$ for $t = 1, 2$.

Taking $\delta_t \equiv \Delta_t$, figure 1 shows that, if one technology is lower cost in both periods, then, as is socially optimal, all consumers purchase that technology. The question of technology choice is richer when the consumers disagree about which technology is the more desirable one. Given the network externalities, each consumer wants all other consumers to purchase his favored technology.

Because each consumer group ignores the negative effects on the other when choosing not to standardize, the private outcome may induce "stranding" or nonstandardization when standardization would be optimal. Figure 3 compares the market outcome and the socially optimal outcome. The diagonally shaded region in the figure represents cases in which there is excessive nonstandardization. Second-period consumers ignore any adverse effects that their technology choice may have on first-period consumers. Similarly, first-period consumers ignore the adverse effects that nonmatching has on later consumers and may choose a technology despite knowing that it will induce stranding.

When standardization does occur, the socially optimal technology may not be selected; there are biases in the market's choice of a standard. The technology that is cheaper in period 1 has a first-mover advantage. When $|\Delta_t| < v_b - v_t$ for both periods, either group of

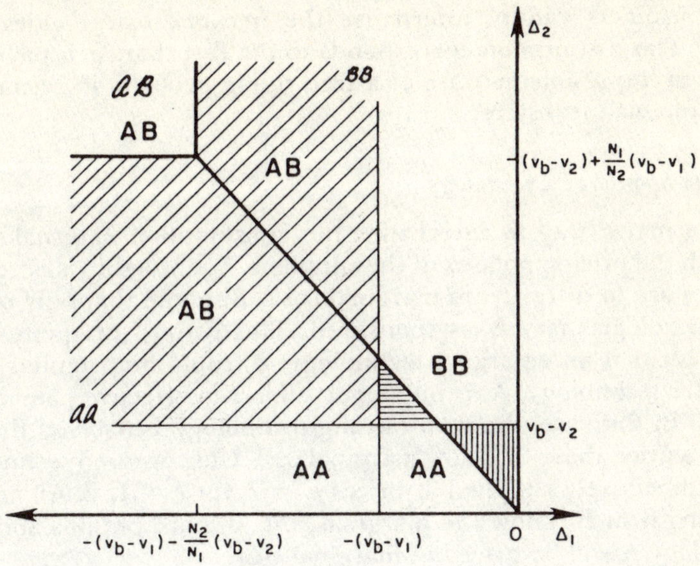

FIG. 3.—Equilibrium vs. optimum: no sponsors. The script characters denote the socially optimal outcomes, with the regions divided by the thick lines. The block characters and thin lines refer to market equilibrium outcomes. The same convention applies to figs. 5 and 7.

consumers would match the other. First-period consumers choose first and lock in their preferred technology, even when it is socially optimal to standardize on the other technology. The range of costs over which this misallocation occurs is given by the vertically shaded area in figure 3. This analysis shows the often-noted tendency to standardize on the technology that initially is superior even though it is the "wrong" technology. Our model shows that this outcome need not be the result of irrationality or impatience.

There is a second bias in the choice of standard. The preferred technology of the larger consumer group can prevail even when it is socially optimal for the preferred technology of the smaller group to do so. The larger group can more easily go it alone. Thus there are cases in which the smaller group, knowing that the larger group will go its own way, caves in and adopts its own less preferred technology. In this way, even the second-period consumers can act as leaders if they are more numerous. In the horizontally shaded region of figure 3 (which is drawn for $N_2 > N_1$), period 2 consumers' preferred technology, B, is adopted as the standard, although total benefits would be larger if A were the standard.

Note that in this subsection, and throughout our analysis, we are assuming that intergenerational transfers are not feasible and hence that consumers cannot internalize the network externalities over time.[12] This assumption corresponds to the fact that such payments are often impossible because of a free-riding problem in raising the funds for such transfers.

B. One Sponsored Technology

An alternative way to internalize the consumption externalities is through the pricing policies of the suppliers. Sponsorship is necessary if prices are to differ from marginal costs. Suppose that only one of the technologies (say B) is sponsored. This pattern of sponsorship would occur if an emerging technology, B, could be patented while the older technology, A, is no longer subject to effective patent protection. In this case, the firm owning technology B (labeled firm B) can set a price above or below marginal cost. Unsponsored technology A is competitively supplied at prices $p_t = c_t$ for $t = 1, 2$. All agents, including firm B, know the marginal costs in both periods and that technology A will be priced at marginal cost.

As usual, we begin the analysis with the second period. The technology adopted by first-period consumers has an installed-base advantage of $v_b - v_2$. Technology B has a (possibly negative) cost advantage of Δ_2. The technology with the net advantage will prevail in the second period. If first-period consumers adopted technology A, then B will win in the second period if and only if $\Delta_2 > v_b - v_2$. If B won in the initial period, then it will also win in the second period if and only if $\Delta_2 \geq v_2 - v_b$.

Now consider the first period. First-period consumers must project second-period sales in making their technology choice. There are several cases to consider. First, suppose that technology B is sufficiently superior that it will be chosen by second-period consumers even if all first-period consumers adopt technology A (i.e., $\Delta_2 > v_b - v_2$). In making their purchases, first-period consumers compare $v_1 - c_1$ with $v_b - q_1$. By setting $q_1 = v_b - v_1 + c_1$, firm B can win both periods' sales and earn total profits of

$$N_1(v_b - v_1 + \Delta_1) + N_2(v_b - v_2 + \Delta_2), \tag{9}$$

where the second-period price is $v_b - v_2 + c_2$. If, instead, firm B loses the first-period competition, it earns total profits (from second-period sales) of

$$N_2(v_2 - v_b + \Delta_2). \tag{10}$$

[12] This assumption is consistent with our earlier assumption that consumers purchasing at a given date coordinate on their Pareto-preferred equilibrium. Unlike intertemporal coordination, within-period coordination requires no transfers since all consumers purchasing at a given date have identical interests.

Comparing expressions (9) and (10), we find that when $\Delta_2 > v_b - v_2$, technology B wins the first-period (as well as second-period) sales if and only if

$$N_1(v_b - v_1 + \Delta_1) + 2N_2(v_b - v_2) \geq 0. \quad (11)$$

When firm B wins in the first period, its price is $q_1 = v_b - v_1 + c_1$. This price may be below its marginal cost, d_1. Firm B is willing to make below-cost sales to first-period consumers because their presence on its network raises the value of technology B to second-period consumers. The sponsor can internalize the external benefits generated when first-period consumers choose the sponsor's technology.

Unlike the first-period price, the second-period price *cannot* fall below marginal cost. Firm B cannot credibly promise first-period consumers that it will subsidize sales in the second period. Thus, in the absence of the ability to commit itself to future subsidies, the sponsor may be unable to internalize the external benefits accruing to first-period consumers.

We turn now to a second case, in which firm B wins the second-period sales if and only if it wins the first-period sales (i.e., $v_2 - v_b \leq \Delta_2 \leq v_b - v_2$). If firm B sets the maximal prices that allow it to win in both periods, its profits are

$$N_1\Delta_1 + N_2(v_b - v_2 + \Delta_2). \quad (12)$$

As long as these profits are positive, selling in both periods is superior to dropping out. Hence, in equilibrium all consumers will purchase technology B if and only if the expression in (12) is positive. Surprisingly, this expression may be positive even when technology B is more costly in *both* periods (i.e., $\Delta_1 < 0$ and $\Delta_2 < 0$). The reason is that firm B is willing to make investments in its technology in the form of penetration pricing in order to build up a customer base, while firms selling the unsponsored technology A are unwilling to do so. Given that firm B has a base of first-period customers, it has an advantage over technology A in the second period, and firm B can earn a return on its investment by making second-period sales at a price above its marginal cost.

The market outcome is illustrated in figure 4.[13] Since only one technology is sponsored, the figure is not symmetric. Because of its ability to invest in a customer base, the sponsored technology B is more likely to be adopted.

[13] The figure uses the fact that, when technology A wins the second-period sales no matter what choice first-period consumers make (i.e., when $\Delta_2 < v_2 - v_b$), first-period consumers compare $v_b - c_1$ with $v_1 - q_1$. In this case, technology B will have positive first-period sales if and only if $\Delta_1 > v_b - v_1$.

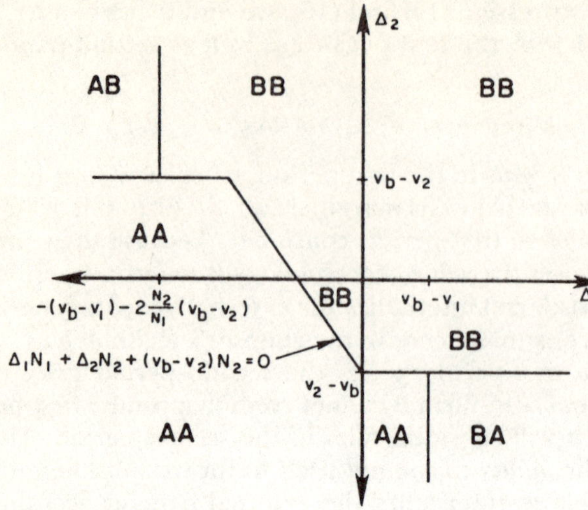

FIG. 4.—Adoption with a single sponsor (B)

The market and the welfare-optimal outcomes are compared in figure 5. With a single sponsor, the market equilibrium still may entail excessive nonstandardization, as is illustrated in the vertically shaded region of the figure; the sponsor ignores period 1 consumers when it wins in period 2. The sponsor internalizes only some of the externalities among consumers. The market outcome also may entail standardization on the sponsored technology in markets where nonstandardization is optimal. This occurs in the cross-hatched area in figure 5.[14]

More generally, there is a bias toward the sponsored technology. In the diagonally shaded region of figure 5, the sponsored technology is adopted although the social optimum entails adoption of the other technology. The most striking case of bias is the one in which technology B wins in both periods even though it has higher costs in each one. The bias is a manifestation of the fact that the sponsor overpromotes its technology. The reason is the "weakened-rival effect": by winning in period 1, firm B faces a rival that is less attractive to consumers in period 2, and thus firm B can appropriate more of the second-period consumer surplus. This transfer is a private, but not a social, benefit of B's winning in the first period.

[14] When the sponsored technology cannot win in the second period ($\Delta_2 \leq v_2 - v_b$), the market may again involve excessive stranding (see the horizontally shaded portion of fig. 5). It is also possible that the *un*sponsored technology is selected as the standard when the sponsored technology would generate higher benefits (the shaded area in fig. 5). Both of these market biases would be corrected if the sponsor were able to commit to second-period subsidies.

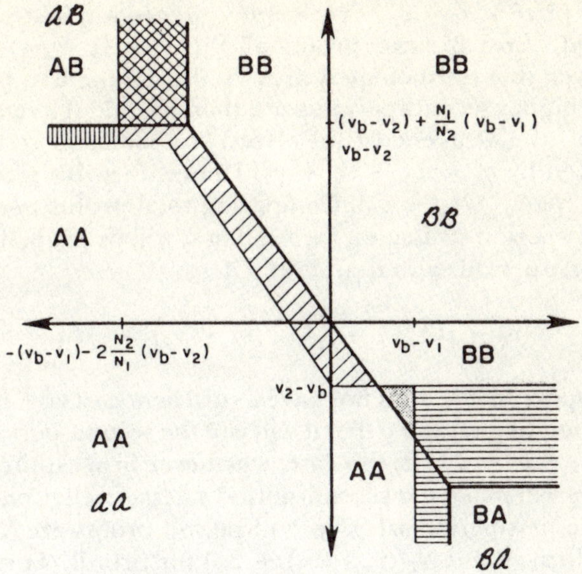

FIG. 5.—Equilibrium vs. optimum with a single sponsor (B)

C. *Both Technologies Sponsored*

When the technologies competing for adoption are both new, each one is likely to enjoy patent protection and hence be sponsored. In this case, the problem is once again symmetric. Hence, we can assume without loss of generality that $\Delta_2 \geq 0$, so that technology B is the cheaper one in period 2.

Consider competition in the second period given first-period sales. The firms play Nash in second-period prices, so either firm is willing to go down as far as its marginal costs in order to undercut its rival. The technology that can offer second-period consumers the greater level of surplus wins all second-period sales and is priced so that it marginally beats the losing one.

Technology B's second-period cost advantage may be so large that it will be chosen even if technology A is priced at cost in the second period and all first-period consumers purchase A (i.e., $\Delta_2 > v_b - v_2$). Consider first-period competition in this case. If technology A wins the first-period sales, then in the second period the maximal price that B can charge satisfies $v_b - c_2 = v_2 - q_2$, or $q_2 = v_2 - v_b + c_2$. Firm B will earn total profits of $N_2[\Delta_2 - (v_b - v_2)]$. Alternatively, firm B can undercut firm A in the first period. First-period consumers recognize that second-period consumers will purchase technology B. Thus first-period consumers compare $v_1 - p_1$ and $v_b - q_1$. Using the fact that firm A is willing to price as low as c_1 to win first-period sales, firm B must set its price at $q_1 = v_b - v_1 + c_1$ in order to win first-period sales. In this event, firm B earns profits of $N_1(\Delta_1 + v_b - v_1)$ in the first period. Given that technology B prevails during the first period, firm B can set a higher second-period price than it could if it had lost in the first period (the weakened-rival effect). Its maximal second-period price is given by $q_2 = v_b - v_2 + c_2$. Hence, firm B's second-period profits are $N_2(\Delta_2 + v_b - v_2)$. Comparing total profits under each of the two first-period strategies, we see that it will be profitable for firm B to win first-period sales if and only if

$$(v_b - v_1) + \left(\frac{2N_2}{N_1}\right)(v_b - v_2) > -\Delta_1. \tag{13}$$

Now suppose that B does not have a sufficient cost advantage in the second period to guarantee that it will win the second-period competition (i.e., $\Delta_2 \leq v_b - v_2$). In this case, whichever firm captures the first-period sales will make the second-period sales as well. Conditional on winning the first-period sales, second-period profits are $N_2[(v_b - v_2) - \Delta_2]$ for firm A and $N_2[(v_b - v_2) + \Delta_2]$ for firm B. Turning to the first period, we investigate the lowest first-period price that would be profitable for each technology. If firm A loses the first-period competition, then it earns no profits in either period. If firm A wins the first-period sales at price p_1, its profits are

$$N_1(p_1 - c_1) + N_2(v_b - v_2 - \Delta_2). \tag{14}$$

Solving (14) for the price that yields zero profits, we obtain

$$\hat{p}_1 = c_1 - \left(\frac{N_2}{N_1}\right)(v_b - v_2 - \Delta_2). \tag{15}$$

Similar calculations show that the minimal price at which firm B would seek first-period sales is

$$\hat{q}_1 = d_1 - \left(\frac{N_2}{N_1}\right)(v_b - v_2 + \Delta_2). \tag{16}$$

The firm that has the lower minimal price is the one that will serve all consumers in both periods. Comparing equations (15) and (16), we see that firm B will win if and only if

$$N_1\Delta_1 + 2N_2\Delta_2 > 0. \tag{17}$$

The equilibrium market outcome is illustrated in figure 6. Recalling that the social decision rule calls for standardization on B instead of A when $N_1\Delta_1 + N_2\Delta_2 > 0$, we see that there is a "second-mover advantage" among firms. Suppose that $N_1 = N_2$, that firm A has the cheaper technology in period 1, and that firm B has the cheaper technology in period 2, with $0 < -\Delta_1 = \Delta_2$. As long as the cost

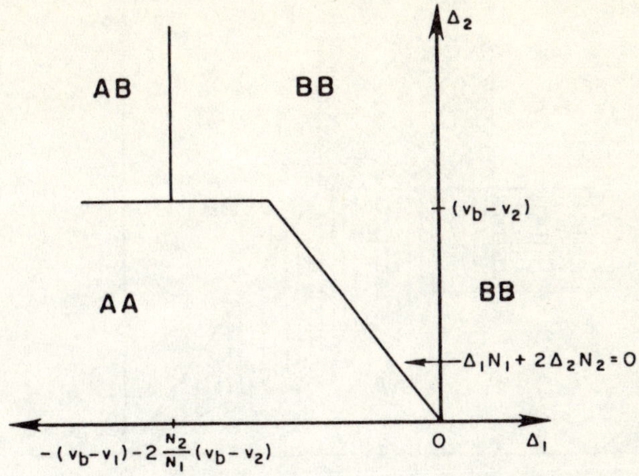

FIG. 6.—Equilibrium adoption with two sponsors

differences are not too large, the market equilibrium entails all consumers' purchasing technology B. The reason is that firm A cannot credibly promise to price below cost in the second period, but firm B *can* price below cost in the first period. Of course, at some point the cost differences become so large that they dwarf the network effects, and consumers in each period choose the technology that is cheaper at the time.

Figure 7 compares the market equilibrium and the second-best outcomes. In the horizontally shaded region of the figure, consumers split their technology choices across periods even though matching would be socially optimal. This excessive splitting occurs only in cases in which the socially optimal pattern of technology choice is AA but the private equilibrium is AB. Because of the timing of play, firm A is unable to internalize the stranding externality that second-period consumers impose on first-period consumers.

With two sponsors, the wrong technology may be chosen when standardization does occur. In the light of the second-mover advantage, we find that the market equilibrium is biased toward adoption of the technology that is cheaper in the second period. The vertically shaded area in figure 7 represents markets in which the socially optimal pattern of technology choice is AA but the market outcome entails BB. Firm B is able to win in the first period despite its higher first-period costs because consumers expect firm A to drop out of the market in period 2, even if it wins in the first period. In the first period, firm A cannot credibly promise to make second-period sales. Firm B's incentives to win in the first period are further strengthened by the weakened-rival effect. Hence, there is excessive choice of technology B in the first period.

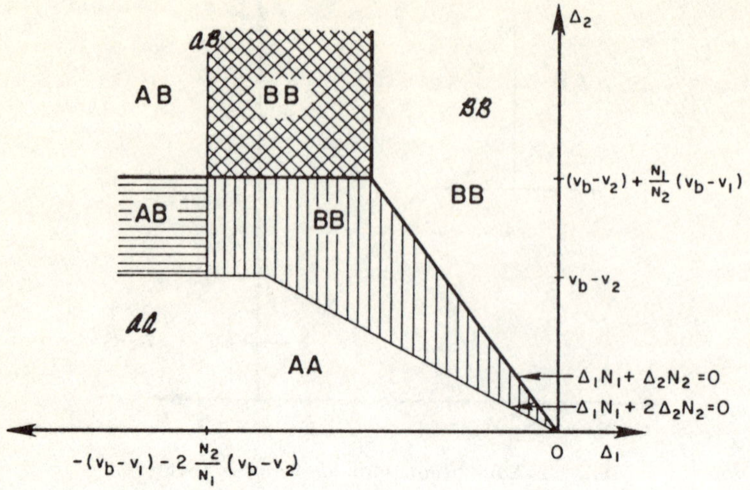

FIG. 7.—Equilibrium vs. optimum with two sponsors

This tendency to standardize on the emerging technology is the opposite of the story one typically hears about industry evolution. With typewriter keyboards, for example, the technology that was "cheaper" in the first period won, to the detriment of later consumers.[15] The typewriter example conforms to our first-mover advantage in the absence of sponsors; there was no sponsor for any given keyboard. Our analysis here suggests that rather different dynamics may arise in the presence of well-defined property rights.

V. The Durable Goods Monopolist on His Head

Suppose now that demands in the second period are elastic, and consider the case of a single sponsor, say firm B. If firm B has market power in the second period, first-period consumers rationally will expect B to exercise this market power, giving rise to a smaller expected network than if marginal cost pricing were practiced in the second period.[16] One can show by example that, in the first period, the sponsored technology may be at a disadvantage relative to the unsponsored one because the latter can credibly commit to marginal cost pricing. This effect is simply a variant of the time consistency

[15] "Cost" refers here to the hedonic cost, where adjustment is made for ease of use. The first period corresponds to early mechanical machines that were subject to jamming. The second period corresponds to modern, nonjamming equipment operated by touch-typists. For the latter, the Dvorak keyboard is superior to QWERTY.

[16] This point is one example of the fact that sponsorship creates market power, which leads to a resource misallocation when demands are elastic.

problem that gave rise to a second-mover advantage in the case of two sponsored technologies that we analyzed above.

The sponsor's problem is the opposite of the durable good monopolist's problem, as posed by Coase (1972) and studied further by Bulow (1982). A durable good monopolist cannot commit itself to restricting future production and maintaining its price over time. Consumers anticipate the falling prices and thus are willing to pay less for early units since their asset value is lower. In our model, the link between prices in different periods comes through the positive consumption externalities. Here, early consumers benefit from lower prices and larger sales in later periods. Thus the time consistency problem for the sponsor is that it cannot commit to a second-period price other than the "high" one that will maximize profits at the time.

There may be mechanisms by which a sponsor can commit himself to setting low second-period prices. For example, the sponsor could make research and development expenditures or physical capital investments that lower its marginal costs. Alternatively, the firm might make its product design public, or the sponsor might advertise in advance that low prices will be offered in the second period to some or all consumers. Given rational first-period consumers, it will be in the firm's interest to limit its own ability to exercise second-period market power.

This point leads to a warning for antitrust enforcement. An industry observer could mistake investments that commit the firm to lower prices for entry deterrence or predation. Similarly, below-cost pricing in the first period need not be indicative of predatory behavior. Finally, our analysis shows that a cartel of producers may be socially desirable to the extent that it allows them to set later prices above marginal cost and thus generates incentives to invest in a new technology in the presence of network externalities. Pricing at the marginal cost of production in each period may not be socially optimal.

VI. Extensions

There are several directions in which to extend our analysis. One could allow for within-period consumer heterogeneity. Within-period heterogeneity differs from our between-period heterogeneity in terms of the timing of purchase decisions; there is no first or second mover within a single period. Within-period heterogeneity gives rise to benefits from variety and to the possibility of multiple equilibria, but the basic forces pushing toward de facto standardization clearly would remain.

Another extension is to consider the case of ex ante uncertainty about the relative attractiveness of competing technologies. There may be uncertainty about second-period costs or about the preferences of consumers over the rival technologies.[17] A consumer purchasing the technology early is unlikely to be certain of future costs or of the adoption decisions by consumers who will make purchases later. This uncertainty may have the greatest effect if consumers can choose when to purchase the technology. With intertemporal substitution, some consumers may choose to wait for cost and demand uncertainty to be resolved before they commit themselves to a specific technology.

References

Arthur, Brian. "Competing Technologies and Lock-in by Historical Small Events: The Dynamics of Allocation under Increasing Returns." Policy paper. Stanford, Calif.: Stanford Univ., Center Econ. Policy Res., January 1985.

Bulow, Jeremy I. "Durable-Goods Monopolists." *J.P.E.* 90 (April 1982): 314–32.

Carlton, Dennis, and Klamer, J. Mark. "The Need for Coordination among Firms, with Special Reference to Network Industries." *Univ. Chicago Law Rev.* 50 (Spring 1983): 446–65.

Coase, Ronald H. "Durability and Monopoly." *J. Law and Econ.* 15 (April 1972): 143–49.

David, Paul A. "Clio and the Economics of QWERTY." *A.E.R. Papers and Proc.* 75 (May 1985): 332–37.

Farrell, Joseph, and Saloner, Garth. "Standardization, Compatibility, and Innovation." *Rand J. Econ.* 16 (Spring 1985): 70–83.

Katz, Michael L., and Shapiro, Carl. "Network Externalities, Competition, and Compatibility." *A.E.R.* 75 (June 1985): 424–40.

———. "Product Compatibility Choice in a Market with Technological Progress." *Oxford Econ. Papers* (1986), in press.

Oren, Shmuel S., and Smith, Stephen A. "Critical Mass and Tariff Structure in Electronic Communications Markets." *Bell J. Econ.* 12 (Autumn 1981): 467–87.

Rohlfs, Jeffrey. "A Theory of Interdependent Demand for a Communications Service." *Bell J. Econ. and Management Sci.* 5 (Spring 1974): 16–37.

[17] In Katz and Shapiro (1986) we examine compatibility choice in a model with uncertainty about the second-period production costs of the two technologies.

[19]

Review of Economic Studies (1981) XLVIII, 395–405 0034-6527/80/00340395$02.00

On the Diffusion of New Technology: A Game Theoretic Approach

JENNIFER F. REINGANUM
California Institute of Technology

1. INTRODUCTION

While much has been written recently on the economics of research and development, most theoretical studies have focused on investment in research when the feasibility and profitability of the innovation may be stochastic (for references see Reinganum (1980*a*, 1980*b*). These authors are especially concerned with the relationship between market structure and the pace of inventive activity. However, the perfection of a new and superior technology confers neither private nor social benefit until that technology is adopted and employed by potential users. In an industry with substantial entry costs, perfection and adoption of an innovation are not necessarily conterminous. While there are many industry-specific and innovation-specific case studies of the diffusion of new technology, the theoretical literature is extremely sparse.

This paper is an attempt at a rigorous (albeit not exceedingly general) analysis of the diffusion of new technology. In particular, consider an industry composed of two firms, each using the current best-practice technology. The firms are assumed to be operating at Nash equilibrium output levels, generating a market price (given demand) and profit allocations. When a cost-reducing innovation is announced, each firm must determine when (if ever) to adopt it, based in part upon the discounted cost of implementing the new technology, and in part upon the behavior of the rival firm. If either firm adopts before the other, it can expect to make substantial profits at the expense of the other firm. On the other hand, the discounted sum of purchase price and adjustment costs may decline with the lengthening of the adjustment period as various quasi-fixed factors become more easily variable. Therefore, although waiting costs the firm more in terms of foregone profits, it may save money on the cost of purchasing the new technology. Thus the firm must weigh the costs and benefits of delaying adoption, as well as take account of its rival's strategic behavior. This problem is formalized and discussed in Section 2. In Section 3 it is shown that there exist two asymmetric Nash equilibria in pure strategies. That is, at a Nash equilibrium, one firm will adopt the innovation at a relatively early date, the other relatively later. Thus even in the case of identical firms and complete certainty, there is a "diffusion" of innovation over time. Section 4 extends the analysis to the case of non-identical firms. In this case, it is shown that a Nash equilibrium exists and is asymmetric. The basic model, though independently developed, closely parallels that of Flaherty (1980). For a discussion of the relationship between this work and that of Flaherty, see Section 4. Section 5 concludes the paper.

2. PROBLEM FORMULATION

Consider a pair of identical firms engaged in a game of duopoly. They are making Nash equilibrium profits[1] of (π_0, π_0) per period. At $t = 0$ a technological improvement designed to reduce production costs is announced and offered for sale.

If firm 1 purchases the new technology before firm 2, then the profit allocations are (π_1, π_2) during each period before firm 2 adopts the innovation. Similarly, if firm 2 adopts first, then the profits are (π_2, π_1) per period until firm 1 adopts. After both firms have adopted the innovation, profits are (π_3, π_3). The industry environment is assumed stationary except for the introduction of the new technology, and no further technical advance is anticipated. For an analysis of a single firm's reaction to anticipated technical advance, see Kamien and Schwartz (1972).

Letting T_1 and T_2 denote the adoption dates of firms 1 and 2, respectively, we can summarize the profit opportunities described above in tabular form.

t	Player 1	Player 2
$0 \leqq t \leqq \min\{T_1, T_2\}$	π_0	π_0
$T_1 \leqq t \leqq T_2$	π_1	π_2
$T_2 \leqq t \leqq T_1$	π_2	π_1
$\infty > t \geqq \max\{T_1, T_2\}$	π_3	π_3

The following assumptions describe the relative magnitudes of the profit rates.

Assumption 1. $\pi_i > 0 \; \forall i \in \{0, 1, 2, 3\}$.

Assumption 2. $\pi_1 > \pi_3 > \pi_2$; $\pi_1 > \pi_0 > \pi_2$.

Assumption 3. $\alpha = \pi_1 - \pi_0 + \pi_2 - \pi_3 > 0$.

Assumption 1 states that in any event, both firms make positive profits. Assumption 2 implies that profits to firm i are greatest when i has adopted the innovation but j has not; next greatest profits occur either when both have adopted or when no firm has yet adopted; finally profit opportunities for i are worst when j has adopted the innovation but i has not. Assumption 3 states that the increment to revenue when one is first exceeds the increment to revenue when one is second. Alternatively, the net value of being first, $\alpha = (\pi_1 - \pi_0) - (\pi_3 - \pi_2)$, is positive.[2]

Define $p(t)$ to be the discounted price of the innovation at t. This may be assumed to include all relevant costs of adjustment. Assume that this function exists and is C^2 for all $t \in [0, \infty)$, though purchases may take place at selected dates rather than continuously.

Let r represent the market rate of interest. The firms' payoffs are given in Definition 1 below.

Definition 1. The payoff to firm 1 is

$$V^1(T_1, T_2) \equiv \begin{cases} g^1(T_1, T_2) & \text{if } T_1 \leqq T_2 \\ g^2(T_1, T_2) & \text{if } T_1 \geqq T_2 \end{cases}$$

where

$$g^1(T_1, T_2) = \int_0^{T_1} \pi_0 e^{-rt} dt + \int_{T_1}^{T_2} \pi_1 e^{-rt} dt + \int_{T_2}^{\infty} \pi_3 e^{-rt} dt - p(T_1)$$

and

$$g^2(T_1, T_2) = \int_0^{T_2} \pi_0 e^{-rt}\, dt + \int_{T_2}^{T_1} \pi_2 e^{-rt}\, dt + \int_{T_1}^{\infty} \pi_3 e^{-rt}\, dt - p(T_1).$$

The payoff to firm 2 is simply $V^2(T_1, T_2) \equiv V^1(T_2, T_1)$. To avoid confusion (and without loss of generality) I will deal almost exclusively with firm 1's decision problem. The corresponding results for firm 2 can be deduced by symmetry.

Notice that while $V^1(T_1, T_2)$ is continuous in T_1 for fixed T_2, it is not differentiable at $T_1 = T_2$. The left-hand derivative at T_2 is

$$g_1^1(T_2, T_2) = (\pi_0 - \pi_1)\, e^{-rT_2} - p'(T_2)$$

while the right-hand derivative at T_2 is

$$g_1^2(T_2, T_2) = (\pi_2 - \pi_3)\, e^{-rT_2} - p'(T_2).$$

By Assumption 3, $g_1^1(T_2, T_2) < g_1^2(T_2, T_2)\ \forall T_2 \in [0, \infty)$.

Assumption 4. Assume $p(t) \geqq 0$ and $p''(t) > r[\pi_1 - \pi_0]e^{-rt}\ \forall t \in [0, \infty)$. Further assume that $\lim_{t\to\infty} p'(t) > 0$.

Since $p(t)$ represents the present value cost of implementing the new technology over the adjustment period t, Assumption 4 states that cost saving can't continue forever. Rather, there is some optimal adjustment period beyond which any further prolongation of the adjustment process begins to increase costs. Without this assumption, firms may postpone adoption forever. Since one is primarily interested in the timing of adoption by strategic players (rather than circumstances under which nonadoption is optimal), Assumption 4 will be maintained throughout. Assumptions 1–3 will also be maintained throughout. One of the versions of Assumption 5 below will be specified as a hypothesis in the results which follow.

Assumption 5a. $p'(0) < \pi_2 - \pi_3$.

Assumption 5b. $p'(0) \geqq \pi_2 - \pi_3$.

Under Assumption 4 it is easy to show that $g^1(T_1, T_2)$ and $g^2(T_1, T_2)$ are strictly concave[3] in T_1 (for fixed T_2). To see this, note that $g_{11}^1 = r(\pi_1 - \pi_0)\, e^{-rt} - p''(t) < 0$ by Assumption 4. Since $g_{11}^2 = r(\pi_3 - \pi_2)\, e^{-rt} - p''(t)$ and $\pi_3 - \pi_2 < \pi_1 - \pi_0$ by Assumption 3, $g_{11}^2 < 0$. In addition, one can prove the following preliminary lemmas.

Lemma 1. *$\exists$ unique $\hat{T} \in [0, \infty)$ and $\hat{\hat{T}} \in [0, \infty)$ which maximize $g^1(T_1, T_2)$ and $g^2(T_1, T_2)$, respectively, independent of T_2. Furthermore*

(a) *if Assumption 5a holds, then $0 \leqq \hat{T} < \infty$ and $0 < \hat{\hat{T}} < \infty$.*

(b) *if Assumption 5b holds, then $\hat{T} = \hat{\hat{T}} = 0$.*

Proof. $\forall T_2[\lim_{t\to\infty} g_1^1(t, T_2) = \lim_{t\to\infty} (\pi_0 - \pi_1)\, e^{-rt} - p'(t) < 0]$ by Assumption 4.[4] By strict concavity and continuity of g^1, there exists a unique $\hat{T} \in [0, \infty)$ which maximizes $g^1(T_1, T_2)$. $\hat{T}$ is defined (independent of T_2) by

$$(\pi_0 - \pi_1)\, e^{-r\hat{T}} - p'(\hat{T}) \leqq 0, \quad \hat{T} \geqq 0, \quad \text{and} \quad [(\pi_0 - \pi_1)\, e^{-r\hat{T}} - p'(\hat{T})]\hat{T} = 0. \tag{3}$$

If Assumption 5b holds, then $\forall T_2[g_1^1(0, T_2) < 0]$, so $\hat{T} = 0$.

$\forall T_2[\lim_{t\to\infty} g_1^2(t, T_2) = \lim_{t\to\infty} (\pi_2 - \pi_3)\, e^{-rt} - p'(t) < 0]$ by Assumption 4.

By strict concavity and continuity of g^2, there exists a unique $\hat{\hat{T}} \in [0, \infty)$ which maximizes $g^2(T_1, T_2)$. $\hat{\hat{T}}$ is defined (independent of T_2) by

$$(\pi_2 - \pi_3) e^{-r\hat{\hat{T}}} - p'(\hat{\hat{T}}) \leqq 0, \quad \hat{\hat{T}} \geqq 0, \quad \text{and} \quad [(\pi_2 - \pi_3) e^{-r\hat{\hat{T}}} - p'(\hat{\hat{T}})]\hat{\hat{T}} = 0. \qquad (4)$$

If Assumption 5a holds, then $\forall T_2[g_1^2(0, T_2) = (\pi_2 - \pi_3) - p'(0) > 0]$, so $\hat{\hat{T}} > 0$. If Assumption 5b holds, then $\forall T_2[g_1^2(0, T_2) = (\pi_2 - \pi_3) - p'(0) \leqq 0]$, and strict concavity ($g_{11}^2 < 0$) implies that $\hat{\hat{T}} = 0$. ||

Condition (3) states that if the first adopter is to adopt after some delay ($\hat{T} > 0$), then the marginal (discounted) opportunity cost of waiting, $(\pi_1 - \pi_0)e^{-r\hat{T}}$, must be exactly off-set by the marginal (discounted) cost savings due to waiting, $-p'(\hat{T})$. If the marginal (discounted) opportunity cost of waiting exceeds the marginal (discounted) cost savings for all T, then immediate adoption is optimal ($\hat{T} = 0$). Condition (4) has a similar interpretation for the second adopter.

Lemma 2. *If Assumption 5a holds, then* $\hat{\hat{T}} > \hat{T}$.

Proof. Note that $\forall T_2[g_1^2(t, T_2) > g_1^1(t, T_2) \forall t \in [0, \infty)]$. By Lemma 1, $\infty > \hat{\hat{T}} > 0$ and $\hat{T} < \infty$. Thus the corner case is obvious: if $\hat{T} = 0$, then $\hat{\hat{T}} > 0$ implies $\hat{\hat{T}} > \hat{T}$. Suppose $\hat{T} > 0$. Now $\forall T_2[(g_1^2(\hat{T}, T_2) > g_1^1(\hat{T}, T_2) = 0]$. Since $g_1^2(\hat{\hat{T}}, T_2) = 0$ and $g^2(\cdot, T_2)$ is strictly concave, $\hat{\hat{T}} > \hat{T}$. ||

3. NASH EQUILIBRIUM IN PURE STRATEGIES

The problem set out in Section 2 (that of determining an optimal adoption date when a competitor exists) can be modeled and solved in a game theoretic framework.

Definition 2. The strategy space for player i is $S_i = [0, \infty)$. A pure strategy for i is a scalar $T_i \in S_i$.

Definition 3. The set of best responses for i to T_j is $\phi_i(T_j) = \{T_i \in S_i \mid V^i(T_i, T_j) \geqq V^i(T'_i, T_j) \forall T'_i \in S_i\}$.

The mapping $\phi_i : S_j \Rightarrow S_i$ is i's best response correspondence.

Definition 4. A strategy pair (T_1^N, T_2^N) is a Nash equilibrium for the game $G = (V^1, V^2, S^1, S^2)$ if

(a) $T_i^N \in S_i, i = 1, 2;$

(b) $V^1(T_1^N, T_2^N) \geqq V^1(T_1, T_2^N) \forall T_1 \in S_1$; and

(c) $V^2(T_1^N, T_2^N) \geqq V^2(T_1^N, T_2) \forall T_2 \in S_2$.

Alternatively, the pair (T_1^N, T_2^N) is a Nash equilibrium if $T_1^N \in \phi_1(T_2^N)$ and $T_2^N \in \phi_2(T_1^N)$; that is, each strategy is a best response to the other.

Theorem 1.

(a) *If Assumption 5a holds, then there exist 2 Nash equilibria in pure strategies*[5] $(T_1^N, T_2^N) = (\hat{T}, \hat{\hat{T}})$ and $(T_1^N, T_2^N) = (\hat{\hat{T}}, \hat{T})$.

(b) *If Assumption 5b holds, then there exists a unique* $(T_1^N, T_2^N) = (0, 0)$.

Proof

(a) See below

(b) It is clear from Lemma 1 that immediate adoption is a dominant strategy for each firm. ||

Before proceeding to prove part (a), consider the implications of Theorem 1. There is the degenerate case (b). In this case, adjustment costs do not decline sufficiently rapidly as to warrant waiting. Thus, regardless of rival behavior, the best action a firm can take is to adopt immediately.

Excepting this degenerate case, we see that, if there is a net value to being first (Assumption 3), then at a Nash equilibrium one firm adopts early (at $\hat{T}$), while the other adopts late at $\hat{\hat{T}}$). It is *never* a Nash equilibrium for identical firms to bring the new technology on line at the same date. Hence even in the case of identical firms and complete certainty, there will be a diffusion of the innovation over time.

The proof of Theorem 1(a) is accomplished via a series of lemmas.

Lemma 3. *$g^1(T_1, T_2) \gtreqless g^2(T_1, T_2)$ as $T_1 \lesseqgtr T_2$.*

Proof. $g^1(T_1, T_2) - g^2(T_1, T_2) = (\alpha/r)(e^{-rT_1} - e^{-rT_2}) \gtreqless 0$ as $T_1 \lesseqgtr T_2$. ‖

Lemma 4. $g^2(\hat{\hat{T}}, \hat{T}) > g^1(\hat{T}, \hat{T})$.

Proof. $g^2(\hat{\hat{T}}, \hat{T}) > g^2(\hat{T}, \hat{T}) = g^1(\hat{T}, \hat{T})$, where the inequality follows from Lemma 1. The equality follows from Lemma 3. ‖

Lemma 5. $g^1(\hat{T}, \hat{\hat{T}}) > g^2(\hat{\hat{T}}, \hat{\hat{T}})$.

Proof. $g^1(\hat{T}, \hat{\hat{T}}) > g^1(\hat{\hat{T}}, \hat{\hat{T}}) = g^2(\hat{\hat{T}}, \hat{\hat{T}})$, where the inequality follows from Lemma 1. The equality follows from Lemma 3. ‖

Lemma 6. *$\exists \tilde{T} \in (\hat{T}, \hat{\hat{T}})$ such that $g^1(\hat{T}, T_2) \lesseqgtr g^2(\hat{\hat{T}}, T_2)$ as $T_2 \lesseqgtr \tilde{T}$.*

Proof. Let $\gamma(\hat{T}, \hat{\hat{T}}, T_2) = g^1(\hat{T}, T_2) - g^2(\hat{\hat{T}}, T_2)$. By Lemma 4, $\gamma(\hat{T}, \hat{\hat{T}}, \hat{T}) < 0$. By Lemma 5, $\gamma(\hat{T}, \hat{\hat{T}}, \hat{\hat{T}}) > 0$. Since $\partial\gamma/\partial T_2 = \alpha\, e^{-rT_2} > 0$, it follows by the intermediate value theorem and the monotonicity of γ in T_2 that $\exists$ a unique $\tilde{T} \in (\hat{T}, \hat{\hat{T}})$ such that $\gamma(\hat{T}, \hat{\hat{T}}, T_2) \lesseqgtr 0$ as $T_2 \lesseqgtr \tilde{T}$. ‖

Figures 1–3 illustrate Lemmas 3–6.

Lemma 7.

$$\phi_1(T_2) = \begin{cases} \hat{\hat{T}} & \text{for } T_2 < \tilde{T} \\ \{\hat{T}, \hat{\hat{T}}\} & \text{for } T_2 = \tilde{T} \\ \hat{T} & \text{for } T_2 > \tilde{T} \end{cases}$$

Proof. Recall Definition 1

$$V^1(T_1, T_2) = \begin{cases} g^1(T_1, T_2) & T_1 \leqq T_2 \\ g^2(T_1, T_2) & T_1 \geqq T_2 \end{cases}.$$

The reasons for the various relations are found in parentheses immediately below the relation itself.

Case 1. $T_2 < \tilde{T}$.

$$\forall T_1 \leqq T_2 [V^1(\hat{\hat{T}}, T_2) = g^2(\hat{\hat{T}}, T_2) > g^1(\hat{T}, T_2) \geqq g^1(T_1, T_2) = V^1(T_1, T_2)]$$

(Def. 1) (Lemma 6) (Lemma 1) (Def. 1)

$$\forall T_1 \geqq T_2, T_1 \neq \hat{\hat{T}} [V^1(\hat{\hat{T}}, T_2) = g^2(\hat{\hat{T}}, T_2) > g^2(T_1, T_2) = V^1(T_1, T_2)].$$

(Def. 1) (Lemma 1) (Def. 1)

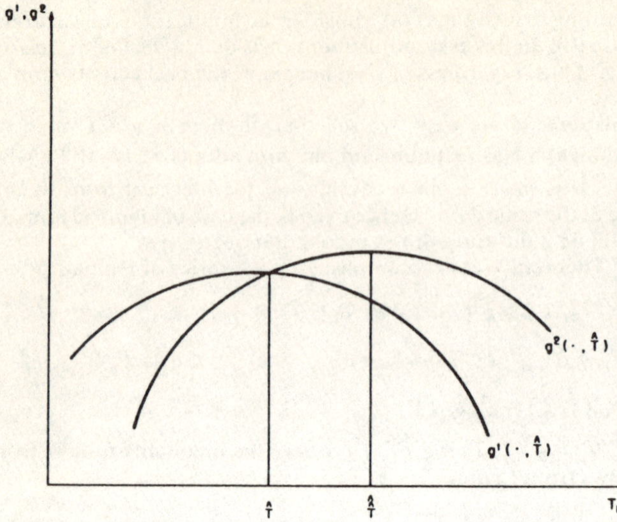

FIGURE 1

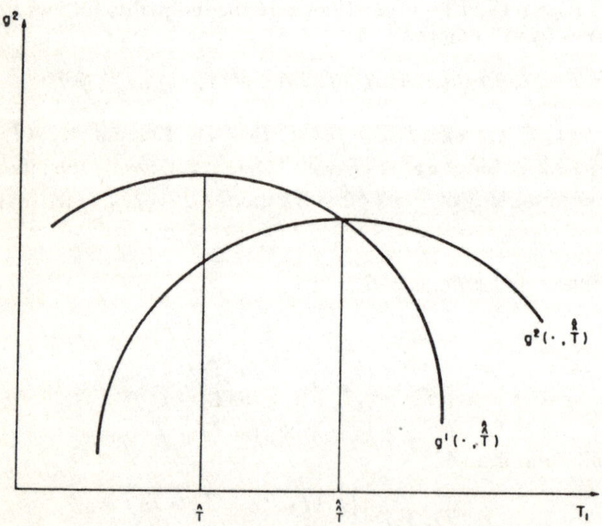

FIGURE 2

Thus $\forall T_1 \neq \hat{\tilde{T}}[V^1(\hat{\tilde{T}}, T_2) > V^1(T_1, T_2)]$. Hence $\phi_1(T_2) = \hat{\tilde{T}}$ for all $T_2 < \tilde{T}$.

Case 2. $T_2 = \tilde{T}$.

$$\forall T_1 \leqq \tilde{T}, T_1 \neq \hat{T}[V^1(\hat{T}, \tilde{T}) \underset{\text{(Def. 1)}}{=} g^1(\hat{T}, \tilde{T}) \underset{\text{(Lemma 1)}}{>} g^1(T_1, \tilde{T}) \underset{\text{(Def. 1)}}{=} V^1(T_1, \tilde{T})];$$

$$\forall T_1 \geqq \tilde{T}, T_1 \neq \hat{\tilde{T}}[V^1(\hat{\tilde{T}}, \tilde{T}) \underset{\text{(Def. 1)}}{=} g^2(\hat{\tilde{T}}, \tilde{T}) \underset{\text{(Lemma 1)}}{>} g^2(T_1, \tilde{T}) \underset{\text{(Def. 1)}}{=} V^1(T_1, \tilde{T})].$$

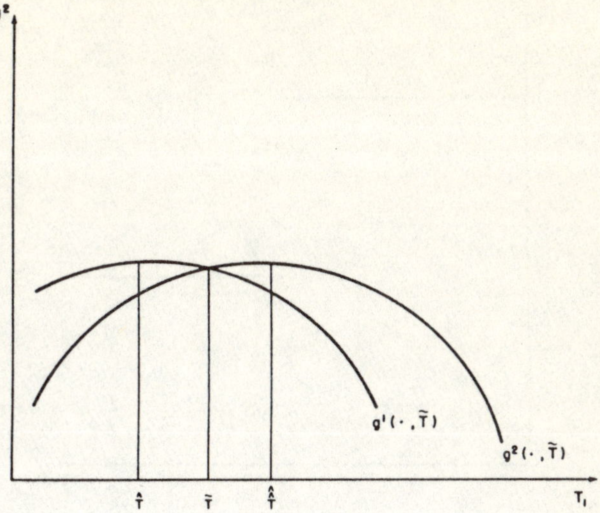

FIGURE 3

But since $g^1(\hat{T}, \tilde{T}) = g^2(\hat{\hat{T}}, \tilde{T})$ by Lemma 6, $V^1(\hat{T}, \tilde{T}) = V^1(\hat{\hat{T}}, \tilde{T}) > V^1(T_1, \tilde{T})$ $\forall T_1 \notin \{\hat{T}, \hat{\hat{T}}\}$. Thus $\phi_1(\tilde{T}) = \{\hat{T}, \hat{\hat{T}}\}$.

Case 3. $T_2 > \tilde{T}$.

$$\forall T_1 \leqq T_2,\ T_1 \neq \hat{T}[V^1(\hat{T}, T_2) \underset{\text{(Def. 1)}}{=} g^1(\hat{T}, T_2) \underset{\text{(Lemma 1)}}{>} g^1(T_1, T_2) \underset{\text{(Def. 1)}}{=} V^1(T_1, T_2)].$$

$$\forall T_1 \geqq T_2[V^1(\hat{T}, T_2) \underset{\text{(Def. 1)}}{=} g^1(\hat{T}, T_2) \underset{\text{(Lemma 6)}}{>} g^2(\hat{\hat{T}}, T_2) \underset{\text{(Lemma 1)}}{\geqq} g^2(T_1, T_2) \underset{\text{(Def. 1)}}{=} V^1(T_1, T_2)];$$

Thus $\forall T_1 \neq \hat{T}[V^1(\hat{T}, T_2) > V^1(T_1, T_2)]$. Hence $\phi_1(T_2) = \hat{T}$ for all $T_2 > \tilde{T}$. ‖

The best response correspondence $\phi_1(T_2)$ is shown in Figure 4.

Proof of Theorem 1(a). By symmetry,

$$\phi_2(T_1) = \begin{cases} \hat{\hat{T}} & \text{for } T_1 < \tilde{T} \\ \{\hat{T}, \hat{\hat{T}}\} & \text{for } T_1 = \tilde{T}. \\ \hat{T} & \text{for } T_1 > \tilde{T} \end{cases}$$

It is apparent that the best response correspondences intersect at two distinct points, $(\hat{T}, \hat{\hat{T}})$ and $(\hat{\hat{T}}, \hat{T})$, as shown in Figure 5. ‖

Recall that interior $\hat{T}$ and $\hat{\hat{T}}$ are determined by equations (3) and (4), repeated here for convenience:

$$g_1^1 = -(\pi_1 - \pi_0)\, e^{-r\hat{T}} - p'(\hat{T}) = 0 \tag{5}$$

$$g_1^2 = -(\pi_3 - \pi_2)\, e^{-r\hat{\hat{T}}} - p'(\hat{\hat{T}}) = 0. \tag{6}$$

Notice that the adoption dates depend on the profit rates only through the relevant opportunity costs of delaying adoption one period—$(\pi_1 - \pi_0)$ for $\hat{T}$ and $(\pi_3 - \pi_2)$ for $\hat{\hat{T}}$.

Comparative statics results are summarized in a corollary to Theorem 1.

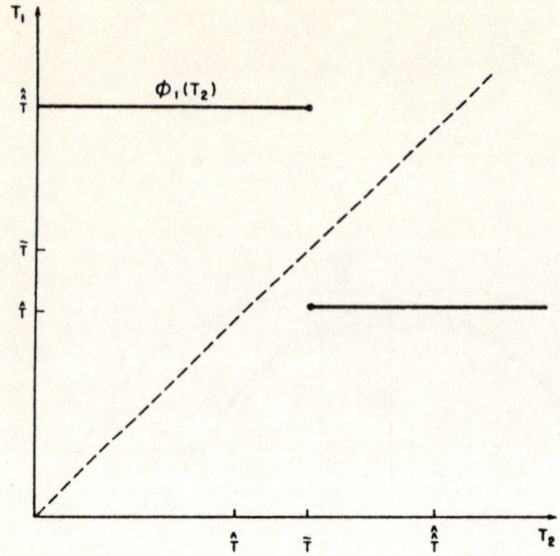

FIGURE 4

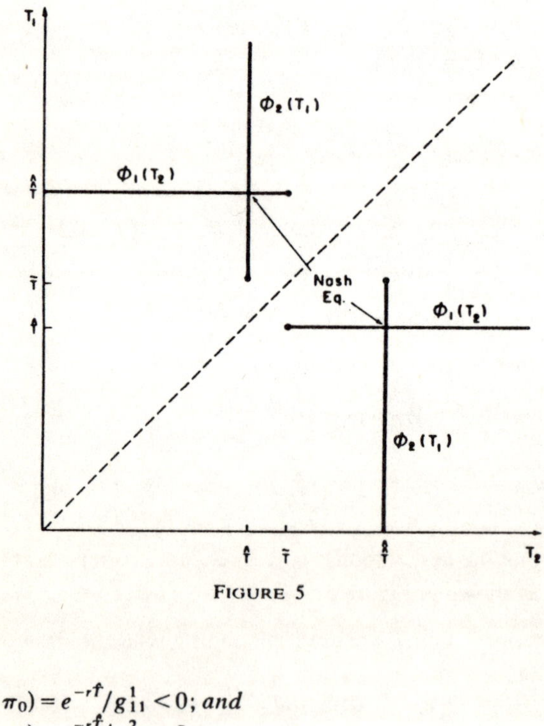

FIGURE 5

Corollary 1.
(a) $\partial\hat{T}/\partial(\pi_1-\pi_0)=e^{-r\hat{T}}/g^1_{11}<0$; *and*
(b) $\partial\hat{\hat{T}}/\partial(\pi_3-\pi_2)=e^{-r\hat{T}}/g^2_{11}<0$.

That is, as the opportunity costs of waiting (foregone revenues) increase, adoption occurs earlier.

4. GENERALIZATIONS

This model may be generalized in at least three directions. First, one may extend the analysis to n identical firms. It is conjectured that this will result in $n!$ Nash equilibria obtained from each other by permuting the indices of the players. While rigorous analysis of the n-player game would probably not be difficult, it would be extremely messy and is not likely to add substantially to the implications of the 2-player case. It would be interesting, however, to examine comparative statics in n.

One can extend the analysis to include uncertainty regarding π_1, π_2 and π_3 simply by interpreting these as mean post-adoption profit opportunities. Finally, one can generalize the model to allow for asymmetric firms. Define π_0^i, π_i^i and π_3^i in the obvious way; also define π_j^i $(j \neq i)$ to be i's profit rate when j has adopted but i has not. Add a subscript to $p_i(T_i)$, the discounted price (plus adjustment costs) to indicate possible scale advantages. One need not indicate which firm is the more efficient—either at production or at technology implementation—at this point. However, the appropriate analogues of Assumptions 1–4 and Assumption 5a will be retained.

The payoff to player i is now

$$V^i(T_1, T_2) = \begin{cases} g^{i1}(T_1, T_2) & T_i \leqq T_j \\ g^{i2}(T_1, T_2) & T_i \geqq T_j \end{cases}$$

where

$$g^{i1}(T_1, T_2) = \int_0^{T_i} e^{-rt}\pi_0^i\,dt + \int_{T_i}^{T_j} e^{-rt}\pi_i^i\,dt + \int_{T_j}^{\infty} e^{-rt}\pi_3^i\,dt - p_i(T_i)$$

and

$$g^{i2}(T_1, T_2) = \int_0^{T_j} e^{-rt}\pi_0^i\,dt + \int_{T_j}^{T_i} e^{-rt}\pi_j^i\,dt + \int_{T_i}^{\infty} e^{-rt}\pi_3^i\,dt - p_i(T_i).$$

Lemmas 1–7 in no way depended on the symmetry of the players. Thus the best response correspondences become

$$\phi_i(T_j) = \begin{cases} \hat{\hat{T}}_i & T_j < \tilde{T}_i \\ \{\hat{T}_i, \hat{\hat{T}}_i\} & T_j = \tilde{T}_i \\ \hat{T}_i & T_j > \tilde{T}_i \end{cases}$$

where $\hat{T}_i$ and $\hat{\hat{T}}_i(> \hat{T}_i)$ maximize g^{i1} and g^{i2}, respectively, independent of T_j $(j \neq i)$; and $\tilde{T}_i \in (\hat{T}_i, \hat{\hat{T}}_i)$ is the unique solution of $g^{i1}(\hat{T}_i, t) = g^{i2}(\hat{\hat{T}}_i, t)$.

Now although it is easily shown that $\hat{\hat{T}}_i > \hat{T}_i$, $i = 1, 2$, one has no idea of the rankings of j's optimal dates relative to i's, especially since I have not specified which firm is "more efficient," nor even any notion of efficiency. However, one can still prove the following theorem characterizing Nash equilibrium with non-identical firms.

Theorem 2. *Under the appropriate analogues of Assumptions* 1–5*a*,

(a) *There is at least* 1 (*and no more than* 2) *Nash equilibrium.*

(b) *All Nash equilibria are asymmetric in the sense that firms will never adopt the new technology simultaneously.*

Proof.

(a) $\phi_1(T_2)$ is non-increasing and, when graphed in (T_1, T_2) space, consists of two horizontal half-lines which cover $[0, \infty)$; $\phi_2(T_1)$ is non-increasing and, when graphed in (T_1, T_2) space, consists of two vertical half-lines which cover $[0, \infty)$. They must intersect at least once and can intersect no more than twice.

(b) It is sufficient to show that matching is never a best response: $\forall T_j$, $T_j \notin \phi_i(T_j)$. Now $\forall T_j < \tilde{T}_i$, $\phi_i(T_j) = \hat{\hat{T}}_i > \tilde{T}_i$; for $T_j = \tilde{T}_i$, $\phi_i(\tilde{T}_i) = \{\hat{\hat{T}}_i, \hat{T}_i\}$ where $\hat{T}_i < \tilde{T}_i$ and $\hat{\hat{T}}_i > \tilde{T}_i$; and $\forall T_j > \tilde{T}_i$, $\phi_i(T_j) = \hat{T}_i < \tilde{T}_i$. Thus $\forall T_j$, $T_j \notin \phi_i(T_j)$. ‖

Let us now relate this to Flaherty (1980). Using a slightly different model, Flaherty focuses on the question: will a larger (lower cost) firm commercialize a new technology earlier than a smaller (higher cost) firm? If the larger firm has a sufficiently large comparative advantage in implementing the new technology, then there exists a Nash equilibrium in which it will adopt sooner than a small firm. If the large firm has no comparative advantage, then there exists an equilibrium in which the smaller firm adopts first. However, in either case there may exist a second equilibrium in which the pattern is reversed. Analysis of the present model with asymmetric payoffs yields similar results (Reinganum, 1980*c*).

5. CONCLUSIONS

The choice of the date of adoption of a new technology has been modeled as a 2-person nonzero-sum game. Under Nash behavior we find that there exists a pair of equilibria in pure strategies (for identical firms). Each Nash equilibrium involves one firm adopting relatively early with the other adopting at a relatively late date, despite the facts that information is perfect and players are identical. For non-identical firms, a Nash equilibrium always exists and is asymmetric so long as the net value of being first is positive.

It is clear that a substantial modeling job remains—that is, to integrate optimal behavior (or strategic behavior) on the part of a seller of the new technology (see Stokey, 1979). Furthermore, there may be information regarding the value of the innovation to be gained by waiting. For an analysis of a single firm's decision problem in the face of such uncertainty, see Jensen (1979).

First version received April 1980; *final version accepted February* 1981 (*Eds.*).

I would like to thank E. Green, J. Jordan and participants in the California Institute of Technology Economic Theory Workshop for helpful comments.

NOTES

1. In light of the results on repeated games, I should specify that these are to be regarded as stage game, rather than supergame, equilibrium profits.

2. It is easy to verify that Assumptions 1–3 hold for simple demand curves (e.g. $P = a - bQ$, $P = a + b/Q$, $P = a - b \ln Q$) when marginal costs are constant. Whether Assumptions 1–3 hold for general cost and demand functions, and for a more general specification of cost reduction, is an open question at this time.

3. If $\lim_{t\to\infty} p'(t) \leqq 0$ and strict concavity is maintained, then $\forall T_2[\lim_{t\to\infty} g_1^1(t, T_2) \geqq 0]$ so $\forall T_2[g_1^1(t, T_2) > 0\ \forall t < \infty]$ and $\exists$ a maximizing value $\hat{T}$. Similarly, $\forall T_2[\lim_{t\to\infty} g_1^2(t, T_2) \geqq 0]$ so $\forall T_2[g_1^2(t, T_2) > 0\ \forall t < \infty]$ and $\exists$ a maximizing value $\hat{\hat{T}}$. If strict concavity is relaxed, optimal adoption dates $\hat{T}$ and $\hat{\hat{T}}$ may exist regardless of the sign of $\lim_{t\to\infty} p'(t)$. However, the existence of such dates would be difficult to verify without specifying the functional form of $p(t)$.

4. The notation $\forall T_i[\cdot]$ means "for all values of T_i in $[0, \infty)$, the bracketed statement is true".

5. It is shown in Reinganum (1980*c*) that there also exist symmetric Nash equilibria in both discrete and continuous mixed strategies. A Nash equilibrium in discrete mixed strategies is to play $\hat{T}$ with probability λ^* and $\hat{\hat{T}}$ with the complementary probability, where

$$\lambda^* = (\pi_1 - \pi_0)/\alpha - r[p(\hat{T}) - p(\hat{\hat{T}})]/\alpha[e^{-r\hat{T}} - e^{-r\hat{\hat{T}}}].$$

A Nash equilibrium in continuous mixed strategies is

$$F_i^N(t) = \Pr\{T_i \leqq t\} = (1/\alpha)[p'(t)\, e^{rt} + \pi_1 - \pi_0],$$

for $t \in [\hat{T}, \hat{\hat{T}}]$.

REFERENCES

FLAHERTY, M. T. (1980) "Timing Patterns of New Technology Adoption in Duopolistic Industries" (Manuscript).

JENSEN, R. A. (1979), "On the Adoption and Diffusion of Innovations Under Uncertainty" (Discussion Paper No. 410, Northwestern University).

KAMIEN, M. I. and SCHWARTZ, N. L. (1972), "Some Economic Consequences of Anticipating Technical Advance", *Western Economic Journal*, **10**, 123–138.

REINGANUM, J. F. (1980*a*), "Dynamic Games of Innovation", *Journal of Economic Theory* (forthcoming).

REINGANUM, J. F. (1980*b*), "A Dynamic Game of R & D: Patent Protection and Competitive Behavior", *Econometrica* (forthcoming).

REINGANUM, J. F. (1980*c*), "On the Diffusion of New Technology: A Game Theoretic Approach" (Social Science Working Paper No. 312 Pasadena, California Institute of Technology).

STOKEY, N. L. (1979), "Intertemporal Price Discrimination", *Quarterly Journal of Economics*, **93**, 355–371.

Part VI
International Technology Transfer

[20]

INTERNATIONAL INVESTMENT AND INTERNATIONAL TRADE IN THE PRODUCT CYCLE *

RAYMOND VERNON

Anyone who has sought to understand the shifts in international trade and international investment over the past twenty years has chafed from time to time under an acute sense of the inadequacy of the available analytical tools. While the comparative cost concept and other basic concepts have rarely failed to provide some help, they have usually carried the analyst only a very little way toward adequate understanding. For the most part, it has been necessary to formulate new concepts in order to explore issues such as the strengths and limitations of import substitution in the development process, the implications of common market arrangements for trade and investment, the underlying reasons for the Leontief paradox, and other critical issues of the day.

As theorists have groped for some more efficient tools, there has been a flowering in international trade and capital theory. But the very proliferation of theory has increased the urgency of the search for unifying concepts. It is doubtful that we shall find many propositions that can match the simplicity, power, and universality of application of the theory of comparative advantage and the international equilibrating mechanism; but unless the search for better tools goes on, the usefulness of economic theory for the solution of problems in international trade and capital movements will probably decline.

The present paper deals with one promising line of generalization and synthesis which seems to me to have been somewhat neglected by the main stream of trade theory. It puts less emphasis upon comparative cost doctrine and more upon the timing of innovation, the effects of scale economics, and the roles of ignorance and uncertainty in influencing trade patterns. It is an approach

* The preparation of this article was financed in part by a grant from the Ford Foundation to the Harvard Business School to support a study of the implications of United States foreign direct investment. This paper is a by-product of the hypothesis-building stage of the study.

with respectable sponsorship, deriving bits and pieces of its inspiration from the writings of such persons as Williams, Kindleberger, MacDougall, Hoffmeyer, and Burenstam-Linder.[1]

Emphases of this sort seem first to have appeared when economists were searching for an explanation of what looked like a persistent, structural shortage of dollars in the world. When the shortage proved ephemeral in the late 1950's, many of the ideas which the shortage had stimulated were tossed overboard as prima facie wrong.[2] Nevertheless, one cannot be exposed to the main currents of international trade for very long without feeling that any theory which neglected the roles of innovation, scale, ignorance and uncertainty would be incomplete.

Location of New Products

We begin with the assumption that the enterprises in any one of the advanced countries of the world are not distinguishably different from those in any other advanced country, in terms of their access to scientific knowledge and their capacity to comprehend scientific principles.[3] All of them, we may safely assume, can secure access to the knowledge that exists in the physical, chemical and biological sciences. These sciences at times may be difficult, but they are rarely occult.

It is a mistake to assume, however, that equal access to scientific principles in all the advanced countries means equal probability of the application of these principles in the generation of new products. There is ordinarily a large gap between the knowledge of a scientific principle and the embodiment of the principle in

1. J. H. Williams, "The Theory of International Trade Reconsidered," reprinted as Chap. 2 in his *Postwar Monetary Plans and Other Essays* (Oxford: Basil Blackwell, 1947); C. P. Kindleberger, *The Dollar Shortage* (New York: Wiley, 1950); Erik Hoffmeyer, *Dollar Shortage* (Amsterdam: North-Holland, 1958); Sir Donald MacDougall, *The World Dollar Problem* (London: Macmillan, 1957); Staffan Burenstam-Linder, *An Essay on Trade and Transformation* (Uppsala: Almqvist & Wicksells, 1961).

2. The best summary of the state of trade theory that has come to my attention in recent years is J. Bhagwati, "The Pure Theory of International Trade," *Economic Journal*, LXXIV (Mar. 1964), 1–84. Bhagwati refers obliquely to some of the theories which concern us here; but they receive much less attention than I think they deserve.

3. Some of the account that follows will be found in greatly truncated form in my "The Trade Expansion Act in Perspective," in *Emerging Concepts in Marketing*, Proceedings of the American Marketing Association, December 1962, pp. 384–89. The elaboration here owes a good deal to the perceptive work of Se'ev Hirsch, summarized in his unpublished doctoral thesis, "Location of Industry and International Competitiveness," Harvard Business School, 1965.

a marketable product. An entrepreneur usually has to intervene to accept the risks involved in testing whether the gap can be bridged.

If all entrepreneurs, wherever located, could be presumed to be equally conscious of and equally responsive to all entrepreneurial opportunities, wherever they arose, the classical view of the dominant role of price in resource allocation might be highly relevant. There is good reason to believe, however, that the entrepreneur's consciousness of and responsiveness to opportunity are a function of ease of communication; and further, that ease of communication is a function of geographical proximity.[4] Accordingly, we abandon the powerful simplifying notion that knowledge is a universal free good, and introduce it as an independent variable in the decision to trade or to invest.

The fact that the search for knowledge is an inseparable part of the decision-making process and that relative ease of access to knowledge can profoundly affect the outcome are now reasonably well established through empirical research.[5] One implication of that fact is that producers in any market are more likely to be aware of the possibility of introducing new products in that market than producers located elsewhere would be.

The United States market offers certain unique kinds of opportunities to those who are in a position to be aware of them.

First, the United States market consists of consumers with an average income which is higher (except for a few anomalies like Kuwait) than that in any other national market — twice as high as that of Western Europe, for instance. Wherever there was a chance to offer a new product responsive to wants at high levels of income, this chance would presumably first be apparent to someone in a position to observe the United States market.

Second, the United States market is characterized by high unit labor costs and relatively unrationed capital compared with practically all other markets. This is a fact which conditions the demand for both consumer goods and industrial products. In the case of consumer goods, for instance, the high cost of laundresses contributes to the origins of the drip-dry shirt and the home washing machine. In the case of industrial goods, high labor cost leads to the early

4. Note C. P. Kindleberger's reference to the "horizon" of the decision-maker, and the view that he can only be rational within that horizon; see his *Foreign Trade and The National Economy* (New Haven: Yale University Press, 1962), p. 15 *passim.*

5. See, for instance, Richard M. Cyert and James G. March, *A Behavioral Theory of the Firm* (Englewood Cliffs, N.J.: Prentice-Hall, 1963), esp. Chap. 6; and Yair Aharoni, *The Foreign Investment Decision Process*, to be published by the Division of Research of the Harvard Business School, 1966.

development and use of the conveyor belt, the fork-lift truck and the automatic control system. It seems to follow that wherever there was a chance successfully to sell a new product responsive to the need to conserve labor, this chance would be apparent first to those in a position to observe the United States market.

Assume, then, that entrepreneurs in the United States are first aware of opportunities to satisfy new wants associated with high income levels or high unit labor costs. Assume further that the evidence of an unfilled need and the hope of some kind of monopoly windfall for the early starter both are sufficiently strong to justify the initial investment that is usually involved in converting an abstract idea into a marketable product. Here we have a reason for expecting a consistently higher rate of expenditure on product development to be undertaken by United States producers than by producers in other countries, at least in lines which promise to substitute capital for labor or which promise to satisfy high-income wants. Therefore, if United States firms spend more than their foreign counterparts on new product development (often misleadingly labeled "research"), this may be due not to some obscure sociological drive for innovation but to more effective communication between the potential market and the potential supplier of the market. This sort of explanation is consistent with the pioneer appearance in the United States (conflicting claims of the Soviet Union notwithstanding) of the sewing machine, the typewriter, the tractor, etc.

At this point in the exposition, it is important once more to emphasize that the discussion so far relates only to innovation in certain kinds of products, namely to those associated with high income and those which substitute capital for labor. Our hypothesis says nothing about industrial innovation in general; this is a larger subject than we have tackled here. There are very few countries that have failed to introduce at least a few products; and there are some, such as Germany and Japan, which have been responsible for a considerable number of such introductions. Germany's outstanding successes in the development and use of plastics may have been due, for instance, to a traditional concern with her lack of a raw materials base, and a recognition that a market might exist in Germany for synthetic substitutes.[6]

6. See two excellent studies: C. Freeman, "The Plastics Industry: A Comparative Study of Research and Innovation," in *National Institute Economic Review*, No. 26 (Nov. 1963), p. 22 *et seq.*; G. C. Hufbauer, *Synthetic Materials and the Theory of International Trade* (London: Gerald Duckworth, 1965). A number of links in the Hufbauer arguments are remarkably similar to

Our hypothesis asserts that United States producers are likely to be the first to spy an opportunity for high-income or labor-saving new products.[7] But it goes on to assert that the first producing facilities for such products will be located in the United States. This is not a self-evident proposition. Under the calculus of least cost, production need not automatically take place at a location close to the market, unless the product can be produced and delivered from that location at lowest cost. Besides, now that most major United States companies control facilities situated in one or more locations outside of the United States, the possibility of considering a non-United States location is even more plausible than it might once have been.

Of course, if prospective producers were to make their locational choices on the basis of least-cost considerations, the United States would not always be ruled out. The costs of international transport and United States import duties, for instance, might be so high as to argue for such a location. My guess is, however, that the early producers of a new product intended for the United States market are attracted to a United States location by forces which are far stronger than relative factor-cost and transport considerations. For the reasoning on this point, one has to take a long detour away from comparative cost analysis into areas which fall under the rubrics of communication and external economies.

By now, a considerable amount of empirical work has been done on the factors affecting the location of industry.[8] Many of these studies try to explain observed locational patterns in conventional cost-minimizing terms, by implicit or explicit reference to labor cost and transportation cost. But some explicitly introduce problems of communication and external economies as powerful locational forces. These factors were given special emphasis in the analyses which were a part of the New York Metropolitan Region Study of the 1950's. At the risk of oversimplifying, I shall try to summarize what these studies suggested.[9]

some in this paper; but he was not aware of my writings nor I of his until after both had been completed.

7. There is a kind of first-cousin relationship between this simple notion and the "entrained want" concept defined by H. G. Barnett in *Innovation: The Basis of Cultural Change* (New York: McGraw-Hill, 1953) p. 148. Albert O. Hirschman, *The Strategy of Economic Development* (New Haven: Yale University Press, 1958), p. 68, also finds the concept helpful in his effort to explain certain aspects of economic development.

8. For a summary of such work, together with a useful bibliography, see John Meyer, "Regional Economics: A Survey," in the *American Economic Review*, LIII (Mar. 1963), 19–54.

9. The points that follow are dealt with at length in the following publications: Raymond Vernon, *Metropolis, 1985* (Cambridge: Harvard Uni-

In the early stages of introduction of a new product, producers were usually confronted with a number of critical, albeit transitory, conditions. For one thing, the product itself may be quite unstandardized for a time; its inputs, its processing, and its final specifications may cover a wide range. Contrast the great variety of automobiles produced and marketed before 1910 with the thoroughly standardized product of the 1930's, or the variegated radio designs of the 1920's with the uniform models of the 1930's. The unstandardized nature of the design at this early stage carries with it a number of locational implications.

First, producers at this stage are particularly concerned with the degree of freedom they have in changing their inputs. Of course, the cost of the inputs is also relevant. But as long as the nature of these inputs cannot be fixed in advance with assurance, the calculation of cost must take into account the general need for flexibility in any locational choice.[1]

Second, the price elasticity of demand for the output of individual firms is comparatively low. This follows from the high degree of production differentiation, or the existence of monopoly in the early stages.[2] One result is, of course, that small cost differences count less in the calculations of the entrepreneur than they are likely to count later on.

Third, the need for swift and effective communication on the part of the producer with customers, suppliers, and even competitors is especially high at this stage. This is a corollary of the fact that a considerable amount of uncertainty remains regarding the ultimate dimensions of the market, the efforts of rivals to preempt that market, the specifications of the inputs needed for production, and the specifications of the products likely to be most successful in the effort.

All of these considerations tend to argue for a location in which communication between the market and the executives directly concerned with the new product is swift and easy, and in which a wide

versity Press, 1960), pp. 38–85; Max Hall (ed.), *Made in New York* (Cambridge: Harvard University Press, 1959), pp. 3–18, 19 *passim*; Robert M. Lichtenberg, *One-Tenth of a Nation* (Cambridge: Harvard University Press, 1960), pp. 31–70.

1. This is, of course, a familiar point elaborated in George F. Stigler, "Production and Distribution in the Short Run," *Journal of Political Economy*, XLVII (June 1939), 305, *et seq.*

2. Hufbauer, *op. cit.*, suggests that the low price elasticity of demand in the first stage may be due simply to the fact that the first market may be a "captive market" unresponsive to price changes; but that later, in order to expand the use of the new product, other markets may be brought in which are more price responsive.

variety of potential types of input that might be needed by the production unit are easily come by. In brief, the producer who sees a market for some new product in the United States may be led to select a United States location for production on the basis of national locational considerations which extend well beyond simple factor cost analysis plus transport considerations.

The Maturing Product [3]

As the demand for a product expands, a certain degree of standardization usually takes place. This is not to say that efforts at product differentiation come to an end. On the contrary; such efforts may even intensify, as competitors try to avoid the full brunt of price competition. Moreover, variety may appear as a result of specialization. Radios, for instance, ultimately acquired such specialized forms as clock radios, automobile radios, portable radios, and so on. Nevertheless, though the subcategories may multiply and the efforts at product differentiation increase, a growing acceptance of certain general standards seems to be typical.

Once again, the change has locational implications. First of all, the need for flexibility declines. A commitment to some set of product standards opens up technical possibilities for achieving economies of scale through mass output, and encourages long-term commitments to some given process and some fixed set of facilities. Second, concern about production cost begins to take the place of concern about product characteristics. Even if increased price competition is not yet present, the reduction of the uncertainties surrounding the operation enhances the usefulness of cost projections and increases the attention devoted to cost.

The empirical studies to which I referred earlier suggest that, at this stage in an industry's development, there is likely to be considerable shift in the location of production facilities at least as far as internal United States locations are concerned. The empirical materials on international locational shifts simply have not yet been analyzed sufficiently to tell us very much. A little speculation, however, indicates some hypotheses worth testing.

Picture an industry engaged in the manufacture of the high-income or labor-saving products that are the focus of our discussion. Assume that the industry has begun to settle down in the United States to some degree of large-scale production. Although the first

3. Both Hirsch, *op. cit.*, and Freeman, *op. cit.*, make use of a three-stage product classification of the sort used here.

mass market may be located in the United States, some demand for the product begins almost at once to appear elsewhere. For instance, although heavy fork-lift trucks in general may have a comparatively small market in Spain because of the relative cheapness of unskilled labor in that country, some limited demand for the product will appear there almost as soon as the existence of the product is known.

If the product has a high income elasticity of demand or if it is a satisfactory substitute for high-cost labor, the demand in time will begin to grow quite rapidly in relatively advanced countries such as those of Western Europe. Once the market expands in such an advanced country, entrepreneurs will begin to ask themselves whether the time has come to take the risk of setting up a local producing facility.[4]

How long does it take to reach this stage? An adequate answer must surely be a complex one. Producers located in the United States, weighing the wisdom of setting up a new production facility in the importing country, will feel obliged to balance a number of complex considerations. As long as the marginal production cost plus the transport cost of the goods exported from the United States is lower than the average cost of prospective production in the market of import, United States producers will presumably prefer to avoid an investment. But that calculation depends on the producer's ability to project the cost of production in a market in which factor costs and the appropriate technology differ from those at home.

Now and again, the locational force which determined some particular overseas investment is so simple and so powerful that one has little difficulty in identifying it. Otis Elevator's early proliferation of production facilities abroad was quite patently a function of the high cost of shipping assembled elevator cabins to distant locations and the limited scale advantages involved in manufacturing elevator cabins at a single location.[5] Singer's decision to invest in Scotland as early as 1867 was also based on considerations of a sort sympathetic with our hypothesis.[6] It is not unlikely that the

4. M. V. Posner, "International Trade and Technical Change," *Oxford Economic Papers*, Vol. 13 (Oct. 1961), p. 323, *et seq.* presents a stimulating model purporting to explain such familiar trade phenomena as the exchange of machine tools between the United Kingdom and Germany. In the process he offers some particularly helpful notions concerning the size of the "imitation lag" in the responses of competing nations.

5. Dudley M. Phelps, *Migration of Industry to South America* (New York: McGraw-Hill, 1963), p. 4.

6. John H. Dunning, *American Investment in British Manufacturing Industry* (London: George Allen & Unwin, 1958), p. 18. The Dunning book

overseas demand for its highly standardized product was already sufficiently large at that time to exhaust the obvious scale advantages of manufacturing in a single location, especially if that location was one of high labor cost.

In an area as complex and "imperfect" as international trade and investment, however, one ought not anticipate that any hypothesis will have more than a limited explanatory power. United States airplane manufacturers surely respond to many "noneconomic" locational forces, such as the desire to play safe in problems of military security. Producers in the United States who have a protected patent position overseas presumably take that fact into account in deciding whether or when to produce abroad. And other producers often are motivated by considerations too complex to reconstruct readily, such as the fortuitous timing of a threat of new competition in the country of import, the level of tariff protection anticipated for the future, the political situation in the country of prospective investment and so on.

We arrive, then, at the stage at which United States producers have come around to the establishment of production units in the advanced countries. Now a new group of forces are set in train. In an idealized form, Figure I suggests what may be anticipated next.

As far as individual United States producers are concerned, the local markets thenceforth will be filled from local production units set up abroad. Once these facilities are in operation, however, more ambitious possibilities for their use may be suggested. When comparing a United States producing facility and a facility in another advanced country, the obvious production-cost differences between the rival producing areas are usually differences due to scale and differences due to labor costs. If the producer is an international firm with producing locations in several countries, its costs of financing capital at the different locations may not be sufficiently different to matter very much. If economies of scale are being fully exploited, the principal differences between any two locations are likely to be labor costs.[7] Accordingly, it may prove wise for the international firm to begin servicing third-country markets from the new location. And if labor cost differences are large enough to offset transport

is filled with observations that lend casual support to the main hypotheses of this paper.

7. Note the interesting finding of Mordecai Kreinin in his "The Leontief Scarce-Factor Paradox," *The American Economic Review*, LV (Mar. 1965), 131–39. Kreinin finds that the higher cost of labor in the United States is not explained by a higher rate of labor productivity in this country.

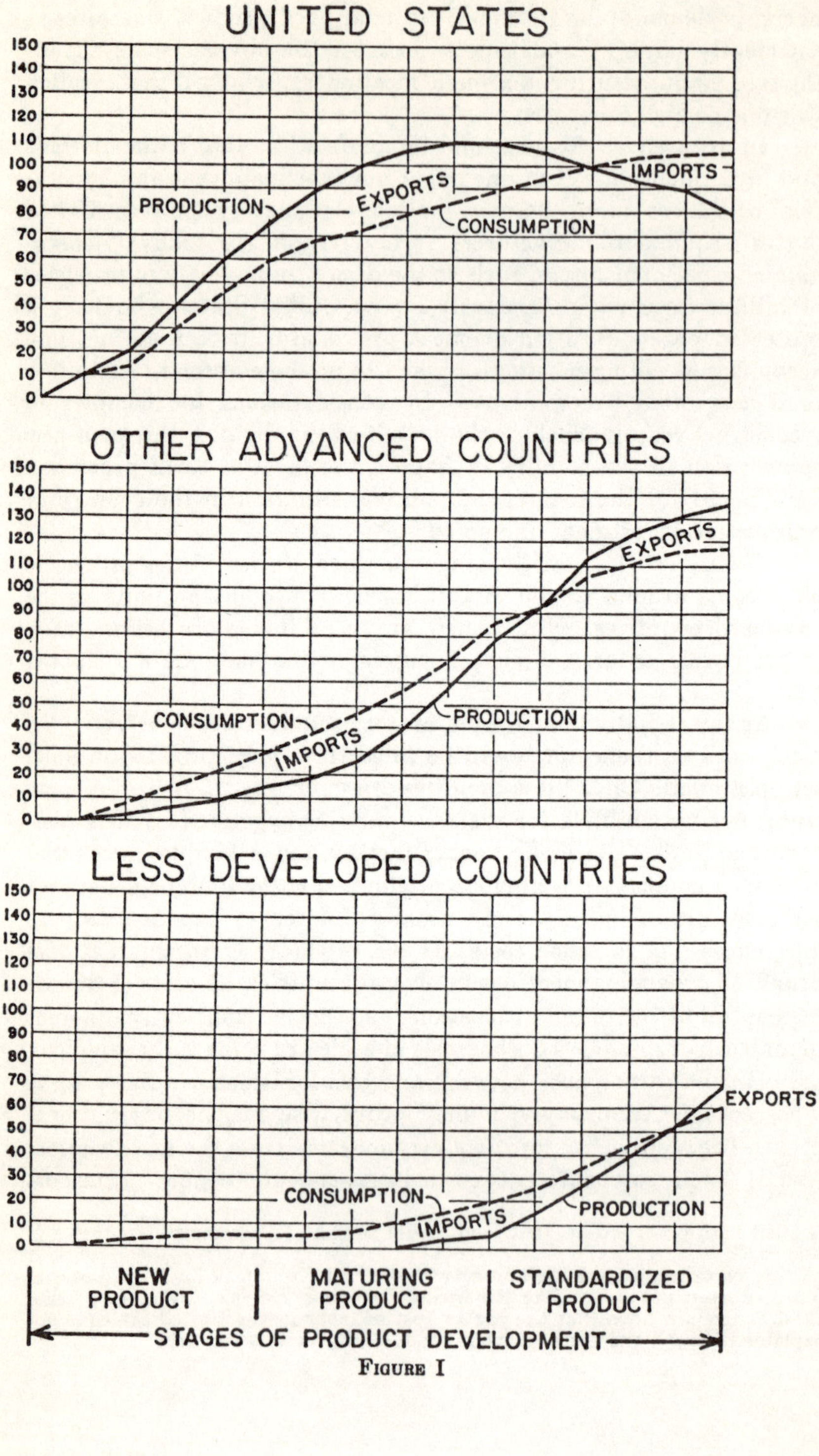
UNITED STATES
150
140
130
120
110
100
90
80
70
60
50
40
30
20
10
0
PRODUCTION
EXPORTS
IMPORTS
CONSUMPTION
OTHER ADVANCED COUNTRIES
150
140
130
120
110
100
90
80
70
60
50
40
30
20
10
0
EXPORTS
CONSUMPTION
IMPORTS
PRODUCTION
LESS DEVELOPED COUNTRIES
150
140
130
120
110
100
90
80
70
60
50
40
30
20
10
0
EXPORTS
CONSUMPTION
IMPORTS
PRODUCTION
NEW PRODUCT
MATURING PRODUCT
STANDARDIZED PRODUCT
STAGES OF PRODUCT DEVELOPMENT

FIGURE I

costs, then exports back to the United States may become a possibility as well.

Any hypotheses based on the assumption that the United States entrepreneur will react rationally when offered the possibility of a lower-cost location abroad is, of course, somewhat suspect. The decision-making sequence that is used in connection with international investments, according to various empirical studies, is not a model of the rational process.[8] But there is one theme that emerges again and again in such studies. Any threat to the established position of an enterprise is a powerful galvanizing force to action; in fact, if I interpret the empirical work correctly, threat in general is a more reliable stimulus to action than opportunity is likely to be.

In the international investment field, threats appear in various forms once a large-scale export business in manufactured products has developed. Local entrepreneurs located in the countries which are the targets of these exports grow restive at the opportunities they are missing. Local governments concerned with generating employment or promoting growth or balancing their trade accounts begin thinking of ways and means to replace the imports. An international investment by the exporter, therefore, becomes a prudent means of forestalling the loss of a market. In this case, the yield on the investment is seen largely as the avoidance of a loss of income to the system.

The notion that a threat to the status quo is a powerful galvanizing force for international investment also seems to explain what happens after the initial investment. Once such an investment is made by a United States producer, other major producers in the United States sometimes see it as a threat to the status quo. They see themselves as losing position relative to the investing company, with vague intimations of further losses to come. Their "share of the market" is imperiled, viewing "share of the market" in global terms. At the same time, their ability to estimate the production-cost structure of their competitors, operating far away in an unfamiliar foreign area, is impaired; this is a particularly unsettling state because it conjures up the possibility of a return flow of products to the United States and a new source of price competition, based on cost differences of unknown magnitude. The uncertainty can be reduced by emulating the pathfinding investor and by investing in the same area; this may not be an optimizing investment

8. Aharoni, *op. cit.*, provides an excellent summary and exhaustive bibliography of the evidence on this point.

pattern and it may be costly, but it is least disturbing to the status quo.

Pieces of this hypothetical pattern are subject to empirical tests of a sort. So far, at any rate, the empirical tests have been reassuring. The office machinery industry, for instance, has seen repeatedly the phenomenon of the introduction of a new product in the United States, followed by United States exports,[9] followed still later by United States imports. (We have still to test whether the timing of the commencement of overseas production by United States subsidiaries fits into the expected pattern.) In the electrical and electronic products industry, those elements in the pattern which can be measured show up nicely.[1] A broader effort is now under way to test the United States trade patterns of a group of products with high income elasticities; and, here too, the preliminary results are encouraging.[2] On a much more general basis, it is reassuring for our hypotheses to observe that the foreign manufacturing subsidiaries of United States firms have been increasing their exports to third countries.

It will have occurred to the reader by now that the pattern envisaged here also may shed some light on the Leontief paradox.[3] Leontief, it will be recalled, seemed to confound comparative cost theory by establishing the fact that the ratio of capital to labor in United States exports was lower, not higher, than the like ratio in the United States production which had been displaced by competitive imports. The hypothesis suggested in this paper would have the United States exporting high-income and labor-saving products in the early stages of their existence, and importing them later on.[4] In the early stages, the value-added contribution of industries engaged in producing these items probably contains an

9. Reported in U.S. Senate, Interstate and Foreign Commerce Committee, *Hearings on Foreign Commerce*, 1960, pp. 130–39.

1. See Hirsch, *op. cit.*

2. These are to appear in a forthcoming doctoral thesis at the Harvard Business School by Louis T. Wells, tentatively entitled" International Trade and Business Policy."

3. See Wassily Leontief, "Domestic Production and Foreign Trade: The American Capital Position Re-examined," *Proceedings of the American Philosophical Society*, Vol. 97 (Sept. 1953), and "Factor Proportions and the Structure of American Trade: Further Theoretical and Empirical Analysis," *Review of Economics and Statistics*, XXXVIII (Nov. 1956).

4. Of course, if there were some systematic trend in the inputs of new products — for example, if the new products which appeared in the 1960's were more capital-intensive than the new products which appeared in the 1950's — then the tendencies suggested by our hypotheses might be swamped by such a trend. As long as we do not posit offsetting systematic patterns of this sort, however, the Leontief findings and the hypotheses offered here seem consistent.

unusually high proportion of labor cost. This is not so much because the labor is particularly skilled, as is so often suggested. More likely, it is due to a quite different phenomenon. At this stage, the standardization of the manufacturing process has not gotten very far; that is to come later, when the volume of output is high enough and the degree of uncertainty low enough to justify investment in relatively inflexible, capital-intensive facilities. As a result, the production process relies relatively heavily on labor inputs at a time when the United States commands an export position; and the process relies more heavily on capital at a time when imports become important.

This, of course, is an hypothesis which has not yet been subjected to any really rigorous test. But it does open up a line of inquiry into the structure of United States trade which is well worth pursuing.

The Standardized Product

Figure I, the reader will have observed, carries a panel which suggests that, at an advanced stage in the standardization of some products, the less-developed countries may offer competitive advantages as a production location.

This is a bold projection, which seems on first blush to be wholly at variance with the Heckscher-Ohlin theorem. According to that theorem, one presumably ought to anticipate that the exports of the less-developed countries would tend to be relatively labor-intensive products.

One of the difficulties with the theorem, however, is that it leaves marketing considerations out of account. One reason for the omission is evident. As long as knowledge is regarded as a free good, instantaneously available, and as long as individual producers are regarded as atomistic contributors to the total supply, marketing problems cannot be expected to find much of a place in economic theory. In projecting the patterns of export from less-developed areas, however, we cannot afford to disregard the fact that information comes at a cost; and that entrepreneurs are not readily disposed to pay the price of investigating overseas markets of unknown dimensions and unknown promise. Neither are they eager to venture into situations which they know will demand a constant flow of reliable marketing information from remote sources.

If we can assume that highly standardized products tend to have a well-articulated, easily accessible international market and to sell largely on the basis of price (an assumption inherent in the definition), then it follows that such products will not pose the problem of market information quite so acutely for the less-developed countries. This establishes a necessary if not a sufficient condition for investment in such industries.

Of course, foreign investors seeking an optimum location for a captive facility may not have to concern themselves too much with questions of market information; presumably, they are thoroughly familiar with the marketing end of the business and are looking for a low-cost captive source of supply. In that case, the low cost of labor may be the initial attraction drawing the investor to less-developed areas. But other limitations in such areas, according to our hypothesis, will bias such captive operations toward the production of standardized items. The reasons in this case turn on the part played in the production process by external economies. Manufacturing processes which receive significant inputs from the local economy, such as skilled labor, repairmen, reliable power, spare parts, industrial materials processed according to exacting specification, and so on, are less appropriate to the less-developed areas than those that do not have such requirements. Unhappily, most industrial processes require one or another ingredient of this difficult sort. My guess is, however, that the industries which produce a standardized product are in the best position to avoid the problem, by producing on a vertically-integrated self-sustaining basis.

In speculating about future industrial exports from the less-developed areas, therefore, we are led to think of products with a fairly clear-cut set of economic characteristics.[5] Their production function is such as to require significant inputs of labor; otherwise there is no reason to expect a lower production cost in less-developed countries. At the same time, they are products with a high price elasticity of demand for the output of individual firms; otherwise, there is no strong incentive to take the risks of pioneering with production in a new area. In addition, products whose production process did not rely heavily upon external economies would be more obvious candidates than those which required a more elaborate industrial environment. The implications of remoteness also would be critical; products which could be precisely described by standardized specifications and which could be produced for inventory without fear of obsolescence would be more relevant than those

5. The concepts sketched out here are presented in more detail in my "Problems and Prospects in the Export of Manufactured Products from the Less-developed Countries," U.N. Conference on Trade and Development, Dec. 16, 1963 (mimeo.).

which had less precise specifications and which could not easily be ordered from remote locations. Moreover, high-value items capable of absorbing significant freight costs would be more likely to appear than bulky items low in value by weight. Standardized textile products are, of course, the illustration par excellence of the sort of product that meets the criteria. But other products come to mind such as crude steel, simple fertilizers, newsprint, and so on.

Speculation of this sort draws some support from various interregional experiences in industrial location. In the United States, for example, the "export" industries which moved to the low-wage south in search of lower costs tended to be industries which had no great need for a sophisticated industrial environment and which produced fairly standardized products. In the textile industry, it was the grey goods, cotton sheetings and men's shirt plants that went south; producers of high-style dresses or other unstandardized items were far more reluctant to move. In the electronics industry, it was the mass producers of tubes, resistors and other standardized high-volume components that showed the greatest disposition to move south; custom-built and research-oriented production remained closer to markets and to the main industrial complexes. A similar pattern could be discerned in printing and in chemicals production.[6]

In other countries, a like pattern is suggested by the impressionistic evidence. The underdeveloped south of Italy and the laggard north of Britain and Ireland both seem to be attracting industry with standardized output and self-sufficient process.[7]

Once we begin to look for relevant evidence of such investment patterns in the less-developed countries proper, however, only the barest shreds of corroboratory information can be found. One would have difficulty in thinking of many cases in which manufacturers of standardized products in the more advanced countries had made significant investments in the less-developed countries with a view of exporting such products from those countries. To be sure, other

6. This conclusion derives largely from the industry studies conducted in connection with the New York Metropolitan Region study. There have been some excellent more general analyses of shifts in industrial location among the regions of the United States. See e.g., Victor R. Fuchs, *Changes in the Location of Manufacturing in the United States Since 1929* (New Haven: Yale University Press, 1962). Unfortunately, however, none has been designed, so far as I know, to test hypotheses relating locational shifts to product characteristics such as price elasticity of demand and degree of standardization.

7. This statement, too, is based on only impressionistic materials. Among the more suggestive, illustrative of the best of the available evidence, see J. N. Toothill, *Inquiry into the Scottish Economy* (Edinburgh: Scottish Council, 1962).

types of foreign investment are not uncommon in the less-developed countries, such as investments in import-replacing industries which were made in the face of a threat of import restriction. But there are only a few export-oriented cases similar to that of Taiwan's foreign-owned electronics plants and Argentina's new producing facility, set up to manufacture and export standard sorting equipment for computers.

If we look to foreign trade patterns, rather than foreign investment patterns, to learn something about the competitive advantage of the less-developed countries, the possibility that they are an attractive locus for the output of standardized products gains slightly more support. The Taiwanese and Japanese trade performances are perhaps the most telling ones in support of the projected pattern; both countries have managed to develop significant overseas markets for standardized manufactured products. According to one major study of the subject (a study stimulated by the Leontief paradox), Japanese exports are more capital-intensive than is the Japanese production which is displaced by imports; [8] this is what one might expect if the hypothetical patterns suggested by Figure I were operational. Apart from these cases, however, all that one sees are a few provocative successes such as some sporadic sales of newsprint from Pakistan, the successful export of sewing machines from India, and so on. Even in these cases, one cannot be sure that they are consistent with the hypothesis unless he has done a good deal more empirical investigation.

The reason why so few revelant cases come to mind may be that the process has not yet advanced far enough. Or it may be that such factors as extensive export constraints and overvalued exchange rates are combining to prevent the investment and exports that otherwise would occur.

If there is one respect in which this discussion may deviate from classical expectations, it is in the view that the overall scarcity of capital in the less-developed countries will not prevent investment in facilities for the production of standardized products.

There are two reasons why capital costs may not prove a barrier to such investment.

First, according to our hypotheses, the investment will occur in industries which require some significant labor inputs in the production process; but they will be concentrated in that subsector of the

8. M. Tatemoto and S. Ichimura, "Factor Proportions and Foreign Trade: the Case of Japan," *Review of Economics and Statistics*, XLI (Nov. 1959), 442–46.

industry which produces highly standardized products capable of self-contained production establishments. The net of these specifications is indeterminate so far as capital-intensiveness is concerned. A standardized textile item may be more or less capital-intensive than a plant for unstandardized petro-chemicals.

Besides, even if the capital requirements for a particular plant are heavy, the cost of the capital need not prove a bar. The assumption that capital costs come high in the less-developed countries requires a number of fundamental qualifications. The reality, to the extent that it is known, is more complex.

One reason for this complexity is the role played by the international investor. Producers of chemical fertilizers, when considering whether to invest in a given country, may be less concerned with the going rate for capital in that country than with their opportunity costs as they see such costs. For such investors the alternatives to be weighed are not the full range of possibilities calling for capital but only a very restricted range of alternatives, such as the possibilities offered by chemical fertilizer investment elsewhere. The relevant capital cost for a chemical fertilizer plant, therefore, may be fairly low if the investor is an international entrepreneur.

Moreover, the assumption that finance capital is scarce and that interest rates are high in a less-developed country may prove inapplicable to the class of investors who concern us here.[9] The capital markets of the less-developed countries typically consist of a series of water-tight, insulated, submarkets in which wholly different rates prevail and between which arbitrage opportunities are limited. In some countries, the going figures may vary from 5 to 40 per cent, on grounds which seem to have little relation to issuer risk or term of loan. (In some economies, where inflation is endemic, interest rates which in effect represent a negative real cost are not uncommon.)

These internal differences in interest rates may be due to a number of factors: the fact that funds generated inside the firm usually are exposed to a different yield test than external borrowings; the fact that government loans are often floated by mandatory levies on banks and other intermediaries; and the fact that funds borrowed by governments from international sources are often re-

9. See George Rosen, *Industrial Change in India* (Glencoe, Ill.: Free Press, 1958). Rosen finds that in the period studied from 1937 to 1953, "there was no serious shortage of capital for the largest firms in India." Gustav F. Papanek makes a similar finding for Pakistan for the period from 1950 to 1964 in a book about to be published.

loaned in domestic markets at rates which are linked closely to the international borrowing rate, however irrelevant that may be. Moreover, one has to reckon with the fact that public international lenders tend to lend at near-uniform rates, irrespective of the identity of the borrower and the going interest rate in his country. Access to capital on the part of underdeveloped countries, therefore, becomes a direct function of the country's capacity to propose plausible projects to public international lenders. If a project can plausibly be shown to "pay its own way" in balance-of-payment and output terms at "reasonable" interest rates, the largest single obstacle to obtaining capital at such rates has usually been overcome.

Accordingly, one may say that from the entrepreneur's viewpoint certain systematic and predictable "imperfections" of the capital markets may reduce or eliminate the capital-shortage handicap which is characteristic of the less-developed countries; and, further, that as a result of the reduction or elimination such countries may find themselves in a position to compete effectively in the export of certain standardized capital-intensive goods. This is not the statement of another paradox; it is not the same as to say that the capital-poor countries will develop capital-intensive economies. All we are concerned with here is a modest fraction of the industry of such countries, which in turn is a minor fraction of their total economic activity. It may be that the anomalies such industries represent are systematic enough to be included in our normal expectations regarding conditions in the less-developed countries.

* * * * *

Like the other observations which have preceded, these views about the likely patterns of exports by the less-developed countries are attempts to relax some of the constraints imposed by purer and simpler models. Here and there, the hypotheses take on plausibility because they jibe with the record of past events. But, for the most part, they are still speculative in nature, having been subjected to tests of a very low order of rigorousness. What is needed, obviously, is continued probing to determine whether the "imperfections" stressed so strongly in these pages deserve to be elevated out of the footnotes into the main text of economic theory.

Harvard Graduate School of Business Administration

[21]

Economic Development and the Transfer of Technology: Some Historical Perspectives

NATHAN ROSENBERG

I

Back in an earlier, more naïve day, we managed to allow ourselves to believe that there was a purely technological solution—a cheap "technological fix"—to the problems of poverty and economic backwardness which beset most of the human race. In the Point Four of his 1949 inaugural address President Truman spoke optimistically of the incalculable benefits which technical assistance could bring to improving the desperate plight of the poor throughout the world "We must embark," he said, "on a bold new program for making the benefits of our scientific advances and industrial progress available for the improvement and growth of underdeveloped areas."

Unfortunately we exaggerated from the outset what could be accomplished solely by making Western techniques available. In some measure the present mood of disillusion with foreign aid in general is due to the unrealistically high expectations which we once attached to such programs (including their effectiveness as weapons of politics and diplomacy). We now realize that there are a host of difficulties—institutional and otherwise—which hamper the successful adoption of foreign technology. I would like to contribute some observations of my own on the obstacles in the way of a successful transfer of technology. I will also, before I am done, go one pessimistic step further, and suggest that there are serious problems, not just in facilitating the transfer but in the very nature of the technology which we presently have to offer and its possibly limited relevance (especially in agriculture) to the problems of poor countries. But it should be emphasized at the outset that the main thrust of this paper is to pose problems for the international transfer of technology which emerge from an examination of earlier—primarily 19th-century—mechanisms through which such transfers have taken

Dr. Rosenberg, of the University of Wisconsin, was a visitor at Nuffield College, Oxford, during the spring of 1970 and is a visiting professor at the University of the Philippines during 1970–71. He has written widely on economic history, including *The American System of Manufactures*.

place. I do not propose to offer authoritative conclusions or recommendations but rather—and much more modestly—to use historical experience to sharpen and to enlarge our appreciation for the difficulties which may still lie ahead.

I propose, first, to discuss the role of the capital goods industries in the 19th century as centers for the creation and diffusion of new techniques. I start from the proposition that technical change is not, and never has been, a random phenomenon. In both the United States and the United Kingdom in the 19th century, technological change became institutionalized in a very special way—that is, in the emergence of a group of specialized firms which were uniquely oriented toward the solution of certain kinds of technical problems. The rapid rate of technological change was completely inseparable from these capital goods firms. In fact, I would regard the emergence of such firms as the fundamental institutional innovation of the 19th century from the point of view of the industrialization process. I am particularly interested in the factors which affected the success and failure of these industries in the context of a developing industrial economy. In general, I shall argue that these industries performed certain critical functions which accounted for the speed of the growth process in the 19th century. But I am anxious to free myself of the charge of suggesting that poor countries today will have to tread exactly the same path. The most important reason why poor countries may *not* have to tread the same path as their industrial predecessors is precisely that industrial countries have already done so. One of the advantages of not taking the lead in economic development is that, once an objective has been reached and clearly demarcated, other and easier routes to attain that objective may become obvious. Or, to put the point a little differently, there is no reason to believe that the optimal path in the development of a new technology is the same as the optimal path for transferring and adapting that technology, once it has been developed.[1] In fact, I want to insist on this point. Economic growth has never been a process of mere replication.

[1] For a more formal treatment of the problems of technological leadership and "followership," see Edward Ames and Nathan Rosenberg, "Changing Technological Leadership and Economic Growth," *Economic Journal*, vol. 73, no. 289 (March 1963). A recent study of the diffusion of new technologies in six West European countries suggested that, for five important new processes, "countries which are pioneers tend to have *slower* speeds of diffusion. This result is consistent with the hypothesis that the pioneer faces all sorts of teething troubles, new problems associated with the new technique; these are likely to be—partly and gradually—solved by the time others adopt it. It is therefore not necessarily desirable to be the first to introduce a new technique" (G. F. Ray, "The Diffusion of New Technology: A Study of Ten Processes in Nine Industries," *National Institute Economic Review* [May 1969], p. 82).

Although one can identify certain broad phenomena and functions common to all rapidly growing economies (changes in the composition of output and labor force, urbanization) the actual paths to development show very wide institutional variations. Certainly the latecomers on the development scene (e.g., Germany, Japan, Russia) followed very different paths and sequences than did pioneering England.[2]

Perhaps it should also be said that the policy implications of what follows are not at all clear—at least not to me. It is not my intention to press the analysis of technology transfer in the last third of the 20th century onto a 19th-century Procrustean bed. Rather, I want to draw upon the experience of the 19th and early 20th centuries as a way of suggesting what are likely to be problems of considerable interest and importance.

II

I have been very much interested in problems of the transmission of technology in the 19th century, not only from one country to another but from one industry to another. The problems involved are not entirely unrelated. Simply because transmitting technologies from advanced to underdeveloped countries presents many unique problems, we should not ignore the fact that transmitting technology from one use to another always presents certain elements of novelty.[3]

In both the United States and United Kingdom, to a very surprising degree, innovations in the area of machine technology were transmitted from existing uses to new uses by a very personal mechanism. Impersonal forms of communication—such as trade journals—were often important in generating interest but hardly ever provided enough of the highly specific information required for the successful transmission of a new productive technique.

There are remarkable similarities, in the United States and the United Kingdom, in the pattern of transmission. A dominant pattern in both countries went as follows: A firm develops a new technique or process in response to a problem in a particular industry—say firearms manufacture. It later becomes apparent that this technique is applicable to typewriters. The technique is "transferred" to the typewriter industry

[2] See Alexander Gerschenkron, "Economic Backwardness in Historical Perspective," in *The Progress of Underdeveloped Areas*, ed. B. F. Hoselitz (Chicago, 1952).

[3] See A. P. Usher, *A History of Mechanical Inventions*, rev. ed. (Cambridge, Mass., 1954), esp. chap. 4; Nathan Rosenberg, "Technological Change in the Machine Tool Industry, 1840–1910," *Journal of Economic History* 23, no. 4 (December 1963): 414–46; Nathan Rosenberg, ed., *The American System of Manufactures* (Edinburgh, 1969).

by the firm which developed the technique actually undertaking the production of typewriters (Remington). In other words, a very common mode of technology transfer *between industries* took place as a result of firms adding to, or switching, their product lines.

An alternative way of stating this is to say that, in the area of machine technology, firms became increasingly specialized by *process* rather than product. In both countries we find firms undertaking a combination or a sequence of products which can be understood only in terms of their common underlying technical processes. We can find many examples of firms in each country undergoing the same product sequences over time—for example, firearms, sewing machines, bicycles—or undertaking the same product combinations. If we examine historically the operation of multiproduct firms in both countries, we find that they frequently arrived at the combination through the development of a specialized skill at a particular process—for example, hydraulics (Armstrong, Tangyes, Fawcetts)—or an unusual expertise in the property of a material—for example, William Fairbairn in Manchester and William Sellers in Philadelphia both undertook activities where the structural properties of iron were critical to success—bridges, iron ships, multistory dwellings, cranes, etc. The center for the transmission of relevant knowledge and techniques from one industry to another, and for the application of known techniques to new uses, was, to a very considerable degree, the individual firm.

Where the transfer of technology involved places geographically distant from one another, the reliance upon the migration of trained personnel (at least temporarily) was very strong. This was true even within the United States. It would be possible to trace the diffusion of machine tool technology in America in the 19th century in terms of a genealogical table showing the movement of a very small number of highly skilled mechanics from the places where they received their training (apprenticeship and early "on-the-job" learning) to the firms where they eventually brought their valuable skills; often these were new firms established by the mechanic himself.[4]

I have been impressed by the extent to which the transfer of technological skills—even between two countries so apparently "close together" as Britain and the United States in the mid-19th century—was dependent upon the transfer of skilled personnel.[5] This was demonstrated

[4] Joseph W. Roe, *English and American Tool Builders* (New Haven, Conn., 1916), in fact contains several such genealogies.

[5] See the illuminating account of a 16th-century transfer in Warren C. Scoville, "Minority Migrations and the Diffusion of Technology," *Journal of Economic His-*

by several striking episodes. For example, when the British government purchased a large quantity of American gun-making machinery in the 1850s and introduced it into the Enfield Arsenal, it was found to be absolutely essential to employ a large number of American machinists and supervisory personnel. At about the same time, Samuel Colt, impressed with the backwardness of British firearms technology as compared with his own plant in Hartford, set up a factory in London for the purpose of producing firearms with American machinery and British workers. Although Colt had been fantastically successful in the United States—where he ran the largest private armory in the world—his London plant had to be closed down, apparently because of the difficulties encountered in employing British workers with unfamiliar American machinery.[6]

There is much evidence that the transmission of industrial technology from England to the Continent in the first half of the 19th century was also heavily dependent upon the same sort of personal mechanism. David Landes has suggested, for example, that in spite of legal prohibitions until 1825, there were at least 2,000 skilled British workers on the Continent providing indispensable assistance in the adoption of the newly developed techniques.

> The best of the British technicians to go abroad were usually entrepreneurs in their own right, or eventually became industrialists with the assistance of continental associates or government subventions. Many of them came to be leaders of their respective trades: one thinks of the Waddingtons (cotton), Job Dixon (machine-building), and James Jackson (steel) in France; James Cockerill (machine construction) and William Mulvany (mining) in Germany; Thomas Wilson (cotton) in Holland; Norman Douglas (cotton) and Edward Thomas (iron and engineering) in the Austrian empire; above all, John Cockerill in Belgium, an aggressive, shrewd businessman of supple ethical standards, who took all man-

tory 11, no. 4 (Fall 1951): 347–60. A. R. Hall has concluded: "It seems fairly clear that in most cases in the 16th century—and indeed long afterwards—the diffusion of technology was chiefly effected by persuading skilled workers to emigrate to regions where their skills were not yet plentiful" (A. R. Hall, "Early Modern Technology, to 1600," in *Technology in Western Civilization*, ed. Melvin Kranzberg and Carroll W. Pursell, Jr. [New York, 1967], 1:85). For a similar conclusion with respect to mining and metallurgy in an earlier period, see Bertrand Gille, "Technological Developments in Europe: 1100 to 1400," in *The Evolution of Science*, ed. Guy S. Metraux and François Crouzet (New York, 1963), p. 201.

[6] These episodes are described at length in the introductory essay in Rosenberg, ed., *The American System of Manufactures*.

ufacturing as his province and with the assistance of first the Dutch and then the Belgium governments made a career of exploiting the innovations of others.[7]

These episodes suggest an important point with respect to the transfer of technology. The notion of a production function as a "set of blueprints" comes off very badly if it is taken to mean a body of techniques which is available independently of the human inputs who utilize it. The point has been well expressed by Svennilson:

> It would be far too crude to assume, as often seems to be the case, that there is a common fund of technical knowledge, which is available to anybody to use by applying his individual skill. We must take into account that only a part, and mainly the broad lines, of technical knowledge is codified by non-personal means of intellectual communication or communicated by teaching outside the production process itself. The technical knowledge of persons who have been trained in actual operations has a wider scope, especially as regards the application of more broad knowledge. The "common fund," thus, covers only part of the technical knowledge to which individuals or groups of persons apply their personal skill.[8]

If, with Svennilson, we define "know-how" as "the capacity to use technical knowledge," then it is apparent that such know-how is essential for the successful utilization of the technical information incorporated in the economist's production function. And such know-how has been transmitted personally in the past simply because it is a noncodified kind of skill which has been largely acquired by direct exposure to and participation in the work process.[9] A somewhat amusing, yet apposite,

[7] David Landes, *The Unbound Prometheus* (Cambridge, 1969), pp. 148–49. Landes adds: "Perhaps the greatest contribution of these immigrants was not what they did but what they taught. . . . The growing technological independence of the Continent resulted largely from man-to-man transmission of skills on the job" (p. 150). For a more detailed discussion of this subject, see W. O. Henderson, *Britain and Industrial Europe, 1750–1870* (Liverpool, 1954).

[8] I. Svennilson, in *Economic Development with Special Reference to East Asia*, ed. K. Berrill (New York, 1964), p. 407.

[9] Some of the recent work of Derek Price lends indirect support to this position. Price argues that there exist powerful professional incentives for the scientist to publish the results of his work quickly and authoritatively, whereas the incentive structure of the working technologist is one which (with some exceptions) encourages concealment and a reluctance to publish. Science is *papyrocentric*, whereas technology is *papyrophobic* (see Derek J. De Solla Price, "Is Technology Historically Independent of Science? A Study in Statistical Historiography," *Technology and Culture* 6, no. 4 [Fall 1965]: 553–68).

instance of the role of know-how has recently been cited by Lucius Ellsworth, who points out that although we possess a complete set of the appropriate craft tools, no one in the United States today *knows how* to use coopers' tools for making wooden barrels—a once-honored and widely practiced craft.[10]

The extent of specialization in the production of machinery was of critical importance in the development of industrial skills. To appreciate this importance we must think in dynamic terms as well as in terms of static allocative efficiency. For there is a crucial learning process involved in machinery production, and a high degree of specialization is conducive both to a more effective learning process and to a more effective *application* of that which is learned. This highly developed facility in the designing and production of specialized machinery is, perhaps, the most important single characteristic of a well-developed capital goods industry and constitutes an external economy of enormous importance to other sectors of the economy.[11]

It is a common practice to look upon industrialization as involving not only growing specialization but also growing complexity and differentiation. While this is certainly true in the sense that there takes place a proliferation of new skills, facilities, commodities, and services, it also overlooks some very important facts. The most important for present purposes is that industrialization was characterized by the introduction of a relatively small number of broadly similar productive processes to a large number of industries.[12] This follows from the familiar fact that industrialization in the 19th century involved the growing adoption of a metal-using technology employing decentralized sources of power.

The use of machinery in the cutting of metal into precise shapes involves, to begin with, a relatively small number of operations (and therefore machine types): turning, boring, drilling, milling, planing, grinding, polishing, etc. Moreover, all machines performing such operations confront a similar collection of technical problems, dealing with such matters as power transmission, control devices, feed mechanisms, friction reduction, and a broad array of problems connected

[10] Lucius F. Ellsworth, "A Directory of Artifact Collections," in *Technology in Early America*, ed. Brooke Hindle (Chapel Hill, N.C., 1966), p. 96.

[11] This paragraph and the three following are taken, with minor modification, from Nathan Rosenberg, "Technological Change in the Machine Tool Industry, 1840–1910."

[12] For more recent evidence on such trends from the point of view of the input-output structure of the economy, see Anne P. Carter, "The Economics of Technological Change," *Scientific American* 214, no. 4 (April 1966): 25–31.

with the properties of metals (such as ability to withstand stresses and heat resistance). It is because these processes and problems became common to the production of a wide range of disparate commodities that industries which were apparently unrelated from the point of view of the nature and uses of the final product became very closely related (technologically convergent) on a technological basis—for example, firearms, sewing machines, and bicycles.

Because of this technological convergence, the machine tool industry, in particular, played a unique role both in the initial solution of technological problems and in the rapid transmission and application of newly learned techniques to other uses. I have found it useful to look at the machine tool industry in the 19th and early 20th centuries as a center for the acquisition and diffusion of new skills and techniques in a "machinofacture" type of economy. Its chief importance lay in its strategic role in the learning process associated with industrialization. This role was a dual one: (1) new skills and techniques were developed or perfected there in response to the demands of specific customers, and (2) once they were acquired, the machine tool industry was the main transmission center for the transfer of new skills and techniques to the entire machine-using sector of the economy.

The questions which currently trouble me are: How do we institutionalize such activities now? What current substitutes are available for the performance of the vital activities of the production and diffusion of industrial skills—activities which were largely performed by the capital goods industries in the 19th century? The experience of successfully industrializing countries in the 19th century indicates that the learning experiences in the design and use of machinery were vital sources of technological dynamism, flexibility, and vitality. Countries which rely upon the importation of a foreign technology are thereby largely cut off from this experience, but other institutional alternatives may well be available. The contemporary rise of the multinational firm in such areas as chemicals, chemical process plants, plastics, and electronic capital goods may reflect, more than anything else, an attempt to fill this gap: to provide institutional and organizational mechanisms which will facilitate the transfer of know-how. This occurs largely by institutional innovations which make the services of the personnel with the appropriate knowledge and skills readily available, on short notice, anywhere in the world.[13]

[13] In a study of the world market in chemical processing plants, Freeman has pointed out: "The main European 'design-engineering' offices of the large contractors typically employ from 300 to 700 total office-based staff. They are concentrated mainly in London (where half a dozen of them have such offices) and to a lesser

III

Anglo-American experience with the introduction and diffusion of technology in the 19th century points strongly to the importance of the composition of consumer demand and the malleability of public tastes. The willingness of the public to accept a homogeneous final product was a decisive factor in the transition from a highly labor-intensive handicraft technology to one involving a sequence of highly specialized machines. Across a whole range of commodities we find evidence that British consumers imposed their tastes on the producer in a manner which seriously constrained him with respect to the exploitation of machine technology. English observers often noted with some astonishment that American products were designed to accommodate, not the consumer, but the machine. Lloyd noted, for example, of the American cutlery trade, that "where mechanical devices cannot be adjusted to the production of the traditional product, the product must be modified to the demands of the machine. Hence, the standard American table knife is a rigid, metal shape, handle and blade forged in one piece, the whole being finished by electroplating—an implement eminently suited to factory production."[14] Even with respect to an object so ostensibly utilitarian as a gun, the British civilian market was dominated by peculiarities of taste which essentially precluded machine techniques. English civilians were in the habit of having their guns made to order, like their clothing, and treating their gunsmith in much the same manner as their tailor. In fact, we hear a Colonel Hawkins instructing his upper-class readers in the 1830s that "the length, bend and casting of a stock [gunstock] must, of course, be fitted to the shooter, who should have his measure for them as carefully entered in a gunmaker's books, as that for a suit of clothes on those of his tailor."[15]

---

extent in The Hague, Paris, and Frankfurt. London is the principal European base of the American contractors and the largest world centre for chemical 'design-engineering' work. But the large American contractors show great ingenuity and flexibility in switching work from one office to another anywhere in the world, in order to achieve an optimum loading of their total resources. The London office, even though employing 500 or fewer, is able to take on several very large contracts because it is part of a world-wide organization and can call upon process design and know-how from the parent company, as well as specialists to deal with crises or bottlenecks. A particular individual, a scale model, or a set of drawings may be flown out at a few hours' notice" (C. Freeman, "Chemical Process Plant: Innovation and the World Market," *National Institute Economic Review* [August 1968], p. 37).

[14] G. I. H. Lloyd, *The Cutlery Trades* (London, 1913), pp. 394–95.

[15] As quoted in *A Treatise on the Progressive Improvement and Present State of the Manufactures in Metal* (London, 1833), 2:105.

On the other hand, Siegfried Giedion has pointed out that "in the United States, where department stores had slowly been developing since the 'forties, ready-made clothes, in contrast to Europe, were produced from the start."[16] Giedion's point is doubly interesting in the present context because, in fact, the United States, although beginning industrialization later than England, took an early lead in the manufacture of both guns and ready-made clothing (as well as in new techniques of merchandising).

There is, it seems to me, an intimate interrelationship between composition of demand and homogeneity of product, on the one hand, and the range of technological possibilities open to society on the other. Certainly the producer must have some minimum degree of freedom to design his product in order to make it suitable for mass production techniques. There is very little doubt that, in this respect, the American producer had several degrees of freedom more than did his English counterpart. This difference had a great deal to do with the fact that mass production technology essentially originated in the United States rather than England, with early American leadership in products which were particularly well suited for mass production—firearms, sewing machines, typewriters, and later automobiles—and with the slow acceptance of many of these techniques in England.

But I am anxious not to leave the impression that Anglo-American differences can all be explained in terms of differences in consumer tastes, although I do think these were important. They seem to have been part of a much more widespread phenomenon in Britain as compared with the United States of "customer initiative" as opposed to "producer initiative." If we examine the relation between producers of capital equipment and their purchasers in both countries, we also find analogous differences. That is, in America the producer of capital goods took the initiative in matters of machine design and successfully suppressed variations in product design which served no clearly defined purpose. He brought about, in other words, a high degree of standardization in the machinery, which very much simplified his own production problems and in turn reduced the price of capital goods. Producer initiative was a very important factor in developing patterns of efficient specialization in American capital goods production.

In England, on the other hand, the capital goods producer remained, to a surprising degree, what Landes has aptly called a "custom tailor working in metal." The explanation for this difference in roles is not at all clear. In the case of many kinds of capital goods, the variations in machine design were quite unrelated to any functional justification.

[16] Siegfried Giedion, *Mechanization Takes Command* (New York, 1948), p. 90.

Buyers of machinery were in the habit of drawing up blueprints with highly detailed specifications which the machine producer had to agree to provide as a condition of fulfillment of the contract. Many of the specifications were often nothing more than the result of historical accident. In England the initiative in matters of machine design was emphatically in the hands of the buyer, and this had serious implications for the machine-producing sector both as a transmission center for new techniques and as a producer of low-cost machinery.[17]

The problems encountered by the machinery-producing sector were further intensified by the role of that strange British institution, the consulting engineer. These engineers were imbued with a professional tradition which often led to an obsession with technical perfection in a purely engineering sense, and they imposed their own tastes and idiosyncrasies upon product design. In America, by contrast, the engineer and engineering skills were more effectively subordinated to business discipline and commercial criteria and did not dominate them.[18] The result was to perpetuate, in Great Britain, a preoccupation with purely technical aspects of the final *product* rather than with the productive *process*. This is reflected in the history of the automobile, where, of course, British engineers developed an early reputation for high-priced, high-quality engineering. In 1912, when Henry Ford was preparing to demonstrate to the world the great possibilities of standardized, high volume production, an influential British trade journal commented: "It is highly to the credit of our English makers that they choose rather to maintain their reputation for high grade work than cheapen that reputation by the use of the inferior material and workmanship they would be obliged to employ to compete with American manufacturers of cheap cars."[19]

[17] At the turn of the 20th century, some evidence of change began to appear. A British observer in 1902 made the following highly revealing comments: "Some few years ago no machine-tool maker ventured to offer advice to a manufacturing engineer as to the tools he might buy or the methods he should follow. He would have been told to mind his own business. Today a good deal of help and advice is often requested and the tool maker and user, as they should be, are often in consultation as to tools, methods and organization. Machine tool making in America has been, I take it, considered a reputable business for some years, whereas here for a time it certainly was but little considered" (William H. Booth, "An English View of American Tools," *American Machinist*, no. 45, November 6, 1902, p. 1580).

[18] See, for example, Daniel H. Calhoun, *The American Civil Engineer: Origin and Conflict* (Cambridge, Mass., 1960), esp. chap. 6.

[19] *Autocar*, September 21, 1912, as quoted in S. B. Saul, "The Motor Industry in Britain to 1914," *Business History* 5, no. 1 (December 1962): 41. The belief long persisted in British industry that high quality was incompatible with mass produc-

The earlier design of locomotives was a classic case of a rampant and pointless individualism. By 1850, after the early period of experimentation (when there was an important justification for variations in design),[20] the basic features of a locomotive stabilized.[21] But innumerable minor variations of this basic design persisted. "Probably five distinct classes of locomotives would afford a variety sufficiently accommodating to suit the varied traffic of railways," an expert wrote in 1855, "whereas I suppose the varieties of locomotives in actual operation in this country and elsewhere are very nearly five hundred."[22] There is much authoritative evidence that, by comparison with American practice, a needless proliferation of designs and specifications prevailed in Great Britain throughout a broad range of iron and steel products and engineering works generally.[23]

This absence of standardization vastly complicated the process of adopting the technology and organization of mass production. It seriously inhibited the growth of specialization by firms which was so characteristic of the American machinery industry in the second half of the 19th century. For example, British railroad companies built their rolling stock in their own workshops. Private locomotive producers

tion. Much evidence on this point for the late 1920s may be found in Committee on Industry and Trade, *Survey of Metal Industries*, pt. 4 (London, 1928), pp. 227–28, 220–21, and passim.

[20] See Burton H. Klein, "A Radical Proposal for R. and D.," *Fortune* 57, no. 5 (May 1958): 112 ff., for a statement of the case for the desirability of parallel research efforts under conditions of uncertainty.

[21] It is interesting to note that, as early as the 1840s, American engineers were playing a vital role in bringing the railroad to Russia. See the account of the work of Major George Washington Whistler in Albert Parry, *Whistler's Father* (Indianapolis, 1939). Of Major Whistler, the awed muzhiks are supposed to have exclaimed: "How smart the foreign gentleman, who has harnessed the samovar and made it run." The book is unfortunately thin on technological history. It does, however, provide a minimal account of Whistler's role in directing the construction of the Saint Petersburg to Moscow railway and of his establishment of a factory at Alexandrovsky for building locomotives and other railroad equipment.

[22] J. H. Clapham, *An Economic History of Modern Britain*, 3 vols. (Cambridge, 1938), 2:76. English orientation toward export markets doubtless increased the difficulties of standardization of locomotive design (see *Survey of Metal Industries*, pp. 168–72). Nevertheless, the high degree of diversification had developed early in the history of the industry when it was certainly preoccupied with the needs of domestic users.

[23] J. Stephen Jeans, ed., *American Industrial Conditions and Competition*, Reports of the Commissioners Appointed by the British Iron Trade Association to Enquire into the Iron, Steel and Allied Industries of the United States (London, 1902), esp. pp. 255–59.

worked essentially for the export trade, and therefore large numbers of firms in Britain were engaged in small-batch production of a wide range of railroad equipment. During the 1920s one locomotive manufacturer was making equipment for twenty-four different gauges, varying from 18 inches to 5 feet 6 inches, and locomotive sizes ranging from 6 to 60 tons.[24] Nothing comparable to the Baldwin Locomotive Company of Philadelphia ever developed, under these circumstances, in Britain. And what was characteristic of British locomotive production was broadly true of all British engineering, where general engineering firms predominated—that is, firms which undertook to produce a range of products which would have seemed incredibly diverse by American standards. British engineering and equipment-producing firms were typically engaged in "batch" or "small-order" production. At the same time, of course, the absence of standardization further increased capital costs by raising the costs of repair and necessitating larger inventories.

Similarly, one of the most characteristic features of automobile production in Britain was the extent to which each firm produced its own components. In part this was due to the failure to standardize, but cause and effect are very hard to disentangle in these matters. There is much evidence suggesting that, in the early days of the automobile industry, British firms found it impossible to make satisfactory arrangements for the production of components by other firms and therefore fell back upon the expedient of producing such components themselves. When William Morris (later Lord Nuffield) made up his mind to purchase the main components of his automobiles from specialist producers (1913), he found that they could not meet his requirements, and he finally had to import components from America.[25]

All this was a very serious handicap because it meant that individual firms were responsible for producing a range of products which was beyond their organizational and technical competence.[26] From the per-

[24] *Survey of Metal Industries*, p. 172.

[25] H. J. Habakkuk, *American and British Technology in the 19th Century* (Cambridge, 1962), p. 203.

[26] A good deal of valuable material on Anglo-American differences in engineering methods at the beginning of the 20th century will be found in the paper by H. F. L. Orcutt, "Modern Machine Methods," and the subsequent discussion, in *Proceedings of the Institution of Mechanical Engineers* (1902), pts. 1–2, pp. 9–112. The greater degree of specialization of both American firms and American machinery as compared with their English counterparts received much attention. J. R. Richardson, who operated a general engineering workshop, justified his unwillingness to purchase American molding machines in the following way: "It was not that English engineers did not understand American methods, but that Americans did not as a rule

spective of American industrialization, where highly elaborate patterns of interfirm specialization developed, this was one of the most striking features of British industry. Not only did railways produce their own locomotives and automobile companies their own components, but users of capital equipment such as machine tools often made the tools themselves—for example, sewing machine companies and textile machinery firms—long after such production had been taken over by specialist firms in the United States.[27]

IV

Although I have already suggested several factors which may account for differences in the degree of specialization in the United States and Britain, I am far from convinced that such factors as differences in taste, standardization, and producer initiative constitute a complete explanation. To put the point in its most general way: American firms showed a much greater talent than British firms for coordinating successfully their relationships with other firms upon whom they were dependent for the supply of essential inputs. American firms learned to integrate their own operations with those of their main suppliers in a way which enabled them to confine their own productive operations to a limited number of specializations. Clearly the British failure to subcontract more efficiently was closely related to the factors which we have already discussed. But the failure seems also to have included an

understand the conditions which obtained in large engineering works in England having a big general practice. There must be a large run of work. Even the most enthusiastic Americans had told him that a large quantity was not needed, that it could be done perfectly well with a dozen, but very often a dozen was a large quantity. Not only was it necessary to have on his catalogue 500 different types and sizes of steam-engines, but an infinite variety of mining and general machinery; and in addition his firm was expected to do anything required, and had to do it even if it only had to be done once" (pp. 72–73). The important unanswered—indeed, unposed—question is why the general engineering firm persisted as it did in England (see S. B. Saul, "The American Impact upon British Industry," *Business History* 3 [1960]: 21–27).

[27] America's superiority in the contriving of highly specialized machinery was clearly recognized as early as the 1850s. A parliamentary committee of distinguished engineers which visited America in 1853 stated: "As regards the class of machinery usually employed by engineers and machine makers, they are upon the whole behind those of England, but in the adaptation of special apparatus to a single operation in almost all branches of industry, the Americans display an amount of ingenuity, combined with undaunted energy, which as a nation we would do well to imitate, if we mean to hold our present position in the great market of the world" (*Report of the Committee on the Machinery of the U.S.*, as reprinted in Rosenberg, ed., *The American System of Manufactures*, pp. 128–29).

inability of British firms to rely more confidently upon the fulfillment of contracts with the degree of precision in engineering specifications which is so important to the assembly of components into a complex final product. F. W. Lanchester, in looking back upon his early experiences in the auto industry, where he attempted to persuade craftsmen to work to standard instructions so as to achieve interchangeability, commented: "In those days, when a body-builder was asked to work to drawings, gauges or templates, he gave a sullen look such as one might expect from a Royal Academician if asked to colour an engineering drawing."[28] The difficulties of successful subcontracting of high-precision components in the face of such attitudes can be readily imagined.

Furthermore, I wonder if the failure might not also involve the absence of some more subtle managerial or administrative talents which may have been available in greater abundance in the United States. For example, after discussing vertical integration as a possible solution to the problems of cooperation and communication among chemical firms, contractors, and component makers, Freeman has pointed out:

> The trend in the United States has been in the opposite direction. The oil and chemical firms make increasing use of specialist contractors and specialist component-makers. The contractors tend to divest themselves of manufacturing and fabricating activities. Studies of the aircraft industry and large weapon systems have found that prime United States contractors make far more use of specialist sub-contractors and external economies than European aircraft firms. When the prime contractor (whether in aircraft or chemical plant) uses large numbers of specialist sub-contractors, this means that he must acquire considerable skills in systems management and must be capable of enforcing very high technical standards on sub-contractors.[29]

[28] As quoted in Habakkuk, p. 203.

[29] Freeman, p. 49. Earlier, Freeman had stated: "We found in our case studies that a number of successful process innovations arose from intimate technical cooperation between chemical firms and contractors or chemical firms and component makers, or all three. This is particularly evident in the United States; much less so in Europe" (p. 48). In an article on the plastics industry, Freeman states, in connection with the machinery for making plastics: "Generally speaking a technically advanced and progressive machine industry will principally benefit the country in which it is located, because material-suppliers, machine-makers and fabricators can more easily co-operate there in experiment, development and design. This tripartite co-operation appears to be closer and more satisfactory in Germany than in Britain" (C. Freeman, "The Plastics Industry: A Comparative Study of Research and Innovation," *National Institute Economic Review* [November 1963], pp. 42–43; see also p. 46).

In another study, of electronic capital goods, Freeman emphasizes that the innovative process is now highly dependent upon a successful collaboration between producers of the final product and specialist makers of components.

> The development process in the electronic capital goods industry consists largely of devising methods of assembling components in new ways, or incorporating new components to make a new design, or developing new components to meet new design requirements. (This is not quite so simple as it may sound. There are more than a hundred different components in a TV set, more than 100 thousand components in a large computer, and more than a million in a big electronic telephone exchange.) Consequently, there must be close collaboration between end-product makers and component makers.[30]

There seem to be some kinds of talents involved here which economists have not yet identified very precisely but which may be important determinants of success in the utilization of certain kinds of technology. The Japanese, it is worth noting, have been extremely successful (especially since the 1920s) at subcontracting, and this ability is almost certainly closely connected with the important role which small firms continue to play in Japanese industry.[31] The Japanese, as Strassmann points out, have been highly successful in the "inter-firm coordination of schedules and standards in subcontracts."[32] On the other hand, the Russians have been singularly unsuccessful in introducing subcontracting arrangements. This failure has been closely connected with many of the organizational failures of Russian central planning.[33]

V

The existence of a well-developed domestic capital goods industry is important to the transmission of technology for a reason which is not yet sufficiently explicit. I have discussed the role of these industries so far in terms of their *capacities*, but we must recognize also their *motivations*. It is the producers of capital goods who have the financial incentive and therefore provide the pressures (marketing, demonstra-

30 C. Freeman, "Research and Development in Electronic Capital Goods," *National Institute Economic Review* (November 1965), p. 63.

31 Paul Strassmann, *Technological Change and Economic Development* (Ithaca, N.Y., 1968), p. 168.

32 Ibid., p. 274.

33 David Granick, *Soviet Metal-fabricating and Economic Development* (Madison, Wis., 1967), chap. 5.

tion) to persuade firms to adopt the innovation (which they produce). Creating a capital goods industry is, in effect, a major way of *institutionalizing* internal pressures for the adoption of new technology. In America the producers of capital goods have always played a major role in persuading and educating machinery users about the superiority and feasibility of new techniques.

This is an extremely important activity in overcoming the inevitable combination of inertia, ignorance, and genuine uncertainty which surrounds an untried product. The introduction of the diesel locomotive by General Motors is a classic case in point. In the United States both the railroad companies and the locomotive producers were extremely skeptical of the diesel engine and resisted its introduction. It took great promotional effort on the part of GM, which developed the diesel, to induce the railroads even to consider and experiment with the innovation. This kind of promotional activity, on the part of capital goods industries with a strong personal motive to gain acceptance for their product, seems to have been a critical factor in the American experience. Here again it has to be asked: what mechanisms or institutions can be substituted for the motivational pressure, provided in the past by domestic capital goods producers, on the part of poor countries which rely on distant foreign producers for their capital equipment?

In general the relationship between machinery producers and their customers was more successful in the United States than in England or elsewhere in Europe. In the confrontations that took place between these two groups, there seems to have been an interchange of information and communication of needs to which the machinery producers responded in a highly creative way. They learned to deal with the requirements of their customers at the same time that the machinery user learned to rely heavily on the judgment and initiative of the machinery supplier. It was, in part, the relative harmony and mutual confidence of these relationships which made it possible for machinery makers to eliminate customer preferences that were technically nonessential or irrelevant and therefore to design more highly standardized machinery. This was a process which involved an intimate knowledge of customer activities and needs and which presupposed frequent face-to-face confrontations and exchange of information (at which the British were much less successful than the Americans).[34]

[34] Strassmann makes the interesting point that "face-to-face contact to establish trust appears to be a need throughout the technical transfer network. Lack of full trust invariably means some concealment of information which impairs perception of needs at subsequent stations. Moreover, a sense of insecurity may blur the trans-

Here the extent of machinery standardization intrudes itself again as a significant variable. I think it is reasonable to argue that, once the technical characteristics of a new machine have become stabilized, the amount of initiative which it is possible for capital goods suppliers to exert with their customers will depend on the extent of standardization. With extreme heterogeneity (British railroads) the role of the capital goods producer tends to become passive and adaptive rather than active and initiatory. Initiative will then reside with the user of equipment, and it will be difficult for anyone to anticipate and cater to his needs (tailor syndrome). This may have a profound effect not only on the ability of the capital goods industries to generate innovations but also on the efficiency with which they are able to produce their machines. This certainly seems to have been an important factor in accounting for differences in performance between British and American capital goods industries.

In the 19th century, a major source of capital saving for the economy as a whole came through improvements in the efficiency of capital goods production. The important analytical point is that any cost reduction in the capital goods sector—whether it is labor saving or capital saving in its immediate factor-proportion bias—is a capital-saving innovation to the economy as a whole. Such a capital-saving stage has been an inevitable but later stage of the sequence by which industrial economies have accommodated themselves to an innovation. When a new machine was introduced, there existed, by definition, no established system of organization for producing the machine. As Marx once stated: "There were mules [spinning mules] and steam-engines before there were any laborers whose exclusive occupation it was to make mules and steam-engines; just as men wore clothes before there were such people as tailors."[35] The fact of the matter is that a very high fraction of new inventions, new products, or new processes, once conceived, are

mitter's vision as well, making him incapable of stating his problems, even on paper, and of reaching decisions. To see faces is to see confusion, satisfaction, or antagonism; to gauge whether status is being, or should be, granted or withheld. The conversational setting from time to time allows the release of tensions through small talk, jokes, and laughter. The urge to conceal as well as the fear of concealment weakens" (Strassmann, pp. 32–33). For a highly suggestive treatment of the spatial diffusion of innovations which turns, in large measure, on the role of person-to-person contacts and personal communications between pairs of individuals, see Torsten Hagerstrand, "Quantitative Techniques for Analysis of the Spread of Information and Technology," in *Education and Economic Development*, ed. C. Arnold Anderson and Mary Jean Bowman (Chicago, 1965), chap. 12.

[35] Karl Marx, *Capital* (New York, 1936), p. 417.

of no economic relevance until the capital goods industries have successfully solved the technical and mechanical problems or developed the new machines which the inventions require. This creative process on the part of the capital goods industries has been badly slighted, and a recognition of this process may help us understand the widely-observed failure of poor countries to develop techniques with the factor-saving bias which they need. In the currently developed economies, the growing efficiency of the capital goods sector was a major source of capital saving for the economy, *both* because a large proportion of all innovations originated in that sector and because it developed a capacity for producing capital goods at lower and lower cost.[36]

From this perspective, some of my concerns about the prospects for poor countries which rely on the importation of foreign capital equipment are obvious. Of course it is of enormous benefit to them to be able to import this equipment, even where the equipment is not optimally factor-biased. But if new techniques are regularly transferred from industrial countries, how will the learning process in the design and the production of capital goods take place? Reliance on borrowed technology perpetuates a posture of dependency and passivity. It deprives a country of the development of precisely those skills which are needed if she is to design and construct capital goods that are properly adapted to her own needs. What, then, are the prospects for underdeveloped countries ever becoming efficient producers of capital goods and, in particular, developing a technology with factor-saving biases more appropriate to their own factor endowments? In the past, as I have argued, the appropriate skills were acquired through an intimate association between the user and the producer of capital goods. In the absence of these experiences, what substitute mechanisms or institutions can be established to provide the necessary skills?

The problem takes on additional importance when we adopt a more realistic conception of the nature of technical change than we have inherited from Schumpeter. Schumpeter accustomed economists to thinking of technical change as involving major breaks, giant discontinuities or disruptions with the past. This rather melodramatic conception fitted in well with his charismatic approach to entrepreneurship. But technological change is also (and perhaps even more importantly) a continuous stream of innumerable minor adjustments, modifications, and adaptations by skilled personnel, and the technical vitality of an economy employing a machine technology is critically affected by its

[36] See Nathan Rosenberg, "Capital Goods, Technology and Economic Growth," *Oxford Economic Papers*, n.s. 15, no. 3 (November 1963): 217–27.

capacity to make these adaptations.[37] The necessary skills, in the past, were developed and diffused in large measure by the capital goods sector. The skills are, inevitably, embodied in the human agent and not in the machine, and unless these skills are somehow made available, the prospects for the viability of a machine technology may not be very good.

As part of our Schumpeterian heritage, we still look upon the transformation of an "invention" into an "innovation" as the work of entrepreneurs. But from a *technological* perspective, it is much more the work of the capital goods industry. This is most obviously the case where the invention is a cost-reducing *process* and does not have to be marketed to final consumers. But in making new products and processes practicable, there is a long adjustment process during which the invention is improved, bugs ironed out, the technique modified to suit the specific needs of users, and the "tooling up" and numerous adaptations made so that the new product (process) can not only be produced but can be produced at low cost. The idea that an invention reaches a stage of commercial profitability *first* and is then "introduced" is, as a matter of fact, simple-minded. It is during a (frequently protracted) shakedown period in early introduction that it becomes obviously worthwhile to *bother* making the improvements. Improvements in the *production* of a new product occur *during* the commercial introduction.

Alternatively put, there has been a tendency to think of a long precommercial period when an invention is treated as somehow shaped and modified by exogenous factors until it is ready for commercial introduction. This is not only unrealistic; it is a view which has also been responsible for the neglect of the critical role of capital goods firms in the innovation process.

We are now coming to realize that modern technology has a long

[37] In the shaping of technology, many more kinds of adjustments are involved than the factor-proportions adjustment, with which economists are typically preoccupied. It may be possible, for example, to incorporate certain kinds of features into a machine which will compensate for (or simply bypass) deficiencies on the part of the labor force. If workers cannot be relied upon to perform appropriate maintenance procedures on expensive machinery, machines can be designed which *require* less maintenance. This occurred in the United States early in the 20th century when it was found that immigrant labor could not be relied upon to attend to the numerous separate lubrication points of their machinery, thus causing frequent breakdowns. This problem was solved by redesigning machine tools to incorporate the centralized, self-acting lubrication system of the automobile. Since this system went into operation automatically when the machine was activated, reliance upon the worker was, in this respect, bypassed (S. Einstein, "Machine-Tool Milestones, Past and Future," *Mechanical Engineering* 52, no. 11 [November 1930]: 961).

umbilical cord. Innumerable unsuccessful foreign aid projects in the past twenty years–including Russian-sponsored as well as American-sponsored projects–have confirmed that when modern technology is carried to points remote from its source, without adequate supportive services, it will often shrivel and die. This is partly because the technology emerged in a particular context, often in response to highly narrow and specific problems, such as may have been defined by a particular natural resource deposit. But, more important, the technology functions well only when it is maintained and nourished by an environment offering it a range of services which are essential to its continued operation. These would include the ability to diagnose correctly the causes of machine breakdown or other sources of inferior performance, the availability of facilities and personnel to perform repair work and to provide routine maintenance and repairs, and the provision of spare parts. In the absence of the kinds of skills produced and embodied in a capital goods industry, repair and maintenance costs in the use of machinery are likely to remain high. A major reason for a domestic capital goods industry, therefore, is that the ability to utilize complex machinery effectively–whatever the country of origin of the machinery–depends upon the kinds of skills which such an industry uniquely makes available.

Moreover, physical proximity between the producer and user of machinery seems to have been indispensable in the past for reasons which we do not really understand but which seem to be rooted basically in the problem of communications. Successful technological change seems to involve a kind of interaction that can best be provided by direct, personal contact. Successful instances of technological change in the past have involved a subtle and complex network of contacts and communication between people, a sharing of interests in similar problems, and a direct confrontation between the user of a machine, who appreciates problems in connection with its use, and the producer of machinery, who is thoroughly versed in problems of machinery production and who is alert to possibilities of reducing machinery (and therefore capital) costs.

VI

With all of the difficulties attaching to the transfer of industrial technology, such technology is nevertheless much easier to transfer than agricultural technology because industrial technology is at least very much self-contained. It tends to be a relatively closed system. Even its dependence upon human inputs can be reduced by making the technology more capital-intensive. But where ecological relationships are involved, as they are in agriculture, everything is quite different. Here there are important interactions between the human enterprise and specific features of the natural environment. Here natural phenomena participate in a much more active way, and productive activity must be more highly responsive to even minor variations and peculiarities of the environment. Agricultural activity is much more closely enmeshed with the natural environment, and this, in turn, has important consequences. It means that solutions to agricultural problems and the ability to make appropriate adaptations to local conditions will hinge upon kinds of knowledge which are not ordinarily possessed by the cultivator: biology, botany, biochemistry, genetics, etc. Major breakthroughs in agriculture are likely to come from the application to local circumstances of scientific disciplines which do not ordinarily flourish in poor countries. I would therefore argue that a major goal of institution building in poor countries should be to equip them with the kind of research abilities and facilities that will enable them to produce the new knowledge required for their own peculiar agricultural resources.

Agriculture's close involvement with nature has another important consequence for the acceptance of new techniques: because of the importance of even minor variations in rainfall, sunshine, soil content, topography, plant diseases, etc., there is a much higher degree of uncertainty concerning the application of new agricultural techniques. Institution building in agricultural environments will have to take this element of uncertainty prominently into account. Institutional arrangements which serve to reduce this uncertainty or to protect the cultivator against a loss that could be personally disastrous may prove to be extremely important in speeding the adoption of new techniques.

In examining the sources of productivity improvement in both industry and agriculture, we must pay attention to the fact that the performance of individual industries will frequently depend not only on resources within that industry but on the availability and the effectiveness of industries which stand in an important complementary relationship with it. The technological inputs (including knowledge) which crucially affect the success of industry *A* are produced by industries *B*, *C*, and *D*. Much of the discussion of the prospects and possibilities for technical improvement in poor countries has suffered from ignoring such interindustry relationships. We have to take account of these relationships and learn how to exploit them.

This point is particularly pertinent to the behavior of agriculture. The history of countries with highly productive agricultural sectors indicates clearly that the major sources of improvements in agricultural productivity have been generated *outside* the agricultural sector. Unless an economy is well equipped with these complementary sources, agriculture is not likely to experience rapid improvements in efficiency. In the American experience the sources of productivity growth in agriculture have come from the machinery-producing sector, which developed a mechanical technology appropriate to agriculture; from research at agricultural experiment stations and other educational institutions; from the fertilizer industry; and increasingly in the 20th century from the study of genetics and from chemistry (including soil chemistry). In fact, in American agriculture the industries which supply agriculture with certain of its inputs have played a role in the introduction of new technologies entirely analogous to that of the capital goods industries in relation to the manufacturing sector.[38] It is the absence of these complementary sectors and resources which is often so damaging to underdeveloped agriculture.[39] If, for example, the fertilizer industry can produce fertilizer only at a very high cost, then all the extension work and exhortation in the world will not induce the peasant to use it—he will reject its use on perfectly rational cost-benefit grounds. What is needed is research oriented toward reducing the cost of the inputs which agriculture receives from the industrial sector. An extension service will not perform an important social function until it has valuable information to extend.[40]

In broad terms, the history of Japanese agricultural development

[38] As Vernon Ruttan has pointed out: "The ratio of purchased inputs to total output has risen steadily over the last several decades in American agriculture. The farm supply industries have played an important role in channeling technological advances into agriculture. Adoption of new technology has usually increased agriculture's dependence on inputs produced in the nonfarm sectors of the economy and has lessened agriculture's dependence upon land inputs" (Vernon Ruttan, "Research on the Economics of Technological Change in American Agriculture," *Journal of Farm Economics* 42, no. 4 [November 1960]: 740).

[39] For an absorbing account of the Soviet Union's heavy reliance upon American technical personnel in her attempt to achieve a rapid mechanization of agriculture, see Dana G. Dalrymple, "The American Tractor Comes to Soviet Agriculture: The Transfer of a Technology," *Technology and Culture* 6, no. 2 (Spring 1964): 191–214. The nature and extent of early Soviet dependence upon the importation of foreign skills is explored in sector-by-sector detail in Anthony C. Sutton, *Western Technology and Soviet Economic Development, 1917 to 1930* (Stanford, Calif., 1968).

[40] On the other hand, the "feedback" role of extension services may also have an important effect upon the productivity of agricultural research. Colleges of agriculture and agricultural experiment stations have frequently suffered from the absence of direct extension-work responsibilities because this has isolated them from a valuable flow of feedback originating at the farm level. (I am grateful to Professor Vernon Ruttan for this observation.)

seems to confirm the importance of these complementary relationships. Japan represents perhaps the most remarkable success story of the 20th century—a story that began 100 years ago. The great improvements in Japanese agricultural productivity, which provided the basis for the growth of the rest of the economy, rested firmly upon inputs provided from outside the agricultural sector. The growth of her agricultural productivity was due to a long series of dynamic interactions between the needs of the farm sector, on the one hand, and the combined ingenuity of industry and the educational establishments to supply these needs, on the other. The outcome of these interactions included the provision of low-cost commercial fertilizer, new farm implements, high-quality education and extension work, and, at a later stage, the beginnings of mechanization. Although it is relatively easy to point out the things which were done well in Japan's spectacular success story, the really critical question is how to incorporate such functions in a set of institutions which will perform successfully in a drastically different environment.[41]

VII

The exact institutional arrangements which will make for success will always be difficult to specify a priori. I would like to suggest that this is so mainly because success or failure always depends on the environment in which any particular institution is immersed—that is, on the complex of *other* institutions as well as widely shared values and traditions which critically affect its operation.

If history teaches us anything in this regard, it is that a wide diversity of institutional forms have proved to be successful under differing conditions. Similarly, history indicates also the very variable experience of the *same* institution when placed in different contexts. Thus, while on the American scene it has been customary to celebrate the virtues of the family farm, the family farm in France has been a favorite whipping boy, held responsible for the continuation of outmoded farm practices, for excessively small farm units, and for an insufficient use

[41] Nor should one ignore social changes which can accelerate the rate of diffusion of existing "best-practice" techniques. Thomas C. Smith (*Agrarian Origins of Modern Japan* [Stanford, Calif., 1959], esp. chap. 7) has pointed to the numerous improvements in farming methods which had been developed under the Tokugawa Shogunate. However, the spread of these techniques had been seriously inhibited by feudal restrictions upon travel and communication. One of the important achievements of the Meiji Restoration was the elimination of barriers to the spread of the best techniques. Consequently, agricultural productivity was raised by narrowing the gap between "best-practice" and "average-practice" techniques. See also Saburo Yamada, "Changes in Output and in Conventional and Nonconventional Inputs in Japanese Agriculture since 1880," *Food Research Institute Studies* 7, no. 3 (1967): 371–413.

of capital. And, while the American owner-operated farm has indeed been a remarkable success, it has been critically dependent for this success upon a wide range of organizational arrangements which absolutely defy simple ideological classification.[42]

The kinds of institutions which will function successfully in a particular environment and in the pursuit of particular objectives is something which it is not easy to generalize about a priori.[43] Certainly this is confirmed by our own recent history and experience, which suggests that a wide diversity of organizational tactics have proven to be successful. Consider the changing mix of government participation in various economic activities in recent years. Not only do we have public production of the TVA variety, but we also have public regulation and control of private utilities, public subsidies to private enterprise, public highways financed on a user-cost basis, public and private institutions of higher education living happily together (or almost so), public support of medical research which is financed partly from private sources, and public sponsorship of specialized activities contracted to private organizations—the Atomic Energy Commission and even the RAND Corporation. If we look specifically at the realm of technology, we find that varying mixes of the public and private sectors have been responsible for a broad range of technological breakthroughs in areas such as the development of atomic energy and its application to peaceful uses, medical research, jet propulsion, electronics, and agriculture. I point this out not to suggest that there is anything optimal about the public-private mix in the American economy but rather to indicate the wide range of institutional devices and combinations which have proved to be successful. Indeed, one of the most spectacularly successful kinds of institutions in our "capitalist" economy in recent years has been the nonprofit corporation, which has been a major source of both new knowledge and new technology.

42 "Organization for research and development in the farm supply industries varies widely. At one extreme are such industries as hybrid seed corn, where a substantial share of research and development has been conducted at USDA and land-grant college experiment stations. At the other extreme is the farm equipment industry where public contributions have apparently been considerably smaller relative to the contributions by private industry. A third variation is represented by the fertilizer industry where the TVA has provided what is in effect an industry research institute for the fertilizer industry with a research program ranging from fundamental chemical and biological research to applied engineering and economics" (Ruttan, p. 741).

43 Similarly, the identification of obstacles to economic development is a hazardous enterprise. For a valuable and provocative discussion, see Albert O. Hirschman, "Obstacles to Development: A Classification and a Quasi-vanishing Act," *Economic Development and Cultural Change* 13, no. 4, pt. 1 (July 1965): 385–93.

Clearly there is no single "best-way-of-doing-things" to which we have rigidly adhered in all sectors of our economic life. If the American experience suggests anything, it is that there has been no single institutional formula for success. It is reasonable to expect that the success formulas will be both different and equally variable in poor countries. Without question we ought to attach the highest research priority to finding out what combinations are likely to prove successful in the special environments of individual countries. This is an area where history can provide some guidance and valuable insights but certainly no authoritative answers.

[22]

The Economic Journal, 87 (*June* 1977), 242–261
Printed in Great Britain

TECHNOLOGY TRANSFER BY MULTINATIONAL FIRMS: THE RESOURCE COST OF TRANSFERRING TECHNOLOGICAL KNOW-HOW*

I. INTRODUCTION

The essence of modern economic growth is the increase in the stock of useful knowledge and the extension of its application. Since the origins of technical and social innovations have never been confined to the borders of any one nation, the economic growth of all countries depends to some degree on the successful application of a transnational stock of knowledge (Kuznets, 1966). In other words, the economic growth of every nation is inextricably linked to the successful international transfer of technology. Nevertheless, economists have been remarkably slow in addressing themselves to the economics of international technology transfer. The result is that "at both the analytic and factual level very little is known about the international transfer of knowhow" (Reynolds, 1966).

This paper addresses itself to this need. The starting-point is Arrow's suggestion that the cost of communication, or information transfer, is a fundamental factor influencing the world-wide diffusion of technology (Arrow, 1969).[1] The purpose of the paper is to examine the level and determinants of the costs involved in transferring technology. The value of the resources which have to be utilised to accomplish the successful transfer of a given manufacturing technology is used as a measure of the cost of transfer. The resource cost concept is therefore designed to reflect the ease or difficulty of transferring technological know-how from manufacturing plants in one country to manufacturing plants in another.

II. TECHNOLOGY TRANSFER AND THE PRODUCTION OF KNOWLEDGE

The literature on technological change recognises that it takes substantial resources to make a new process or product feasible (Mansfield, 1968). However, it is common to assume that the cost of transferring the innovation to other firms is very much less, so that the marginal costs of successive application is trivial compared to the average cost of research, development, and application. This paradigm is sometimes extended to international as well as domestic technology transfer (Rodriguez, 1975).[2] Buttressing this view is a common belief that

* The findings described in this paper resulted from research undertaken for my Ph.D. dissertation, "The Multinational Corporation and the Resource Cost of International Technology Transfer" (Cambridge, Mass.: Ballinger, 1976). The trenchant comments of Professor Edwin Mansfield were much appreciated during all phases of the study. My particular gratitude goes to the participating firms, without whose co-operation this paper would not have been possible. I should also like to acknowledge the financial support provided for this study by the National Science Foundation, under a grant to Professor Edwin Mansfield of the University of Pennsylvania.

[1] Arrow asks: "If one nation or class has the knowledge which enables it to achieve high productivity, why is not the other acquiring that information?...The problem turns on the differential between costs of communication within and between classes" (or nations). P. 33.

[2] "Transmission of technology between countries is assumed costless. Thus, it is possible for the

technology is nothing but a set of blueprints that is usable at nominal cost to all. Nevertheless, it has been pointed out that generally "only the broad outlines of technical knowledge are codified by non-personal means of intellectual communication, or communication by teaching outside the production process itself" (Berrill, 1964). The cost of transfer, which can be defined to include both transmission and absorption costs, may therefore be considerable when the technology is complex and the recipient firm does not have the capabilities to absorb the technology. The available evidence is unfortunately very sketchy. From the case studies of Mueller and Peck, Arrow inferred that transfer costs must be high (Arrow, 1962). From the Hall and Johnson study of the transfer of aerospace technology from the United States to Japan, it is not clear that this is true (Hall and Johnson, 1970). Robinson believes that economists' views on transfer costs are exaggerated (Robinson, 1973) while Mansfield and Freeman take the opposite view (Freeman, 1965; Mansfield, 1973). The lack of compelling evidence is apparent, and the appeals for further research (Mansfield, 1974; UNCTAD, 1970) seem to be well founded.

III. THE SAMPLE

The domain of this study is the transfer of the capability to manufacture a product or process from firms in one country to firms in another. Consequently the transfers can be considered as horizontal,[1] and in the design phase.[2] Data on 26 fairly recent international technology transfer projects were obtained. The proprietary nature of much of the data meant that sampling costs were high, which in turn severely limited the size of the sample that could be collected. All 26 transfers were conducted by firms which were multinational in the scope of their manufacturing activity, although they varied considerably in sales value (10–20 billion U.S. dollars) and R & D expenditures (1·2–12·5% of sales value). All had headquarters in the United States. The transferees were on the average much smaller and less research-intensive. In 12 instances they were wholly owned subsidiaries of the transferor, in 8 instances the transferor and transferee were joint ventures partners, in 4 instances transfers were to wholly independent private enterprises, and the remaining 3 were to government enterprises. Table 1 shows that 17 of the projects fall into a broad category which will be labelled "chemicals and petroleum refining." The remaining 9 projects fall into a category which will be labelled "machinery".[3] Table 1 also indicates the wide geographical dispersion of the transferees.

---

country which owns the technology to operate a plant in a foreign country without any transfer of factors." P. 122.

[1] Horizontal transfer refers to the transfer of technical information from one project to another. It can be distinguished from vertical transfer, which refers to the transfer of technical information within the various stages of a particular innovation process, e.g. from the basic research stage to the applied research stage.

[2] For the distinctions between materials transfer, design transfer, and capacity transfer, see Hayami and Ruttan (1971).

[3] Chemicals and petroleum refining thus embrace ISIC categories (United Nations, 1968), 351, 353, and 356, while "machinery" embraces categories 381, 382, and 383.

Table 1

Twenty-six Technology Transfer Projects: 3 Digit ISIC Category and Transferee Location

Location	"Chemicals and Petroleum Refining"			"Machinery"			Total
	351: industrial chemicals	353: petroleum refineries	356: plastic products	381: fabricated metal products machinery and equipment	382: machinery except electrical	383: electrical machinery, appliances, and supplies	
Canada	1	1	0	1	0	0	3
Northern and Western Europe	3	1	0	0	4	1	9
Australia	0	0	1	0	0	0	1
Japan	3	0	0	0	1	0	4
Eastern Europe	2	0	0	0	0	0	2
Latin America	3	0	0	0	0	1	4
Asia (excluding Japan)	0	1	0	0	0	1	2
Africa	1	0	0	0	0	0	1
Total	13	3	1	1	5	3	26

IV. DEFINITION OF TECHNOLOGY TRANSFER COSTS

An economic definition of transfer cost is developed below. The emphasis is on the resources which must be utilised to transfer technological know-how. Of course royalty costs or rents must be incurred merely to secure access to the technology, but these costs are not the focus of attention of this paper.[1] In order to appreciate the import of the definition that will be presented, a distinction must first be made between two basic forms in which technology can be transferred. The first form embraces physical items such as tooling, equipment, and blue prints. Technology can be embodied in these objects. The second form of technology is the information that must be acquired if the physical equipment or "hardware" is to be utilised effectively. This information relates to methods of organisation and operation, quality control, and various other manufacturing procedures. The effective conveyance of such "peripheral" support constitutes the crux of the process of technology transfer, and it typically generates the associated information flows. It is towards discovery of the cost of transfer of this "unembodied"[2] knowledge that the attention is directed.

Technology transfer costs are therefore defined as the costs of transmitting and absorbing all of the relevant unembodied knowledge. The costs of performing the various activities which have to be conducted to ensure the transfer of the necessary technological know-how will represent the cost of technology transfer.[3] Clearly, a great many skills from other industries (e.g. design engineering) will be needed for plant design, plant construction, and equipment installation. However, not all of these skills will have to be transferred to ensure the success of the project. As defined, the costs of transfer clearly do not include all of the costs of establishing a plant abroad and bringing it on stream.

The definition of transfer costs presented at the conceptual level can be translated into operational measures by considering the nature of a given project activity. At the operational level the subset of project costs identified as transfer costs fall into four groups. The first group is the cost of pre-engineering technological exchanges. During these exchanges the basic characteristics of the technology are revealed to the transferee, and the necessary theoretical insights are conveyed. The second group of costs included are the engineering costs associated with transferring the process design and the associated process engineering[4] in the case of process innovations, or the product design and production engineering[5] in the case of product innovations. If the technology has already

[1] Many observers equate the cost of technology with royalty fees (Mason, 1973; Gillette, 1973). Royalty costs are considered in the dissertation from which this paper was taken.

[2] Unembodied knowledge is the term used here to denote knowledge not embodied in capital goods, blueprints, and technical specifications, etc.

[3] All of the relevant costs are included, irrespective of which entity initially or eventually incurs them.

[4] Process engineering for continuous flow technology involves the compilation of flow diagrams, heat balances, control instrumentation, etc. It can be distinguished from detailed engineering which involves the translation and elaboration of the process engineering into a manufacturing facility.

[5] Production engineering for a specified item can be divided into two phases: production design and process planning. Production design is the modification of the functional design in order to reduce manufacturing costs. (Functional product design is the design of a product to fulfil certain specifications and requirements.) Given the design, process planning for manufacture must be carried out to specify,

been commercialised,[1] transmission may simply involve transferring existing drawings and specifications with the minimum of modification. However, the process of absorption may be more difficult, requiring the utilisation of considerable consulting or advisory resources. "Engineering" costs not falling into the specified categories[2] are excluded from transfer costs. The excluded engineering costs are essentially the plant or detailed engineering costs, net of advisory or consulting costs. This residual is assumed to correspond with routine drafting costs. Routine drafting is generally performed by technicians under the supervision of engineers. Drafting skills do not have to be transferred for the viability of the project to be assured. Accordingly, drafting is not considered to represent a transfer activity.[3]

The third group of costs are those of R & D personnel (salaries and expenses) during all phases of the transfer project. These are not the R & D costs associated with developing the underlying process or product innovations. Rather, they are the R & D costs associated with solving unexpected problems and adapting or modifying the technology. For instance, research scientists may be utilised during the transfer if new and unusual technical problems are encountered[4] with the production inputs. These R & D costs are generally small or non-existent for international transfers falling into the "design transfer" category.

The fourth group of costs are the pre-start-up training costs and the "excess manufacturing costs". The latter represent the learning and debugging costs incurred during the start-up phase, and before the plant achieves the design performance specifications. It is quite possible that no marketable output will be produced during the initial phases of the start-up. Nevertheless, normal labour, materials, utilities, and depreciation costs will be incurred, together with the costs of the extra supervisory personnel that will inevitably be required to

in careful detail, the processes required and their sequence. The production design first sets the minimum possible costs that can be achieved through the specification of materials, tolerances, basic configurations, methods of joining parts, etc. Process planning then attempts to achieve that minimum through the specification of processes and their sequence to meet the exacting requirements of the design specifications. The accepted end-point for production design is manifested by the drawing release. Process planning takes over from this point and develops the broad plan of manufacture of the part or product. A distinction can also be drawn between process planning and the layout of the physical facilities. Some process planning will take place during the layout phases of the design of a production system, Process plans can be regarded as inputs to the development of a layout. (*McGraw-Hill Encyclopedia.* 1960.)

[1] An innovation is said to have been commercialised if it has already been applied in a facility of economic size which is essentially non-experimental in nature. Thus pilot plant or prototype application is not considered to represent commercialisation.

[2] These categories are (*a*) process or design engineering costs and related consultation for process innovations or (*b*) production engineering expenses for product innovations; and (*c*) costs of engineering supervision and consultation (salaries plus travel and living) for the plant engineering.

[3] Drafting costs can be considered an implementation cost rather than a transfer cost, the implication being that if the host country does not have these skills, the viability and cost of the project is unlikely to be affected. The advisory and consulting costs, on the other hand, represent transfer costs since these activities are necessary if the technology is to be adjusted to the local circumstance and requirements. Clearly, if an existing plant was to be duplicated in its own environment, consulting costs could be expected to go to zero, whereas routine drafting would still have to be performed.

[4] Referring to process technologies, it is possible that differences in feedstocks amongst various locations may create problems that only research scientists can effectively handle. Similarly, changes in atmospheric conditions or water supply could have unexpected consequences for some highly complex processes.

assist in the start-up. The operating losses incurred during initial production are very often a close approximation to excess manufacturing costs.[1]

V. TRANSFER COSTS: DATA AND HYPOTHESES

1. *The Level of Transfer Costs*

The above definition was used to calculate the transfer costs for 26 projects. The results are presented in Table 2. The costs are given in absolute dollars, and then normalised by total project costs.[2] For the sample as a whole, transfer costs average 19 % of total project costs. Clearly, the data do not support the notion that technology is a stock of blueprints usable at nominal cost to all. Nevertheless, there is considerable variation in the sample data, with transfer costs ranging from 2 % to 59 % of total project costs. The number of factors influencing transfer costs is undoubtedly very great,[3] but some factors are likely to have a more pervasive influence than others. The discussion to follow is restricted to hypotheses for which statistical testing is feasible, given the available data. Two groupings of testable hypotheses can be identified: characteristics of the technology/transferor, and characteristics of the transferee/host country.

2. *Technology/Transferor Characteristics*

A critical factor in the transfer of technology is the extent to which the technology is completely understood by the transferor. The number of manufacturing start-ups[4] or applications which the transferor has already conducted with a specific technology can be used as an index of this knowledge.[5] An increase in the number of applications is likely to lower transfer costs since with each start-up additional knowledge about the technology is acquired. Since no two manufacturing start-ups are identical, each start-up provides the firm with the opportunity to observe the effects of different operating parameters and differences in equipment design. Each application can be regarded as a new experiment which

[1] An important consideration is the extent to which excess manufacturing costs correctly reflect technology transfer costs rather than the costs of discovering and overcoming the idiosyncrasies of a new plant. One way to confront this issue is to consider the level of excess manufacturing costs when an absolutely identical plant is constructed in a location adjacent to an existing plant. Further, assume the second plant embodies the same technology as the first plant, and the labour force from the first is transferred to the second for the purpose of performing the manufacturing start-up. The assumption is that under these circumstances excess manufacturing costs in the second plant will be zero, or very nearly so. The correctness of this assumption was corroborated by a subsample of project managers subsequently questioned about this matter. The postulated circumstance would be identical to shutting down the first plant and then starting it up again. Some excess manufacturing costs might be incurred during the initial hours of operation if the plant embodies flow process technology. (For the projects in the sample the average duration of the manufacturing start-up was 8·2 months.) However, these costs are unlikely to be of sufficient magnitude to challenge the validity of classifying excess manufacturing costs as a component of technology transfer costs.

[2] Total project costs are measured according to the inside boundary limits definition commonly employed by project accountants. Installations outside the plant perimeter are thereby excluded.

[3] For a broader view of the spectrum of hypotheses, see the author's Ph.D. dissertation.

[4] Manufacturing start-ups are synonymous with the number of applications of the technology. If a new plant is built for each application, it would also by synonymous with the number of plants that are built which utilise the technology.

[5] Corporations engaged in technology transfer ventures not grounded on their own technology are known to have encountered massive transfer problems and costs.

yields new information and new experience.[1] Transfer will be facilitated the more fully the technology is understood. Besides these engineering economies, additional applications provide expanded opportunities for the pre-start-up training of the labour force. Clearly, if identical or similar plants exist elsewhere,

Table 2

Sample Data on the Resource Costs of Technology Transfer: 26 International Projects

Chemicals and Petroleum Refining		Machinery	
Transfer costs: dollar amount (thousands)	Transfer costs: dollar amount / total project cost	Transfer costs: dollar amount (thousands)	Transfer costs: dollar amount / total project cost
49	18	198	26
185	8	360	32
683	11	1,006	38
137	17	5,850	45
449	8	555	10
362	7	1,530	42
643	6	33	59
75	10	968	24
780	13	270	45
2,142	6		
161	2		
586	7		
877	7		
66	4		
2,850	19		
7,425	22		
3,341	4		

then experienced operators from these plants can be used to assist the start-up in the new plant. In addition, untrained operators can be brought into existing plants for pre-start-up training.

The second variable to be considered is the age of the technology. The age of the technology is defined as the number of years since the beginning of the first commercial application of the technology[2] anywhere in the world, and the end[3] of the technology transfer programme. The age of an innovation will determine the stability of the engineering designs and the transferor's knowledge of the manufacturing procedures. The older the technology, *ceteris paribus*, then the greater have been the opportunities for interaction between the development groups and the manufacturing and operating groups within the firm. Problems

[1] The first application represents first commercialisation of the technology. This will result in the creation of a set of basic engineering drawings and specifications. Duplication and alteration of these for subsequent start-ups will involve a modest cost compared to the initial cost of constructing them.

[2] If there is more than one key innovation embodied in the technology, then the date of commercial application of the most recent key innovation is the reference date.

[3] Age is defined up to the end of the transfer programme since any knowledge about the technology acquired up to this point is potentially useful for the transfer. For the very first start-up, age will be the length of the transfer minus the development overlap.

stand a better chance of already being ironed out, and the drawings are likely to be more secure. Further, since technology is not embodied in drawings alone, there is a great deal of uncodified information – the relevant "art". This kind of knowledge is carried by the supervisors, engineers, and operators. As the age of the technology increases, more individuals in the firm have the opportunity to acquire this non-codified information, and hence are potentially available to assist in the transfer. There will, however, be some point after which greater age will begin to increase the cost of transfer. When the length of stay of corporate personnel begins to be outstripped by the age of technology, then the non-codified dimensions of design knowledge may be lost to the firm.[1]

It is necessary to distinguish the cost reductions resulting from additional start-ups from the cost reductions resulting from greater age of the technology. For continuous flow technologies, additional applications of an innovation in entirely new plants will allow experimentation with scale and with the basic parameters of the design. This will generate a greater understanding of the technology. On the other hand, greater age, given the number of applications or start-ups, generally permits experimentation only with operating parameters, the design of the plant remaining fixed throughout.

The third technology variable to be considered is the number of firms utilising the technology, or one that is "similar and competitive". This is taken to represent the degree to which the innovation and the associated manufacturing technology is already diffused throughout the industry. The greater the number of firms with the same or similar and competitive technology, then the greater the likelihood that technology is more generally available, and can therefore be acquired at lower cost.[2]

These technology variables and the attendant hypotheses begin to take on some extra significance when viewed together. Taken singly they define the technology to only a limited degree. Together, they hypothesise, *ceteris paribus*, that the most difficult and hence costly technology to transfer is characterised by very few previous applications, a short elapsed time since development, and limited diffusion. Technology displaying such characteristics can be termed "leading-edge" technology. "Leading-edge" technology is likely to be in a state of flux; the engineering drawings will be constantly altering, thus frustrating the transfer. In comparison, state-of-the-art technology is hypothesised, *ceteris paribus*, to involve lower transfer costs since the engineering drawings are more likely to be finalised and the fundamentals of the technology stand a better chance of being more fully understood.

[1] In the limit, the firm could terminate its utilisation of a particular technology, and the non-codified information associated with it could be gradually lost for ever as the technology becomes historic. Further, the drawings associated with technology that is very old may suffer from so many small alterations that the very essence of even the codified technology may become quite obscure. Since none of the technology transfer projects in the sample were historic in the above sense, the relevant range of the hypothesised age–transfer cost function involves an inverse relationship between the age of the technology and the cost of transfer.

[2] An identification problem may exist here because more firms may have applied the technology because the transfer cost is low.

3. *Transferee and Host Country Characteristics*

The technical and managerial competence of the transferee will be presented as an important determinant of the ease with which technology can be absorbed. The years of manufacturing experience of the transferee in a given 4-digit ISIC industry (United Nations, 1968) is used as an index of the extent to which managers, engineers, and operators have command over the general manufacturing skills of an industry. A firm skilled in the manufacture of a group of products is likely to have less difficulty absorbing a new innovation in that industry group than is the firm which has had no previous experience manufacturing products in a particular industry group (Rawski, 1975). Older enterprises, with their skilled manufacturing personnel, seem more likely to be able to understand and apply codified knowledge to the manufacture of a new product, or the utilisation of a new process.[1]

Another variable to be considered is the size of the transferee. Although less compelling, the reasoning behind the hypothesis that transfer costs decline with firm size is that larger firms generally have a wider spectrum of technical and managerial talent which can be called on for assistance during the transfer. A small firm may be technically and managerially quite competent yet unable to absorb new technology easily because of the extra demands placed on its scarce managerial and technical manpower. Consultants may have to be engaged by the smaller firms to perform tasks that are typically handled internally in larger firms.

A third variable considered is the R & D activity of the transferee. When unusual technical problems are unexpectedly encountered, an in-house R & D capability is likely to be of value. Oshima has argued that the R & D capability of Japanese firms facilitated the low-cost importation of foreign technology by Japanese firms (Oshima, 1973). The R & D to sales ratio of the transferee is taken as an index of its R & D capability, and an inverse relationship between this and transfer cost is postulated.

The final variable considered is designed to reflect the level of development of the host country infrastructure, which is hypothesised to be a determinant of the cost of transfer. For example, the level of skill formation in the host country will influence the amount and type of training that the labour force will require. Similarly, if the new venture is to acquire its inputs domestically, the quality of the inputs available will undoubtedly influence the level of start-up costs. There are many other considerations of similar kind which could be discussed. However, the high degree of cross-sectional collinearity between indices of development (Kuznets, 1966) makes the identification of separate effects statistically difficult. However, GNP *per capita*, a measure of productive capacity, can

[1] According to Rawski, recent experience of the People's Republic of China shows that during at least some phases of industrialisation, production experience may be a key determinant of the level and fungibility of industrial skills. Rawski notes that "with their skilled veteran workers and experienced technical persons, old industrial bases and old enterprises find it easier to tackle complicated technical problems than new enterprises and new industrial bases. With these advantages, it is the established centers which are best able to copy foreign equipment samples, to extract useful information from foreign technological publications, and to apply it to current domestic problem areas." (Rawski (1975), p. 386.)

be expected to capture some of the above considerations, and it will be used in this study as an index of economic development. A negative relationship between transfer cost and GNP/*per capita* is postulated.[1]

VI. DETERMINANTS OF THE COST OF INTERNATIONAL TECHNOLOGY TRANSFER: TESTS AND RESULTS

1. *The Model*

The basic model to be tested is

$$C_i = f(U_i, G_i, E_i, R_i, S_i, N_i, P_i, Z_i), \tag{1}$$

where C_i is the transfer cost divided by the total project cost for the ith transfer; U_i is the number of previous applications or start-ups that the technology of the ith transfer has undergone by the transferor;[2] G_i is the age of the technology in years; E_i is the number of years of manufacturing experience that the recipient of the ith transfer has accumulated; R_i is the ratio of research and development to sales for the recipient of the ith transfer, calculated for the year the transfer commenced; S_i is the volume of sales, measured in millions of dollars, of the recipient of the ith transfer; N_i is the number of firms identified by the transferor as having a technology that is identical or "technically similar and economically competitive" to the technology underlying the ith transfer; P_i is the level of GNP *per capita* of the host country (International Bank, 1973); Z_i is the random error term for the ith transfer. The expected derivatives are:

$$\frac{\partial C_i}{\partial U_i} < 0, \quad \frac{\partial C_i}{\partial G_i} < 0, \quad \frac{\partial C_i}{\partial E_i} < 0, \quad \frac{\partial C_i}{\partial R_i} < 0, \quad \frac{\partial C_i}{\partial S_i} < 0, \quad \frac{\partial C_i}{\partial N_i} < 0, \quad \frac{\partial C_i}{\partial P_i} < 0.$$

Since one of the best tests of any hypothesis is to look for the convergence of independent lines of evidence, the testing of this model will proceed in two phases. First, cross-section data on 26 completed projects is utilised in a linear version of the model estimated by ordinary least-squares procedures. Secondly, cost estimates provided by project managers for comparable projects are pooled to test a more specific non-linear version of the model.

2. *Statistical Tests: Phase 1*

The model to be tested is

$$C_i = \alpha_0 + \alpha_1 \bar{U}_i + \alpha_2 G_i + \alpha_3 E_i + \alpha_4 R_i + \alpha_5 S_i + \alpha_6 N_i + \alpha_7 P_i + Z_i, \tag{2}$$

where $\bar{U}_i$ is a dummy variable taking the value 1 if the transfer represents the first manufacturing start-up, and zero otherwise. $\bar{U}_i$ is used rather than U_i for empirical reasons, since the first start-up is often of critical importance. The sample was dichotomised because of the large differences between continuous flow process technology, and product technology. One category includes chemicals and petroleum refining and the other includes machinery (see Table 1).

[1] The sample did not include countries where high GNP statistics were grossly dependent on oil revenues.

[2] The number of previous manufacturing start-ups was significant in Phase 1 only when it was included as a dummy variable taking the value 1 if there had been no previous manufacturing start-ups of this technology by the transferring firm, and zero otherwise.

The results in Table 3 indicate that in chemicals and petroleum refining $\bar{U}_i$, N_i, and E_i are significant at the 0·05 level and carry the expected signs. In the machinery category the variables N_i, G_i, and E_i all carry the expected signs and are significant at the 0·05 level. N_i and E_i are thus significant in both industry

Table 3

Regression of Coefficients and t Statistics in Regression Equations to Explain C (The Cost of Transfer)

	Chemicals and petroleum refining		Machinery	
Independent variable	Equation (1)*	Equation (2)*	Equation (1)*	Equation (2)*
Constant	12·79	13·42	16·67	65·98
	(6·82)	(6·98)	(8·27)	(6·60)
Novelty dummy variable $\bar{U}$†	6·73	6·11	—	1·62
	(1·92)	(1·75)		(0·15)
Number of firms variable	−0·37	−0·39	−1·29	−1·26
	(−2·06)	(−2·22)	(−2·28)	(−1·95)
Age of technology variable (years)	—	—	−2·43	−2·35
			(−3·53)	(−2·51)
Experience of transferee variable (years in 4-digit ISIC)	−0·09	−0·08	−0·84	−0·85
	(−1·66)	(−1·42)	(−3·37)	(−2·95)
Size of transferee variable (thousands of dollars of sales)	—	−0·0009	—	—
		(−1·18)		
Number of observations	17	17	9	9
R^2	0·56	0·61	0·78	0·78
F	5·66	4·73	6·00	3·22
Significance level of F	0·01	0·02	0·04	0·12

* Omitted coefficient indicates variable dropped from the regression equation.
† Note 2, p. 251.

groupings, strongly supporting the hypothesis that transfer costs decline as the number of firms with identical or "similar and competitive" technology increases, and as the experience of the transferee increases. However, R_i and P_i were not significant in any of the equations, and although S_i carries the expected sign and approaches significance in one of the regressions it is not possible to be more than 85 % sure that the sign is correct or that the coefficient is different from zero.[1]

[1] Multicollinearity does not appear to be a serious problem in any of the equations. Correlations amongst pairs of the independent variables were never significant at the 0·05 level. The stability of the regression coefficients further suggests that multicollinearity is not a serious problem. Dummy variables were introduced to test for the effects of the organisational relationship between transferor and transferee (affiliate/non affiliate, public enterprise/private enterprise), but they were not found to be statistically significant determinant of transfer costs. Application of a forward step by step procedure did not reveal a preferred subset of variables. However, it is possible that the correct model is the simultaneous equation model $Ci = f(N_i, \ldots)$, $N_i = f(C_i, \ldots)$. To eliminate simultaneous equation bias it would be desirable to use a two-stage procedure. A predictor of N could first be obtained by regressing N_i on arguments other than C_i. This could then be used as an argument in the transfer cost regression. It was not possible to obtain a good predictor of N using the available cross-section data, so this procedure was not employed. Consistency was sacrificed for efficiency. It is therefore possible that simultaneous equation bias remains in the model. Therefore, the estimates of the parameters may not be consistent.

The results therefore generally support the hypotheses advanced earlier, but there are differences in the size of coefficients as well as in the specification of the equations between the industry groups. In particular, the novelty variable $\bar{U}_i$ is significant in chemicals and petroleum refining, but insignificant in machinery. The converse is true for the age variable G_i. The reason may be that there exists relatively less latitude for production experimentation with continuous flow process technology than with product technology. Once the plant is constructed, the extent to which the design parameters can be changed is rather minimal because of the degree of interdependence in the production system. In comparison, many product technologies allow greater design flexibility. Innumerable small changes to the technology are very often possible without massive reconstruction of the plant. It is also of interest that the coefficient of the experience variable E_i is considerably larger in machinery than in chemicals and petroleum refining. This is consistent with other findings that reveal important learning economies in fabrication and assembling (Tilton, 1971).

3. *Statistical Tests: Phase II*

The above analysis is handicapped by the small sample size and the very high costs of adding additional observations. Limited variation in exogenous variables coupled with the problem of omitted variables can imply difficulties with bias and identification. For the projects in the sample, a procedure was therefore devised to hold the missing variables constant while generating large variation in the exogenous variables. The respondent firms were asked to estimate how the total transfer costs would vary for each project if one particular exogenous variable happened to take a different value, assuming all other variables remain constant. The responses were taken into account only if the exercise generated circumstances within the bounds of an executive's experience. Given these limitations, the change specified was quite large in order to provide a robust sample. Generally the actual value of a selected variable was hypothesised first to halve and then to double. The estimated impact on transfer costs was noted. The exercise was performed for the following independent variables: the number of applications or start-ups that the technology has undergone; the age of the technology; the number of years of previous manufacturing experience possessed by the transferee in a given four-digit industry; the research and development expenditures to sales ratio for the transferee; the size (measured by sales value) of the transferee. For each variable this exercise generated at most three observations (including the actual) or transfer costs for each project. Pooling across projects produces enough observations for ordinary least-squares regression analysis.

The estimation procedure is commenced by assuming that the shape of the cost function can be represented by the following relatively simple but quite specific equation

$$C_j = V e^{\phi/X_j}. \tag{3}$$

C is the estimated transfer cost as a percentage of total project cost, X represents the value of various independent variables, j refers to the jth observation.

With this specification, the transfer cost for a project asymptotically approaches a minimum non-zero value as the value of each X increases. That is, as X goes to infinity, C goes to V. Therefore V is the minimum transfer cost with respect to the X variable. However, there is no maximum cost asymptote for the range of the data. The expression for the elasticity of transfer cost with respect to X is given by

$$\frac{-X}{C}\frac{dC}{dX} = \frac{\phi}{X}. \tag{4}$$

Thus for a specified value of X, the elasticity of transfer cost with respect to X is determined by ϕ. Hence the elasticity depends only on ϕ and X. In order to estimate the function, the log of the arguments in (3) are taken:

$$\log C_j = \log V + \frac{\phi}{X_j}. \tag{5}$$

Dummy variables are used to pool the observations across projects. Inclusion of dummy variables allows the minimum cost asymptote to vary from project to project. It is assumed that ϕ is constant across projects. These assumptions provide a pooled sample with intercepts which vary across projects.

Ordinary least-squares regressions of log C_j on the dummy variables and $1/X_j$ then proceeded for five different X variables, and for five data sets. These were: total transfers; transfers within the chemical and petroleum refining category; transfers in the machinery category; transfers of chemicals and petroleum refining technology to developed countries; and transfers of chemical and petroleum refining technology to less developed countries.[1] The Chow test (Chow, 1960) of equality between sets of coefficients in two linear regressions revealed that the separation of the sample along industry lines was valid, except for the research and development variable. However, there was no statistically valid reason for disaggregating the chemical and petroleum refining subsample according to differences in GNP *per capita* in the host countries.

The results of the estimation are contained in Table 4. The high R^2 values are partly because the large across-project variation in costs is being captured by the project dummies. The intercept term was always highly significant and the coefficients on all the dummies were significantly different from each other. All of the coefficients are significantly greater than zero at the 0·20 level and the age of the technology, the number of manufacturing start-ups, transferee size and experience achieve at least the 0·05 significance level in one or other of the subsamples. In several cases the coefficients are highly significant, providing strong statistical support for the hypotheses that have been advanced. The number of previous applications once again has a sizeable impact. Diffusion and manufacturing experience are particularly important in the machinery category.

The calculation of elasticities allows interpretation and comparisons of estimated effects. Average or point elasticities for some typical sample values of X are presented in Table 5. These estimates suggest that in the chemicals and petroleum-refining category, the second start-up could lower transfer costs by

[1] A purely arbitrary classification was used where less developed countries were defined as those with GNP/*per capita* less than $1,000.

34% over the first start-up, other variables held constant. The corresponding change for conducting a third start-up is 19%. The other elasticities can be interpreted similarly.

Table 4

Estimated values of ϕ (obtained from regressing $\log C_j$ on $\log V + \phi/X_j$) *together with corresponding t-statistics, sample size, degrees of freedom, and coefficient of determination* R^2

Variable	ϕ	*t*-statistic	Sample size	Degrees of freedom	R^2
Start-ups					
Chemicals and petroleum refining	0·46	4·23	45	25	0·92
Machinery	0·19	1·76	20	10	0·91
Age					
Chemicals and petroleum refining	0·04	1·29	47	30	0·89
Machinery	0·41	2·19	21	13	0·94
Experience					
Chemicals and petroleum refining	0·007	0·85	52	33	0·78
Machinery	0·57	6·08	23	14	0·91
Size					
Chemicals and petroleum refining	0·008	1·17	54	35	0·88
Machinery	0·081	5·18	17	10	0·99
R & D sales					
Total sample	0·06	1·58	59	30	0·90

VII. DIFFERENCES BETWEEN INTERNATIONAL AND DOMESTIC TECHNOLOGY TRANSFER

Although this is primarily a study of international technology transfer, it is apparent that many of the characteristics of international technology transfer are also characteristic of the technology transfer that occurs within national borders, but there are differences. For instance, distance and communication costs very often differentiate international from domestic transfers. Although the communications revolution of the twentieth century has enormously reduced the barriers imposed by distance,[1] the costs of international communication are often significant.[2] Language differences can also add to communication costs, especially if the translation of engineering drawings is required. The experience of Polyspinners Ltd at Mogilev in the Soviet Union (Jones, 1973) is ample testimony to the extra costs that can be encountered.[3] International differences in units of measurements and engineering standards can compound the problems encountered (Meursinge, 1971). Additional sources of difficulty are rooted in

[1] Facsimile equipment exists which can be used to transmit messages and drawings across the Atlantic instantaneously.

[2] One of the participating companies indicated that travel, telegraph, freight, and insurance added about 10% to the total cost of a project established in New Zealand.

[3] The project manager estimated that documentation alone cost £500,000, and the translation a similar amount.

the cultural and attitudinal differences between nations, as well as differences in the level of economic development and the attendant socioeconomic structure.

It is of interest to know the magnitude and determinants of the "international component" of the transfer cost. Unfortunately, foreign and domestic transfers

Table 5

Arc or Point Elasticity of Transfer Costs With Respect to Number of Start-ups, Age of Technology, Experience, Size and R & D/Sales of Transferee

Independent variable	Chemicals and petroleum refining	Machinery
	Arc elasticity	
Number of start-ups		
1–2	0·34	0·14
2–3	0·19	0·08
3–4	0·13	0·05
9–10	0·05	0·02
14–15	0·03	0·01
Age of technology (years)	Point elasticity	
1	0·04	0·41
2	0·02	0·20
3	0·01	0·14
10	0·00	0·04
20	0·00	0·02
Experience of transferee (years)	Point elasticity	
1	0·007	0·57
2	0·003	0·28
3	0·002	0·19
10	0·001	0·06
20	0·000	0·03
Size of transferee (millions of sales dollars)	Point elasticity	
1·0	0·008	0·081
10	0·001	0·008
20	0·000	0·004
100	0·000	0·001
1000	0·000	0·000
R & D/Sales of transferee (%)	Total sample point elasticity	
1	0·06	
2	0·03	
3	0·02	
4	0·01	
5	0·01	
6	0·01	

are rarely identical in scope or in timing, and so it is not possible to gather comparative data on implemented projects at home and abroad. It was therefore found necessary to rely on estimates provided by the firms involved in international transfers. For the projects in the sample, project managers were asked to estimate the dollar amount by which transfer costs would be different if the international transfers in the sample had occurred domestically, holding firm and technology characteristics constant. The procedure was designed to highlight the effects of country characteristics such as differences in language, differences in engineering and measurement standards, differences in economic infrastructure and business environment, and geographical distance from the transferor. The international component of the transfer cost for the projects in the sample could be obtained by subtracting the estimated transfer cost from the

Table 6

International Component of Transfer Cost*

Chemicals and petroleum refining		Machinery	
Dollar amount (thousands)	As % of actual transfer cost	Dollar amount (thousands)	As % of actual transfer cost
3·03	6·07	35·55	17·88
0·00	0·00	−399·37	−110·93
−12·81	−1·87	50·06	4·93
43·90	31·00	830·70	14·20
0·00	0·00	−4·59	−0·02
5·17	1·42	226·80	14·82
132·75	20·63	0·67	1·99
0·00	0·00	−134·40	13·87
342·00	43·84	34·98	12·95
0·00	—		
0·00	—		
0·00	0·00		
−10·77	−6·66		
−50·16	−8·52		
0·00	—		
637·32	72·60		
−1·33	−1·99		
1,723·81	60·48		
1,370·25	18·45		
524·25	15·69		

* Amount of actual transfer costs attributable to the fact that transfer was international rather than domestic. (Accordingly negative values indicate that firms estimated that transfer costs would be higher had the transfer been domestic.) In general, these numbers were derived from taking the weighted average of estimated changes in the various identifiable components of transfer costs.

actual transfer cost. The data, contained in Table 6, reveal that the difference in cost is not always positive. This indicates that in at least some of the cases, the international transfer of an innovation was estimated to cost less than a comparable domestic transfer. This may seem paradoxical at first, given that international technology transfer generally augments the transfer activities that have to be performed.[1] An analysis of the determinants of the international component of transfer costs may yield an explanation.

Several hypotheses are presented and tested. The first is that the difference is

[1] The source of the apparent paradox may be differences in labour costs. Nevertheless, the identification of the transfer for which international transfer costs less than domestic transfer is an issue of importance.

large and positive when the technology has not been previously commercialised. National boundaries are often surrogates for cultural and language barriers, differences in methods and standards of measurement, and distance from the home country. During first commercialisation of a product or process, there are generally enormous information flows across the development-manufacturing interface. The hypothesis is that placing a national boundary at this interface can complicate matters considerably, and escalate the costs enormously. The second hypothesis is that transfers to government enterprises in centrally planned economies will involve higher transfer costs. Transferors can expect numerous delays and large documentation requirements (Jones, 1973). The third hypothesis is that the less the diffusion of the technology, measured as before by the number of firms utilising the innovation, the greater the positive differential associated with international technology transfer. The fourth hypothesis is that whereas, in general, low levels of economic development are likely to add to transfer costs because of inadequacies in the economic infrastructure, this may be more than offset, in some circumstances, by low labour costs. Labour costs can have a substantial impact on excess manufacturing costs, especially in relatively labour intensive industries. Since machinery manufacture is relatively labour intensive, the hypothesis is that the GNP *per capita* in the host country is positively related to the transfer cost differential in this classification, but is negatively associated with the differential in the chemicals and petroleum refining category.

To test these hypotheses it is assumed that

$$D_i = \alpha_0 + \alpha_1 d_i + \alpha_2 \bar{U}_i + \alpha_3 N_i + \alpha_4 P_i + Z_i,$$

where D_i is the "international component" as a percentage of actual transfer cost for the ith transfer. d_i is a dummy variable which takes the value 1 if the recipient of the ith transfer is a government enterprise in a centrally planned economy, and zero otherwise. The other variables carry the same definitions as previously. The expected derivations are:

$$\frac{\partial D_i}{\partial d_i} > 0, \quad \frac{\partial D_i}{\partial \bar{U}_i} > 0, \quad \frac{\partial D_i}{\partial N_i} < 0;$$

$\partial D_i / \partial P_i \gtrless 0$ according to the industry category (the partial is postulated positive for the machinery category, and negative otherwise). Least-squares estimates of the α's were obtained, the results being:

Chemicals and petroleum:

$$D_i = \underset{(0.91)}{0.285} + \underset{(5.01)}{3.84 d_i} + \underset{(4.89)}{4.46 \bar{U}_i} \quad (n = 17,\ r^2 = 0.71).$$

Machinery:[1]

$$D_i = \underset{(-1.96)}{-8.59} \quad \underset{(-5.98)}{-1.39 N_i} + \underset{(3.90)}{0.005 P_i} \quad (n = 9,\ r^2 = 0.94).$$

[1] O_i was omitted from the machinery regression since none of the actual transfers in this category were to government enterprises in centrally planned economies.

The hypotheses are to some extent borne out by the data, but the small sample size must counsel caution in the interpretation of these results.[1] In chemicals and petroleum, the results indicate that transfers to government enterprises, and transfers before first commercialisation, involve substantial extra costs. Furthermore, both N_i and P_i are significant in the machinery category, despite the small number of observations, yet they are insignificant in chemicals and petroleum refining, where there are more than twice as many degrees of freedom. Apparently, the level of host country development and the degree of diffusion of an innovation have no bearing on the international–domestic transfer cost differential in the chemicals and petroleum grouping. This calls for an explanation. The diffusion variable N_i is taken to indicate the degree to which the requisite skills are generally available. The statistical results suggest that the relevant skills for highly capital intensive industries, such as chemicals and petroleum refining, are more easily transferred internationally than are the requisite skills in the machinery category.[2] Furthermore, P_i was not significant in chemicals and petroleum refining, suggesting that costs of transfer are independent of the level of economic development in this category. This is consistent with speculation that international transfer is no more difficult than domestic transfer when the underlying technology is highly capital intensive. The perceived reluctance of multinational firms to adapt technology to suit the capital–labour endowments of less developed countries could well be rooted in the desire to avoid escalating transfer costs to unacceptable levels.

VIII. CONCLUSION

The resources required to transfer technology internationally are considerable. Accordingly, it is quite inappropriate to regard existing technology as something that can be made available to all at zero social cost. Furthermore, transfer costs vary considerably, especially according to the number of previous applications of the innovation, and how well the innovation is understood by the parties involved. It is equally inappropriate, therefore, to make sweeping generalisations about the process of technology transfer and the costs involved. For instance, technology transfer in chemicals and petroleum refining displayed relatively low transfer costs, presumably because it is possible to embody sophisticated process technology in capital equipment, which in turn facilitates the transfer process.

The analysis of the determinants of technology transfer costs provided some interesting findings with development implications. The success of the more experienced enterprises, indicated by lower transfer costs, points towards economic models which emphasise the accumulation of skills, rather than fixed

[1] If the second observation on D_i in the machinery category is excluded, and the regression results recomputed, the estimates of the coefficients exhibit considerable instability and the "goodness of fit" deteriorates. The estimated equation is

$$D_i = \underset{(1\cdot 14)}{-4\cdot 96} \underset{(2\cdot 40)}{-0\cdot 66 N_i} + \underset{(1\cdot 94)}{0\cdot 003 P_i} \quad (n = 8,\ r^2 = 0\cdot 45).$$

These estimates are nevertheless significant at the 0·05 level for a one-tail test.

[2] This is consonant with the views expressed by several project managers in the chemical industry. It was asserted that technology could be transferred with equal facility to almost anywhere in the world, including less developed countries, assuming host government interference is held constant.

assets or capital, in facilitating the technology transfer process. This seems consonant with the findings of several economic historians (Rosenberg, 1970; Rawski, 1975).

The results also provide some managerial implications for the multinational firm. Consider the costs associated with separating production from development (Arditti, 1968). The results indicate that the international transfer of technology is most likely to be viable when production runs are long enough to allow second sourcing. The especially high cost of transfer before first application favours the development location, at least for production of initial units. However, transfer costs will be lowered once the first production run has been commenced, and international transfer then becomes more likely, a finding consistent with the product cycle model (Vernon, 1966). However, inter-industry differences are important, and the costs involved in separating first production from development did not prove to be an insurmountable transfer barrier for an important subset of the sample projects.

A second implication is that since transfer costs decline with each application of a given innovation, technology transfer is a decreasing cost activity. This can be advanced as an explanation for the specialisation often exhibited by engineering firms in the design and installation of particular turnkey plants,[1] a characteristic particularly noteworthy of the petrochemical industry.

A third set of managerial implications relate to the criteria which might be used for the selection of a joint venture or licensing partner to utilise the innovating firms' technology abroad. While the manufacturing experience, size, and R & D to sales ratio of the transferee were identified as statistically significant determinants of transfer costs for the sample, there was also evidence to suggest that, *ceteris paribus*, any firm moderately matured in these dimensions is a good candidate to absorb the technology at the minimum possible transfer cost. It is not clear, therefore, that super giant firms have any advantage in this respect over moderately sized firms. Nor is it clear that highly research intensive firms have more than a slight cost advantage in absorbing technology over firms with a minimal commitment to research and development activity. However, manufacturing experience is important, especially for transferring machinery technology. In addition, there is evidence that transfers to governments in centrally planned economies involve substantial extra costs, perhaps because of high documentation requirements, or differences in language and managerial procedures.

Technology transfer by multinational firms is clearly a complex matter. Collection and analysis of proprietary data has provided some helpful insights. Few issues have been settled although many have been raised. Further analytic research and more extensive data collection is required if our understanding of international technology transfer is to be improved.

D. J. TEECE

Stanford University, California

Date of receipt of final typescript: November 1976

[1] Turnkey plants generally embody state-of-the-art technology.

References

Arrow, K. (1969). "Classificatory Notes on the Production and Transmission of Technological Knowledge." *Amercian Economic Review; Papers and Proceedings*, vol. 52 (May), pp. 29–35.

—— (1962). Comment in Universities-National Bureau Committee for Economic Research. *The Rate and Direction of Inventive Activity*. Princeton: Princeton University Press.

Arditti, F. (1968). "On the Separation of Production from the Developer." *Journal of Business*, vol. 41 (July), pp. 317–28.

Baranson, J. (1967): *Manufacturing Problems in India: The Cummings Diesel Experience*. Syracuse, N.Y.: Syracuse University Press.

Berrill, K. (ed.) (1964). *Economic Development with Special Reference to East Asia*. New York: St Martins Press.

Chow, G. C. (1960). "Tests of Equality between Sets of Coefficients in Two Linear Regressions." *Econometrica*, vol. 28 (July), pp. 591–605.

Freeman, C. (1965). "Research and Development in Electronic Capital Goods." *National Institute Economic Review*, no. 34, vol. 34 (November), pp. 1–70.

Gillette, R. (1973). "Latin America: Is Imported Technology Too Expensive?" *Science*, vol. 191 (6 July), pp. 4–44.

Hall, G. R. and Johnson, R. E. (1970). "Transfers of United States Aerospace Technology to Japan." In *The Technology Factor in International Trade* (ed. R. Vernon). N.Y.: National Bureau of Economic Research.

Hayami, Y. and Ruttan, V. (1971). *Agricultural Development and International Perspective*. Baltimore: Johns Hopkins.

International Bank for Reconstruction and Development (1973). *World Bank Atlas*. Washington, D.C.: I.B.R.D.

Jones, D. (1973). "The 'Extra Costs' in Europe's Biggest Synthetic Fiber Complex at Mogilev, U.S.S.R." *Worldwide Projects and Installations*, vol. 7 (May/June), pp. 30–5.

Kuznets, S. (1966). *Modern Economic Growth: Rate, Structure, Spread*. New Haven: Yale University Press.

Mansfield, E. (1974). "Technology and Technical Change." In *Economic Analysis and the Multinational Enterprise* (ed. J. Dunning). London: Allen and Unwin.

—— (1968). *The Economics of Technological Change*. New York: Norton.

—— (1973). "Discussion of the Paper by Professor Griliches." In *Science and Technology in Economic Growth* (ed. B. R. Williams). New York: John Wiley.

Mansfield, E., Rapoport, J., Schnee, J., Wagner, S. and Hamburger, M. (1971). *Research and Innovation In the Modern Corporation*. New York: W. W. Norton.

Mason, R. Hal (1973). "The Multinational Firm and the Cost of Technology to Developing Countries." *California Management Review*, vol. 15 (Summer), pp. 5–13.

Meursinge, J. (1971). "Practical Experience in the Transfer of Technology." *Technology and Culture*, vol. 12 (July), pp. 469–70.

McGraw-Hill Encyclopedia of Science and Technology (1960). Vols. 4, 10, pp. 639–44. New York: McGraw-Hill.

Oshima, K. (1973). "Research and Development and Economic Growth in Japan." In *Science and Technology in Economic Growth* (ed. B. R. Williams). New York: John Wiley.

Rawski, T. (1975). "Problems of Technology and Absorption in Chinese Industry." *American Economic Review*, vol. 65 (May), pp. 363–88.

Reynolds, L. (1966). Discussion. *American Economic Review*, vol. 56 (May), pp. 112–14.

Robinson, E. A. G. (1973). "Discussion of the Paper by Professor Hsia." In *Science and Technology in Economic Growth* (ed. B. R. Williams). New York: John Wiley.

Rodriguez, C. A. (1975). "Trade in Technical Knowledge and the National Advantage." *Journal of Political Economy*, vol. 93 (February), pp. 121–35.

Rosenberg, N. (1970). "Economic Development and the Transfer of Technology: Some Historical Perspectives." *Technology and Culture*, vol. 11 (October), pp. 550–75.

Teece, D. (1976). *The Multinational Corporation and the Resource Cost of International Technology Transfer*. Cambridge: Ballinger.

—— (1977). "Time-Cost Tradeoffs: Elasticity Estimates and Determinants for International Technology Transfer Projects." *Management Science*, vol. 23 (April).

Tilton, J. (1971). *International Diffusion of Technology: The Case of Semiconductors*. Washington, D.C.: Brookings Institution.

United Nations (1968). *International Standard Industrial Classification of all Economic Activities*. United Nations Statistical Papers, Series M., Number 4. New York: United Nations.

UNCTAD (1970). "The Transfer of Technology." *Journal of World Trade Law*, vol. 4 (September/October), pp. 692–718.

Vernon, R. (1966). "International Investment and International Trade in the Product Cycle. *Quarterly Journal of Economics*, vol. 80 (May), pp. 190–207.

Part VII
The Management of Technology

[23]

The Economic Implications of Learning by Doing

It is by now incontrovertible that increases in per capita income cannot be explained simply by increases in the capital-labor ratio. Though doubtless no economist would ever have denied the role of technological change in economic growth, its overwhelming importance relative to capital formation has perhaps only been fully realized with the important empirical studies of Abramovitz [1] and Solow [11]. These results do not directly contradict the neo-classical view of the production function as an expression of technological knowledge. All that has to be added is the obvious fact that knowledge is growing in time. Nevertheless a view of economic growth that depends so heavily on an exogenous variable, let alone one so difficult to measure as the quantity of knowledge, is hardly intellectually satisfactory. From a quantitative, empirical point of view, we are left with time as an explanatory variable. Now trend projections, however necessary they may be in practice, are basically a confession of ignorance, and, what is worse from a practical viewpoint, are not policy variables.

Further, the concept of knowledge which underlies the production function at any moment needs analysis. Knowledge has to be acquired. We are not surprised, as educators, that even students subject to the same educational experiences have different bodies of knowledge, and we may therefore be prepared to grant, as has been shown empirically (see [2], Part III), that different countries, at the same moment of time, have different production functions even apart from differences in natural resource endowment.

I would like to suggest here an endogenous theory of the changes in knowledge which underlie intertemporal and international shifts in production functions. The acquisition of knowledge is what is usually termed " learning," and we might perhaps pick up some clues from the many psychologists who have studied this phenomenon (for a convenient survey, see Hilgard [5]). I do not think that the picture of technical change as a vast and prolonged process of learning about the environment in which we operate is in any way a far-fetched analogy; exactly the same phenomenon of improvement in performance over time is involved.

Of course, psychologists are no more in agreement than economists, and there are sharp differences of opinion about the processes of learning. But one empirical generalization is so clear that all schools of thought must accept it, although they interpret it in different fashions: Learning is the product of experience. Learning can only take place through the attempt to solve a problem and therefore only takes place during activity. Even the Gestalt and other field theorists, who stress the role of insight in the solution of problems (Köhler's famous apes), have to assign a significant role to previous experiences in modifying the individual's perception.

A second generalization that can be gleaned from many of the classic learning experiments is that learning associated with repetition of essentially the same problem is subject to sharply diminishing returns. There is an equilibrium response pattern for any given stimulus, towards which the behavior of the learner tends with repetition. To have steadily increasing performance, then, implies that the stimulus situations must themselves be steadily evolving rather than merely repeating.

The role of experience in increasing productivity has not gone unobserved, though the relation has yet to be absorbed into the main corpus of economic theory. It was early observed by aeronautical engineers, particularly T. P. Wright [15], that the number of labor-hours expended in the production of an airframe (airplane body without engines) is a decreasing function of the total number of airframes of the same type previously produced. Indeed, the relation is remarkably precise; to produce the Nth airframe of a given type, counting from the inception of production, the amount of labor required is proportional to $N^{-1/3}$. This relation has become basic in the production and cost planning of the United States Air Force; for a full survey, see [3]. Hirsch (see [6] and other work cited there) has shown the existence of the same type of " learning curve " or " progress ratio," as it is variously termed, in the production of other machines, though the rate of learning is not the same as for airframes.

Verdoorn [14, pp. 433-4] has applied the principle of the learning curve to national outputs; however, under the assumption that output is increasing exponentially, current output is proportional to cumulative output, and it is the former variable that he uses to explain labor productivity. The empirical fitting was reported in [13]; the estimated progress ratio for different European countries is about ·5. (In [13], a neo-classical interpretation in terms of increasing capital-labor ratios was offered; see pp. 7-11.)

Lundberg [9, pp. 129-133] has given the name " Horndal effect " to a very similar phenomenon. The Horndal iron works in Sweden had no new investment (and therefore presumably no significant change in its methods of production) for a period of 15 years, yet productivity (output per manhour) rose on the average close to 2% per annum. We find again steadily increasing performance which can only be imputed to learning from experience.

I advance the hypothesis here that technical change in general can be ascribed to experience, that it is the very activity of production which gives rise to problems for which favorable responses are selected over time. The evidence so far cited, whether from psychological or from economic literature is, of course, only suggestive. The aim of this paper is to formulate the hypothesis more precisely and draw from it a number of economic implications. These should enable the hypothesis and its consequences to be confronted more easily with empirical evidence.

The model set forth will be very simplified in some other respects to make clearer the essential role of the major hypothesis; in particular, the possibility of capital-labor substitution is ignored. The theorems about the economic world presented here differ from those in most standard economic theories; profits are the result of technical change; in a free-enterprise system, the rate of investment will be less than the optimum; net investment and the stock of capital become subordinate concepts, with gross investment taking a leading role.

In section 1, the basic assumptions of the model are set forth. In section 2, the implications for wage earners are deduced; in section 3 those for profits, the inducement to invest, and the rate of interest. In section 4, the behavior of the entire system under steady growth with mutually consistent expectations is taken up. In section 5, the divergence between social and private returns is studied in detail for a special case (where the subjective rate of discount of future consumption is a constant). Finally, in section 6, some limitations of the model and needs for further development are noted.

1. The Model

The first question is that of choosing the economic variable which represents " experience ". The economic examples given above suggest the possibility of using cumulative output (the total of output from the beginning of time) as an index of experience, but this does not seem entirely satisfactory. If the rate of output is constant, then the stimulus to learning presented would appear to be constant, and the learning that does take place is a gradual approach to equilibrium behavior. I therefore take instead cumulative gross investment (cumulative production of capital goods) as an index of experience. Each new machine produced and put into use is capable of changing the environment in which production takes place, so that learning is taking place with continually new stimuli. This at least makes plausible the possibility of continued learning in the sense, here, of a steady rate of growth in productivity.

The second question is that of deciding where the learning enters the conditions of production. I follow here the model of Solow [12] and Johansen [7], in which technical change is completely embodied in new capital goods. At any moment of new time, the new capital goods incorporate all the knowledge then available, but once built their productive efficiency cannot be altered by subsequent learning.

To simplify the discussion we shall assume that the production process associated with any given new capital good is characterized by fixed coefficients, so that a fixed amount of labor is used and a fixed amount of output obtained. Further, it will be assumed that new capital goods are better than old ones in the strong sense that, if we compare a unit of capital goods produced at time t_1 with one produced at time $t_2 > t_1$, the first requires the co-operation of at least as much labor as the second, and produces no more product. Under this assumption, a new capital good will always be used in preference to an older one.

Let G be cumulative gross investment. A unit capital good produced when cumulative gross investment has reached G will be said to have *serial number* G. Let

$\lambda(G)$ = amount of labor used in production with a capital good of serial number G,
$\gamma(G)$ = output capacity of a capital good of serial number G,
x = total output,
L = total labor force employed.

It is assumed that $\lambda(G)$ is a non-increasing function, while $\gamma(G)$ is a non-decreasing function. Then, regardless of wages or rental value of capital goods, it always pays to use a capital good of higher serial number before one of lower serial number.

It will further be assumed that capital goods have a fixed lifetime, $\bar{T}$. Then capital goods disappear in the same order as their serial numbers. It follows that at any moment of time, the capital goods in use will be all those with serial numbers from some G' to G, the current cumulative gross investment. Then

$$x = \int_{G'}^{G} \gamma(G)\,dG, \qquad (1)$$

$$(2) \qquad L = \int_{G'}^{G} \lambda(G)dG.$$

The magnitudes x, L, G, and G' are, of course, all functions of time, to be designated by t, and they will be written $x(t)$, $L(t)$, $G(t)$, and $G'(t)$ when necessary to point up the dependence. Then $G(t)$, in particular, is the cumulative gross investment up to time t. The assumption about the lifetime of capital goods implies that

$$(3) \qquad G'(t) \geqq G(t - \bar{T}).$$

Since $G(t)$ is given at time t, we can solve for G' from (1) or (2) or the equality in (3). In a growth context, the most natural assumption is that of full employment. The labor force is regarded as a given function of time and is assumed equal to the labor employed, so that $L(t)$ is a given function. Then $G'(t)$ is obtained by solving in (2). If the result is substituted into (1), x can be written as a function of L and G, analogous to the usual production function. To write this, define

$$\Lambda(G) = \int \lambda(G)dG,$$

$$(4) \qquad \Gamma(g) = \int \gamma(G)dG.$$

These are to be regarded as indefinite integrals. Since $\lambda(G)$ and $\gamma(G)$ are both positive, $\Lambda(G)$ and $\Gamma(G)$ are strictly increasing and therefore have inverses, $\Lambda^{-1}(u)$ and $\Gamma^{-1}(v)$, respectively. Then (1) and (2) can be written, respectively,

$$(1') \qquad x = \Gamma(G) - \Gamma(G'),$$

$$(2') \qquad L = \Lambda(G) - \Lambda(G').$$

Solve for G' from (2′).

$$(5) \qquad G' = \Lambda^{-1}[\Lambda(G) - L].$$

Substitute (5) into (1′).

$$(6) \qquad x = \Gamma(G) - \Gamma\{\Lambda^{-1}[\Lambda(G) - L]\},$$

which is thus a production function in a somewhat novel sense. Equation (6) is always valid, but under the full employment assumption we can regard L as the labor force available.

A second assumption, more suitable to a depression situation, is that in which demand for the product is the limiting factor. Then x is taken as given; G' can be derived from (1) or (1′), and employment then found from (2) or (2′). If this is less than the available labor force, we have Keynesian unemployment.

A third possibility, which, like the first, may be appropriate to a growth analysis, is that the solution (5) with L as the labor force, does not satisfy (3). In this case, there is a shortage of capital due to depreciation. There is again unemployment but now due to structural discrepancies rather than to demand deficiency.

In any case, except by accident, there is either unemployed labor or unemployed capital; there could be both in the demand deficiency case. Of course, a more neoclassical model, with substitution between capital and labor for each serial number of capital good, might permit full employment of both capital and labor, but this remains a subject for further study.

In what follows, the full-employment case will be chiefly studied. The capital shortage case, the third one, will be referred to parenthetically. In the full-employment case, the depreciation assumption no longer matters; obsolescence, which occurs for all capital goods with serial numbers below G', becomes the sole reason for the retirement of capital goods from use.

The analysis will be carried through for a special case. To a very rough approximation, the capital-output ratio has been constant, while the labor-output ratio has been declining. It is therefore assumed that

$$(7) \qquad \gamma(G) = a,$$

a constant, while $\lambda(G)$ is a decreasing function of G. To be specific, it will be assumed that $\lambda(G)$ has the form found in the study of learning curves for airframes.

$$(8) \qquad \lambda(G) = bG^{-n},$$

where $n > 0$. Then

$$\Gamma(G) = aG, \Lambda(G) = cG^{1-n}, \text{ where } c = b/(1-n) \text{ for } n \neq 1.$$

Then (6) becomes

$$(9) \qquad x = aG[1 - \left(1 - \frac{L}{cG^{1-n}}\right)^{1/(1-n)}] \text{ if } n \neq 1.$$

Equation (9) is always well defined in the relevant ranges, since from (2′),

$$L = \Lambda(G) - \Lambda(G') \leqq \Lambda(G) = cG^{1-n}.$$

When $n = 1$, $\Lambda(G) = b \log G$ (where the natural logarithm is understood), and

$$(10) \qquad x = aG(1-e^{-L/b}) \text{ if } n = 1.$$

Although (9) and (10) are, in a sense, production functions, they show increasing returns to scale in the variables G and L. This is obvious in (10) where an increase in G, with L constant, increases x in the same proportion; a simultaneous increase in L will further increase x. In (9), first suppose that $n < 1$. Then a proportional increase in L and G increases L/G^{1-n} and therefore increases the expression in brackets which multiplies G. A similar argument holds if $n > 1$. It should be noted that x increases more than proportionately to scale changes in G and L in general, not merely for the special case defined by (7) and (8). This would be verified by careful examination of the behavior of (6), when it is recalled that $\lambda(G)$ is non-increasing and $\gamma(G)$ is non-decreasing, with the strict inequality holding in at least one. It is obvious intuitively, since the additional amounts of L and G are used more efficiently than the earlier ones.

The increasing returns do not, however, lead to any difficulty with distribution theory. As we shall see, both capital and labor are paid their marginal products, suitably defined. The explanation is, of course, that the private marginal productivity of capital (more strictly, of new investment) is less than the social marginal productivity since the learning effect is not compensated in the market.

The production assumptions of this section are designed to play the role assigned by Kaldor to his " technical progress function," which relates the rate of growth of output per worker to the rate of growth of capital per worker (see [8], section VIII). I prefer to think of relations between rates of growth as themselves derived from more fundamental relations between the magnitudes involved. Also, the present formulation puts more stress on gross rather than net investment as the basic agent of technical change.

Earlier, Haavelmo ([4], sections 7.1 and 7.2) had suggested a somewhat similar model. Output depended on both capital and the stock of knowledge; investment depended on output, the stock of capital, and the stock of knowledge. The stock of knowledge was either simply a function of time or, in a more sophisticated version, the consequence of investment, the educational effect of each act of investment decreasing exponentially in time.

Verdoorn [14, pp. 436-7] had also developed a similar simple model in which capital and labor needed are non-linear functions of output (since the rate of output is, approximately, a measure of cumulative output and therefore of learning) and investment a constant fraction of output. He notes that under these conditions, full employment of capital and labor simultaneously is in general impossible—a conclusion which also holds for for the present model as we have seen. However, Verdoorn draws the wrong conclusion: that the savings ratio must be fixed by some public mechanism at the uniquely determined level which would insure full employment of both factors; the correct conclusion is that one factor or the other will be unemployed. The social force of this conclusion is much less in the present model since the burden of unemployment may fall on obsolescent capital; Verdoorn assumes his capital to be homogeneous in nature.

2. Wages

Under the full employment assumption the profitability of using the capital good with serial number G' must be zero; for if it were positive it would be profitable to use capital goods with higher serial number and if it were negative capital good G' would not be used contrary to the definition of G'. Let

$$w = \text{wage rate with output as numéraire.}$$

From (1′) and (7)

$$(11) \qquad G' = G - (x/a)$$

so that

$$(12) \qquad \lambda(G') = b\left(G - \frac{x}{a}\right)^{-n}.$$

The output from capital good G' is $\gamma(G')$ while the cost of operation is $\lambda(G')w$. Hence

$$\gamma(G') = \lambda(G')w$$

or from (7) and (12)

$$(13) \qquad w = a\left(G - \frac{x}{a}\right)^{n}/b.$$

It is interesting to derive labor's share which is wL/x. From (2′) with $\Lambda(g) = cG^{1-n}$ and G' given by (11)

$$L = c\left[\; G^{1-n} - \left(G - \frac{x}{a}\right)^{1-n} \right],$$

for $n \neq 1$ and therefore

$$\text{(14)} \qquad wL/x = a\left[\left(\frac{G}{x} - \frac{1}{a}\right)^{n}\left(\frac{G}{x}\right)^{1-n} - \left(\frac{G}{x} - \frac{1}{a}\right)\right]/(1-n) \text{ for } n \neq 1,$$

where use has been made of the relation, $c = b/(1-n)$. It is interesting to note that labor's share is determined by the ratio G/x.

Since, however, x is determined by G and L, which, at any moment of time, are data, it is also useful to express the wage ratio, w, and labor's share, wL/x, in terms of L and G. First, G' can be found by solving for it from (2′).

$$\text{(15)} \qquad G' = \left(\; G^{1-n} - \frac{L}{c}\right)^{1/(1-n)} \text{ for } n \neq 1.$$

We can then use the same reasoning as above, and derive

$$\text{(16)} \qquad w = a\left(G^{1-n} - \frac{L}{c}\right)^{n/(1-n)}/b,$$

$$\text{(17)} \qquad \frac{wL}{x} = \frac{\left[\left(\frac{L}{G^{1-n}}\right)^{(1-n)/n} - \frac{1}{c}\left(\frac{L}{G^{1-n}}\right)^{1/n}\right]^{n/(1-n)}}{b\left[1 - \left(1 - \frac{L}{cG^{1-n}}\right)^{1/(1-n)}\right]}.$$

Labor's share thus depends on the ratio L/G^{1-n}; it can be shown to decrease as the ratio increases.

For completeness, I note the corresponding formulas for the case $n = 1$. In terms of G and x, we have

$$\text{(18)} \qquad w = (aG - x)/b,$$

$$\text{(19)} \qquad wL/x = \left(\frac{aG}{x} - 1\right) \log \frac{G/x}{(G/x) - (1/a)}.$$

In terms of G and L, we have

$$\text{(20)} \qquad G' = Ge^{-L/b},$$

$$\text{(21)} \qquad w = \frac{aG}{be^{L/b}},$$

$$\text{(22)} \qquad wL/x = \frac{L}{b(e^{L/b} - 1)}.$$

In this case, labor's share depends only on L, which is indeed the appropriate special case ($n=1$) of the general dependence on L/G^{1-n}.

The preceding discussion has assumed full employment. In the capital shortage case, there cannot be a competitive equilibrium with positive wage since there is necessarily unemployment. A zero wage is, however, certainly unrealistic. To complete the model, it would be necessary to add some other assumption about the behavior of wages. This case will not be considered in general; for the special case of steady growth, see Section 5.

3. Profits and Investment

The profit at time t from a unit investment made at time $v \leqq t$ is

$$\gamma[G(v)] - w(t)\ \lambda[G(v)].$$

In contemplating an investment at time v, the stream of potential profits depends upon expectations of future wages. We will suppose that looking ahead at any given moment of time each entrepreneur assumes that wages will rise exponentially from the present level. Thus the wage rate expected at time v to prevail at time t is

$$w(v)\ e^{\theta(t-v)},$$

and the profit expected at time v to be received at time t is

$$\gamma[G(v)]\ [1-W(v)\ e^{\theta(t-v)}],$$

where

(23) $$W(v) = w(v)\ \lambda[G(v)]/\gamma[G(v)],$$

the labor cost per unit output at the time the investment is made. The dependence of W on v will be made explicit only when necessary. The profitability of the investment is expected to decrease with time (if $\theta > 0$) and to reach zero at time $T^* + v$, defined by the equation

(24) $$We^{\theta T^*} = 1.$$

Thus T^* is the expected economic lifetime of the investment, provided it does not exceed the physical lifetime, $\bar{T}$. Let

(25) $$T = \min(\bar{T}, T^*).$$

Then the investor plans to derive profits only over an interval of length T, either because the investment wears out or because wages have risen to the point where it is unprofitable to operate. Since the expectation of wage rises which causes this abandonment derives from anticipated investment and the consequent technological progress, T^* represents the expected date of obsolescence. Let

$$\rho = \text{rate of interest.}$$

If the rate of interest is expected to remain constant over the future, then the discounted stream of profits over the effective lifetime, T, of the investment is

$$(26)\qquad S = \int_0^T e^{-\rho t}\,\gamma[G(v)]\,(1 - W e^{\theta t})dt,$$

or

$$(27)\qquad \frac{S}{\gamma[G(v)]} = \frac{1 - e^{-\rho T}}{\rho} + \frac{W(1 - e^{-(\rho-\theta)T})}{\theta - \rho}.$$

Let

$$(28)\qquad V = e^{-\theta T} = \max\,(e^{-\theta \bar{T}}, W),\ \alpha = \rho/\theta.$$

Then

$$(29)\qquad \frac{\theta S}{\gamma[G(v)]} = \frac{1 - V^{\alpha}}{\alpha} + \frac{W(1 - V^{\alpha-1})}{1-\alpha} = R(\alpha).$$

The definitions of $R(\alpha)$ for $\alpha = 0$ and $\alpha = 1$ needed to make the function continuous are:

$$R(0) = -\log V + W(1 - V^{-1}),\ R(1) = 1 - V + W \log V.$$

If all the parameters of (26), (27), or (29) are held constant, S is a function of ρ, and, equivalently, R of α. If (26) is differentiated with respect to ρ, we find

$$dS/d\rho = \int_0^T (-t)e^{-\rho t}\,\gamma[G(v)]\,(1 - W e^{\theta t})dt < 0.$$

Also

$$S < \gamma[G(v)] \int_0^T e^{-\rho t}dt = \gamma[G(v)]\,(1 - e^{-\rho T})/\rho$$

$$< \gamma[G(v)]/\rho.$$

Since obviously $S > 0$, S approaches 0 as ρ approaches infinity. Since R and α differ from S and ρ, respectively, only by positive constant factors, we conclude

$$dR/d\alpha < 0,\ \lim_{\alpha \to +\infty} R(\alpha) = 0.$$

To examine the behavior of $R(\alpha)$ as α approaches $-\infty$, write

$$R(\alpha) = -\frac{(1/V)^{1-\alpha}}{(1-\alpha)^2}\,[(1-\alpha)V + \alpha W]\left(\frac{1-\alpha}{\alpha}\right) + \frac{1}{\alpha} + \frac{W}{1-\alpha}.$$

The last two terms approach zero. As α approaches $-\infty$, $1 - \alpha$ approaches $+\infty$. Since $1/V > 1$, the factor

$$\frac{(1/V)^{1-\alpha}}{(1-\alpha)^2}$$

approaches $+\infty$, since an exponential approaches infinity faster than any power. From (28), $V \geqq W$. If $V = W$, then the factor,

$$(1 - \alpha)V - \alpha W = \alpha(W - V) + V,$$

is a positive constant; if $V > W$, then it approaches $+\infty$ as α approaches $-\infty$. Finally,

$$\frac{1 - \alpha}{\alpha}$$

necessarily approaches -1. Hence,

(30) $R(\alpha)$ is a strictly decreasing function, approaching $+\infty$ as α approaches $-\infty$ and 0 as α approaches $+\infty$.

The market, however, should adjust the rate of return so that the discounted stream of profits equals the cost of investment, i.e., $S = 1$, or, from (29),

(31) $R(\alpha) = \theta/\gamma[G(v)]$.

Since the right-hand side of (31) is positive, (30) guarantees the existence of an α which satisfies (31). For a given θ, the equilibrium rate of return, ρ, is equal to $\alpha\,\theta$; it may indeed be negative. The rate of return is thus determined by the expected rate of increase in wages, current labor costs per unit output, and the physical lifetime of the investment. Further, if the first two are sufficiently large, the physical lifetime becomes irrelevant, since then $T^* < \bar{T}$, and $T = T^*$.

The discussion of profits and returns has not made any special assumptions as to the form of the production relations.

4. Rational Expectations in a Macroeconomic Growth Model

Assume a one-sector model so that the production relations of the entire economy are described by the model of section 1. In particular, this implies that gross investment at any moment of time is simply a diversion of goods that might otherwise be used for consumption. Output and gross investment can then be measured in the same units.

The question arises, can the expectations assumed to govern investment behavior in the preceding section actually be fulfilled? Specifically, can we have a constant relative increase of wages and a constant rate of interest which, if anticipated, will lead entrepreneurs to invest at a rate which, in conjunction with the exogenously given rate of interest to remain at the given level? Such a state of affairs is frequently referred to as " perfect foresight," but a better term is " rational expectations," a term introduced by J. Muth [10].

We study this question first for the full employment case. For this case to occur, the physical lifetime of investments must not be an effective constraint. If, in the notation of the last section, $T^* > \bar{T}$, and if wage expectations are correct, then investments will disappear through depreciation at a time when they are still yielding positive current profits. As seen in section 2, this is incompatible with competitive equilibrium and full employment. Assume therefore that

(32) $T^* \leqq \bar{T}$;

then from (28), $W = V$, and from (29) and (31), the equilibrium value of ρ is determined by the equation,

(33) $$\frac{1 - W^\alpha}{\alpha} + \frac{W - W^\alpha}{1 - \alpha} = \frac{\theta}{a},$$

where, on the right-hand side, use is made of (7).

From (16), it is seen that for the wage rate to rise at a constant rate θ, it is necessary that the quantity,

$$G^{1-n} - \frac{L}{c},$$

rise at a rate $\theta(1 - n)/n$. For θ constant, it follows from (33) that a constant ρ and therefore a constant α requires that W be constant. For the specific production relations (7) and (8), (23) shows that

$$W = a \frac{\left(G^{1-n} - \frac{L}{c}\right)^{n/(1-n)}}{b} \frac{bG^{-n}}{a} = \left(1 - \frac{L}{cG^{1-n}}\right)^{n/(1-n)},$$

and therefore the constancy of W is equivalent to that of L/G^{1-n}. In combination with the preceding remark, we see that

(34) L increases at rate $\theta(1 - n)/n$, G increases at rate θ/n.

Suppose that

σ = rate of increase of the labor force,

is a given constant. Then

(35) $\theta = n\,\sigma/(1-n)$,

(36) the rate of increase of G is $\sigma/(1-n)$.

Substitution into the production function (9) yields

(37) the rate of increase of x is $\sigma/(1-n)$.

From (36) and (37), the ratio G/x is constant over time. However, the value at which it is constant is not determined by the considerations so far introduced; the savings function is needed to complete the system. Let the constant ratio be

(38) $G(t)/x(t) = \mu$.

Define

$g(t)$ = rate of gross investment at time $t = dG/dt$.

From (36), $g/G = \sigma/(1 - n)$, a constant. Then

$$(39) \qquad g/x = (g/G)(G/x) = \mu\, \sigma/(1 - n).$$

A simple assumption is that the ratio of gross saving (equals gross investment) to income (equals output) is a function of the rate of return, ρ; a special case would be the common assumption of a constant savings-to-income ratio. Then μ is a function of ρ. On the other hand, we can write W as follows, using (23) and (13):

$$(40) \qquad W = a\,\frac{\left(G - \frac{x}{a}\right)^n}{b}\;\frac{bG^{-n}}{a} = \left(1 - \frac{x}{aG}\right)^n = \left(1 - \frac{1}{a\mu}\right)^n.$$

Since θ is given by (35), (33) is a relation between W and ρ, and, by (40) between μ and ρ. We thus have two relations between μ and ρ, so they are determinate.

From (38), μ determines one relation between G and X. If the labor force, L, is given at one moment of time, the production function (9) constitutes a second such relation, and the system is completely determinate.

As in many growth models, the rates of growth of the variables in the system do not depend on savings behavior; however, their levels do.

It should be made clear that all that has been demonstrated is the existence of a solution in which all variables have constant rates of growth, correctly anticipated. The stability of the solution requires further study.

The growth rate for wages implied by the solution has one paradoxical aspect; it increases with the rate of growth of the labor force (provided $n < 1$). The explanation seems to be that under full employment, the increasing labor force permits a more rapid introduction of the newer machinery. It should also be noted that, for a constant saving ratio, g/x, an increase in σ decreases μ, from (39), from which it can be seen that wages at the initial time period would be lower. In this connection it may be noted that since G cannot decrease, it follows from (36) that σ and $1-n$ must have the same sign for the steady growth path to be possible. The most natural case, of course, is $\sigma > 0$, $n < 1$.

This solution is, however, admissible only if the condition (32), that the rate of depreciation not be too rapid, be satisfied. We can find an explicit formula for the economic lifetime, T^*, of new investment. From (24), it satisfies the condition

$$e^{-\theta T^*} = W.$$

If we use (35) and (40) and solve for T^*, we find

$$(41) \qquad T^* = \frac{-(1-n)}{\sigma} \log \left[1 - \frac{1}{a\mu}\right]$$

and this is to be compared with $\bar{T}$; the full employment solution with rational expectations of exponentially increasing wages and constant interest is admissible if $T^* \leq \bar{T}$.

If $T^* > \bar{T}$, then the full employment solution is inadmissible. One might ask if a constant-growth solution is possible in this case. The answer depends on assumptions about the dynamics of wages under this condition.

We retain the two conditions, that wages rise at a constant rate θ, and that the rate of interest be constant. With constant θ, the rate of interest, ρ, is determined from (31); from (29), this requires that

(42) W is constant over time.

From the definition of W, (23), and the particular form of the production relations, (7) and (8), it follows that the wage rate, w, must rise at the same rate as G^n, or

(43) G rises at a constant rate θ/n.

In the presence of continued unemployment, the most natural wage dynamics in a free market would be a decreasing, or, at best, constant wage level. But since G can never decrease, it follows from (43) that θ can never be negative. Instead of making a specific assumption about wage changes, it will be assumed that any choice of θ can be imposed, perhaps by government or union or social pressure, and it is asked what restrictions on the possible values of θ are set by the other equilibrium conditions.

In the capital shortage case, the serial number of the oldest capital good in use is determined by the physical lifetime of the good, i.e.,

$G' = G(t - \bar{T})$. From (43),

$$G(t - \bar{T}) = e^{-\theta \bar{T}/n} G.$$

Then, from (1′) and (7),

$$x = aG(1 - e^{-\theta \bar{T}/n}),$$

so that the ratio, G/x, or μ. is a constant,

(44) $\mu = 1/a(1 - e^{-\theta \bar{T}/n})$.

From (43), $g/G = \theta/n$; hence, by the same argument as that leading to (39),

(45) $g/x = \theta/na(1 - e^{-\theta \bar{T}/n})$.

There are three unknown constants of the growth process, θ, ρ, and W. If, as before, it is assumed that the gross savings ratio, g/x, is a function of the rate of return, ρ, then, for any given ρ, θ can be determined from (45); note that the right-hand side of (45) is a strictly increasing function of θ for $\theta \geqq 0$, so that the determination is unique, and the rate of growth is an increasing function of the gross savings ratio, contrary to the situation in the full employment case. Then W can be solved for from (31) and (29).

Thus the rate of return is a freely disposable parameter whose choice determines the rate of growth and W, which in turn determines the initial wage rate. There are, of course, some inequalities which must be satisfied to insure that the solution corresponds to the capital shortage rather than the full employment case; in particular, $W \leqq V$ and also the labor force must be sufficient to permit the expansion. From (2′), this means that the labor force must at all times be at least equal to

$$cG^{1-n} - c(G')^{1-n} = cG^{1-n}(1 - e^{-\theta(1-n)\bar{T}/n});$$

if σ is the growth rate of the labor force, we must then have (46)

$$(46) \qquad \sigma \geqq \theta(1-n)/n,$$

which sets an upper bound on θ (for $n < 1$). Other constraints on ρ are implied by the conditions $\theta \geqq 0$ and $W \geqq 0$ (if it is assumed that wage rates are non-negative). The first condition sets a lower limit on g/x; it can be shown, from (45). that

$$(47) \qquad g/x \geqq 1/a\bar{T};$$

i.e., the gross savings ratio must be at least equal to the amount of capital goods needed to produce one unit of output over their lifetime. The constraint $W > 0$ implies an interval in which ρ must lie. The conditions under which these constraints are consistent (so that at least one solution exists for the capital shortage case) have not been investigated in detail.

5. Divergence of Private and Social Product

As has already been emphasized, the presence of learning means that an act of investment benefits future investors, but this benefit is not paid for by the market. Hence, it is to be expected that the aggregate amount of investment under the competitive model of the last section will fall short of the socially optimum level. This difference will be investigated in detail in the present section under a simple assumption as to the utility function of society. For brevity, I refer to the *competitive solution* of the last section, to be contrasted with the *optimal* solution. Full employment is assumed. It is shown that the socially optimal growth rate is the same as that under competitive conditions, but the socially optimal ratio of gross investment to output is higher than the competitive level.

Utility is taken to be a function of the stream of consumption derived from the productive mechanism. Let

$$c = \text{consumption} = \text{output} - \text{gross investment} = x - g.$$

It is in particular assumed that future consumption is discounted at a constant rate, β, so that utility is

$$(48) \qquad U = \int_0^{+\infty} e^{-\beta t}c(t)dt = \int_0^{+\infty} e^{-\beta t}x(t)dt. - \int_0^{+\infty} e^{-\beta t}g(t)dt.$$

Integration by parts yields

$$\int_0^{+\infty} e^{-\beta t}g(t)dt = e^{-\beta t}G(t)\Big|_0^{+\infty} + \beta\int_0^{+\infty} e^{-\beta t}G(t)dt.$$

From (48),

$$U = U_1 - \lim_{t \to +\infty} e^{-\beta t} G(t) + G(0), \tag{49}$$

where

$$U_1 = \int_0^{+\infty} e^{-\beta t}[x(t) - \beta\, G(t)]dt. \tag{50}$$

The policy problem is the choice of the function $G(t)$, with $G'(t) \geqq 0$, to maximize (49), where $x(t)$ is determined by the production function (9), and

$$L(t) = L_0 e^{\sigma t}. \tag{51}$$

The second term in (49) is necessarily non-negative. It will be shown that, for sufficiently high discount rate, β, the function $G(t)$ which maximizes U_1 also has the property that the second term in (49) is zero; hence, it also maximizes (49), since $G(0)$ is given.

Substitute (9) and (51) into (50).

$$U_1 = \int_0^{+\infty} e^{-\beta t} G(t) \left[a - \beta - a\left(1 - \frac{L_0 e^{\sigma t}}{cG^{1-n}}\right)^{1/(1-n)}\right] dt.$$

Let $\bar{G}(t) = G(t)\, e^{-\sigma t/(1-n)}$.

$$U_1 = \int_0^{+\infty} e^{-\left(\beta - \frac{\sigma}{1-n}\right)t}\, \bar{G}(t) \left[a - \beta - a\left(1 - \frac{L_0}{c\bar{G}^{1-n}}\right)^{1/(1-n)}\right] dt.$$

Assume that

$$\beta > \frac{\sigma}{1-n}; \tag{52}$$

otherwise an infinite utility is attainable. Then to maximize U_1 it suffices to choose $\bar{G}(t)$ so as to maximize, for each t,

$$\bar{G}\left[a - \beta - a\left(1 - \frac{L_0}{c\bar{G}^{1-n}}\right)^{1/(1-n)}\right]. \tag{53}$$

Before actually determining the maximum, it can be noted that the maximizing value of $\bar{G}$ is independent of t and is therefore a constant. Hence, the optimum policy is

$$G(t) = \bar{G}\, e^{\sigma t/(1-n)}, \tag{54}$$

so that, from (36), the growth rate is the same as the competitive. From (52), $e^{-\beta t}G(t) \longrightarrow 0$ as $t \longrightarrow +\infty$.

To determine the optimal $\bar{G}$, it will be convenient to make a change of variables. Define

$$v = \left(1 - \frac{L_0}{c\bar{G}^{1-n}}\right)^{n/(1-n)}.$$

so that

(55) $$\bar{G} = \left[\frac{L_0}{(1 - v^{(1-n)/n})}\right]^{1/(1-n)}.$$

The analysis will be carried through primarily for the case where the output per unit capital is sufficiently high, more specifically, where

(56) $$a > \beta.$$

Let

(57) $$\gamma = 1 - \frac{\beta}{a} > 0.$$

The maximizing $\bar{G}$, or v, is unchanged by multiplying (53), the function to be maximized, by the positive quantity, $(c/L_0)^{1/(1-n)}/a$ and then substituting from (55) and (57). Thus, v maximizes

$$(1 - v^{(1-n)/n})^{-1/(1-n)} (\gamma - v^{1/n}).$$

The variable v ranges from 0 to 1. However, the second factor vanishes when $v = \gamma^n < 1$ (since $\gamma < 1$) and becomes negative for larger values of v; since the first factor is always positive, it can be assumed that $v < \gamma^n$ in searching for a maximum, and both factors are positive. Then v also maximizes the logarithm of the above function, which is

$$f(v) = -\frac{\log (1 - v^{(1-n)/n})}{1 - n} + \log (\gamma - v^{1/n}),$$

so that

$$f'(v) = \frac{v^{\frac{1}{n} - 2}}{n} \left[\frac{\gamma - v}{(1 - v^{(1-n)/n}) (\gamma - v^{1/n})}\right].$$

Clearly, with $n < 1$, $f'(v) > 0$ when $0 < v < \gamma$ and $f'(v) < 0$ when $\gamma < v < \gamma^n$, so that the maximum is obtained at

(58) $$v = \gamma.$$

The optimum $\bar{G}$ is determined by substituting γ for v in (55).

From (54), L/G^{1-n} is a constant over time. From the definition of v and (58), then,

$$\gamma = \left(1 - \frac{L}{cG^{1-n}}\right)^{n/(1-n)}$$

for all t along the optimal path, and, from the production function (9),

(59) $$\gamma = \left(1 - \frac{x}{aG}\right)^n \text{ for all } t \text{ along the optimal path.}$$

This optimal solution will be compared with the competitive solution of steady growth studied in the last section. From (40), we know that

(60) $$W = \left(1 - \frac{x}{aG}\right)^n \text{ for all } t \text{ along the competitive path.}$$

It will be demonstrated that $W < \gamma$; from this it follows that *the ratio G/x is less along the competitive path than along the optimal path.* Since along both paths,

$$g/x = [\sigma/(1 - n)] \, (G/x),$$

it also follows that *the gross savings ratio is smaller along the competitive path than along the optimal path.*

For the particular utility function (48), the supply of capital is infinitely elastic at $\rho = \beta$; i.e., the community will take any investment with a rate of return exceeding β and will take no investment at a rate of return less than β. For an equilibrium in which some, but not all, income is saved, we must have

(61) $\rho = \beta.$

From (35), $\theta = n\sigma/(1 - n)$; hence, by definition (28),

(62) $\alpha = (1 - n)\beta/n\sigma.$

Since $n < 1$, it follows from (62) and the assumption (52) that (63)

(63) $\alpha > 1.$

Equation (33) then becomes the one by which W is determined. The left-hand side will be denoted as $F(W)$.

$$F'(W) = \frac{1 - W^{\alpha-1}}{1 - \alpha}.$$

From (63), $F'(W) < 0$ for $0 \geqq W < 1$, the relevant range since the investment will never be profitable if $W > 1$. To demonstrate that $W < \gamma$, it suffices to show that $F(W) > F(\gamma)$ for that value of W which satisfies (33), i.e., to show that

(64) $F(\gamma) < \theta/a.$

Finally, to demonstrate (64), note that $\gamma < 1$ and $\alpha > 1$, which imply that $\gamma^{\alpha} < \gamma$, and therefore

$$(1 - \alpha) - \gamma^{\alpha} + \alpha\gamma > (1 - \alpha)(1 - \gamma).$$

Since $\alpha > 1$, $\alpha(1 - \alpha) < 0$. Dividing both sides by this magnitude yields

$$\frac{1 - \gamma^{\alpha}}{\alpha} + \frac{\gamma - \gamma^{\alpha}}{1 - \alpha} < \frac{1 - \gamma}{\alpha} = \frac{\theta}{a}$$

where use is made of (57), (28), and (61); but from (33), the left-hand side is precisely $F(\gamma)$, so that (64) is demonstrated.

The case $a \leqq \beta$, excluded by (56), can be handled similarly; in that case the optimum v is 0. The subsequent reasoning follows in the same way so that the corresponding competitive path would have $W < 0$, which is, however, impossible.

6. Some Comments on the Model

(1) Many writers, such as Theodore Schultz, have stressed the improvement in the quality of the labor force over time as a source of increased productivity. This interpretation can be incorporated in the present model by assuming that σ, the rate of growth of the labor force, incorporates qualitative as well as quantitative increase.

(2) In this model, there is only one efficient capital-labor ratio for new investment at any moment of time. Most other models, on the contrary, have assumed that alternative capital-labor ratios are possible both before the capital good is built and after. A still more plausible model is that of Johansen [7], according to which alternative capital-labor ratios are open to the entrepreneur's choice at the time of investment but are fixed once the investment is congealed into a capital good.

(3) In this model, as in those of Solow [12] and Johansen [7], the learning takes place in effect only in the capital goods industry; no learning takes place in the use of a capital good once built. Lundberg's Horndal effect suggests that this is not realistic. The model should be extended to include this possibility.

(4) It has been assumed here that learning takes place only as a by-product of ordinary production. In fact, society has created institutions, education and research, whose purpose it is to enable learning to take place more rapidly. A fuller model would take account of these as additional variables.

REFERENCES.

[1] Abramovitz, M., " Resource and Output Trends in the United States Since 1870," *American Economic Review, Papers and Proceedings of the American Economic Associations*, 46 (May, 1956): 5-23.

[2] Arrow, K. J., H. B. Chenery, B. S. Minhas, and R. M. Solow, " Capital-Labor Substitution and Economic Efficiency," *Review of Economics and Statistics*, 43 (1961): 225-250.

[3] Asher, H., *Cost-Quantity Relationships in the Airframe Industry*, R-291, Santa Monica, Calif.: The RAND Corporation, 1956.

[4] Haavelmo, T. *A Study in the Theory of Economic Evolution*, Amsterdam: North Holland, 1954.

[5] Hilgard, E. R., *Theories of Learning*, 2nd ed., New York: Appleton-Century-Crofts, 1956.

[6] Hirsch, W. Z., " Firm Progress Radios," *Econometrica*, 24 (1956): 136-143.

[7] Johansen, L., " Substitution vs. Fixed Production Coefficients in the Theory of Economic Growth: A Synthesis," *Econometrica*, 27 (1959): 157-176.

[8] Kaldor, N., " Capital Accumulation and Economic Growth," in F. A. Lutz and D. C. Hague (eds.), *The Theory of Capiatl*, New York: St. Martin's Press, 1961, 177-222.

[9] Lundberg, E., *Produktivitet och räntabilitet*, Stockholm: P. A. Norstedt and Söner, 1961.

[10] Muth, J., " Rational Expectations and the Theory of Price Movements," *Econometrica* (in press).

[11] Solow, R. M., " Technical Change and the Aggregate Production Function," *Review of Economics and Statistics*, 39 (1957): 312-320.

[12] Solow, R. M., " Investment and Technical Progress," in K. J. Arrow, S. Karlin, and P. Suppes (eds.), *Mathematical Methods in the Social Sciences*, 1959, Stanford, Calif.: Stanford University Press, 1960, 89-104.

[13] Verdoorn, P. J., " Fattori che regolano lo sviluppo della produttività del lavoro," *L'Industria*, 1 (1949).

[14] Verdoorn, P. J., " Complementarity and Long-Range Projections," *Econometrica*, 24 (1956): 429-450.

[15] Wright, T. P., " Factors Affecting the Cost of Airplanes," *Journal of the Aeronautical Sciences*, 3 (1936): 122-128.

Stanford. KENNETH J. ARROW.

[24]

UNCERTAINTY, LEARNING, AND THE ECONOMICS OF PARALLEL RESEARCH AND DEVELOPMENT EFFORTS *

Richard R. Nelson

IN recent years, there has been a very marked rise in the interest economists have shown in the process of invention, and in industrial Research and Development — the institutionalization of inventive activities by business enterprises. The developing interest in invention has stemmed from several roots. One important one is the growing body of research findings on productivity,[1] which turned the attention of economists interested in economic growth toward the process of technological change. These studies showed that only a small fraction of the total increase in output per worker which had occurred in the American economy since the late 19th century could be explained by increased capital per worker. The lion's share had to be attributed to something else, to increased productivity or efficiency. The term "increased productivity" covers a wide number of different elements and the operation by which increased productivity is defined and measured obscures a variety of economic phenomena. Better allocation of existing factor supplies (the process of dynamic adjustment) and capital formation in humans (education, health, etc.) are two of the most important. But it seems obvious that technological change also has been an important ingredient.

A second source of the interest in invention is the changing way that economists are coming to look at the competitive process. Increasingly the focus is on competition through new products, rather than on direct price competition. And concurrently, normative considerations are shifting toward conditions of long run growth rather than fixing on short term Pareto optimality. In a sense these developments represent a renaissance of Schumpeter.

A third source of the heightened interest in inventive activity has been the cold war, and the growing awareness that our national security may depend on the output of our military research and development effort — organized inventive effort for the purpose of creating more effective weapons. Closely related to the interest in military R and D [2] is the growing concern with the technological race to which the Soviets have challenged us. The interest in inventive activity which stems from interest in defense and the space program has tended to be more micro-oriented than the interest stemming from concern with problems of economic growth. The studies generated have tended to be normative — analyses of conditions of efficiency (how resources should be allocated), rather than analyses of factors determining the actual allocation of inventive effort.

This paper is focused on one aspect of R and D efficiency: the problem of achieving a given objective at minimum cost. Thus the paper is directly concerned with the third area of interest discussed above. However, some of the implications of the model seem to have relevance to more macro issues of policy, and to a positive analysis of the allocation of inventive effort.

In Section I, the basic efficiency problem is discussed and the concept of a parallel path strategy is introduced. Section II presents a model. Lest the reader expect too much, let me state at the outset that the model is much too crude to provide reliable quantitative answers. Rather, the purpose of the model is to provide a framework for analysis and to explore some implications of two aspects of the R and D process, the uncertainty which exists

* This paper is one product of a continuing RAND study of Research and Development management. The idea of a parallel development strategy has a long history at RAND. Burton Klein, William Meckling, Emmanuel Mesthene, Leland Johnson, Thomas Marschak, Armen Alchian, William Capron, and others all have contributed to its evolution.

[1] See, for example: J. Schmookler, "The Changing Efficiency of the American Economy, 1869–1934," this Review, XXXIV (Aug. 1952); M. Abramowitz, "Resource and Output Trends in the United States since 1870," American Economic Review, XLVI (May 1956); Robert Solow, "Technical Change and the Associate Production Function," this Review, XXXIX (Aug. 1957); B. F. Massell, "Capital Formation and Technological Change in United States' Manufacturing," this Review, XLII (May 1960).

[2] Research and Development will henceforth be abbreviated as R and D.

at the outset of any R and D effort, and the learning which takes place during the course of the effort.[3] Section III presents some implications for efficient R and D management. And in Section IV some more general implications with respect to a positive theory of invention, and to problems of macro policy, are discussed.

I. The Choice among Alternative Means

Consider the following problem. Assume that the Air Force is interested in developing an advanced fighter aircraft and that a certain performance must be achieved if the plane is to have a capability significantly greater than planes currently in the force. There are several competing designs. All of them have at least some promise, though considering expected cost, expected performance, and expected development time, some proposals are more promising than others. However, the estimate of the relevant parameters are known to be subject to considerable error. What should the Air Force decide to do?

Substitute the phrase "industrial research laboratory" for "Air Force", the phrase "new product" for "fighter aircraft", and the problem is a very common one facing R and D management.

The problem of choosing among alternative paths to a given objective is a difficult one. It is very easy to make choices which, ex-post, turn out to be the wrong ones. For scientists and research administrators with long experience in R and D are in general agreement that there is considerable uncertainty as to the outcome of any development program, the uncertainty, of course, being closely related to the degree of knowledge in the relevant fields and the advance being sought. For example, a recent RAND study showed that early cost estimates of proposed airframes, engines, and electronic equipment for military aircraft have tended to be wrong often by a factor of five or more. Performance estimates and development-time estimates have also been extremely unreliable. For example, one guidance system turned out to have only one-quarter the forecasted accuracy, and to achieve even this required a guidance package weighing four times the originally planned weight. One missile program achieved half the forecasted accuracy in the course of a program lasting three times longer and costing many times more than originally estimated. These examples are not extreme.[4]

Yet despite the unreliability of estimates, choices must be made. Given the riskiness of R and D, what is a good way to make choices? One answer to this question is that on the basis of the best available estimates the decision maker should pick the alternative with the greatest expected net value, that is, if the objective is fixed, with the smallest expected cost. But this approach neglects one important characteristic of R and D — estimates of cost, performance, and development time tend to improve as development proceeds and information accumulates. For an outlay relatively small compared with total development cost, a very great increase in the accuracy of estimates can often be achieved. For example, Robert Schlaifer, in his book on the development of airplane engines, reports:

> "Although the complete development of any engine . . . is necessarily an extremely expensive business, only a relatively small amount of preliminary development . . . is needed to show whether an innovation has promise or not." [5]

RAND studies have tended to confirm Schlaifer's findings. By the time a fighter development program has proceeded to the point of a flight test or an engine program to the fifty-hour test, estimates of cost, performance, and development time are greatly improved. Many of the problems involved in radar developments were discovered quite early in the programs. Similarly with missiles. And the early stages of development are usually the cheapest stages.

[3] In a number of ways the framework has much in common with some earlier work of Albert Hart, particularly his "Risk, Uncertainty, and the Unprofitability of Compounding Probabilities," reprinted in *Readings in the Theory of Income Distribution*, Blakiston, 1946.

[4] B. H. Klein, "A Radical Proposal for R and D," *Fortune*, LVII (May 1958); R. R. Nelson, "The Economics of Invention," *Journal of Business of University of Chicago*, XXXII (April 1959); A. W. Marshall and W. H. Mecklins, "Predictability of the Costs, Time, and Success of Development," P-1821 (the RAND Corporation, 1960); B. H. Klein, "The Nature of Military R + D," P-1818 (the RAND Corporation, 1960).

[5] R. Schlaifer and S. D. Heron, *The Development of Aircraft Engines and Fuels* (Cambridge, Mass., 1949), 90.

The possibility of acquiring significantly improved estimates at relatively low cost suggests that it may be economical *not* to choose one design or contractor for an R and D job on the basis of first estimates, which experience has shown to be notoriously unreliable, but rather to initiate parallel development efforts, cutting down the list of competing projects as estimates improve. The best decision to make on the basis of some information may be to delay a final decision until more information is obtained.

In its most general form the logic of this strategy is simply that choices among alternatives are likely to be better, the more information the decision maker has when he makes his final choice. This is almost tautological. What makes the argument somewhat special and differentiates it from more standard sequential-decision theory formulations is, first, the way that information is to be acquired by the decision maker and, second, the kind of information that is acquired. In the parallel-path strategy, information about a development alternative is acquired by doing almost exactly the same things that would be done were the alternative actually chosen, that is, design and development work. And the information acquired in doing early development work on an alternative design is likely to relate almost exclusively to that design. This way of getting information is in contrast to general background scientific research to enable better predictions to be made about a wide range of alternative technical possibilities, the kind of information gathering activity that statistical sequential-decision theory treats. The parallel-development strategy implies, first, that the designs considered for development are novel and push the frontiers of knowledge, and, second, that the decision maker is too pressed for time to postpone a decision to start development until more general background research is done, or that no line of general research appears particularly promising as a source of the information required to permit better development choices.

The use of a parallel-path strategy for the solution of difficult development problems is standard practice in several of our outstanding industrial laboratories.[6] It is extremely common in agricultural and medical research. And in the atomic-bomb project, one of the most spectacularly successful military projects the United States has ever undertaken, the parallel-path strategy was employed. James Conant wrote in a letter to Bush on May 14, 1942:

> "All five methods will be entering very expensive pilot plant development during the next six months. . . . [But] while all five methods now appear to be about equally promising, clearly the time to production . . . by the five routes will certainly not be the same but might vary by six months or a year because of unforeseen delays. Therefore, if one discards one or two or three of these methods now, one may be betting on the slower horse unconsciously."

All this is reasonable enough, but strictly qualitative. How many competing projects should one run? What are the factors that affect the answer to this question? To this we now turn.

II. The Model

Let us return to the choice problem facing the Air Force: to select a strategy for the efficient development of a modern fighter aircraft, when several competing designs have been proposed and there is considerable uncertainty with respect to which is best.

A. *Uncertainty, Learning, and the Time Path of Estimates*

The arguments for a parallel-path development strategy in such a situation rest on two separate assumptions. First, at the start of a development program, estimates of the cost, time, and performance of a proposed weapon system are subject to great uncertainty. Second, these estimates improve and become more reliable as development proceeds.

What do we mean when we say that estimates tend to get better as development proceeds? Taking cost as an example, I think we mean that the actual final cost is likely to be closer to a later estimate than to an earlier one. One way to formalize this is to say that the

[6] For an interesting case study see T. Marschak, "*Strategy and Organization in a System Development Project*," P-1901 (the RAND Corporation, 1960).

expected squared error of a cost estimate decreases with time, or:

$$E(\check{C}_t - C)^2 > E(\check{C}_{t+\theta} - C)^2 \quad (1)$$

where C is actual ex post cost, $\check{C}_t$ is the estimate of cost made at time t, and $\check{C}_{t+\theta}$ is the estimate of cost made at a later date, $t + \theta$.

If this is what we mean when we say that estimates tend to improve as development proceeds, we are implying that given a cost estimate $\check{C}_t$ at time t, or $\check{C}_{t+\theta}$ at time $t + \theta$, a probability distribution of final cost can be defined. What do we mean by this? One interpretation is that the probability distribution of cost is subjective. The estimator knows that many things can happen in a development program, and for every sequence of possible events he assigns both a probability and a cost. $C_{t+\theta}$ is the mean of this subjective probability distribution. His estimates improve through time because as development proceeds some of these events occur and others are logically excluded; thus the range of possibilities is narrowed down.[7]

Cost estimates tend to improve as development proceeds because we gain more data and because some events occur and others are logically excluded. We would expect, therefore, the average squared *change* in a cost estimate between time t and time $t + \theta$ to be an increasing function of θ. If θ is small not very much is likely to happen between t and $t + \theta$ which will warrant a change in the best estimate. But if θ is large a great deal can happen which will change the estimate. And the larger θ, the greater the number of events that are likely to occur, and thus, the more likely it is that the estimate will change greatly.

One implication of this is that if θ is less than the total development period, the expected squared change in the cost estimate between time t and time $t + \theta$ should be less than the expected squared change in the cost estimate between time t and the end of development. But since estimated cost at the end of the development *is* actual ex post cost, C, the expected squared change in the cost estimate between time t and the end of development *is* the expected squared *error* of the cost estimate made at time t. Thus the expected squared change in a cost estimate between time t and time $t + \theta$ is smaller than the expected squared error of the cost estimate made at time t.

$$E(\check{C}_t - \check{C}_{t+\theta})^2 < E(C_t - C)^2 \quad (2)$$

Let us tighten the assumption of equation (2). Assume that the change in the cost estimate between time t and time $t + \theta$ and the change between time $t + \theta$ and the end of the development program are statistically independent.[8] Then we can write:

$$E(\check{C}_t - C)^2 = E(\check{C}_t - \check{C}_{t+\theta})^2 + E(\check{C}_{t+\theta} - C)^2 \quad (3)$$

Equation (3) states that the expected squared error in a cost estimate made at time t equals the expected squared *change* in the cost estimate between time t and time $t + \theta$, *plus* the expected squared error in the cost estimate made at this later time $t + \theta$.

Equation (3) is extremely important in our model. The first term of (3) is a measure of the total uncertainty regarding cost at time t, the start of a development program. The third term of (3) is a measure of the uncertainty that will still remain at time $t + \theta$. The second term, the expected squared change in the cost estimate between time t and time $t + \theta$, must therefore be a measure of how much uncertainty is reduced between time t and time

[7] Another interpretation is to state that for any specific case final cost is, to be sure, a number, not a random variable. Assume that cost estimates of many systems have been made at a given stage of development, say θ months after project initiation, and that later the final cost for each system has been observed. Consider the subset of systems with the same estimated cost $\check{C}_{t+\theta}$. For this subset, final cost, C, is a random variable; and our probability distribution of C, given $\check{C}_{t+\theta}$, is defined in this sense. In this interpretation, for each subset of all cost estimates with the same $\check{C}_{t+\theta}$ there is an average final cost $\acute{C}$. It is assumed that $\check{C} = \acute{C}$

[8] In other words assume that the time path of estimates is Markovian. In terms of sampling theory, the cost estimation problem is assumed to be similar to the following more conventional problem. The statistician wants to estimate what will be the mean of a sample of size 100 that he will draw from a known population. He will draw his sample one by one. After he has drawn, say n elements, he can estimate what the mean of his distribution will be when he has drawn 100-n more. The time path of such estimates (an estimate made for each value of n) will be Markovian, if the draws from the population are independent.

$t + \theta$. It must be a measure of how much is learned by running a project for θ time units. Thus $E(\check{C}_t - \check{C}_{t+\theta})^2$, henceforth denoted σ^2_θ, is a measure of three things: (1) the expected squared change in the cost estimate between time t and time $t + \theta$; (2) the variance of the probability distribution of what the cost estimate will be at time $t + \theta$, as computed at time t; (3) a measure of how much uncertainty will be reduced, that is, how much will be learned between time t and time $t + \theta$. The third interpretation is, of course, the most interesting one for the purposes of this paper.

The preceding discussion has focused on cost estimates. The same arguments apply to performances and development-time estimates, though it is theoretically obvious and empirically true that the time paths of cost, time, and performance estimates are not statistically independent. If a project runs into difficulties, it will cost more money to achieve a given performance in a given time, or if time and money are fixed, either performance will be lower than expected earlier, or it will take longer to achieve a given performance at given cost. Given the performance-cost-time trade-off, unexpected obstacles will usually degrade estimated net military worth in all three dimensions, though they need not. This is a matter of choice, given a trade-off surface, which has shifted adversely.

B. *The Optimum Number to Run in Parallel*

Consider the following class of strategies for the Air Force. Now, at time t, select n (where n is the number to be determined) design proposals, and spend M_θ dollars on *each* of them over a period of θ months. At the end of θ months, pick the project which looks best *then*, and run it to completion, dropping the $n - 1$ others.

To simplify exposition let us fix utility, all costs other than R and D cost, and development time, and focus on development cost. The logic of the analysis obviously does not depend on our eliminating all relevant variables save development cost, and in the simple numerical example in Section IIC, development time is explicitly considered in the analysis. To the extent that cost, performance, and development time can be substituted for each other in a development program, if a strategy saves money for a given performance and time, it can, in general, also save some time and gain some performance, with a somewhat smaller cost saving. This is a matter of choice.

A SPECIAL CASE. Let us begin by assuming that at time t, the Air Force has absolutely no way to rank the alternative designs. Given the assumptions listed above, this is equivalent to assuming that at time t the estimated cost, $\check{C}_{i,t}$, of each alternative, i, is the same, say $\acute{C}$. We shall also assume that the probability distribution of what the cost estimate will be at time t, as estimated at time $t + \theta$, is the same for all projects with variance σ^2_θ, and that the time paths of cost estimates are statistically independent.

If n projects are run in parallel until time $t + \theta$ at a cost of M_θ each, and if at time $t + \theta$ the project with the lowest estimated cost (unbiased) is selected for completion and the $n - 1$ others are abandoned, then clearly total expected cost is a function of n, the number of projects run in parallel.

$$E(C)_n = (n - 1)M_\theta + E_n min_i \check{C}_{i,\, t+\theta} \qquad (4)$$

The first term, $(n - 1)\ M_\theta$, is the cost of the $(n - 1)$ abandoned projects. The second term, $E_n min_i\ \check{C}_{i,t+\theta}$, is the expected cost of the project judged best at time $t + \theta$. $\check{C}_{i,\,t+\theta}$, is the cost estimate of project i made at time $t + \theta$. The index i runs from 1 to n. In terms of probability theory, $E_n min_i\ \check{C}_{i,\,t+\theta}$ is the expected value of the smallest element of a sample of size n drawn from a given population with mean $\acute{C}$ and variance σ^2_θ.

For notational convenience, let us set the date at which we choose our n projects as $t = 0$, and thus let $t + \theta = \theta$.

$$E(C)_n = (n - 1)M_\theta + E_n min_i \check{C}_{i\theta} \qquad (4a)$$

Holding θ constant, we wish to find an n so as to minimize $E(C)_n$. The first difference with respect to n of (4a) is:

$$\triangle_n E(C)_n = M_\theta + \triangle_n E_n min_i \check{C}_{i\theta} \qquad (5)$$

M_θ is a positive constant. $\triangle_n E_n min_i C_{i\theta}$ is negative for all probability distributions relevant to our analysis, that is, the expected value of the smallest member of a sample of size n

decreases as n increases.[9] Therefore, equation (5) has both a negative term and a positive term. Depending on which term is larger, marginal expected cost will be negative or positive.

The second difference with respect to n of equation (4a) can be written as:

$$\Delta_n^2 E(C)_n = \Delta_n^2 E_n min_i \check{C}_{i\theta} \qquad (6)$$

It can be proved that (6) is positive.[10] The expected value of the smallest member of a sample of size n decreases as n increases, but at a diminishing rate. Thus marginal expected cost must be an increasing function of n. Therefore, optimum n, the n that minimizes expected cost, must be the last n for which $\Delta_n E(C)_n$ is negative. This is the last n for which the negative (but increasing) $\Delta_n E_n min_i C_{i\theta}$ outweighs the positive (and constant) M_θ. In the example presented in Section IIC, optimum n is three.

On what does n^*, the number of projects that should be started and run in parallel for θ time units, depend? Intuition should tell us that the number depends both on the cost of running the projects and on the amount learned. First, n^* should be greater, the smaller is M_θ, the cost of running a project for θ time units. Second, n^* should be greater, the greater is σ^2_θ, the expected squared change in cost estimates in θ time units and a measure of the amount learned in θ months.

A glance at equation (5), the marginal expected cost equation, verifies the first statement. Clearly, if $\Delta_n E_n min_i \check{C}_{i\theta}$ is negative and an increasing function of n, the last n for which equation (5) is negative will be greater, the smaller is M_θ.

To prove the second statement requires an assumption about the probability distribution of changes in cost estimates. In particular, let Z be a random variable with a zero mean and a standard deviation of one, and let $\check{C}_{i\theta}$ be a linear function of Z_i. That function must be $\check{C}_{i\theta} = \acute{C} + \sigma_\theta Z_i$ where σ^2_θ is the variance of the $\check{C}_{i\theta}$ distribution and $E_n min_i \check{C}_{i\theta} = \acute{C} + \sigma_\theta E_n min_i Z_i$. Therefore:

$$\Delta_n E_n min_i \check{C}_{i\theta} = \sigma_\theta \Delta_n E_n min_i Z_i \qquad (7)$$

Thus, given a mean, and the underlying "shape" of a distribution, the expected value of the smallest member of a sample of size n is greater, the greater is σ^2_θ. And the greater σ^2_θ, the greater will be the decrease, as n increases by one, in the expected value of the smallest of a sample of size n, for every value of n. Thus, given M_θ, the larger will be the value of n at which M_θ first outweighs $\Delta_n E_n min_i \check{C}_{i\theta}$. This verifies the second statement.

A MORE GENERAL ANALYSIS. Our analysis thus far has rested on the following assumptions: (a) At time t, each of the alternative development projects being considered has the same estimated cost, $\acute{C}$; (b) The probability distribution of changes in the cost estimate between time t and time $t + \theta$ is the same for all projects, with mean zero and variance σ^2_θ; (c) Project costs are statistically independent; (d) It costs M_θ to run a project for θ time units.

The assumptions are quite stringent, but they have been useful for developing the logic of the parallel-path strategy. The assumptions can be relaxed, and if they are, it is possible to show two things. First, if, at time t, different projects have different estimated costs, then the greater the estimated cost of the nth ranked project, the less the gain from adding

[9] Proof: Assuming X can take on only positive values and has an upper bound, b:

$$E(X) = \int_0^b X f(X)\, dX = \int_0^b X df(X)$$
$$= \int_0^b X d[F(X) - 1]. \qquad (1)$$

Integrating by parts

$$E(X) = X[F(X) - 1]\Big|_0^b - \int_0^b [F(X) - 1]\, dX \qquad (2)$$

The first term is zero. Thus:

$$E(X) = \int_0^b [1 - F(X)]\, dX = \int_0^b G(X)\, dX \qquad (3)$$

where $G(X)$ is the probability that a variable will equal or exceed X. Let X be the smallest of a sample of size n. $G(X)$ thus must equal $G(C)^n$, the probability that each member of a sample of size n will equal or exceed C. Thus the expected value of the smallest element of a sample of size n can be written

$$E_n min_i C_i = \int_0^b G(C)^n\, dC. \qquad (4)$$

The first difference with respect to n of equation (4) is

$$\Delta_n E_n min_i C_i = \int_0^b [G(C)^n - G(C)^{n-1}]\, dC$$
$$= \int_0^b [G(C)^{n-1} (G(C) - 1)]\, dC \qquad (5)$$

(5) is definitely negative. The second difference of equation (4) is

$$\Delta_n^2 E_n min_i C_i = \int_0^b [G(C)^n - 2\, G(C)^{n-1}$$
$$+ G(C)^{n-2}]dC = \int_0^b [G(C)^{n-2}[(G(C) - 1)^2]dC \qquad (6)$$

(6) is definitely positive.

[10] Fn. 9.

it to the competing program set. For assumptions (b) and (c) imply that if $(n - 1)$ projects are run in parallel, these projects should be the projects ranked highest at time t. Adding the nth ranked project to the set run in parallel will pay off if, and only if, at time $t + \theta$, its estimated cost is lower than each of the $(n - 1)$ projects that were ranked higher at time t. Given the estimated costs of the $(n - 1)$ highest-ranked projects, the probability of this happening is smaller, the greater the estimated cost of the nth ranked project.

Second, again with relaxed assumptions, it is possible to show that if the costs of the several designs depend in part on the same factors so that there is a positive relationship between project costs, then the greater the positive relationship between the cost of the nth ranked project and the $(n - 1)$ higher-ranked projects, the less the gain from adding it to the competing program set. For if the costs of the various competing systems are highly related, then, even if the cost estimate for each system made at time t is quite unreliable, it is still not likely that the *relative* costs will differ greatly from the estimates. Thus it is unlikely that the project ranked n^{th} at time t will be ranked first at time $t + \theta$ and this possibility is the sole reason for adding the nth ranked project to the set run in parallel.[11]

The analysis can be extended to permit θ, the length of the competition period, to vary, but the conclusions are not surprising. Given the constraint upon space and attention, it seems more worthwhile to sketch a related but more promising line of generalization. Clearly θ, the time that n projects are run in competition, is a variable in most development-decision problems. But equally clearly, there is no reason to select n projects at time t, run them all to time $t + \theta$, and then select just one for continuing development. A more gradual weeding out of alternatives would almost certainly be a better strategy. At time $t + \theta_1$, several of the n projects started at time t might be dropped; at time $t + \theta_2$, several more of the remaining projects might be dropped; and so on, until one project finally is chosen.

The preceding analysis might seem adequate to tackle this more general strategy. The analysis could be made at each decision time, and if optimum n turned out to be less than the number then in competition, the number would be reduced. If optimum n turned out to be equal to or greater than the number in competition, all projects would be continued. But it is not this simple. The facts that the selection process can be gradual, that a final decision need not be made at time $t + \theta$, and that the number of projects kept in competition at time $t + \theta$ cannot exceed the number begun in competition at time t affect the number started at time t. This sequential weeding problem is essentially a large dynamic-programming problem and I hope to be able to formulate it as such sometime in the future. However, I doubt very seriously that any of the major qualitative conclusions of the simple model will be changed by such a formulation.

---

[11] Let i be the rank of a project at time t. Thus $\hat{C}_i \leq \hat{C}_j$ for i less than j, where $\hat{C}_i$ is the cost of project ranked i^{th} as estimated at time t. To prove the first proposition it is sufficient to assume that $G(C)$ is greater, the greater is $\hat{C}_i$, for all values of C. That is, assume that the probability that the cost of a project, as estimated at time $t + \theta$, will exceed any C, is greater, the greater is $\hat{C}_i$, the cost of that project as estimated at time t. Continuing the argument of the preceding footnote, the expected cost of the project ranked first at time $t + \theta$ can be written:

$$E_n min_i \check{C}_i = \int_0^b G_1(C) G_2(C) \ldots G_n(C) dC. \qquad (1)$$

Equation (1) is the expected value of the smallest member of a sample of size n, where the first member of the sample is drawn from population G_1, the second from population G_2, and so on.

Taking the first difference with respect to n

$$\Delta_n E_n min_i \check{C}_i = \int_0^b G_1(C) G_2(C) \ldots G_{n-1}(C) \; [G_n(C) - 1] \, dC. \qquad (2)$$

This expression is negative, but the decrease in expected cost is smaller, the larger is $G_n(C)$. And we have assumed that $G_n(C)$ is greater, the greater is $\hat{C}_n$, the estimated cost of the nth ranked project at time t. This proves the first proposition.

To prove the second proposition, note that if project costs (and estimates of cost) are not statistically independent, then each $G_i(C)$ above must be replaced with the conditional expression, $G_i(C; \; C_j \geqq C)$ with $j<i$. Let us measure the positive relationship between project i and the other projects as some increasing functional of $G_i[C; \; C_j \geqq C]_{j<i} - G_i(C)$, $0 \leqq C \leqq C_{max}$. Let us replace $[G_n(C) - 1]$ in equation (2) above by $[G_n(C; \; C_j \geqq C)_{j<i} - 1]$. Clearly, the greater the positive relationship (defined as above) between project n and the other projects, the less negative is (2), other things equal. This proves the second proposition.

C. A Simple Numerical Example

In this section a very simple numerical example is presented to illustrate the model. No new theoretical material is presented. The purpose of this section is to make the analysis a little more concrete. To add a bit of variety, development time as well as development cost will be considered.

Returning to the problem of developing a new fighter plane, assume that at present there is absolutely no information for ranking the alternative designs. The estimates of development time and development cost are exactly the same for all projects. Estimated development cost for all projects is $\hat{C}$ = $90 million accounting period; for example witness the ects is $\overline{T}$ = 58 months. The variances of the probability distributions, however, are large; hence it is likely that there will be significant differences in cost and time among the competing projects.

In the past the most successful programs were marked by an early first flight (prior to twenty months after contract initiation), while in few unsuccessful development programs (expensive and long) was there an early first flight. Furthermore, expenditure during the first twenty months was not, on the average, greater in programs marked by an early first flight than in programs not so marked. Thus, if the Air Force runs several projects in parallel for twenty months and if by that time one of the projects has produced a flying vehicle, the chances of selecting a "good" design will be significantly improved.

Assume that past history yields the following data:

1. M_θ, the average expenditure on a fighter program during the first twenty months = $10 million.
2. M_1, the average total expenditure on a fighter development program from contract initiation to development completion for those programs in which the first flight occurs prior to twenty months = $60 million. These programs will henceforth be referred to as class I programs. M_2, the average total expenditure on fighter development programs in which first flight occurs *after* twenty months = $110 million. Call these class II programs. Thus, on the average, class I programs cost $(M_2 - M_1)$, $50 million less than class II programs.
3. T_1, the average number of months taken by class I programs from contract initiation to completion = 40 months. T_2, the average number of months taken by class II programs = 70 months. Thus, on the average, class I programs took $(T_2 - T_1)$, 30 months less than class II programs.
4. P, the proportion of past fighter development programs in class II = .6. $1 - P$, the proportion of past fighter development programs in class I = .4. In terms of our time path of estimates model, $\hat{C}_{i,t} = \hat{C}$ and $T_{i,t} = \overline{T}$ for all i, at time t. At time $t + \theta$ C_i, $t + \theta$ can take on one of two possible values, M_1 or M_2. The probability that C_i, $t + \theta$ will equal M_1 is $(1 - P)$, and the probability that $\hat{C}_i$, $t + \theta$ will equal M_2 is P. $\overline{T}_i$, $t + \theta$ has a similar distribution. Remember, in this example we are considering both cost and time as variables (though we are fixing performance). And in our example cost and time estimates made at time $t + \theta$ are made solely on the basis of whether the vehicle has flown or not. No other data is considered.

Class I projects were not always better than class II projects, though they were better on the average, as the above figures show. The Air Force believes that the alternative designs of the present program can be considered a random sample drawn from an infinite population with the same relevant parameters as the set of all observed past fighter development programs. Given this data, what is a good strategy for the Air Force?

Consider the following set of strategies: At time t, select n (where n is the number to be determined) design proposals and spend money on them for twenty months (until time $t + \theta$). Expected total outlay for the first twenty months will thus be nM_θ (n = $10 million). At $t + \theta$, the end of twenty months, pick one project to run to completion, dropping all of the $(n - 1)$ others, spending no more money on them. If at the end of twenty months one of the development programs in the decision set is in class I, pick it. In this case, expected additional time to completion will be $T_1 = \theta =$ 20 months, and additional money outlay will be $M_1 - M_\theta = \$50$ million. Thus, if at least one of the projects flies before the end of twenty months, expected total cost will be:

$$E(C) = (n-1)M_\theta + M_1 = (n-1)\ \$10 \text{ million} + \$60 \text{ million}$$

where M_1 is the expected total cost of the project that is completed, and $(n-1)\ M_\theta$ is the cost of the $(n-1)$ projects that are abandoned. (It is assumed that at $t+\theta$ there is no information enabling *intra*-class ranking; only *inter*-class ordering is possible. Thus, if there is more than one class I project in the decision set at $t+\theta$, the one to be chosen must be selected at random.) If at the end of twenty months no projects in the decision set are in class I — that is, if no planes have flown — then the Air Force must pick a class II project. In this case, expected additional time to completion will be $T_2 - \theta = 50$ months, and money outlay will be $M_2 - M = \$100$ million. Expected total outlay will be:

$$E(C) = (n-1)M_\theta + M_2 = (n-1)\ \$10 \text{ million} + \$110 \text{ million}.$$

What is the best strategy for the Air Force? That is, how many projects should be brought to $t+\theta$ (at which time one will be selected, the other $(n-1)$ dropped)? To answer this question, we must examine both expected cost and expected time to completion.

First, consider the Air Force's expected total cost as a function of n, the number of projects brought to $t+\theta$.

$$E(C)_n = (n-1)\ M_\theta + M_2P^n + M_1\ (1-P^n) \quad (8)$$

$$E(C)_n = (n-1)\ \$10 \text{ million} + .6^n \cdot \$110 \text{ million} + (1-.6^n)\ \$60 \text{ million} \quad (8a)$$

(See Table I.) Equation (8) is a special case of equation (4a). The term $(n-1)\ M_\theta$ is the expected total cost of the $n-1$ projects that are run in competition for θ months until time $t+\theta$ and then abandoned. $M_2P^n + M_1\ (1-P^n)$ is the expected cost of the project selected as "best" at $t+\theta$, the winner of the project competition. It is the expected value of the smallest member of a sample of size n drawn from a population where C can be M_1 or M_2 with probabilities $1-P$ and P. Remember, the Air Force will always choose a class I project at time $t+\theta$ if it can, a class II project otherwise. P^n is the probability that all n projects will be in class II, which means that the "winner" will be a class II project. $1-P^n$ is the probability that at least one of the n projects in the decision set at $t+\theta$ will be a class I project; hence the "winner" will be a class I project.

TABLE 1. — DEVELOPMENT MONEY COST
(*$ Million*)

	$n=1$	$n=2$	$n=3$	$n=4$	$n=5$
$E(C)$	90	88	91	96.5	104
$\Delta_nE(C)$		− 2	+ 3	+ 5.5	+ 7.5

TABLE 2. — DEVELOPMENT TIME
(*Months*)

	$n=1$	$n=2$	$n=3$	$n=4$	$n=5$
$E(T)$	58	50.8	46.5	43.9	42.3
$\Delta_nE(T)$		− 7.2	− 4.3	− 2.6	− 1.6

TABLE 3. — DEVELOPMENT REAL COST — MONEY PLUS TIME AT $1 MILLION A MONTH

	$n=1$	$n=2$	$n=3$	$n=4$	$n=5$
$E(C^*)$	148	138.8	137.5	140.4	146.3
$\Delta_nE(C^*)$		− 9.2	− 1.3	+ 2.9	+ 5.9
M_θ		10	10	10	10
$(C^*_2 - C^*_1)\ (P^n - P^{n-1})$		−19.2	−11.3	− 7.1	− 4.1

How does expected cost change as n increases? As n increases, the probability that the winner of the project competition will be a class I project increases. But the cost of the projects that are run until $t+\theta$ and then abandoned also increases. These two factors work in opposite directions.

$$\Delta_nE(C) = M_\theta + (M_2 - M_1)\ (P^n - P^{n-1}) \quad (9)^{12}$$

$$\Delta_nE(C) = \$10 \text{ million} + \$50 \text{ million}\ (.6^n - .6^{n-1}) \quad (9a)$$

$\Delta_nE(C)$ is defined as marginal expected cost and is shown in Table I.

The first term of equation (9), M_θ, is positive. The second term, $(M_2 - M_1)\ (P^n -$

[12] Equation (9) is a special case of equation (5).

P^{n-1}), will always be negative. From Table 1 we can see that the second term outweighs the first for $n = 2$, but thereafter is outweighed by the first. Expected total cost is minimized for $n = 2$.

At this point in the argument it should be stated clearly that if the Air Force decides to bring two projects to time $t + \theta$ and then choose between the two, the results will not always prove, after the fact, to be good. It may turn out that neither project will have flown in twenty months, in which case the Air Force will not be in a better position to make up its mind at the end of twenty months than it was initially. Or, perhaps more importantly, if one of the projects has flown and the other not, the class I project, which will then be chosen at time $t + \theta$, may turn out to be more expensive than its class II competitor. All class I projects are *not* better than all class II projects; it is only that their average cost is lower. There is a smaller probability of ending up with a long, expensive program if several projects are brought to time $t + \theta$ and then one is chosen than if the decision is made at time t; but there is a possibility of making a poorer choice at time $t + \theta$ than at time t. The parallel-path strategy improves expectations, but it is not foolproof.

How does expected time to completion vary with n? Using the same analysis as before, it can be seen that:

$$E(T)_n = T_2P^n + T_1(1 - P^n) \quad (10)$$

$$E(T)_n = 70 \text{ months } .6^n + 40 \text{ months } (1 - .6^n) \quad (10a)$$

$$\Delta_n E(T) = (T_2 - T_1)(P^n - P^{n-1}) \quad (11)$$

$$\Delta_n E(T) = (30 \text{ months})(.6^n - .6^{n-1}) \quad (11a)$$

(See Table 2.[13])

Tables 1 and 2 show the cost and development time that the Air Force may expect if it runs n projects competitively in parallel for $\theta = 20$ months and picks the "best" on the basis of data available at $t + \theta$. As n increases, expected development *time* falls, and expected development *cost* first falls and then rises. Clearly, optimum n is two or greater. How much greater depends on the value that the Air Force places on time — the time-money utility trade-off. For example, assume the Air Force is willing to spend an extra \$1 million for each month of expected development time it can save. Or, to put it another way, the "cost" of each month of development *time* is \$1 million.[14]

With this additional assumption, it is possible to compute optimum n, the number of projects the Air Force should run in parallel until time $t + \theta$. Adding time cost, \$1 million a month, to money cost, the total *real* cost of a class I project is $C_1^* = 60 + 40 = 100$, and the total real cost of a class II project is $C_2^* = 110 + 70 = 180$. The expected total cost of development, as a function of n, thus can be written:

$$E(C^*)_n = (n - 1)M_\theta + [C_2^* P^n + C_1^*(1 - P^n)] \quad (12)$$

$$E(C^*)_n = (n - 1)\ \$10 \text{ mill.} + .6^n\ \$180 \text{ mill.} + (1 - .6^n)\ \$100 \text{ mill.} \quad (12a)$$

and

$$\Delta_n E(C^*) = M_\theta + (C_2^* - C_1^*)(P^n - P^{n-1}) \quad (13)$$

$$\Delta_n E(C^*) = \$10 \text{ mill.} + \$80 \text{ mill. } (.6^n - .6^{n-1}) \quad (13a)$$

The second term to the right of the equality sign in equation (12) is the expected total cost (time cost plus money cost) of the "best" of the n projects run in parallel for twenty months. The first term is the cost of the $n - 1$ projects that are abandoned. Clearly, $E(C^*)_n$ decreases as n increases if $\Delta_n E(C^*)$ is negative and increases with n if $\Delta_n E(C^*)$ is positive. The first term of equation (13), M_θ, is a positive constant. The second term is negative, but is an increasing function of n. (See Table 3). $\Delta_n E(C^*)$ is the sum of these two terms and thus must be an increasing function of n. There must be a unique last n for which $\Delta_n E(C^*)$ is negative, and optimum n thus must be unique.

In our example, optimum n is three; that is, the Air Force should run three projects in parallel for twenty months. At that time it should select the "best" project of the three. (In the context of the model this means that if one of the three planes has flown, pick that project. If more than one has flown, pick one of these at random. If none has flown, pick one of these at random. Remember, it is as-

[13] See p. 29.

[14] More generally, we would want to work with indifference curves for $E(C)_n$ and $E(T)_n$.

sumed that the Air Force cannot make *intra*-class rankings). It can be seen from Tables 1 and 2 that the Air Force, if it uses the multiple-path approach instead of running only one project all the way, can expect to pay $.8 million more (about one per cent more — $91 million as compared with $90 million), and can expect the development program to be completed almost twelve months sooner (more than twenty per cent sooner — 46.5 months as compared with 58 months).

As was previously noted, even if not concerned with saving time but only with achieving the objective at the smallest possible expected cost, the Air Force ought to delay final commitment on a project until time $t + \theta$. If expected cost is the only consideration, the Air Force should bring two projects to time $t + \theta$ and then pick the best. This strategy will be cheaper, on the average, than a strategy of commitment to a single project at the start of the program ($88 million as compared with $90 million).

III. Some Implications for Efficient R and D Management

The preceding sections have explored the logic of the parallel-path approach to the solution of development problems. It has been shown that if there is considerable uncertainty regarding the best way to achieve a particular objective, and if it is possible to reduce uncertainty substantially in the early stages of a development effort, the parallel running of several alternatives may be a very good strategy. It has been shown that the factors that determine the number of projects that should be run in parallel are (1) the cost of running a project during the period of competition, (2) the expected improvement in estimates during the period of competition, (3) the difference among the cost and performance estimates of the competing projects, and (4) the design similarities and differences of the competing projects.

If the early stage development costs are small, and the expected improvement in the estimates large, if the competing projects are quite similar in their estimated cost and performance because there is little data on which to base estimates, but quite different in design, then it certainly pays to run several projects in parallel. If, on the other hand, the early stages of development are likely to be expensive and little important additional information is likely to be acquired, and if the designs of the competing projects are similar but one seems to have a clear advantage over the others, then probably the best decision is to pick that one. In short, parallel development of alternative designs seems called for when the technical advances sought are large, when much additional information can be gained from prototype testing, and when the cost of a few prototypes is small relative to total systems' cost.

The discussion of the preceding sections has assumed that each alternative is developed by a separate and independent contractor. Clearly the logic of the analysis is not affected by organizational arrangements. The logic certainly is applicable to a single company seeking to achieve a development objective efficiently, and many companies do try multiple paths to the solution of difficult development problems.[15]

It should be noted again, here, that though our formal analysis has been in terms of reducing costs, the parallel-path strategy also can gain time.[16] Indeed, one might argue that the parallel-path strategy is rational only when time is of importance. When time is not pressing, the uncertainties which call for the parallel-path strategy if development is attempted *now*, might better be interpreted as a signal that development should *not* be attempted until more background research has been done.

With this in mind, certain important difficulties of the parallel-path strategy become clear when it is noted that such a strategy implies a set of complementary policies, particularly in cases where several contractors are involved.[17] Obviously, if projects are to be weeded out, commitments beyond the support of the early stages of development should not be made to the competing contractors.

[15] See Marschak, op. cit.

[16] And increased quality of the first product of development.

[17] See B. Klein and W. Meckling, "An Application of Operations Research to Development Decisions," *Operations Research*, VI (May–June 1958).

Also, since the value of running the projects in parallel lies in the information generated, clearly the competing contractors or project groups should be directed to achieve testable prototypes as quickly as possible. Since a strategy of parallel development calls for, as complementary policies, a stress on early testing rather than a direct effort to design and build the final hardware, and the avoidance of large commitments to any contractor or project early in the program, it may be argued that these complementary aspects of parallel development may cost time and may seriously slow down the rate of progress on any particular project. Thus it might be agreed that the fastest of several projects might be chosen with a higher probability if decision is delayed, but argued that the pace of all projects would be slowed by the policy of limited early expenditure.

It is true that in the atomic-bomb project, the method that actually produced chain-fissionable material for the first bombs was considered relatively unpromising early in the program, and that the method considered early in the program to be most promising did not produce appreciable quantities of material until after the war was over. But it might well be argued that had all our money and effort been allocated to this latter method, it might have produced material not only just sooner, but sooner than the former method.

This argument hinges on the shape of the function relating rate of expenditure on a project to development time. It is my belief, based on a number of case studies, that this curve has sharply diminishing returns; that is, in the early stages of a project, development time may actually be increased, not decreased, as the rate of expenditure climbs above a certain level. In short, a low early rate of expenditure on several projects with a consequent improved choice is a far more effective way to buy time than a faster rate of expenditure on a single project right from the start.

The returns to additional resources allocated to an R and D project are different in kind from the returns to additional resources allocated to production of a good. In R and D, the goal is to produce one good design. As increased resources are directed to R and D, more men can be put to work designing each component, and division of labor can be increased, but it seems likely that sharply diminishing returns to increased rate of expenditure at the early stages of a program might result for a number of reasons. First and most important, the number of engineers who do the important design work on a project is usually quite small, and good communication between the designing engineers is extremely important. If the size of the engineering staff is large, much of the better engineer's time is likely to be taken up with directing and helping his group. Second, a large formal structure may burden creativity and hinder the finding of good, quick solutions to problems by requiring a more formal task assignment than is possible with a smaller staff. Third, case studies strongly suggest that early testing of components and a prototype system contributes to quick and inexpensive solutions to development problems, and that a large engineering staff definitely slows the process of getting a component ready for a test. Someone is always doing a last minute redesign.

This is not to argue that a very small staff and a low rate of expenditure will yield a quicker result than a slightly larger staff. Clearly, if time is important, a good development strategy should reflect this fact both in running more projects in parallel and in spending at a faster rate on each of the projects that is undertaken. There are diminishing returns to both strategies.

IV. Some Broader Normative and Positive Implications

A. *Multiple Decision Making Units, Multiple Inventions, and the Efficient Allocation of Resources*

One of the most striking aspects of the history of technological change is the very common occurrence of multiple inventions; several inventors or groups of inventors coming up with similar solutions to practical problems at about the same time. There are at least half a dozen "inventors" of the steamship, the electric bulb, the internal combustion engine, and the airplane, to name a few. This is scarcely surprising. Demand is an extremely important factor determining what problems inventors try to solve, and the state of knowledge is an extremely important factor determining how, and with what success, they go about solving them. Common perception of demand and common scientific and technical knowledge should be expected to generate similar inventive interests, and similar approaches, among a wide group of inventors.

Because of this, the response to increased profit opportunities tends to be an increase in the number of individuals and organizations attempting to invent in the relevant fields. And one result is often the almost simultaneous appearance of inventions which are very close substitutes.

There are a host of problems, both of efficiency and equity, raised in running the invention industry as a lottery. Nonetheless, the notion that parallel inventive efforts are often the most efficient way to achieve a given objective indicates that we should be wary in damning the wastefulness of independent and competitive efforts. While it is true that one result is ex-post waste, other results are likely to be greater speed in achieving a satisfactory invention, and higher quality of the invention the economy finally accepts.

The argument of section II was carried forth under the assumption that projects would be weeded out before they were completed. The savings were in terms of R and D time and cost. But it would seem that if quality and production cost are not likely to be the same for different designs, then it may be good strategy to develop several designs completely, test their performance, calculate their production cost, and only then make a choice. This, in essence, is what the market economy tends to do in cases of multiple inventions.

If we take as given, first R and D cost, and second the uncertainty regarding quality and cost of the alternative designs, then a simple extension of the model of Section II implies that the number of alternative inventors which should be offered the economic system should be greater, the greater the demand for the invention.[18] Historical accounts are not inconsistent with the notion that the market economy tends to result in just that.

Of course we could obtain the appropriate number of parallel efforts without having a number of independent sources of inventions. We could rationalize the field of invention, and assign a single organization to conduct R and D in certain specified areas. The organization could run parallel approaches to difficult problems. And clearly, if the competing projects were run within one organization, there would be a much smaller likelihood that exactly the same problems would be worked on and solved in the same way by different groups, thus canceling the advantages of the multiple-path approach. Similarly, to the extent that information created in one project turns out to be relevant to the work of a competing project, it is much more likely that this information would be communicated and shared. But the incentives for each of the groups would be smaller than they would be were the stakes greater, and there is some evidence that companies do more imaginative R and D when there is a race for success than when there is no competition. Further, the very substantial number of cases in which existing companies and interests were backward in attempting to develop new technologies, in which outside inventions played the key role, and the birth of new firms was required to complete the innovation, indicates that decentralized decision making in the field of invention yields many social benefits. There would be significant disadvantages as well as advantages from "rationalizing" the invention industry.

B. Economies of Scale in R and D

The notion that the running of parallel inventive efforts may be an efficient strategy in situations where there is considerable uncertainty — generally situations in which major technological break-throughs are being sought, sheds some additional light on the advantages large industrial laboratories have over smaller ones. In addition to the ad-

[18] Consider a given probability distribution of unit production costs. Note that the first terms of equation (4a) can be interpreted as total R and D cost, and the second term as the expected production cost of the invention which the economy finally accepts. The relevant production cost clearly is the total (discounted) cost of all the units produced in the new product's life history. Ignoring discounting, this equals average unit cost times the total quantity produced.

vantages stemming from the ability to make profitable use of what would be external economies in a smaller laboratory[19] — the ability to make profitable use of a wide range of expertise (division of labor), and the ability to reduce overall risk by carrying a large portfolio of projects — the large laboratory is in a much better position to run the "optimizing" number of projects in parallel.

Marschak's[20] paper on the development of a major new microwave relay system at the Bell Telephone Laboratories is convincing evidence on this point. Two facts seem clear from his description. One is that the eventual success of the development was greatly facilitated by the running of multiple approaches to the solution of difficult development problems. The other is that a smaller laboratory probably would not have had the resources to support the multiple efforts.

In a sense, the conclusion of this section — that large laboratories have considerable advantages in R and D — conflicts with the conclusion of the preceding section, that there are many advantages in having multiple decision making units. Perhaps it is not too facetious to suggest that the joint conclusion is that an optimal organization of our economy-wide R and D effort should have a mixture of a few very big units and many smaller units so that we can have our cake and eat it too.

[19] See R. R. Nelson, "The Simple Economics of Basic Scientific Research," *The Journal of Political Economy*, LXVII (June, 1959).

[20] Op. cit.

[25]

THE ECONOMIC JOURNAL

SEPTEMBER 1989

The Economic Journal, 99 (*September* 1989), 569–596
Printed in Great Britain

INNOVATION AND LEARNING: THE TWO FACES OF R&D*

Wesley M. Cohen and Daniel A. Levinthal

Economists conventionally think of R&D as generating one product: new information. We suggest that R&D not only generates new information, but also enhances the firm's ability to assimilate and exploit existing information. In this paper we consider the implications of this dual role of R&D for the firm's incentive to invest in R&D. We argue that recognition of this second role of R&D suggests that the ease and character of learning within an industry will both affect R&D spending and condition the influence of appropriability and technological opportunity conditions on R&D. For example, we show that, contrary to the traditional result, intra-industry spillovers may encourage equilibrium industry R&D investment.

Scholars of technological change have observed that firms invest in own R&D to be able to utilise information which is available externally (e.g., Tilton, 1971; Allen, 1977; Mowery, 1983).[1] Tilton, for example, states that one of the main reasons firms invested in R&D in the semiconductor industry was that, '... an R & D effort provided an in-house technical capability that could keep these firms abreast of the latest semi-conductor developments and facilitate the assimilation of new technology developed elsewhere' (1971, p. 71). Accordingly, we argue that while R&D obviously generates innovations, it also develops the firm's ability to identify, assimilate, and exploit knowledge from the environment—what we call a firm's 'learning' or 'absorptive' capacity. While encompassing a firm's ability to imitate new process or product innovations, absorptive capacity also includes the firm's ability to exploit outside knowledge of a more intermediate sort, such as basic research findings

* Cohen's research for this paper was supported by the Division of Policy Research and Analysis of the National Science Foundation (PRA 83–10664). For their comments, we thank Mark Kamlet, Bruce Kogut, Jonathan Leland, Richard Levin, Richard Nelson, and Sidney Winter, and the participants in the 1988 bienniel Congress of the International Joseph A. Schumpeter Society, especially F. M. Scherer and Jianye Wang. We are also particularly grateful to Steven Klepper and two anonymous referees for their thoughtful and extensive comments. We thank Michael Dodman and Nam Pham for computing assistance, and Maria Viskochil for typing.

The representations and conclusions presented herein are those of the authors. They have not been adopted in whole or in part by the Federal Trade Commission, its Bureau of Economics, or any other entity within the Commission. The FTC's Disclosure Avoidance Officer has certified that the data included in this paper do not identify individual company line of business data.

[1] In a different context, Evenson and Kislev (1973) make a similar point when they observe that the international transfer of agricultural technology depends, in part, upon the recipients' own research efforts.

that provide the basis for subsequent applied research and development. Also, in light of the dependence of industrial innovation upon extramural knowledge,[2] absorptive capacity represents an important part of a firm's ability to create new knowledge. In this regard, the exercise of absorptive capacity represents a sort of learning that differs from learning-by-doing, the focus of industrial economists' work on firm learning in recent years (e.g., Spence, 1981; Lieberman, 1984). Learning-by-doing typically refers to the automatic process by which the firm becomes more practiced, and, hence, more efficient at doing what it is already doing. In contrast, with absorptive capacity a firm may acquire outside knowledge that will permit it to do something quite different.

The role that R&D plays in learning has received little attention in the past because, following Arrow (1962) and Nelson (1959), economists have assumed that technological knowledge which is in the public domain is a public good. Like a radio signal or smoke pollution, its effects are thought to be costlessly realised by all firms located within the neighbourhood of the emission. When economists do think about the costs of knowledge transfer, they typically identify them with immediate information processing or imitation costs. In suggesting technological knowledge is a public good, Arrow and others do not deny the existence of such costs, but argue that they are typically small relative to the cost of creating new knowledge. This argument, however, raises the question of what determines these immediate costs of assimilating technological knowledge. We suggest that if these costs are relatively small, it is by virtue of the considerable R&D already conducted by the firms in the vicinity of the 'emission'; the firm has already invested in the development of its absorptive capacity in the relevant field.[3]

Thus, we are suggesting that the long-run cost of learning may be substantial. Second, most of this cost is borne via the development of a stock of prior knowledge that constitutes the firm's absorptive capacity. Third, a significant benefit of R&D is its contribution to this knowledge base. Therefore, the incentives to learn should influence R&D spending. Those incentives will be shaped by the quantity of knowledge to be assimilated and the ease with which learning may occur. The ease of learning, in turn, depends upon the characteristics of the underlying technological and scientific knowledge upon which innovation depends in a given industry.

To explore the implications of the dual role of R&D for R&D investment, we construct a simple theoretical model of the generation of a firm's

[2] According to Brock (1975) most of the computer industry's innovations originated with developments outside the industry, particularly in semiconductors. In his study of the aluminium industry, Peck (1962) makes the same point. In a study of twenty-seven major inventions made between 1946 and 1955, Hamberg (1963) showed that twenty came from sources outside industrial R&D laboratories. In his study of twenty-five major discoveries introduced into the United States by DuPont, Mueller (1962) indicated that, despite the company's reputation for path-breaking research, fifteen originated with work done outside the company. In their study of information usage in technological innovation, Johnston and Gibbons (1975) even concluded that information obtained from outside the company contributes significantly more to the solution to technical problems than does internally available information.

[3] For example, compare a firm's immediate costs of reverse engineering a product versus the considerable costs, incurred over time, of acquiring the experience and background necessary to execute the task.

technological knowledge. Our model considers the basic sources of technological knowledge utilised by a firm: the firm's own R&D, knowledge which originates with its competitors' R&D spillovers,[4] and knowledge which originates outside the industry. Section I of the paper develops the model structure. In Section II, we explore the determinants of industry equilibrium R&D, focusing on the effect of the ease of learning. In Section III, we test hypotheses suggested by our framework with survey data on technological opportunity and appropriability conditions recently collected by Levin *et al.* (1983, 1987), and data on business unit R&D expenditures from the Federal Trade Commission's Line of Business Programme. In this section, we examine how the ease of learning shapes the influence on R&D spending of appropriability and technological opportunity conditions.

I. MODEL STRUCTURE

Central to our model is the determination of the firm's stock of knowledge. We represent additions to firm i's stock of technological and scientific knowledge by z_i, and assume that z_i increases the firm's gross earnings, Π^i (i.e., $\Pi^i_{z_i} > 0$), but at a diminishing rate (i.e., $\Pi^i_{z_i z_i} < 0$). We characterise the determination of z_i such that,

$$z_i = M_i + \gamma_i(\theta \sum_{j \neq i} M_j + T), \tag{1}$$

where M_i is a firm's investment in R&D; γ_i is the fraction of knowledge in the public domain that the firm is able to assimilate and exploit, and represents the firm's absorptive capacity; θ is the degree of intra-industry spillovers; and T is the level of extra-industry knowledge. Other firms' investments in research and development, represented by M_j for $j \neq i$, also contribute to z_i. The degree to which the research effort of one firm may spill over to a pool of knowledge potentially available to all other firms is characterised by θ where $0 \leqslant \theta \leqslant 1$. A value of θ of one means that the R&D effort of one firm increases the pool of knowledge available to other firms by the total amount of the firm's R&D. A value of zero means that the benefits of research are exclusively appropriated by the firm conducting the research. Exogenous factors such as patent policy shape θ.

Our model postulates that a firm's capacity to absorb externally generated knowledge depends on its R&D effort. We assume that $0 \leqslant \gamma_i \leqslant 1$. When $\gamma_i = 1$, the firm absorbs all knowledge that is in the public domain. Alternatively, when $\gamma_i = 0$, the firm absorbs none. We assume that the firm's own R&D increases absorptive capacity, $\gamma_M > 0$, though at a decreasing rate, $\gamma_{MM} < 0$. Although not formally captured by our static model, we think of γ as being cumulatively built through past as well as current R&D effort.

We consider γ_i to be a function not only of M_i, but also of a variable β (i.e.,

[4] We define spillovers to include any original, valuable knowledge generated in the research process which becomes publicly accessible, whether it be knowledge fully characterising an innovation, or knowledge of a more intermediate sort.

$\gamma_i \equiv \gamma(M_i, \beta)$) that is determined by the characteristics of the underlying scientific and technological knowledge that affect the ease of learning from the environment. The variable, β, reflects the characteristics of outside knowledge that make R&D more or less critical to the maintenance and development of absorptive capacity. Although difficult to specify all the relevant characteristics *a priori*, we suggest they would include the complexity of the knowledge to be assimilated, and the degree to which the outside knowledge is targeted to the needs and concerns of the firm. When outside knowledge is less targeted to the firm's particular needs and concerns, a firm's own R&D becomes more important in permitting it to recognise, assimilate and exploit valuable knowledge. Sources that produce less targeted knowledge include, for example, university laboratories involved in basic research, while more targeted knowledge may be generated by contract research laboratories or input suppliers. In addition, the degree to which a field is cumulative, or a field's pace of advance should also affect how critical R&D is to the development of absorptive capacity. To the extent that findings in a field build upon prior findings, an understanding of prior research is necessary to the assimilation of subsequent findings. The pace of advance of a field affects the importance of R&D to developing absorptive capacity because, the faster is the pace of knowledge generation, the larger will be the staff required to keep abreast of developments.[5]

We define β such that a higher level indicates that the firm's ability to assimilate external knowledge is more dependent on the firm's own R&D. This dependence is reflected in two effects. We assume that the larger is β, the larger is the marginal impact of own R&D on absorptive capacity such that $\gamma_{M\beta} \equiv \gamma_{M\beta}(M_i, \beta) > 0$. In addition, for a given level of M, absorptive capacity decreases with β (i.e., $\gamma_\beta(M_i, \beta) < 0$).[6] Thus, we are assuming that increasing β increases the marginal effect of R&D on absorptive capacity, but diminishes the level of absorptive capacity. This second effect indicates that when external knowledge becomes more difficult to assimilate, the firm will assimilate less of it for a given R&D effort.

Equation (1) reflects the assumption that the appropriation of other firms' research is realised through the interaction of θ and γ, indicating that the firm cannot assimilate what is not spilled out. Also, the firm cannot passively assimilate externally available knowledge. It must invest in its own R&D to absorb any of the R&D output of its competitors. In this sense, not only is absorptive capacity endogenous, but the appropriability of rents due to inventive activity is endogenous as well.

Another critical determinant of R&D is technological opportunity, which may be thought of as how costly it is for the firm to achieve technical advance in a given industry. We incorporate two determinants of technological

[5] Pavitt (1987), in an analysis of policies affecting the utilisation of technology, focuses on some of the same characteristics of knowledge as considered here, including complexity and cumulativeness.

[6] Therefore, as β approaches zero, absorptive capacity is less responsive to the level of own R&D and γ_i approaches 1. Thus, a β of zero is equivalent to the setting in which absorptive capacity is not endogenous and $\gamma \equiv 1$.

opportunity in our model. One is the quantity of extra-industry technological knowledge, represented by the variable T. Examples of T include the output of government or university laboratories, or the knowledge provided by equipment suppliers. Like its assimilation of competitors' R&D output, a firm's assimilation of outside knowledge is constrained by its absorptive capacity.[7] Another dimension of technological opportunity, represented by the variable f, is the degree to which new knowledge, z, improves the technological performance of the firm's manufacturing processes or products. We assume that the more that knowledge contributes to technological performance, the more it increases profit (i.e., $\Pi^i_{z_i f} > 0$).[8]

We assume that there are n firms in the industry with symmetric R&D policies. Each firm chooses its level of R&D to maximise profits, taking the R&D levels of the other firms as given. Thus, we model a symmetric Nash equilibrium in R&D levels. Accordingly, the firms anticipate the effect of changes in their R&D investments on their competitors' knowledge levels and, in turn, the effects of these changes on the firms' own profits. As a result, the firm's profit is a function not only of its own knowledge, z_i, but the technological knowledge of all firms in the industry. To reflect the effect of rivalry, we assume that an increase in a competitor's knowledge diminishes both firm i's profits and firm i's marginal benefit from increasing its knowledge level so that $\Pi^i_{z_j} < 0$, and $\Pi^i_{z_i z_j} < 0$, where z_j represents the change in a competitor's knowledge level. When knowledge has a large impact on technological performance, increases in competitors' knowledge levels are more damaging to own profits (i.e., $\Pi^i_{z_j f} < 0$). In addition to being a function of increments to firms' knowledge levels, we assume that profit due to R&D is also a function of demand conditions.

Differentiating Π^i with respect to M_i yields,

$$R \equiv \Pi^i_{z_i}[1 + \gamma_{M_i}(\theta \sum_{j \neq i} M_j + T)] + \theta \sum_{j \neq i} \gamma_j \Pi^i_{z_j}. \tag{2}$$

The function R is the marginal return to own R&D. Deriving this expression for each firm and setting it equal to one (the per unit cost of R&D) generates a set of equations characterising the firm's optimal R&D policy given its competitors' R&D levels. When solved simultaneously, these equations yield the equilibrium value of each firm's R&D, which we represent as M^*.

II. ANALYSIS

We will consider the direct effect of the ease of learning on R&D, as well as its impact on the influence of spillovers and technological opportunity on R&D spending. In this latter regard, we will show that as learning becomes more dependent on own R&D, increasing technological opportunity or spillovers

[7] For simplicity, we assume that all extra-industry research findings are made public, implying that the spillover parameter applicable to extra-industry knowledge equals unity.

[8] See Cohen and Levin (1989) for a discussion of the two dimensions of technological opportunity represented here as T and f.

will tend to elicit more R&D effort, *ceteris paribus*. Indeed, the presence of an endogenous absorptive capacity may change the qualitative effect of these traditionally considered determinants of inventive activity. Most notably, spillovers may encourage R&D under some conditions.

In the context of our n-firm symmetric equilibrium, we can show that for any arbitrary parameter, k, that influences M^*, the sign of $(\partial M^*/\partial k)$ equals the sign of R_k, where R is defined by equation (2) above.[9] Therefore, to simplify the presentation of our results, we only present the analogues of R_k, including R_β, R_θ, R_T, and R_f to establish the signs of, respectively, $(\partial M^*/\partial \beta)$, $(\partial M^*/\partial \theta)$, $(\partial M^*/\partial T)$ and $(\partial M^*/\partial f)$.[10] In addition to analysing the direct effect of β on R&D spending, we also examine how the ease of learning affects the influence of spillovers and technological opportunity on R&D; that is, the effect of β on $(\partial M^*/\partial \theta)$, $(\partial M^*/\partial T)$ and $(\partial M^*/\partial f)$. Exploring these effects analytically is, however, quite complex. Thus, to provide an intuitive sense of the basic forces at work, we simplify the analysis by considering only the first-order effects of knowledge on firm profit and assume the second-order condition to be constant. Since we therefore cannot claim the analysis to be general, we have confirmed the intuition developed here with numerical analysis.[11] Given our assumption of a constant second-order condition, we can show that the signs of the cross-partial derivatives, $(\partial^2 M^*/\partial \theta\, \partial \beta)$, $(\partial^2 M^*/\partial T\, \partial \beta)$ and $(\partial^2 M^*/\partial f\, \partial \beta)$ are identical to the signs of, respectively, $R_{\theta\beta}$, $R_{T\beta}$, and $R_{f\beta}$. Therefore, to simplify further the presentation of our results, we only present $R_{\theta\beta}$, $R_{T\beta}$, and $R_{f\beta}$.

II.1 *Direct Effect of the Ease of Learning*

As β increases, reflecting, for example, an increase in the complexity or a diminution in the targeted quality of outside knowledge, own R&D becomes more critical to absorptive capacity. However, that also means that the cost per unit of knowledge, z, increases. Despite these seemingly countervailing effects of β, differentiating equation (2) by β reveals that equilibrium R&D increases with β (i.e., $(\partial M^*/\partial \beta) > 0$).[12] We can see the main forces at work if we focus on the first-order effects.[13] Deriving R_β and constraining the second-order terms of

[9] This assumes that the Nash equilibrium is stable. The proof is available in a technical appendix available from the authors.

[10] The complete derivative, $\partial M^*/\partial k$, equals $R_k/[a+(n-1)b]$, where $a \equiv \partial^2 \Pi^i/\partial M_i \partial M_i$ and $b \equiv \partial^2 \Pi^i/\partial M_i \partial M_j$.

[11] Results of the numerical analysis are contained in the technical appendix available from the authors.

[12] This result holds if we assume that the sufficient condition for the reaction function to be downward sloping applies when $\gamma_M = 0$. The reaction function is downward sloping when $\Pi^i_{M_i M_j} < 0$. A sufficient condition for this to hold is that $\Pi^i_{z_j z_j} + \Pi^i_{z_i z_j} + (n-2)\, \Pi^i_{z_j z_k} < 0$, where $k \neq j \neq i$.

[13] The complete characterisation of R_β is:

$$R_\beta = \Pi^i_{z_i} \gamma_{M\beta}[\theta(n-1)M + T] + \theta(n-1)\gamma_\beta \Pi^i_{z_j}$$

$$+ \gamma_\beta[\theta(n-1)M + T]\,([\Pi^i_{z_i z_i} + (n-1)\,\Pi^i_{z_i z_j}]\{1 + \gamma_M[\theta(n-1)M + T]\}$$

$$+ \theta(n-1)\gamma\,[\Pi^i_{z_j z_j} + \Pi^i_{z_j z_i} + (n-2)\,\Pi^i_{z_j z_k}]).$$

Π^i with respect to knowledge to equal zero allow us to characterise the effect of β on R&D spending as

$$\operatorname{sign}\left(\frac{\partial M^*}{\partial \beta}\right) = \operatorname{sign}\left(\Pi^i_{z_i}\{\gamma_{M\beta}[\theta(n-1)\,M + T]\} + \theta(n-1)\frac{\partial \gamma}{\partial \beta}\Pi^i_{z_j}\right). \qquad (3)$$

The first expression, $\Pi^i_{z_i}(\cdot)$, indicates that with a higher β the firm has greater incentive to conduct R&D because its own R&D has become more critical to assimilating both its competitors' spillovers, $\theta(n-1)\,M$, and extra-industry knowledge, T. At the same time, the second expression, $\theta(n-1)\,(\partial\gamma/\partial\beta)\,\Pi^i_{z_j}$, indicates that competitors' levels of absorptive capacity, $(n-1)\,\gamma$, decline with β. As a consequence, competitors are less able to exploit the firm's spillovers. Due to both of these effects, the payoff to the firm's R&D increases and, *ceteris paribus*, more R&D is conducted.

II.2. *Intraindustry Spillovers*

Economists (e.g., Nelson, 1959; Arrow, 1962) have long argued that research spillovers diminish firms' incentives to invest in R&D by undermining the appropriability of returns to inventive activity. The contribution of R&D to a firm's absorptive capacity implies, however, that there is an offsetting incentive associated with spillovers, because only through its own R&D may a firm exploit the knowledge generated by its competitors.

To explore the relationship between spillovers and equilibrium R&D, we differentiate equation (2) with respect to θ. In the absence of an endogenous absorptive capacity, we can verify the traditional claim that higher levels of spillovers reduce R&D investment.[14] With an endogenous absorptive capacity, the result changes. In addition to a negative appropriability incentive, we now also observe a positive absorption incentive. These countervailing forces may be clearly seen if we eliminate the effect of diminishing returns to knowledge (represented by the second-order terms of firm profits with respect to knowledge).[15] In this case, deriving R_θ permits us to sign $(\partial M^*/\partial\theta)$ where,

$$\operatorname{sign}\left(\frac{\partial M^*}{\partial \theta}\right) = \operatorname{sign}\,[\Pi^i_{z_i}\gamma_M(n-1)\,M + (n-1)\,\gamma\,\Pi^i_{z_j}]. \qquad (4)$$

Contrary to the standard proposition that increasing spillovers reduces the incentive to invest in R&D, the sign of $(\partial M^*/\partial\theta)$ is ambiguous. This ambiguity is due to two offsetting effects: the benefit to the firm of increasing its absorptive capacity, represented by $\Pi^i_{z_i}\gamma_M\,(n-1)\,M$, and the loss associated with the diminished appropriability of rents as spillovers increase, represented by

[14] As with the comparative static result with respect to β, we need to assume that the sufficient condition for the reaction function to be downward sloping holds.

[15] The complete characterisation of R_θ is:

$$R_\theta = \Pi^i_{z_i}\gamma_M(n-1)\,M + (n-1)\,\gamma\,\Pi^i_{z_j}$$
$$+\gamma(n-1)M\{[\Pi^i_{z_iz_i} + (n-1)\Pi^i_{z_iz_j}]\{1+\gamma_M[\theta(n-1)M+T]\} + \theta(n-1)\gamma\,[\Pi^i_{z_iz_j} + \Pi^i_{z_jz_j} + (n-2)\,\Pi^i_{z_jz_k}]\}.$$

$(n-1)\gamma\Pi^i_{z_j}$. The former expression indicates that with an endogenous absorptive capacity, a higher spillover rate provides a positive incentive to conduct R&D by increasing the pool of available knowledge. Thus, our introduction of absorptive capacity changes the traditional result in two ways. Most importantly, the desire to assimilate the knowledge generated by other firms provides a positive incentive to invest in R&D as θ increases.[16] Second, the dis-incentive associated with other firms' assimilation of the firm's R&D output is dampened because other firms' absorptive capacities are less than unity.[17]

Equation (4) also suggests that the positive impact of θ is typically greater the larger is β, the impact of own R&D on absorptive capacity. To examine this formally we evaluate the sign of the cross-partial derivative, $(\partial^2 M^*/\partial\beta\,\partial\theta)$, by deriving $R_{\beta\theta}$. If we focus only on first-order effects of z on profit and assume that $\gamma_{MM} = 0$, then we can show that $R_{\beta\theta}$ is positive.[18] There are three distinct effects. First, a higher β enhances the positive absorption incentive to invest in R&D. Second, by reducing competitors' absoptive capacities, a higher β encourages R&D by mitigating the negative appropriability effect of spillovers, *ceteris paribus*. Finally, since a higher β increases the optimal level of R&D, β increases the optimal M_j which, while increasing the pool of knowledge available to firm i, also increases firm j's absorptive capacity to firm i's detriment. However, as long as the magnitude of the marginal return to own knowledge exceeds the marginal effect on profits of a change in a competitor's knowledge level (i.e., $|\Pi^i_{z_i}| > |\Pi^i_{z_j}|$), this last effect is positive.[19]

In light of the importance of spillovers in economists' analysis of R&D incentives, we examine how demand conditions and market structure interact with β to determine the qualitative effect of spillovers on R&D spending. In order to do this, we develop a model of cost-reducing technological change.[20] We specify a cost function with a constant marginal cost of production that is a decreasing exponential function of z_i, the firm's knowledge level. We also assume that the demand curve and the absorptive capacity function have constant elasticities with respect to price and R&D spending, respectively.

[16] While Levin and Reiss (1988) show that high spillovers and high R&D investment may coincide when the productivity of spillovers (i.e., their impact on cost reduction) is high, they do not show, as we do in our theoretical analysis and as the empirical results below suggest, that increasing the extent of spillovers may have a direct positive effect on R&D investment incentives.

[17] Suppose that the appropriability disincentive effect outweighs the absorption incentive effect such that we obtain the standard result that equilibrium R&D diminishes with θ. If R&D declines, the level of absorptive capacity will also decline in equilibrium given our assumption that $\gamma_M > 0$. This decline will, in turn, diminish competitors' abilities to exploit the firm's spillovers, and thus dampens the negative incentive effect with which spillovers are associated. Thus, the interaction of an endogenous absorptive capacity and θ means that even where the negative appropriability effect dominates, its effect is attenuated relative to a world where firm's absorptive capacity is not a function of its own R&D.

[18] Under these conditions,

$$R_{\beta\theta} = (n-1)\left[M_j\gamma_{M\beta}\Pi^i_{z_i} + \gamma_\beta\Pi^i_{z_j} + \gamma_M\frac{\partial M^*}{\partial\beta}(\Pi^i_{z_i} + \Pi^i_{z_j})\right].$$

[19] We can show that $|\Pi^i_{z_i}| > |\Pi^i_{z_j}|$ in the context of the specialised model discussed below.

[20] The development of this model and its analysis is contained in the technical appendix available from the authors. As Spence (1984) suggests, such a model may be interpreted to consider product as well as process innovation if we think of products as delivering services to customers, and product innovation as simply increasing the quantity of service per unit of output.

Again, we analyse Nash equilibria in which the firms' R&D policies are assumed to be symmetric.

In this specialisation of our model, spillovers are more likely to cause an increase in industry equilibrium R&D where β is high, which is indicated as well by the general analysis above. We also find that the higher is the price elasticity of demand or the less concentrated is the industry, the more likely it is that equilibrium R&D investment will rise with spillovers, *ceteris paribus*. Thus, as firms face a more competitive environment, in the sense that they are less inter-dependent due to either an increase in demand elasticity or a lower concentration level, spillovers are more likely to encourage R&D investment. The intuition is that as an industry becomes more competitive, the private loss associated with the public good character of R&D spillovers diminishes relative to the private benefit of being able to exploit competitors' spillovers.

II.3. *Technological Opportunity*

Our model suggests that increased levels of extra-industry knowledge will only substitute for own R&D unless absorptive capacity is endogenous. With an endogenous absorptive capacity, there is also a positive incentive for the firm to conduct R&D in order to exploit the pool of external knowledge. To formalise this intuition, we characterise the sign of $(\partial M^*/\partial T)$ by deriving R_T, and assume $\theta = 0$ to highlight the role of T:[21]

$$\text{sign}\left(\frac{\partial M^*}{\partial T}\right) = \text{sign}\,[\gamma_M \Pi^i_{z_i} + (\Pi^i_{z_i z_i} + (n-1)\,\Pi^i_{z_i z_j})\,\gamma(1+\gamma_M T)]. \qquad (5)$$

In the absence of an endogenous absorptive capacity (i.e., $\gamma_M \equiv 0$), this expression is negative. A higher T merely substitutes for the firm's own R&D. However, with an endogenous absorptive capacity, a higher T also provides an incentive to enhance the firm's absorptive capacity, as reflected in the first expression on the right hand side.

To examine the impact of the second dimension of technological opportunity, f, which reflects the degree to which new knowledge, z, improves technological performance, we derive R_f to sign $(\partial M^*/\partial f)$:

$$\text{sign}\left(\frac{\partial M^*}{\partial f}\right) = \text{sign}\,(\Pi^i_{z_i f}\{1+\gamma_M\,[T+\theta(n-1)\,M]\} + \theta(n-1)\,\Pi^i_{z_j f}). \qquad (6)$$

The net effect of f is ambiguous. There is a positive effect of increasing the payoff to the firm's own knowledge level, and a negative effect of spillovers being more damaging to the firm. Thus, both components of technological opportunity, f and T, have ambiguous effects on optimal R&D spending, although for different reasons. The countervailing negative effect of T results from diminishing returns to knowledge, and that of f reflects the negative appropriability incentive associated with spillovers.

[21] We can derive the same qualitative result as below with a positive θ under the assumption invoked and motivated before, that $\Pi^i_{z_i z_i} + \Pi^i_{z_j z_j} + (n-2)\,\Pi^i_{z_j z_k} < 0$. The complete characterisation of R_T is:

$$R_T = \Pi^i_{z_i}\gamma_M + \gamma\{[\Pi^i_{z_i z_i} + (n-1)\,\Pi^i_{z_i z_j}]\{1+\gamma_M[\theta(n-1)\,M+T]\} + \theta(n-1)\,\gamma^2\,[\Pi^i_{z_i z_j} + \Pi^i_{z_j z_j} + (n-2)\,\Pi^i_{z_j z_k}]\}.$$

We are interested not only in the direct effect of technological opportunity on the firm's R&D level, but, particularly for the empirical analysis, how this effect is mediated by the ease of learning. We find that increasing β has the same qualitative effect on the influence of both T and f. Increasing β causes increases in T or f to call forth higher levels of R&D spending, notwithstanding the sign of their direct effects. If we again focus exclusively on the first-order effects of knowledge on firm profits and assume $\gamma_{MM} = 0$, we sign $(\partial^2 M^*/\partial T \partial \beta)$ by deriving $R_{T\beta}$, which equals $\gamma_{M\beta} \Pi^i_{z_i}$. A higher level of β increases the incentive to build absorptive capacity in order to assimilate the greater level of outside knowledge. Thus, the first-order effect suggests that increasing β causes R&D to increase more (or decline less) with increases in external knowledge.

The analysis of the effect of β on $(\partial M^*/\partial f)$ resembles that of β on $(\partial M^*/\partial \theta)$, involving three similar incentives.[22] The first represents an enhancement of the absorption incentive effect. The second is that increasing β diminishes competitors' absorptive capacities, and thereby reduces the negative appropriability incentive. Finally, as firm j's optimal R&D increases with β, increases in the pool of knowledge are all the more valuable as f increases. However, to the firm's detriment, the competitor's absorptive capacity also increases with M_j. If the positive marginal effect of f on the contribution of own knowledge to profits is greater than the magnitude of its negative marginal effect on the influence of competitors' knowledge on firm i's profits (i.e., $|\Pi^i_{z_i f}| > |\Pi^i_{z_j f}|$), then this latter incentive is positive. Given this assumption, the first-order effects suggest that higher levels of β should cause R&D to increase (or decrease less) as f increases.

III. EMPIRICAL ANALYSIS

Our theoretical analysis suggests that the determinants of the ease of learning (i.e., β), including the targeted quality of knowledge and the characteristics of scientific fields (such as pace of advance and cumulativeness), should have a direct effect on R&D intensity. We also hypothesise that these variables condition the influence on R&D of technological opportunity and appropriability conditions. The Levin *et al.* survey data on technological opportunity and appropriability conditions do not permit construction of direct measures of the ease of learning or its determinants, and thus do not allow analysis of their direct effects. These survey data do, however, permit an examination of the ways in which the effects of technological opportunity and appropriability on R&D intensity are influenced by the ease of learning.

In Section III.1, we describe the data. In Section III.2, we operationalise the

[22] Deriving $R_{\beta f}$ and again focusing on the first-order effects, we have

$$R_{\beta f} = \gamma_{M\beta}[T + \theta(n-1)M]\Pi^i_{z_i f} + \theta(n-1)\gamma_\beta \Pi^i_{z_j f} + \theta(n-1)\gamma_M(\Pi^i_{z_i f} + \Pi^i_{z_j f})\frac{\partial M^*}{\partial \beta}.$$

Consistent with our assumption above that $|\Pi^i_{z_i}| > |\Pi^i_{z_j}|$, we assume that $|\Pi^i_{z_i f}| > |\Pi^i_{z_j f}|$. Again, this property can be shown to hold in the context of the example developed in the technical appendix available from the authors.

hypotheses and discuss the empirical specification. In Section III.3, estimation issues are considered, and in Section III.4, the results are presented.

III.1. Data

Our dependent variable is company-financed business unit research and development expenditures, expressed as a percentage of business unit sales and transfers over the period 1975–7. The variable was averaged to control for possible differences in the impact and timing of business cycles across industries. Business unit level data on firms' R&D expenditures and sales were obtained from the Federal Trade Commission's Line of Business Programme. Data on inter-industry differences in technological opportunity and appropriability conditions were drawn from the survey data collected by Levin *et al.* (1983, 1987).[23] The survey-based variables are defined at the FTC's line of business level, which is between the SIC's three and four digit levels of aggregation. All the survey-based variables are industry (line of business) mean scores computed as an average over all respondents within a given industry.[24]

We used two samples in our analysis. The larger sample includes both R&D performing and non-performing business units, and consists of 1,719 business units representing 318 firms in 151 lines of business. The smaller sample includes only R&D performing business units, and consists of 1,302 business units representing 297 firms in 151 lines of business. Table 1 presents the mean, median, and minimum and maximum values for, respectively, R&D intensity, business unit sales, and firm sales for each of the two samples. As Table 1 suggests, business units performing R&D are on average larger than those that do none, and tend to be operated by larger parent firms. We should note that even our more inclusive sample is not quite representative of the manufacturing sector as a whole, because the sample is restricted to only those lines of business covered by the Levin *et al.* survey.[25]

III.2. *Specification*

Our hypotheses are tested in the context of a simple empirical model of business unit R&D intensity which considers technological opportunity, appropriability, and demand conditions as principal industry-level determinants.

[23] We wish to thank Professors Levin, Klevorick, Nelson, and Winter for permitting us to use these data.

[24] Respondents were R&D laboratory managers.

[25] Both samples are drawn from a larger sample, constructed from the FTC's Line of Business Programme dataset, that consists of 2,494 business units, representing 345 firms operating in 244 manufacturing lines of business. This larger dataset, developed by Cohen and Mowery (1984), excludes all firms in the FTC database that operate mainly in regulated industries, and firms with obvious inter-temporal inconsistencies in reporting methods or other obvious reporting errors. All business units operating outside the manufacturing sector, and those that were not continuously active during the period 1974–7, were excluded as well. See Cohen and Mowery (1984, Appendix V) for a discussion of the screening procedures used to check the validity of the FTC's Line of Business Programme R&D data. Relative to the full sample of 2,494 business units, the firms and business units covered by the Levin *et al.* survey tend to be somewhat larger, and R&D intensity is somewhat higher. For example, in the sample of 2,494 business units, the mean R&D intensity is 1·59%, the mean business unit sales is 0·20 billion dollars, and the mean sales of the parent firm 2·31 billion dollars. In comparison, Table 1 indicates that for our sample of 1,719 business units, the mean R&D intensity is 1·81%, the mean business unit sales is 0·25 billion dollars, and the mean sales of the parent firm is 2·46 billion dollars.

Table 1

Descriptive statistics on R&D intensity, business unit sales and firm sales by sample

	R&D intensity* (%)		Business unit sales ($billion)		Firm sales ($billion)	
	All business units (N = 1,719)	R&D performers only (N = 1,302)	All business units (N = 1,719)	R&D performers only (N = 1,302)	All firms (N = 318)	R&D performers only (N = 297)
Mean	1·81	2·31	0·25	0·30	2·46	2·57
Median	0·81	1·30	0·07	0·08	1·23	1·31
Minimum†	0·00	0·00	0·00	0·01	0·11	0·12
Maximum†	22·35	22·35	16·04	16·04	34·13	34·13

* Business unit R&D spending as a percentage of business unit sales and transfers.

† To preserve the confidentiality of responding firms, we report the mean values of the four smallest and largest business units of firms, respectively.

Technological Opportunity

In assessing the importance for technological progress of what are considered to be two sources of technological opportunity – the relevance of science and the importance of extra-industry sources of knowledge – the Levin *et al.* survey data permit us to examine the effects of the determinants of the ease of learning on the influence of technological opportunity. The survey provides a measure of the relevance to an industry's R&D of each of eleven basic and applied scientific fields for each line of business (Levin *et al.*, 1983, Question III.A). The basic sciences include biology, chemistry, geology, mathematics, and physics. The applied sciences include agricultural science, applied mathematics/operations research, computer science, materials science, medical science, and metallurgy. In the survey questionnaire, relevance is assessed on a seven-point Likert scale in which a score of one indicates no relevance and a seven indicates the field is very relevant.[26] 'Relevance' may be interpreted to reflect either of the two dimensions of technological opportunity represented in our model, T or f. If by the relevance of a field of science, the respondent means the quantity of the research findings of, say, physics that applies to its business, then T, the quantity of outside knowledge, is the appropriate representation. If, alternatively, relevance refers to the impact on profit of the technical advance

[26] As indicated in Cohen *et al.* (1987), there are several statistical problems associated with the use of Likert-scale survey responses as independent variables in regressions. The most important is whether responses along a semantic continuum can be treated as if they were interval data. In the absence of adequate alternative measures, we assume that such treatment is reasonable. Given this assumption, there remain other potential sources of measurement error. First, individual respondents may differ in their use of the seven-point scale. After conducting a preliminary examination of the importance of inter-rater differences in mean responses and in the variance of responses, Levin concluded that the ranking of industry mean responses to particular questions is reasonably insensitive to correction for these individual effects. Another form of measurement error is introduced by using industry means instead of individual responses. We attempt to control for this type of error by including the number of survey responses per industry among the variables used to correct regression results for heteroscedasticity.

generated by using some quantity of research output from a particular field, then f is the appropriate representation.

The survey questionnaire also asks respondents to evaluate the importance (on a seven-point scale) of the contributions of various external sources of knowledge to technical progress within each line of business (Levin *et al.*, 1983, Question III.E). Five sources of outside knowledge are considered: upstream suppliers of raw materials and equipment (*MATERIALTECH* and *EQUIPTECH*, respectively), downstream users of the industry's products (*USERTECH*), government agencies and research laboratories (*GOVTECH*), and universities (*UNIVTECH*). Like our interpretation of the 'relevance' of scientific fields, we interpret the 'importance of the contribution to technical progress' of the external sources of knowledge to refer to either T or f. If 'importance of the contribution' reflects T, it represents how much knowledge from each source is of use. If it reflects f, it represents the impact of a given quantity of knowledge from that source on technical advance and profit.

In our empirical analysis, we are able to approach the ambiguity of the terms 'relevance' and 'importance' with equanimity because, whether these notions represent T or f, our model predicts that β should condition their influence on R&D intensity in the same qualitative fashion. Specifically, for a given change in either T or f, the effect of that change on R&D should be more positive the higher is β.

The question is how do we observe the mediating effect of β on the influence of technological opportunity (i.e., f or T) given no direct measure of β or its determinants. Recall that we have eleven variables indicating the technological opportunity associated with each of eleven scientific fields for each line of business,[27] and five variables indicating the technological opportunity associated with each of five sources of external knowledge. These two sets of variables allow us to test for the influence of β if one assumes that β varies across the elements of each set. Determinants of β, such as cumulativeness, pace of advance, or the targeted quality of knowledge, are all field-specific to some degree. Therefore, one test of the influence of β is to see whether the effect on R&D spending of technological opportunity differs across the field variables. However, we can go beyond this relatively weak proposition.

Our theory suggests that as β increases, the effect of either f or T on R&D spending becomes more positive. Therefore, if we can ordinally rank the value of β across either the field variables or the knowledge sources variables, then we can predict the ordinal rank of the effects of these variables on R&D intensity. We suggest that there are categorical differences across the elements of each set

[27] One might argue that it is not the relevance of every field of science that represents the dimension of technological opportunity associated with an industry's closeness to science. Rather, it may be the relevance of the most relevant field. To represent this notion of technological opportunity, we examined the effect of a variable that represents the maximum of the relevance scores received by the fields of science for each line of business. Given the close conceptual and operational relationship between this variable and the vector of variables representing the relevance of each of the scientific fields, it is not surprising that the coefficient of the variable is never significant when the field vector is included, while it is always positive and significant when the field relevance variables are dropped. Also, when this variable alone is dropped, none of the qualitative results change. We therefore omit it from our specifications.

of variables that roughly correspond to differences in one determinant of β. A characteristic that distinguishes the basic from the applied sciences is the extent to which findings are targeted to the needs and concerns of firms, where the knowledge associated with basic science is less targeted than that associated with applied science. Accordingly, the β value associated with basic science is higher than that associated with applied science. Our analysis of the influence of β then suggests that for a given change in the technological opportunity associated with less targeted basic science, firms will conduct relatively more R&D. Thus, the coefficient values of the technological opportunity variables associated with the basic sciences should exceed those of the applied sciences.

The targeted quality of knowledge not only distinguishes fields; it also distinguishes the knowledge originating from external sources. By distinguishing the five extra-industry sources of knowledge on the basis of how targeted to a firm's needs the knowledge is, we can predict the ordinal ranking of the associated effects on R&D spending. Although we cannot distinguish all five sources of knowledge on the basis of how targeted they are to a firm's needs, we can approximately rank four of the five. First, we would expect suppliers' knowledge to be most targeted since such knowledge is often conveyed to stimulate sales. Between equipment and materials suppliers, the knowledge provided by equipment suppliers is typically more targeted since suppliers tend to differentiate their products more via support service and point-of-sale information. On the other end of the spectrum, we suggest that the knowledge generated by university laboratories is the least targeted because it tends to be the most oriented toward basic science and is typically conducted with little regard for its exploitation by manufacturing firms. The knowledge generated by government laboratories also tends to be basic scientific, although many government laboratories also concern themselves with applied problems. In addition, some government laboratories work quite closely with, and are often managed by, large manufacturing firms. Thus, as the knowledge from each of these four sources is scored as 'more important' to a given industry, the magnitude of their respective effects should be ranked in ascending order as follows: 1. *EQUIPTECH*; 2. *MATERIALTECH*; 3. *GOVTECH*; 4. *UNIVTECH*.

It is difficult to predict the ordinal rank of *USERTECH*, the importance of knowledge from users, for two reasons. First, as von Hippel (1978) suggests, users will often provide a product idea to potential suppliers, but the informativeness of the 'solution concept' is quite variable. Therefore, the targeted quality of the information is variable as well. Second, *USERTECH* may actually reflect something other than technological opportunity because this particular source of knowledge, users, is also an important source of information on market demand. *USERTECH* may, therefore, represent some dimension of demand conditions.

Appropriability

Having considered how the ease of learning affects the influence of technological opportunity on R&D spending, we now consider how an endogenous absorptive capacity shapes the way in which appropriability conditions influence R&D spending. To represent the level of intra-industry spillovers, θ, we use a measure drawn from the Levin *et al.* survey. Respondents were asked to rate (on a seven-point scale) the effectiveness of six mechanisms used by firms to capture and protect the competitive advantages of new processes and new products.[28] For a line of business, *APPROPRIABILITY* is the maximum score received by any one of these mechanisms for either process or product innovations. Thus, if *APPROPRIABILITY* increases, the spillover level declines.[29]

Our general model predicts that as β increases, the negative appropriability incentive associated with spillovers diminishes relative to the positive absorption incentive, implying that the coefficient of *APPROPRIABILITY* should decline. Our more specialised model predicts that the coefficient of *APPROPRIABILITY* should also decline as the price elasticity of demand and the number of firms in the industry increase. To test these hypotheses, we interact measures of each of these three variables with *APPROPRIABILITY*. We should, however, be cautious in interpreting our results, since our prediction of the effects of industry structure and the demand elasticity emerge from our specialised model that, while directly considering process R&D, may be interpreted to also consider product R&D only under restrictive assumptions.

As an index of the number of firms in an industry, we employ four-firm concentration ratios ($C4$) taken from the 1977 Census of Manufactures. When necessary, these concentration ratios are aggregated to the FTC's line of business level using the value of shipments as weights. Since $C4$ will tend to diminish as the number of firms in the industry increases, we hypothesise a positive coefficient on the interaction of $C4$ and *APPROPRIABILITY*. To represent the price elasticity of demand, we employ price elasticities (*PELAS*) developed by Levin (1981), and predict a negative effect of the interaction of *APPROPRIABILITY* and *PELAS*.

The representation of β is a more difficult matter. We distinguish categorically among industries on the basis of whether basic science or applied science is, on average, more relevant, and assume that the targeted quality of extramural knowledge is less in the former case. Two dummy variables, *DUMAPP* and *DUMBAS*, denote those industries for which the applied or the basic sciences are, respectively, the more relevant. Thus *DUMAPP* designates those industries for which β is relatively low, and *DUMBAS* designates those for which β is relatively high. Our analysis therefore suggests that the coefficient of *APPROPRIABILITY* × *DUMAPP* will exceed that of *APPROPRIABILITY* × *DUMBAS*.

Other determinants

To evaluate the role played by our featured variables, we control for other conventionally considered determinants of R&D intensity. Another variable which may reflect a dimension of technological opportunity is industry

[28] These mechanisms are patents to prevent duplication, patents to secure royalty income, secrecy, lead time, moving quickly down the learning curve, and complementary sales and service efforts (Levin *et al.*, 1983, Questions I.A and I.B).

[29] In terms of our theory, *APPROPRIABILITY* represents $(1-\theta)$. The reader should note that while spillover effects, reflecting the interaction of θ and γ in our model, are endogenous, $(1-\theta)$, or *APPROPRIABILITY* in the empirical model, is not.

maturity. We employ *NEWPLANT* to reflect the relative maturity of an industry's technology. It measures the percentage of an industry's property, plant, and equipment installed within the five years preceding 1977, as reported to the FTC's Line of Business Programme.

To represent industry demand conditions, we use industry estimates, previously developed by Levin (1981), of price elasticity (*PELAS*) and income elasticity (*INCELAS*), and a time shift parameter (*DGROWTH*). We expect demand growth and income elasticity to be positively associated with R&D intensity, but the expected effect of the price elasticity of demand is ambiguous. Elastic demand should provide a positive incentive to invest in process R&D, while an inelastic demand may encourage product R&D by magnifying the returns from a rightward shift in the demand curve.[30]

III.3 *Estimation Issues*

To estimate our specifications, three problems are addressed. First, using a Breusch-Pagan statistic to test for heteroscedasticity, we could not always accept the null hypothesis of homoscedasticity. Where heteroscedasticity was evident, we found the error structure to be one of 'multiplicative heteroscedasticity,' in which the logarithm of the error variance is a linear function of the exogenous variables and the number of respondents to the Levin *et al.* survey question in the associated line of business. In these instances we followed the procedure suggested by Harvey (1976) to obtain asymptotically efficient GLS estimates of the parameters. Breusch-Pagan statistics were then calculated for each transformed specification, and in no case can we reject the null hypothesis of homoscedasticity at the 0·05 level.

In addition to heteroscedasticity, there is a second estimation problem. Of the business unit observations in our sample, 24% perform no R&D. If the independent variables in our model affect both the probability of conducting R&D and the amount of R&D spending, then estimating our specification on a sample that includes only R&D performing business units will bias the resulting parameter estimates due to truncation of the error term. The Tobit model addresses this problem, but at the cost of the possible specification error introduced by restricting the way in which the explanatory variables simultaneously determine the probability of performing R&D and the amount of R&D spending. Indeed, it may be the case that the variables reflecting the ease of learning may not affect the probability of engaging in R&D, but only affect R&D performance given that a firm is already conducting R&D. The reason is that a firm may require some initial level of absorptive capacity before it can recognise the characteristics of the information environment relevant to its R&D decision. For example, the findings of university labs may have no effect on a firm's propensity to conduct R&D if, without the requisite

[30] One may think that another variable, industry concentration, should also be considered as an independent determinant of R&D intensity. However, we omit industry concentration because it has been shown (Levin *et al.*, 1985) that once more fundamental determinants, such as appropriability and technological opportunity conditions, are controlled for, industry concentration appears to exercise no influence on R&D intensity. Indeed, when concentration is added to our benchmark specification presented in Table 2 below, we find the same result.

absorptive capacity, it cannot recognise their value. Thus, the firm must already be an R&D performer for such factors to have any effect. In this event, the sample should be confined to only the R&D performers, and OLS (or, if required, GLS), rather than Tobit, is preferred assuming that any unobserved variables affecting the likelihood of performing R&D are not correlated with the error term. In light of the uncertainty with respect to the appropriate estimation model, we explore the robustness of our results by presenting the OLS and GLS coefficient estimates for the R&D performing business units only, and the Tobit coefficient estimates for the full sample which also includes non-performers.[31]

The third estimation problem is posed by the likely endogeneity of market concentration with respect to R&D spending.[32] Although concentration *per se* is not featured, we do examine the role of the interaction of the four-firm concentration ratio ($C4$) with *APPROPRIABILITY*. Expecting this interactive variable to be endogenous, we employ a two stage least squares procedure. We construct an instrument by regressing $APPROPRIABILITY \times C4$ on the right-hand-side variables plus other variables thought to affect concentration.[33] Based on Lee (1981), we also use the predicted value from this regression as an instrument in our Tobit estimation.

III.4 *Results*

Table 2 presents the OLS, GLS, and Tobit estimates of the effects of the nature of knowledge inputs and other industry characteristics on business unit R&D intensity. In general, the results confirm our hypotheses.

Technological Opportunity

First, the coefficients (α_8–α_{18}) of the eleven variables representing the technological opportunity associated with the scientific fields are jointly significant at the 0·01 level across all three estimation methods.[34] With regard to the role of learning, we can reject across all three estimation methods the hypothesis that the effects on R&D spending of the eleven applied and basic science variables are equal. This finding is consistent with our hypothesis that there exist differences in characteristics of the fields, such as pace of advance, cumulativeness, and targetedness, that affect the ease of learning, and, in turn, the influence of technological opportunity on R&D spending.

The results also confirm the related hypothesis that increasing the technological opportunity associated with the relatively less targeted basic

[31] The OLS/GLS coefficients and the Tobit coefficients are not directly comparable because the latter should be interpreted in our model as a weighted average of two effects: (1) the effect of an increase in an independent variable on the probability that the dependent variable, R&D intensity, is greater than zero and (2) the effect on the expected value of R&D intensity given that it is above zero. If one assumes that Tobit is the correct estimation procedure, then the OLS coefficients should provide a biased estimate of the second effect. However, we are making no such assumption here.

[32] Suggested by Schumpeter's (1942) notion of creative destruction, the endogeneity of market concentration is also highlighted by Phillips (1971), and Levin and Reiss (1988), among others.

[33] These include minimum efficient scale and capital intensity.

[34] For the relevant F, Chi-square, and Wald statistics, see the bottom of the table under discussion. In this case, see Table 2.

Table 2

Effects of knowledge and industry characteristics on R&D intensity

Parameter	Variable/hypothesis	Regression coefficient (standard error)		
		OLS (1)	GLS (2)	Tobit (3)
α_1	*INTERCEPT*	−4·226**	−2·016*	−3·391**
		(1·348)	(0·945)	(1·261)
α_2	*APPROPRIABILITY*	0·396*	0·360**	0·260
		(0·156)	(0·104)	(0·161)
α_3	*USERTECH*	0·387**	0·409**	0·510**
		(0·099)	(0·070)	(0·106)
α_4	*UNIVTECH*	0·346**	0·245**	0·321*
		(0·128)	(0·089)	(0·147)
α_5	*GOVTECH*	0·252*	0·170*	0·200*
		(0·100)	(0·076)	(0·100)
α_6	*MATERIALTECH*	−0·315**	−0·198**	−0·369**
		(0·096)	(0·070)	(0·097)
α_7	*EQUIPTECH*	−0·392**	−0·462**	−0·570**
		(0·111)	(0·079)	(0·115)
α_8	*BIOLOGY*	0·176	0·042	0·159
		(0·096)	(0·057)	(0·116)
α_9	*CHEMISTRY*	0·195**	0·095	0·149
		(0·071)	(0·050)	(0·078)
α_{10}	*MATH*	0·146	0·131	0·091
		(0·117)	(0·093)	(0·122)
α_{11}	*PHYSICS*	0·189	0·037	0·156
		(0·109)	(0·082)	(0·109)
α_{12}	*AGRICULTURAL SCIENCE*	−0·373**	−0·253**	−0·259*
		(0·084)	(0·055)	(0·101)
α_{13}	*APPLIED MATH/OPERATIONS RESEARCH*	−0·220	−0·001	−0·325*
		(0·135)	(0·099)	(0·136)
α_{14}	*COMPUTER SCIENCE*	0·336**	0·157	0·446**
		(0·123)	(0·093)	(0·121)
α_{15}	*GEOLOGY*	−0·301**	−0·228**	−0·327**
		(0·082)	(0·058)	(0·095)
α_{16}	*MATERIALS SCIENCE*	−0·005	−0·028	0·231*
		(0·121)	(0·089)	(0·116)
α_{17}	*MEDICAL SCIENCE*	−0·076	−0·042	−0·036
		(0·089)	(0·062)	(0·106)
α_{18}	*METALLURGY*	−0·244**	−0·169**	−0·322**
		(0·075)	(0·052)	(0·085)
α_{19}	*NEWPLANT*	0·055**	0·041**	0·042**
		(0·008)	(0·006)	(0·007)
α_{20}	*PELAS*	−0·180**	−0·071	−0·147*
		(0·061)	(0·044)	(0·060)
α_{21}	*INCELAS*	1·062**	0·638**	1·145**
		(0·170)	(0·136)	(0·180)
α_{22}	*DGROWTH*	0·060	−0·059	−0·018
		(0·090)	(0·052)	(0·106)
	H_0: *FIELDS* = 0	F(11, 1280)	χ^2 (11)	χ^2 (11)
	($\alpha_8, \alpha_9, \alpha_{10}, \ldots, \alpha_{18} = 0$)	9·341**	8·575**	99·059**
	$H_0: \alpha_8 = \alpha_9 = \alpha_{10} = \ldots = \alpha_{18}$	F(10, 1280)	χ^2 (10)	χ^2 (10)
		9·987**	9·287**	98·839**

Continued overleaf

Table 2 (*cont.*)

Variable/hypothesis	OLS (1)	GLS (2)	Tobit (3)
H_0: *EXTRA-INDUSTRY KNOWLEDGE* = 0	F(5, 1280) 17·433**	χ^2 (5) 25·615**	χ^2 (5) 90·988**
$(\alpha_3, \alpha_4, \alpha_5, \alpha_6, \alpha_7 = 0)$			
$H_0: \alpha_3 = \alpha_4 = \alpha_5 = \alpha_6 = \alpha_7$	F(4, 1280) 19·912**	χ^2 (4) 29·194**	χ^2 (4) 89·782**
$H_0: \alpha_3 = \alpha_4$	F(1, 1280) 0·054	χ^2 (1) 1·894	χ^2 (1) 0·948
$H_0: \alpha_4 = \alpha_5$	F(1, 1280) 0·230	χ^2 (1) 0·291	χ^2 (1) 0·322
$H_0: \alpha_5 = \alpha_6$	F(1, 1280) 19·844**	χ^2 (1) 14·403**	χ^2 (1) 20·32**
$H_0: \alpha_6 = \alpha_7$	F(1, 1280) 0·200	χ^2 (1) 4·856*	χ^2 (1) 1·335
R^2	0·278		
n	1,302	1,302	1,719

* Significant at the 0·05 level.
** Significant at the 0·01 level.

sciences elicits more R&D spending than does increasing the technological opportunity associated with the applied sciences. A comparison of the coefficient magnitudes of the basic (α_8, α_9, α_{10}, α_{11}, α_{15}) versus the applied (α_{12}, α_{13}, α_{14}, α_{16}, α_{17}, α_{18}) sciences in Table 2 reveals that, with the sole exceptions of the coefficient of *GEOLOGY* (α_{15}) among the basic sciences and *COMPUTER SCIENCE* (α_{14}) among the applied sciences, the coefficients are uniformly greater for the basic sciences.[35] Furthermore, one could readily argue that, although classed as a basic science by the Levin *et al.* survey, geology is as targeted as the applied fields considering its inductive methodology and intensive use by firms in the extractive sector. The other major exception to the predicted pattern, *COMPUTER SCIENCE*, reminds us that an implicit premise of this hypothesised pattern is that field characteristics affecting β other than their targeted quality do not vary. However, casual empiricism suggests that relative to the other applied fields, the pace of advance of computer science has been rapid. Thus, it is not surprising that computer science is the major exception to the postulated pattern of coefficient values if a field's pace of advance also affects β, as suggested above.

To probe further our observation that the coefficient values of the less targeted, basic science field variables exceed those of the applied fields, we estimated a specification, otherwise identical to that of Table 2, in which we constrained the coefficients of the basic sciences, excluding geology, to be the same, and the coefficients of the applied sciences, including geology, to be the same. Thus, we are examining the effect on R&D spending as the overall technological opportunity associated with basic science and applied science, respectively, change. The constrained GLS coefficient estimate of the effect of

[35] In column (3) presenting the Tobit results, we find one additional exception to this pattern, *MATERIALS SCIENCE*.

the technological opportunity associated with the basic sciences equals 0·130, and that of the applied sciences equals −0·049.[36] The Tobit coefficient estimates are 0·182 and −0·115, respectively. We can always reject the null hypothesis of the equality of these coefficients, confirming that, relative to the effect of an increase in the technological opportunity associated with applied science, an increase in that associated with basic science elicits more R&D.

Our predicted ranking of the coefficient magnitudes associated with the extra-industry sources of knowledge (*UNIVTECH*, *GOVTECH*, *MATERIALTECH*, and *EQUIPTECH*) is largely confirmed. Table 2 shows that, across all estimation methods, the coefficient estimate for *UNIVTECH* is greater than that of *GOVTECH*, which exceeds that of *MATERIALTECH*, which, in turn, exceeds that of *EQUIPTECH*. The appropriate F, Chi-square, and Wald test statistics, presented at the bottom of Table 2, indicate whether the coefficients differ significantly from one another. The coefficients of *UNIVTECH* and *GOVTECH* do not differ significantly. The coefficient estimate of *GOVTECH* is, however, significantly higher than that of *MATERIALTECH* across all estimation methods. In the GLS results the coefficient estimate of *MATERIALTECH* is significantly greater than that of *EQUIPTECH*.

Although we had no priors with respect to the ordinal rank of the coefficient of *USERTECH*, its coefficient estimates are nonetheless surprising. For all three sets of results, *USERTECH* had the highest coefficient estimate, and was always significantly different from zero. *USERTECH* may, however, be representing some dimension of demand conditions, as suggested above. For example, a high score for *USERTECH* may indicate that information from buyers is important when product differentiation is a critical dimension of competition, and that, as Comanor (1967) suggests, when product differentiation is important, firms tend to do more R&D.[37]

The coefficient patterns for both the scientific field variables and the extra-industry knowledge source variables are predicted by our hypothesis that an increase in technological opportunity (i.e., f or T) elicits greater R&D spending the higher is β. Is there, however, another explanation? If we assume that the values of the relevance scores for a given field of science or knowledge source have little effect on own R&D productivity, and the main effect occurs *across* the technological opportunity variables, perhaps the results simply indicate that the marginal productivity of R&D is greater as the underlying

[36] The OLS estimates yield the identical qualitative results. Also, constraining the basic science variables and the applied science variables to have the same coefficient values raises the question whether, beyond this simple distinction, do individual field characteristics add any explanatory power. To address this, we tested whether the coefficient values across the applied science variables equal one another, and the coefficient values across the basic science variables equal one another. With geology considered as a basic science, we can reject both null hypotheses at the 0·05 confidence level. Treating geology as an applied science changes the result. We then cannot reject the null hypothesis that the coefficients of the basic science variables are equal, but can reject, at the 0·01 level, the null hypothesis that the applied science variables exercise the same effect.

[37] To probe the relationship between *USERTECH* and product differentiation, we computed the correlation coefficient between it and two variables, also drawn from the Levin *et al.* survey, that measure on a seven-point Likert scale the importance in each line of business of, respectively, 'improving the physical properties of the product,' and 'improving the performance characteristics of the product.' The correlation coefficients are, respectively, 0·23 and 0·28, and each is significant at the 0·01 level.

knowledge base is more basic scientific in character.[38] First, aside from the hazard of ignoring the variation in technological opportunity associated with each field or knowledge source, it is far from obvious that basic science will have a greater effect on R&D productivity at the margin than applied science. Second, even if this were so, the observed ordinal ranking of coefficients would not necessarily be the prediction. For example, if we equate outside knowledge's productivity impact with f, the theoretical analysis above indicates that the effect of f on own R&D is ambiguous, and consequently predicts no ordinal ranking of the sort observed. Alternatively, since the results indicate that outside applied science tends to substitute for internal R&D while basic science tends to complement it, perhaps they simply show that extramural basic research is more complementary to internal R&D than applied research.[39] We are left, however, with the question of why external applied research acts as a substitute while more basic research acts as a complement. We suggest that because basic science is less targeted to the needs and concerns of the firm, a firm must invest more to assimilate and exploit it.

Appropriability

In Table 2, the OLS and GLS estimate of the coefficient of *APPROPRIABILITY* is positive and significant, suggesting spillovers have a negative effect on R&D. Though also positive, the Tobit coefficient estimate is not significant. These results suggest a negative *net* effect of spillovers. They do not, however, disconfirm our hypothesis that spillovers generate countervailing appropriability and absorption incentives. We therefore consider whether, as predicted, the positive absorption incentive increases relative to the negative appropriability incentive as β, the price elasticity of demand, and the number of firms in the industry rise.

Table 3 presents the coefficient estimates of the interactions between *APPROPRIABILITY* and, respectively, market concentration ($C4$), the price elasticity of demand (*PELAS*), and the two dummy variables representing β (*DUMAPP* and *DUMBAS*). The results largely confirm that the ease of learning conditions the effect of *APPROPRIABILITY* as hypothesised. First, the four interactions are jointly significant across all estimation methods. Second, the hypothesis that the coefficient of *APPROPRIABILITY* × *DUMAPP* should exceed that of *APPROPRIABILITY* × *DUMBAS* holds across all estimation methods, and the difference is always significant. Therefore, the

[38] More formally, this argument assumes that, for example, f varies across fields or knowledge sources, but not with the associated relevance scores, and that the results reflect the direct effect, $(\partial M^*/\partial f)$, not the cross partial, $(\partial^2 M^*/\partial f \partial \beta)$ and therefore do not reflect the influence of the ease of learning.

[39] Because our central hypotheses with respect to the technological opportunity variables concern the ordinal ranking of their coefficients, we have not considered their signs. We do observe, however, that the signs of the applied fields are generally negative and those of the basic fields are generally positive, and the signs of *MATERIALTECH* and *EQUIPTECH* are negative, while those of *UNIVTECH* and *GOVTECH* are positive. Consistent with these findings, our theoretical analysis suggests that the sign of the direct effect of technological opportunity (either f or T in this model) on R&D may vary. If we interpret the respective technological opportunity variables as representing f, then a negative effect may be realised if the negative appropriability incentive effect is relatively large while β is relatively small. Alternatively, if these variables represent T, then diminishing returns to knowledge may account for negative signs, again assuming that β is relatively small. In either case, a higher value of β will tend to be associated with a positive coefficient.

Table 3

Effect of appropriability interactions and knowledge and industry characteristics on R&D intensity

Parameter	Variable/ Hypothesis	Regression coefficient (standard error) OLS (1)	GLS (2)	Tobit (3)
α_1	INTERCEPT	−5·184** (1·522)	−2·355* (1·037)	−4·086** (1·461)
α_2	*APPROPRIABILITY × C4*	0·213 (0·128)	0·342** (0·103)	0·368** (0·130)
α_3	*APPROPRIABILITY × PELAS*	−0·192 (0·106)	−0·200* (0·091)	−0·176 (0·103)
α_4	*APPROPRIABILITY × DUMAPP*	0·448* (0·202)	0·248 (0·143)	0·211 (0·194)
α_5	*APPROPRIABILITY × DUMBAS*	0·302 (0·208)	0·174 (0·144)	0·094 (0·206)
α_6	*USERTECH*	0·470** (0·104)	0·397** (0·069)	0·612** (0·107)
α_7	*UNIVTECH*	0·374** (0·131)	0·318** (0·091)	0·395** (0·147)
α_8	*GOVTECH*	0·221* (0·106)	0·069 (0·079)	0·137 (0·107)
α_9	*MATERIALTECH*	−0·258** (0·098)	−0·074 (0·070)	−0·303** (0·100)
α_{10}	*EQUIPTECH*	−0·401** (0·111)	−0·484** (0·077)	−0·574** (0·117)
α_{11}	*BIOLOGY*	0·314** (0·102)	0·185** (0·071)	0·276* (0·114)
α_{12}	*CHEMISTRY*	0·289** (0·084)	0·081 (0·062)	0·191* (0·088)
α_{13}	*MATH*	0·184 (0·131)	0·151 (0·097)	0·123 (0·143)
α_{14}	*PHYSICS*	0·373** (0·117)	0·323** (0·091)	0·310* (0·128)
α_{15}	*AGRICULTURAL SCIENCE*	−0·441** (0·088)	−0·273** (0·064)	−0·308** (0·099)
α_{16}	*APPLIED MATH/OPERATIONS RESEARCH*	−0·237 (0·148)	−0·117 (0·102)	−0·366* (0·152)
α_{17}	*COMPUTER SCIENCE*	0·294* (0·124)	0·116 (0·090)	0·433** (0·122)
α_{18}	*GEOLOGY*	−0·363** (0·084)	−0·240** (0·061)	−0·365** (0·097)
α_{19}	*MATERIALS SCIENCE*	−0·110 (0·125)	−0·150 (0·095)	0·116 (0·118)
α_{20}	*MEDICAL SCIENCE*	−0·179 (0·093)	−0·133 (0·070)	−0·133 (0·103)
α_{21}	*METALLURGY*	−0·315** (0·077)	−0·195** (0·053)	−0·393** (0·089)
α_{22}	*NEWPLANT*	0·057** (0·008)	0·049** (0·006)	0·045** (0·007)
α_{23}	*PELAS*	0·936 (0·611)	1·082* (0·527)	0·892 (0·573)
α_{24}	*INCELAS*	1·077** (0·170)	0·587** (0·131)	1·112** (0·188)

Table 3 (*cont.*)

Parameter	Variable/ Hypothesis	Regression coefficient (standard error) OLS (1)	GLS (2)	Tobit (3)
α_{25}	*DGROWTH*	0·068	−0·074	0·004
		(0·090)	(0·053)	(0·105)
	$H_0: \alpha_2, \alpha_3, \alpha_4, \alpha_5 = 0$	F(4, 1277)	$\chi^2(4)$	$\chi^2(4)$
		5·773**	7·367**	18·783**
	$H_0: \alpha_4 = \alpha_5$	F(1, 1277)	$\chi^2(1)$	$\chi^2(1)$
		10·495**	5·169*	5·545*
	R^2	0·287		
	n	1,302	1,302	1,719

* Significant at the 0·05 level.
** Significant at the 0·01 level.

positive absorption incentive associated with spillovers appears to increase relative to the negative appropriability incentive in industries where β is higher. Third, as predicted, the effect of *APPROPRIABILITY* × *C4* is positive across all estimation methods, and is significant in the GLS and Tobit estimations. Finally, the coefficient of *APPROPRIABILITY* × *PELAS* is always negative as predicted, but is significant only in the GLS estimate.[40]

As suggested above, the predictions on the effects of the interactions of *APPROPRIABILITY* with concentration and price elasticity are derived from a model of process innovation.[41] The dependent variable in the Table 3 specification, however, is total business unit R&D intensity. Therefore, to explore the robustness of our results, we replace our dependent variable with an approximation of business unit process R&D intensity and re-estimate the equation.[42] We do not feature this specification, however, because adjusting our dependent variable comes at the cost of introducing measurement error into the majority of our right-hand-side variables. While the Levin *et al.* survey permits us to measure *APPROPRIABILITY* for process innovations alone, none of the measures of technological opportunity distinguish between process and product innovation. This introduces measurement error if respondents' evaluations of technological opportunity are weighted largely by their views characterising product, rather than process innovation, which may be expected since most R&D in American manufacturing is product R&D.[43] Moreover, we

[40] It is, however, significant at the 0·10 level in both the Tobit and OLS results.

[41] In contrast, the effect of the interaction of our proxy variable for β is derived from the general model.

[42] To approximate business unit process R&D intensity, we multiply total business unit R&D intensity by the share of each industry's innovations represented by process innovations. The estimates of each industry's share of process innovations were developed by F. M. Scherer using patent data. Scherer (1982, 1984) classified all US patents granted within a 15 month period in the mid-1970's by industry of origin and industry of use, and assumed that process innovations are those represented by patents used in their industry of origin. We use this data to divide an industry's R&D expenditure between process and product R&D assuming that each industry devotes to process innovation a percentage of R&D equal to the percentage of its patents assigned to processes. As noted in Cohen and Levin (1989), this assumption is suspect, however, to the extent that process innovations are less likely to be patented than product innovations.

[43] Employing Scherer's measures of the shares of process and product innovation, we find that approximately 29% of the business unit R&D in our sample is dedicated to process innovation, and 71% is dedicated to product innovation.

cannot reasonably assume that the technological opportunity facing product innovation is the same as that facing process innovation within a line of business.

Despite these difficulties, using process R&D intensity as the dependent variable scarcely changes the qualitative results of interest. The coefficient of *APPROPRIABILITY* × *PELAS* remains negative, but is no longer significant. The coefficient of *APPROPRIABILITY* × *C4* becomes significantly positive across all three estimates. The coefficient of *DUMAPP* remains significantly greater than that of *DUMBAS* in the OLS and GLS specifications.[44] However, consistent with the expected introduction of measurement error into the technological opportunity measures, the F statistics testing the joint significance of the technological opportunity variables decline, and the variance explained by the equation drops by more than half.

As predicted, increases in β, the price elasticity of demand, and the number of firms in the industry appear to diminish the coefficient of *APPROPRIABILITY*, reflecting a relative increase in the positive absorption incentive associated with spillovers. The question remains, however, whether spillovers may, on balance, actually encourage R&D in some industries. To explore this possibility, we examine the effect of *APPROPRIABILITY* in the four two-digit SIC code level industries for which our sample contains enough lines of business to permit separate industry regressions.[45] These include SICs 20 (food processing), 28 (chemicals), 35 (machinery), and 36 (electrical equipment). The OLS and GLS estimates show that in SIC's 28 and 36, the effect of *APPROPRIABILITY* is negative and significant, implying that R&D intensity rises with spillovers.[46] It is interesting to note that SIC 36 is the two-digit industry group in which the semiconductor industry is found, the industry whose coincidence of high spillovers and high R&D perplexed Spence (1984).[47] Most importantly, we obtain these results after controlling for other industry-level variables conventionally thought to drive R&D spending, including technological opportunity and demand conditions. Although a negative effect of spillovers in these two industry groups does not represent a direct test of our model, it does suggest, particularly when considered with our Table 3 results, that the positive absorption incentive associated with spillovers may be

[44] In the Tobit estimation, the difference is significant at the 0·10 level.

[45] The technical appendix available from the authors provides a detailed discussion of the specification, the estimation procedure, and the results for these two-digit level industry regressions.

[46] In the Tobit results, the sign of *APPROPRIABILITY* is also negative for SIC's 28 and 36, but the coefficient estimates are not quite significant at the 0·05 level.

Our explanation for a negative coefficient on *APPROPRIABILITY* is that firms invest in R&D to build the absorptive capacity that permits them to assimilate and exploit spillovers. An alternative explanation is that the immediate R&D effort associated with the process of imitation in these industries is considerable, and we are simply observing a lot of costly imitation. However, if such immediate imitation costs are largely a negative function of prior investment in absorptive capacity, as we would suggest, this second argument is a corollary of our own. Also, this argument alone would not explain the interaction effects observed in Table 3.

[47] Bernstein (1988) also found a positive effect of spillovers on R&D intensity for the Canadian electrical products and chemicals industries. Bernstein and Nadiri (1989), however, found no such positive effect in their study of spillovers in American manufacturing.

sufficiently strong in some cases to more than offset the negative appropriability incentive.

IV. CONCLUSION

We have argued that firms invest in R&D not only to pursue directly new process and product innovation, but also to develop and maintain their broader capabilities to assimilate and exploit externally available information. Recognition of the dual role of R&D suggests that factors that affect the character and ease of learning will affect firms' incentives to conduct R&D. We hypothesised that such factors include the degree to which knowledge is targeted to a firm's needs, and, more generally, the character of knowledge within each of the scientific and technological fields upon which innovation depends. Our analytic model suggests these factors should both exercise a direct effect on inventive activity, and condition the influence of more conventionally considered determinants.

Aside from market demand, applied economists have identified technological opportunity and appropriability as the principal industry-level determinants of the firm's inventive activity (Cohen and Levin, 1989). We suggest, however, that since the effect of both of these classes of variables depend importantly upon the assimilation of external knowledge, their influence is mediated by the firm's capacity to recognise, assimilate, and exploit information. Therefore, those variables that affect the ease and character of learning should, in turn, influence how technological opportunity and appropriability conditions influence R&D spending. The empirical results generally confirm that the influence of both appropriability and technological opportunity conditions is affected by determinants of the ease of learning, particularly the targeted quality of knowledge inputs. This, in turn, suggests that the characteristics of knowledge that affect the ease of firm learning may represent an important class of determinants of R&D investment.

The observation that R&D creates a capacity to assimilate and exploit new knowledge sheds new light on a range of questions. For example, it provides a ready explanation of why some firms may invest in basic research even when the preponderance of findings spill out into the public domain. Specifically, firms may conduct basic research less for particular results than to be able to identify and exploit potentially useful scientific and technological knowledge generated by universities or government laboratories, and thereby gain a first-mover advantage in exploiting new technologies. Likewise, basic research may permit firms to act as a rapid second mover in the face of spillovers from a competitor's innovation. This perspective implies that variables that influence the firm's incentives to learn should affect the incentives to conduct basic research. For example, as automobile manufacturing depends increasingly on fields that draw heavily on basic science, such as microelectronics and ceramics, we expect that manufacturers will expand their basic research efforts in physics and chemistry to evaluate and exploit new findings in these areas. Similarly, as a firm's technological progress depends upon an increasing number of fields of basic science, a firm will increase its basic research as it mounts efforts in each field. Thus, even if a firm is not widely diversified, but the knowledge inputs relevant to technical advance become more varied, we may expect an increase in basic research. In contrast, Nelson (1959) hypothesises that firms which are more diversified in terms of product markets will invest more heavily in basic research because, assuming imperfect markets for information, they will be better able to exploit the potentially wide-ranging findings.

Recognition of the dual role of R&D also offers important implications for the analysis of the adoption and diffusion of innovations. Although the importance of various forms of learning have been highlighted in the diffusion literature, it has not been widely appreciated in this context that a firm's R&D contributes to its ability to learn.[48] Our perspective implies that the ease of learning, and, thus, technology adoption, is affected by the character of the knowledge inputs in question. For example, we would suggest that an innovation which is purely capital embodied is less costly to adopt than more disembodied innovations that require more complementary internal effort, and more pre-existing expertise in an area. We also conjecture that a product innovation developed on the basis of a well established underlying knowledge base will diffuse more rapidly among users than one grounded in a more recently developed body of scientific or technological knowledge. Finally, R&D's contribution to absorptive capacity also implies that, particularly for new intermediate goods that require substantial complementary R&D effort on the part of adopters, the R&D expenditures of the innovating industry and those of the adopting industry are jointly determined in the long run.

In addition to offering theoretical and empirical implications, our analysis suggests a lesson for technology policy. Economists have long cautioned policymakers about the welfare costs of policies, such as patents, that curtail the negative incentive effects of intra-industry spillovers by conferring monopoly power. Recently, Spence (1984) has highlighted a cost of a different sort. By increasing the appropriability of the rents due to new knowledge, society foregoes the socially beneficial efficiency effects of spillovers associated with the diminution of redundant R&D effort. Our analysis of the role that R&D plays in firm learning adds another dimension to the evaluation of the welfare effects of patents and similar policies. In particular, it implies that the negative incentive effects of spillovers and, thus, the benefits of policies designed to mitigate these effects, may not be as great as supposed.

Carnegie Mellon University

Date of receipt of final typescript: February 1989

[48] An important exception is Mansfield *et al.* (1977) whose finding that diffusion occurs more rapidly in more R&D intensive industries supports our general argument. See Thirtle and Ruttan (1987) and Stoneman (1987) for recent reviews of the literature on adoption and diffusion. Both of these works briefly describe the way in which the notion of learning has been treated in this literature.

References

Allen, T. J. (1977). *Managing the Flow of Technology*. Cambridge: MIT Press.

Arrow, K. J. (1962). 'Economic welfare and the allocation of resources for invention,' in (R. R. Nelson, ed.) *The Rate and Direction of Inventive Activity*. Princeton: Princeton University Press.

Bernstein, J. L. (1988). 'Costs of production, intra and inter-industry R&D spillovers: Canadian evidence,' *Canadian Journal of Economics*, vol. 21, pp. 324–47.

—— and Nadiri, M. I. (1989). 'Research and development and intraindustry spillovers: an empirical application of dynamic duality,' *Review of Economic Studies*, vol. 56, pp. 249–69.

Brock, G. W. (1975). *The U.S. Computer Industry*. Cambridge: Ballinger.

Cohen, W. M. and Levin, R. C. (1989). 'Empirical studies of innovation and market structure.' In (R. C. Schmalensee and R. Willig, Eds.) *Handbook of Industrial Organization Vol. II*. Amsterdam: Elsevier Science Publishers, B.V.

——, —— and Mowery, D. C. (1987). 'Firm size and R&D intensity: a re-examination,' *Journal of Industrial Economics*, vol. 35, pp. 543–63.

—— and Mowery, D. C. (1984). 'The internal characteristics of the firm and the level and composition of research and development spending: interim report, NSF grant PRA 83–10664,' Mimeo, Carnegie-Mellon University.

Comanor, W. S. (1967). 'Market structure, product differentiation, and industrial research,' *Quarterly Journal of Economics*, vol. 81, pp. 639–57.

Dixit, A. (1986). 'Comparative statics for oligopoly,' *International Economic Review*, vol. 27, pp. 107–22.

Evenson, R. E. and Kislev, Y. (1973). 'Research and productivity in wheat and maize,' *Journal of Political Economy*, vol. 81, pp. 1309–29.

Hamberg, D. (1963). 'Invention in the industrial research laboratory,' *Journal of Political Economy*, vol. 71, pp. 95–115.

Harvey, A. C. (1976). 'Estimating regression models with multiplicative heteroscedasticity,' *Econometrica*, vol. 44, pp. 461–5.

Johnston, R. and Gibbons, M. (1975). 'Characteristics of information usage in technological innovation,' *IEEE Transactions on Engineering Management*, vol. EM-22, pp. 27–34.

Lee, L. F. (1981). 'Simultaneous equations models with discrete and censored variables.' In (C. F. Manski and D. McFadden, eds.) *Structural Analysis of Discrete Data with Econometric Applications*. Cambridge: The MIT Press.

Levin, R. C. (1981). 'Toward an empirical model of Schumpeterian competition,' Yale University, Working Paper No. 43.

——, Cohen, W. M. and Mowery, D. C. (1985). 'R&D appropriability, opportunity, and market structure: new evidence on some Schumpeterian hypotheses,' *American Economic Review Proceedings*, vol. 75, pp. 20–4.

——, Klevorick, A. K., Nelson, R. R. and Winter, S. G. (1983). 'Questionnaire on industral research and development,' Yale University, mimeo.

——, ——, —— and —— (1987). 'Appropriating the returns from industrial R&D,' *Brookings Papers on Economic Activity*, pp. 783–820.

—— and Reiss, P. C. (1988). 'Cost-reducing and demand-creating R&D with spillovers,' *Rand Journal of Economics*, vol. 19, pp. 538–56.

Lieberman, M. B. (1984).'The learning curve and pricing in the chemical processing industries,' *Rand Journal of Economics*, vol. 15, pp. 213–28.

Mansfield, E., Rapoport, J., Romeo, A., Wagner, S. and Husic, F. (1977). *The Production and Application of New Industrial Technology*. New York: W. W. Norton.

Mowery, D. C. (1983). 'The relationship between intrafirm and contractual forms of industrial research in American manufacturing, 1900–1940,' *Explorations in Economic History*, vol. 20, pp. 351–74.

Mueller, W. F. (1962). 'The origins of the basic inventions underlying DuPont's major product and process innovations, 1920 to 1950.' In (R. R. Nelson, ed.) *The Rate and Direction of Inventive Activity*. Princeton: Princeton University Press.

Nelson, R. R. (1959). 'The simple economics of basic research,' *Journal of Political Economy*, vol. 67, pp. 297–306.

Pavitt, K. (1987). 'The objectives of technology policy,' *Science and Public Policy*, vol. 14, pp. 182–8.

Peck, M. J. (1962). 'Inventions in the postwar American aluminum industry.' In (R. R. Nelson, ed.) *The Rate and Direction of Inventive Activity*. Princeton: Princeton University Press.

Phillips, A. (1971). *Technology and Market Structure*. Lexington, Mass.: D. C. Health.

Scherer, F. M. (1982). 'Inter-industry technology flows in the United States,' *Research Policy*, vol. 11, pp. 227–45.

—— (1984). 'Using linked patent and R&D data to measure interindustry technology flows.' In (Z. Griliches, ed.) *R&D, Patents, and Productivity*. Chicago: University of Chicago Press for the National Bureau of Economic Research.

Schumpeter, J. A. (1942). *Capitalism, Socialism, and Democracy*. New York: Harper.

Spence, A. M. (1981). 'The learning curve and competition.' *The Bell Journal of Economics*, vol. 12, pp. 49–70.

—— (1984). 'Cost reduction, competition, and industry performance,' *Econometrica*, vol. 52, pp. 101–21.

Stoneman, P. (1987). *The Economic Analysis of Technology Policy*. Oxford: Clarendon Press.

Thirtle, C. G. and Ruttan, V. W. (1987). *The Role of Demand and Supply in the Generation and Diffusion of Technical Change*. Chur: Harwood Academic Publishers.

Tilton, J. H. (1971). *International Diffusion of Technology: The Case of Semiconductors*. Washington, D.C.: Brookings Institution.

von Hippel, E. (1978). 'Successful industrial products from customer ideas,' *Journal of Marketing*, vol. 42, pp. 39–49.

Name Index

The International Library of Critical Writings in Economics

1. Multinational Corporations
   Mark Casson
2. The Economics of Innovation
   Christopher Freeman
3. Entrepreneurship
   Mark Casson
4. International Investment
   Peter J. Buckley
5. Game Theory in Economics
   Ariel Rubinstein
6. The History of Economic Thought
   Mark Blaug
7. Monetary Theory
   Thomas Mayer
8. Joint Production of Commodities
   Neri Salvadori and Ian Steedman
9. Industrial Organization
   Oliver E. Williamson
10. Growth Theory (Volumes I, II and III)
    R. Becker and E. Burmeister
11. Microeconomics: Theoretical and Applied (Volumes I, II and III)
    Robert E. Kuenne
12. The Economics of Health (Volumes I and II)
    A.J. Culyer
13. Recent Developments in Macroeconomics (Volumes I, II and III)
    Edmund S. Phelps
14. Urban and Regional Economics
    Paul C. Cheshire and Alan W. Evans
15. Modern Public Finance (Volumes I and II)
    A.B. Atkinson
16. Exchange Rate Economics (Volumes I and II)
    Ronald MacDonald and Mark P. Taylor

17. The Economic Value of Education: Studies in the Economics of Education
Mark Blaug

18. Development Economics (Volumes I, II, III and IV)
Deepak Lal

19. The New Classical Macroeconomics (Volumes I, II and III)
Kevin D. Hoover

20. The Economics of the Environment
Wallace E. Oates

21. Post Keynesian Theory of Growth and Distribution
Carlo Panico and Neri Salvadori

22. Dynamic Labor Demand and Adjustment Costs
Giorgio Galeazzi and Daniel S. Hamermesh

23. The Philosophy and Methodology of Economics (Volumes I, II and III)
Bruce J. Caldwell

24. Public Choice Theory (Volumes I, II and III)
Charles K. Rowley

25. Evolutionary Economics
Ulrich Witt

26. Economics and Psychology
Shlomo Maital and Sharone L. Maital

27. Social Choice Theory (Volumes I, II and III)
Charles K. Rowley

28. Non-Linear Dynamics in Economic Theory
Marc Jarsulic

29. Recent Developments in Experimental Economics (Volumes I and II)
John D. Hey and Graham Loomes

30. Monopoly and Competition Policy (Volumes I and II)
F.M. Scherer

31. The Economics of Technical Change
Edwin Mansfield and Elizabeth Mansfield

32. The Economics of Exhaustible Resources
Geoffrey Heal

Future titles will include:

The Economics of Transport
Herbert Mohring

General Equilibrium Theory
Gerard Debreu

International Trade
J. Peter Neary

The Foundations of Public Finance
Peter Jackson

Labor Economics
Orley Ashenfelter

International Finance
Robert Z. Aliber

Welfare Economics
William J. Baumol and Janusz A. Ordover

Agricultural Economics
G.H. Peters

The Theory of Inflation
Michael Parkin

The Economics of Information
David K. Levine and Steven A. Lippman

Foundations of Analytical Marxism
John E. Roemer

The Theory of the Firm
Mark Casson

The Economics of Inequality and Poverty
A.B. Atkinson

Implicit Contract Theory
Sherwin Rosen

Business Cycle Theory
Finn E. Kydland

The Economics of Housing
John M. Quigley

The Economics of Institutions
G.M. Hodgson

Population Economics
Julian L. Simon

The Economics of Crime
Isaac Ehrlich

The Economics of Integration
Willem Molle

Economics and Discrimination
William A. Darity, Jr

Financial Intermediaries
M.K. Lewis

The Rhetoric of Economics
Donald McCloskey

Ethics and Economics
Amitai Etzioni

Migration
Oded Stark

The Economics of Ageing
John Creedy

The Economics of Privatization and Deregulation
Elizabeth E. Bailey and Janet Rothenberg Pack

Markets and Socialism
Alec Nove and Ian Thatcher

Gender and Economics
Jane Humphries

Location Theory
Melvin L. Greenhut and George Norman

Economic Forecasting
Paul Ormerod

Macroeconomics and Imperfect Competition
Jean-Pascal Benassy

The Economics of Training
Robert J. LaLonde

The Economics of Defence
Keith Hartley